Cell Division
in Higher Plants

EXPERIMENTAL BOTANY

An International Series of Monographs

CONSULTING EDITORS

J. F. Sutcliffe

School of Biological Sciences, University of Sussex, England

AND

P. Mahlberg

Department of Botany, Indiana University, Bloomington, Indiana, U.S.A.

Forthcoming Titles

Cell Division
in Higher Plants

Edited by

M. M. YEOMAN

Department of Botany
University of Edinburgh

1976
ACADEMIC PRESS
London   New York   San Francisco
A Subsidiary of Harcourt Brace Jovanovich, Publishers

ACADEMIC PRESS INC. (LONDON) LTD
24/28 Oval Road,
London NW1

United States Edition published by
ACADEMIC PRESS INC.
111 Fifth Avenue
New York, New York 10003

582.08
C33
106922
ort.1978

Library of Congress Catalog Card Number: 75 34566
ISBN: 0 12 770550 3

PRINTED IN GREAT BRITAIN BY
BUTLER & TANNER LTD
FROME AND LONDON

Contributors

P. A. AITCHISON, *Department of Botany, University of Edinburgh, The King's Buildings, Mayfield Road, Edinburgh EH9 3JH, Scotland.*

M. D. BENNETT, *Plant Breeding Institute, Cambridge, Maris Lane, Trumpington, Cambridge CB2 2LQ, England.*

R. BROWN, F.R.S., *Department of Botany, University of Edinburgh, The King's Buildings, Mayfield Road, Edinburgh EH9 3JH, Scotland.*

F. A. L. CLOWES, *Botany School, University of Oxford, South Parks Road, Oxford OX1 3RA, England.*

J. E. DALE, *Department of Botany, University of Edinburgh, The King's Buildings, Mayfield Road, Edinburgh EH9 3JH, Scotland.*

A. W. DAVIDSON, *Department of Biological Sciences, Thames Polytechnic, Wellington Street, London SE19 6PF, England.*

A. F. DYER, *Department of Botany, University of Edinburgh, The King's Buildings, Mayfield Road, Edinburgh EH9 3JH, Scotland.*

W. A. JENSEN, *Department of Botany, University of California, Berkeley, California 94720, U.S.A.*

R. M. LEECH, *Department of Biology, University of York, Heslington Road, York YO1 5DD, England.*

R. F. LYNDON, *Department of Botany, University of Edinburgh, The King's Buildings, Mayfield Road, Edinburgh EH9 3JH, Scotland.*

I. D. J. PHILLIPS, *Department of Biological Sciences, University of Exeter, Washington Singer Laboratories, Perry Road, Exeter EX4 4QG, England.*

M. M. YEOMAN, *Department of Botany, University of Edinburgh, The King's Buildings, Mayfield Road, Edinburgh EH9 3JH, Scotland.*

Preface

This book is an attempt to present an up-to-date account of cell division in higher plants by bringing together in one volume contributions from specialists distinguished in their particular area of study. The text is arranged in four parts. The introductory chapter considers the significance of cell division, both as a process and as part of overall growth. This is followed by a section dealing solely with the processes of mitotic and meiotic cell division, which in turn is succeeded by a series of chapters concerned with the relationship of division to growth and the generation of form. A short summary containing some speculation about the possible future direction of research in cell division concludes the text.

Differences in approach and style of presentation are inevitable in a multi-author work despite the combined efforts of contributors and editor. However, I believe that any disadvantages which have accrued from this approach are far outweighed by the advantages of employing the combined expertise of specialists. A number of steps have been taken to prevent fragmentation and encourage integration, including the deliberate retention of some overlap, the inclusion of cross-references wherever appropriate, the use of a unified bibliography which also serves as an author index, and a terminal summary in which the relationships between chapters are emphasized. This book is aimed at senior undergraduates, post-graduates and established scientists who wish to discover more about cell division or are contemplating research in this field.

As Editor I should like to thank not only my co-authors but also Professor J. F. Sutcliffe (Consulting Editor of the Experimental Botany Series) and Barbara Renvoize of Academic Press. We also gratefully acknowledge permission to include data previously published from many authors and journals. In the general preparation of the manuscript, assembly of the bibliography and index, I have been greatly helped by Sheena Littledyke, Esmé Mills, Hilary Pritchard, Betty Raeburn and Bill Foster. I also wish to thank Marysia Miedzybrodzka, Paul Aitchison, Allan Gould and Alex McLeod for their valuable assistance.

September, 1975   M. M. YEOMAN

Contents

A. Introduction

1. Significance of Division in the Higher Plant

Department of Botany, University of Edinburgh, Scotland

1. INTRODUCTION

This introductory chapter is concerned with the significance of division for certain phases of the growth of the higher plant. The process of division could of course be analysed in contexts other than that of growth and in systems other than that of the higher plant. But growth is undoubtedly the most comprehensive expression of division activity, and the integrated multicellular system provides the unique context in which the full versatility of the process can be realized. The assessment of significance is attempted in terms of two aspects: firstly by reference to the course of division in the phases of enlargement, development and differentiation; and secondly by reference to the intrinsic characteristics of the process itself. This treatment is justified by the circumstance that effects of division are only intelligible in terms of the nature of the division process itself.

It is necessary to emphasize that the discussion that is developed only refers to certain phases of growth. An analysis of the nature of the involvement of division in the whole compass of growth is certainly beyond the scope of this chapter. The phases of growth that have been selected for treatment are those that constitute the vegetative stage. The restriction may be arbitrary, but it confines discussion to the one area in which knowledge, although inadequate, is nevertheless more extensive than it is in any other. It may be repeated that by growth is meant a complex of enlargement, development and differentiation. It is recognized that the three phenomena are essentially connected expressions of the same continuous process. Ultimately it is claimed that this is derived from the progressive release of information from the genome, and the significance of division in this context is accorded some attention.

II. SIGNIFICANCE OF MITOSIS

Early investigations of cell division in the higher plant were restricted to the study of mitosis (Chapter 2). Available techniques imposed this limitation and the generally spectacular aspects of the process attracted particular attention. The conclusions that could be drawn during this phase were necessarily restricted, but one generalization was elaborated which was held, with some justification, to be of sovereign importance. This incorporated the claim that mitosis ensures that each cell acquires the same genetic equipment, in other words that each cell acquires the same genome. This generalization could of course only be elaborated after three particularly important aspects of mitosis had been described. The first comprehensive description of the process was published by Strasburger in 1880. This report outlined the main features of the process, but it did not draw attention to two important characteristics. It did not emphasize that the number of strands or chromosomes is normally constant, and it did not specify that the fragments or chromatids are generated through longitudinal splits in the chromosomes. For some years after Strasburger's classical work was published, while it was conceded that the chromosome divided, it was not universally accepted that it did so longitudinally. Many workers insisted that the cleavage was transverse. With improvements in technique the longitudinal splitting was established and the attachment of the two chromatids to spindle strands from different poles was in due course demonstrated.

The full significance of the mitotic succession could not of course be appreciated until the status and the structure of the chromosomes had been defined. The general subject of the concepts that have emerged in the study of heredity is not relevant to the present theme. It is necessary only to emphasize the supreme importance of one of these for the interpretation of the significance of mitosis. It was established that the hereditary determinants, or genes as they were subsequently called, are carried on the chromosomes and that each chromosome incorporates a constant linear sequence of genes; clearly these two generalizations endow the mitotic sequence with high significance (Morgan, 1919). For if each chromosome splits longitudinally, then since the genes are linearly disposed and each is transversely aligned, each chromatid must carry the same gene complement. Further, since the number of chromosomes is constant and the two chromatids are dragged to separate poles, with respect to the genome, division is necessarily equational. It secures the formation of two cells each carrying exactly the same nuclear gene complement. This general conclusion is of course one of outstanding importance. Subsequent experience has shown, however, that it requires to be complemented by another that is possibly of comparable significance. The inference has occasionally been made that because two identical nuclei are generated, this implies the elaboration of two identical cells. The course of mitosis does not justify this assumption and important considerations derived from other connections indicate that it is certainly untenable. The evidence shows that while mitosis secures the emergence of identical nuclei it does not invariably lead to the establishment of two cells that carry the same potential (see below).

Since the two nuclei that are the products of division carry identical genomes all the cells in a multicellular system incorporate the same genome. Not only so, but all carry the same genetic information as the fertilized egg of the embryo sac. They do so of course since they are all derived from this unit by division. This circumstance is the basis for an important biological phenomenon. Since the nuclear complement is the same as that of the fertilized egg the possibility is inherent in the situation that every cell can have the same potentiality, that every cell incorporates the capacity to yield a whole organism. Not only is this so, but it can realize this capacity through the same developmental sequence as that which is traversed from the egg. Every cell may in fact be totipotent. The inherent capacity must of course be qualified by the operation of different environmental circumstances and by different structural conditions in the cytoplasm. Nevertheless that totipotency is not simply an inference enshrined in a doctrine but is a realizable phenomenon has now been demonstrated in a surprising variety of situations (Steward *et al.*, 1970). The development of an intact plant from a single vegetative cell has been demonstrated with material from callus cultures (Chapter 12), from endosperm, and from the epidermis of cultured seedlings. Moreover, in each connection the intact plant has been formed through a developmental sequence which is traversed by the normal embryo.

III. CELL DIVISION AND GROWTH

One of the more curious fallacies that has been repeated at intervals over at least the last 100 years is the claim that cell division cannot be considered a phase of growth and that the study of each is irrelevant to that of the other. This oddity is resurrected at intervals and proclaimed with ardour. That the attitude is misguided cannot be doubted, but it must be conceded that it has had the support of high authority, and notably that of Sachs (1887). Indeed it may have been Sachs who inspired the tradition. In the *Lectures on Plant Physiology* published in 1887 he had this to say:

> "Growth—i.e. the increase in volume and change of form—may take place in a plant even without accompanying cell-divisions. In this connection, I have already repeatedly referred to the non-cellular plants, such as *Botrydium, Caulerpa, Vaucheria*, etc., and particularly to the Myxomycetes. It is important to bear this fact in mind; because it proves that the formation of cells is a phenomenon subordinate to, and independent of, growth. The excessive importance for organic life hitherto ascribed to cell-formation found expression in this direction also, in that it was believed that growth depended upon the formation of cells. This is, however, not the case. On the other hand, however, the fact is of course important, that while a few hundred simple forms of plants exist in which growth is not accompanied by cell-division, in all other plants growth and cell-division are intimately connected with one another. In attempting, then, to make clear the relations of the two processes—growth and cell-division—it is above all to be insisted upon that growth is the primary, and cell-division the secondary and independent phenomenon."

Evidently Sachs was persuaded into the position he adopted by the situation he observed in coenocytic systems such as those of *Vaucheria* and *Botrydium*. In these, enlargement occurs without the formation of partitioning walls. The argument no doubt took the form that since in at least some instances growth can continue without the deposition of walls, then clearly the primary process is surface extension and the formation of compartments enclosed by walls must be secondary. Sachs attempted to support his claim by referring to the situation in another alga, *Stypocaulon*. In this the thallus terminates in a large bladder-like cell. The apical cell continues to expand, and as it does so a small cell is segmented from an inconspicuous base. The pattern of events showed, it was claimed, that surface expansion is the more fundamental condition and that segmentation is a derivative process. Clearly to Sachs division was simply a device that established compartments within a space that is created by surface extension.

It is instructive to explore the probable origin of the misconception. The volume in which the passage quoted above occurs was published in 1887. The conclusion which was formulated was presumably developed some years earlier. It was in fact probably developed before Strasburger's description of mitosis had been published. Sachs was therefore probably not in a position to appreciate that the deposition of the wall is preceded by the formation of another nucleus and therefore of another protoplast. If he had been in such a position he would undoubtedly have understood that the formation of the protoplast is the initial step that precedes surface extension, and further that the deposition of the wall has the significance only of being the culmination of a process that incorporates a number of earlier phases. Within the terms of this interpretation the absence of transverse wall formation in the coenocytic algae clearly does not provide grounds for asserting that surface expansion is the primary requirement in growth. Even in these algae the basic process is division. This yields the protoplasts which can sustain surface expansion.

In an immediate sense it cannot be gainsaid that the higher plant is composed of cells and that it cannot grow indefinitely unless the number of component cells is increased continuously. This statement is no doubt a platitude to many, but to some the validity of the converse claim is impressive. It cannot be denied that in certain instances division does not lead to significant enlargement of the whole. The circumstances in which this situation may arise are explained below. But the evidence provided by arresting division through agents that do not interfere with normal metabolic activity is fatal to this objection. Ionizing radiations tend to disrupt nuclear structures and, when a tissue is exposed to sufficiently high intensities of these, growth ultimately ceases. The arrest of growth is not immediate, but the limited enlargement that is observed after treatment is due to the expansion of

immature cells present in the system at the time of exposure to the radiation. The final cessation of growth is undoubtedly a consequence of
the arrest in division that follows from nuclear disintegration. A similar
sequence is observed when certain analogues of nucleic acid bases are
supplied to growing tissues. After treatment with compounds such as
2-thiouracil and 8-azaguanine (Brown, 1963), again some enlargement
may still continue and again this must be attributed to the expansion
of immature cells. In due course growth ceases and this is certainly an
inevitable effect of cessation of division induced by the disturbance to
the synthesis of particular nucleic acids.

It must be conceded that without division growth is not observed.
But nor is it observed when photosynthesis is interrupted. It could be
that for enlargement of the whole plant division is simply a necessary
condition, and not an intrinsic phase of the process. There is little doubt
that it is in fact an integral stage, and it is so in the sense that in normal
circumstances it can be a rate-limiting step and that it moulds the cell
into a state in which it can subsequently expand. It evokes the metabolic
situation in which expansion can subsequently occur. Division and
expansion are thus different phases in a continuous process. The justification for this claim requires further consideration of the mechanism
of division.

IV. MITOSIS AND INTERPHASE

It has already been stressed that in the earliest stages of the investigation of division attention was restricted to mitosis. This restriction is
emphasized by the nature of the early attempts to measure the relative
rate of division. Experience with structures such as staminal hairs had
shown that the formation of mitotic figures preceded the cleavage of
the cell into two units. This experience was accepted as indicating that
division begins with the first stages of mitosis. From this the inference
was inevitably drawn that when a mitotic figure was observed this
showed that the cell had been stimulated into division from a so-called
"resting state". In a system such as a meristem this inference clearly
justified the further conclusion that the frequency of mitotic figures is
a measure of the proportion of cells that are dividing at any one time.
A determination of the proportion of mitotic figures in the population
is therefore a relative measure of the rate of division. A procedure for
determining a value which represented the percentage of cells in mitosis
was developed. The value was termed the mitotic index. Thousands
of such mitotic indices were determined in countless hundreds of investigations. The differences between indices were taken as a relative

measure of corresponding differences in the frequency of division. It may be emphasized that the general procedure must be accepted as valid if the assumption on which it is based is justified. It is now clear that the assumption is not justified and the index as a measure of the relative rate of division is therefore grossly misleading. Another legacy from the phase in which division was considered to be co-extensive with mitosis is the notion of the necessity of a particular stimulator. In a system such as a meristem if the majority of cells are thought of as being in a resting state the transition to active division presumably requires the operation of a stimulant. If the interpretation is accepted then it must be assumed that the incidence of the stimulant is random and sporadic, for otherwise the irregular distribution of mitotic figures would not be intelligible. The notion of a specific stimulant has never been abandoned, and it can be invoked on grounds other than those outlined. But the amplification of the interpretation of the nature of division has certainly diminished the status of the stimulant.

The restriction of the study of division to mitosis led to yet another serious misconception. The course of events in mitosis necessarily implies that the dry mass of the chromosomes in the product nuclei must be half that of those in the parent nucleus. Through this change the quantity of genetic material is reduced to half that in the parent cell. Observation showed that the products of a division may themselves divide. Clearly this second mitosis can only occur if at some stage the mass of chromosomal material is doubled. In the phase in which it was assumed that the cell is impelled into division from a resting state by the operation of a stimulant it was suggested that the doubling occurred at the beginning of mitosis in early prophase. This interpretation is of course entirely consistent with the assumption that division begins with the induction of mitosis. In itself this particular misconception was not any more misleading than several others. Historically, however, it carries particular significance. For it was the modification of the appreciation of the nature of the doubling process that introduced the current phase in the interpretation of division.

The change followed from the identification of the chemical nature of the genetic component of the chromosome and from the elaboration of techniques for estimating it in the cell. The decisive component was identified when Avery demonstrated that the transfer of pathogenicity from one strain of *Pneumococcus* to another was mediated through the transfer of DNA (Avery *et al.*, 1944). The transmission of pathogenicity in effect represented a transfer of genetic information. That this transfer could be secured by an exchange of DNA indicated that this substance was the chemical basis of the gene.

The identification of DNA as the genetic substance was preceded by the elaboration of a cytochemical reaction, the Feulgen reaction, that provided the basis for the quantitative estimation of this material (Feulgen and Rosenbech, 1924). A purple complex is generated, and the quantity of this can be measured by densitometry. The absorption of light of the appropriate wavelength provides a relative measure of the level of DNA in the nucleus. The application of this technique led to a comprehensive reappraisal of the course of division. It was shown that the doubling of DNA that the comprehensive process requires occurs not at the beginning of mitosis but in the preceding interphase (Swift, 1950). Evidently interphase is not a resting stage but it represents an integral part of the comprehensive succession of division. The doubling occurs in a process of replication which raises the DNA content often in a mid-interphase stage from the 2C to the 4C level (Chapter 3).

Typically the interphase is now considered as being composed of three successive phases, G_1, S and G_2. G_1 is the interval, or the gap, between the last telophase and DNA replication which is completed in S. G_2 is the gap between S and the beginning of mitosis. This formulation incorporates one extremely important assumption. It is a characteristic of groups of dividing cells, such as those in meristems, that the products of those that have divided may subsequently traverse another mitosis. In terms of the earlier tradition of interpretation this simply represents a random evocation of activity in dormant cells. The observation that replication occurs in the interphase suggests on the other hand that a current mitosis is in a sense a consequence of an earlier one. It suggests that when telophase has occurred the products are delivered into a state of G_1 from which replication inevitably follows. Evidently division cannot be considered as a linear process which begins with prophase and terminates with the completion of cytokinesis. It must be envisaged as a cyclical process which provides for the return of progeny to the state from which replication can begin. The extent of the conceptual change may be appreciated by considering the implications of one particular technique for measuring the durations of interphase and of the different stages of mitosis. If in the meristem division is cyclical then the situation is similar to that in a culture of microorganisms in which the vigour of division can be assessed in terms of a mean generation time. A similar quantity can be determined with a meristem. In this, dividing cells are usually distinguished by being essentially non-vacuolated. The increment in the total number of cells due to division activity during any interval is readily determined. The number of non-vacuolated and therefore of potentially dividing cells is also readily determined. From the two quantities an average generation time or an average cycle period may be calculated (Brown, 1951). It is true that the average covers a large range of cycle durations, but it cannot be denied that the average is a property of the population as a whole in the same sense as the average respiration rate per cell is such. From the average cycle time the average durations of interphase and mitosis can be calculated by taking the proportion of cells in mitosis as representing a corresponding fraction of the total duration. Thus if the proportion of mitotic figures is 10% and the total duration is 20 h then the duration of mitosis is 2 h and of interphase 18 h. By applying the same principle the average duration of each stage of mitosis may be estimated. This procedure gives values similar to those provided by other methods. It is undeniably valid. Being such the assumptions on which it is based are of immediate significance. The average generation time is meaningful only if it refers to a situation in which all the cells divide and do so repeatedly. The second qualification is particularly significant. It implies a repetitive cycling through G_1, S, G_2 and mitosis. Secondly, in this context the mitotic index carries a different significance. In a cyclical system it can be taken to represent only the relative durations of mitosis and interphase. It is not a relative measure of the rate of division. In a system in which divisions are not synchronized the number of mitoses seen at any one time is determined by the length of mitosis relative to that of interphase.

The reappraisal of the nature of the division process prompts a query of some importance. Above it has been argued that the notion of random stimulation requires a stimulant. In the present context it may be asked, if the process is cyclical and repetitive, why does the process ultimately cease? In the higher plant many of the cells produced from the meristem cease to divide. If in fact it is a characteristic of division that when cells emerge from telophase they are in the state of G_1 and can then begin another cycle, it becomes necessary to ask what the mechanism may be that arrests the repetitive sequence. In this connection the synchrony that has been observed when growth begins in certain inocula for callus cultures may be instructive (Yeoman and Evans, 1967). When inocula from Jerusalem artichoke tubers are transferred to a suitable medium division begins immediately and the first two or three cycles in the dividing cells are approximately synchronous. The evidence indicates that division begins from early G_1 and the cells remain in phase with each other since the tissue is strikingly uniform. The fact that the dividing cells begin in G_1 suggests that they are arrested at that stage and that at least one of the effects of excising the explant and transferring it to the culture medium is the removal of an inhibitor. It must be supposed that all the cells of the tuber at some stage accumulate an inhibitor which acts selectively in G_1 (Clowes, 1965a). The uniformity of response is more easily interpreted in terms of a selective inhibitor than of a generalized promoter. If exhaustion of a promoter had been the basis for the original cessation of division, arrest in all cells at one particular stage could not have been expected. (It may be noted that it has been suggested that the development of the quiescent zone in the roots of certain species is due to arrest in G_1. It has not been proposed that the arrest is due to the operation of a selective inhibitor, but such may be the case.) The evidence is certainly tenuous, but it is not incompatible with the general position that the transition from a dividing to an expanding state is due to the operation of a selective inhibitor which acting on division holds it in the state of G_1. Whenever division is resumed the division cycle begins from this point and if all the dividing cells are exposed to the same environment then they progress synchronously. In most instances the tissue is not uniform and synchrony is not achieved.

V. THE MECHANISM OF DIVISION

The discussion of the division process developed in the preceding paragraphs is based mainly on events in the nucleus with special reference to changes in the DNA component. This treatment is justified by the primary significance of the nucleus and of the DNA complement for the determination of genetic potentiality. It is evident, however, that division must involve not only the nucleus but every single component of the cell (Chapter 2). Since in a typical situation the cell is fragmented into two more or less equal portions, clearly in the sequence of cycles cell components must be amplified, for otherwise a situation could be reached where the nucleus is not supported by an adequate cytoplasmic system. Either in every cycle or with some feature in particular cycles, every single chemical component must be enhanced, the extent of every structural feature must be enlarged and the scope of the metabolic system must be extended. It is important to notice that this growth when it occurs is confined to the interphase. Little or no change in dry mass occurs during mitosis. The accumulation of dry matter is confined to the interphase. The change in activity that is involved in the transition from interphase to mitosis is accompanied by a change in respiration rate. During the interphase the rate remains comparatively high, but in early prophase it begins to decline and continues to decrease until a minimum is reached in later metaphase (Erickson, 1964). Thus when respiration is vigorous and energy is freely available synthetic activity is high, but when respiration is low and the energy supply is restricted major synthetic activity is depressed.

Growth, although almost certainly a necessary condition for the continuation of division, does not seem to be causally linked to it. The average level of protein or any other crucial complex per cell can vary considerably in a dividing population and a causal link between division and interphase growth is therefore almost certainly not involved. The increase in total dry mass in a rapidly dividing population can be negligible; hence one of the arguments referred to earlier in favour of the contention that division is not a component of the growth process. The independence of growth is shown by conditions in the meristem. The average dry weight per cell tends to increase from the extreme apex to the basal limits of the meristem. Over the same length, however, the rate of division can increase, reach a peak and then decline as the base is approached. The independence of the growth process is further demonstrated in the character of the growth that continues after division has been arrested. This seems to be simply a prolongation of the process that was in being before division ceased. An independence of growth from division is shown again by the sequence of events when growth begins in the inoculum for a callus culture. The promotion of growth involves an induction of division and a stimulation of protein accumulation. But division proceeds relatively more rapidly than protein synthesis and as a result the level of protein per cell may decline sharply (Yeoman et al., 1965). The evidence shows that there is no quantitative link between division and cellular growth. On the other hand it does not show that there is no connection between division and the quality of the growth that is made. It does not show that division does not endow the complex that is accumulated with particular properties, which become the basis for particular activities. Evidence that is reviewed in another connection suggests that such a control by the division process does in fact operate.

While division may be independent of the synthetic reactions of growth, it does depend on synthetic processes peculiar to itself. When a cell reaches the state of G_1 it is not equipped with a mechanism that is independent of the genome that drives the division sequence. The necessary mechanism must be elaborated and inevitably this requires a particular set of synthetic reactions. In a formal sense it may be said that the mechanism that is elaborated requires the formation of two types of components—structural elements and particular catalysts. It is recognized that the distinction between the two is arbitrary and that they are both different phases of an integrated system. It is recognized that structural elements are derived from and depend on the catalytic, and that the catalytic depend on the structural for their operation. The full extent of the structural phase certainly cannot be specified. While much of the protein that is synthesized during the interphase is assimilated into the growth component some undoubtedly becomes attached to a structural component of division. It has been suggested that prophase begins with the formation of a protein baffle into which the nucleus migrates. The 4C level of DNA is a requirement for mitotic division to occur and the increment from the 2C level may therefore be considered as part of the structural mechanism. This probably also requires the synthesis of histone. It is significant that while the DNA content doubles so also does the nuclear content of histone (Woodard *et al.*, 1961). The temporal and quantitative coincidence suggests that the two are closely linked. One of the first structural components formed is the nucleolus. This body is dispersed when mitosis begins and is reconstituted when telophase is complete. It is certainly part of the division mechanism of interphase since it has been shown that when certain analogues of the nucleic acids are applied which arrest division they accumulate in the nucleolus (Heyes and Vaughan, 1968). The nucleolus is primarily committed to the elaboration of the components of the ribosome, and the effect of the analogues is presumably to interfere with this process. If this is the position, then the accumulation of ribosomes must be considered as being one of the synthetic processes that is involved in assembling the structural apparatus. The visible feature that emerges during division, and that has probably received most attention, is the spindle. It has been suggested that this is formed from a band of microtubules that condenses within the wall at about the level at which the equator of the spindle is subsequently aligned. It is possible that the band is formed at the periphery of the protein baffle that marks an early phase in the progress of division. The system is elaborated by the movement of the microtubules into the baffle and by the formation of the spindle transversely across it. It is completed when threads from opposite poles become attached to centromeres which must be regarded as important structural elements.

The different structures appear at different stages of the cycle. They are no doubt the product of a particular metabolic state, and the sequential emergence of the different components therefore suggests the promotion of a succession of metabolic states. If this is the case then the whole course of division is presumably controlled by the release of a number of catalysts in a particular temporal order. The phenomenon of what may be called rigid spontaneous synchrony is of particular significance in this connection. One form of natural synchrony has already been considered. In this the tissue consists of intact cells with

intervening walls and the degree of synchrony is only loose. The period
over which mitosis begins, for instance, stretches over about 2 h, and
synchrony is confined to about two cycles. In a second form the system
within which division occurs is without dividing walls, the correspondence within the population of nuclei can be remarkably close, and
synchronous division can be maintained through a large number of
cycles (Erickson, 1964). The closeness of the synchrony can be such
that all the nuclei are in anaphase, although the duration of this stage
is only about 5 min in normal circumstances. It is this form of synchrony
that is particularly relevant here. The phenomenon has been observed
in myxomycetes, in many young endosperms, in the gametophyte of
gymnosperms, in groups of developing sperms in antheridia and in
giant cells formed when certain nematodes infect plant roots. In all
these systems a decisive feature is undoubtedly the absence of continuous walls. The nuclei lie in a more or less continuous cytoplasm.
The case of the endosperm is instructive. In many species after a phase
in which walls are not assembled wall formation begins. As soon as this
stage is reached synchrony is no longer observed. In other species wall
formation begins immediately the endosperm is deposited. In these,
synchrony does not emerge at any stage in the development of the tissue.
It may be suggested that the attainment of any division stage is due
to the release of a particular solute complex. This complex, it may be
supposed, consists of an activator and a characteristic catalytic system.
The activator may be a catalyst, but it is necessary to distinguish it
from others for reasons that are developed below. The whole solute
complex may be characterized as a stimulant system. Close synchrony
is attained when a stimulant system can diffuse freely from one protoplast to the next and this condition is only secured when walls with
their associated membranes are not formed. If, through heterogeneity
in the group, one nucleus tends to produce the next stimulant system
more slowly or another more rapidly than the next, each will nevertheless be exposed to the same stimulant concentration and each will therefore react simultaneously with the rest. The state of close simultaneity
is inferred from the appearance of the system in mitosis. Since the state
is displayed in successive mitotic phases it can be assumed that the synchrony of mitosis follows from a corresponding condition developed in
interphase. Thus the cytological data indicate that the population of
protoplasts traverse the whole cycle with rigid simultaneity as a result
of the release of a succession of stimulant systems which, since they are
not confined within boundary walls, become uniformly distributed
rapidly. It may be argued that the presence of a wall should not affect
the uniformity of distribution of a solute complex since free flow might

be expected to continue along plasmodesmata. Solute movement along plasmodesmata undoubtedly occurs, but the total cross-sectional area of these is small relative to the total area of the wall and is certainly not great enough to secure uniform concentrations within minutes, which is the order of time that the situation requires.

The incorporation of a group of catalysts within each stimulant system is necessitated by the change in structural components that the cycle entails. In mitosis the structural changes are spectacular and they certainly could not be completed without the intervention of appropriate catalysts. The situation suggests that the division process involves the successive release of different groups of enzymes. A direct demonstration that the levels of different enzymes are markedly enhanced at different stages of the division cycle has only recently been provided in tissues of a higher plant; that in fact in such a system the successive formation of particular catalysts constitutes one aspect of the division mechanism. Investigation has been hindered by a lack of suitable experimental material. Recently at least one system has become available and preliminary results with this are highly suggestive (Yeoman and Aitchison, 1973). The evidence indicates that the activity of certain enzymes is enhanced only in G_1 and certain others only in S. Moreover, it has been found that the suppression of enzyme synthesis in both phases leads to the arrest of division. Three enzymes have been shown to be produced in G_1 only. They are malic dehydrogenase, ATP glucokinase and glucose-6-phosphate dehydrogenase. The data, fragmentary though they are, suggest that the increases in the levels of the three enzymes do not occur simultaneously but serially. The level of each is markedly enhanced to a maximum which is maintained at least through a large part of the interphase. The enzymes whose production it has been shown is confined to the S phase are DNA polymerase, thymidine kinase and thymidine monophosphate kinase (Harland *et al.*, 1973). The course through which the activity level of each of these enzymes changes is different from that observed with the G_1 group. The level tends to increase throughout S and the rate of enhancement seems to be closely linked to the change in the level of DNA.

The evidence suggests that the structural mechanism that is elaborated during division is the product of a serial enhancement of the activities of particular groups of enzymes in G_1, S, G_2 and mitosis. If this is the case then the mechanism which promotes change in activity in the different stages is a matter of acute importance. In this connection the nature of the inhibitor which suppresses an increase in activity and at the same time arrests division in G_1 is of particular significance. It has been shown that enhancement of activity can be suppressed by the application of 6-methyl purine. It has also been shown that this compound interferes with the progress of transcription. It depresses the formation of various RNA species, and thus effectively prevents the mobilization of information from the genome. Thus the evidence shows that the enhancement of the activities of the enzymes whose levels in the different stages of G_1 has been examined is a consequence of a stimulation to transcription. The inhibitor also arrests further progress through G_1. The two effects are undoubtedly connected, although it cannot be inferred that they are causally related. The three particular enzymes whose levels in G_1 have been examined may not be immediately significant for the elaboration of the division apparatus. They may represent only a contribution to the growth component. Nevertheless, whether they contribute to the elaboration of the division mechanism or not, they certainly indicate the nature of the processes through which the catalysts that are important in that mechanism are generated. Others may well be more immediately involved, but if they are then undoubtedly they are also the products of transcription, and the effect of the 6-methyl purine on the course of G_1 is due to the suppression of the formation of these and all other enzymes (Chapter 3).

The three enzymes whose enhancement has been examined in S are susceptible to 6-methyl purine, and again the inhibitor affects the progress of this one particular stage. It may be supposed that the action of the inhibitor is the same as it is in G_1. It suppresses, through arresting transcription, the formation of enzymes that are required in the promotion of S. In this instance it is probable that at least two of the enzymes involved are important components of the catalytic phase of the division apparatus. The polymerase undoubtedly contributes to the elaboration of DNA in replication. If the formation of this enzyme is depressed the synthesis of additional DNA does not occur and the S-phase cannot be completed. The thymidine monophosphate kinase mediates the conversion of thymidine monophosphate ultimately to thymidine triphosphate. Again if this reaction is depressed by a restriction of transcription an essential component of DNA is not provided and the progress of replication is necessarily arrested.

The effect of one inhibitor fluordeoxyuridine (FUdR) has been examined which has a selective action on the development of S. When FUdR is introduced into the system it arrests the formation of the polymerase and of the two kinases. At the same time it depresses DNA formation and inhibits the progress of S. In this case the action of the inhibitor is probably not due to an immediate effect on transcription, although a depression of transcription is probably involved as a secondary effect. The most spectacular effects with FUdR are observed when the compound is added in G_1. The progress of G_1 is not affected, but it has been shown that FUdR inhibits the formation of thymidine monophosphate; it may be supposed that the enzyme required in the synthesis of the nucleotide is released in G_1, but after release it is immobilized and consequently the triphosphate is not formed. The suppression of the formation of the triphosphate necessarily restricts the formation of DNA and replication consequently does not occur. Arrest in the S phase inevitably follows. This chain of events does not show how the formation of the polymerase and the two kinases is depressed. This, it may be suggested, is a direct consequence of the failure of this process. It is not inconceivable that certain forms of transcription depend on the progress of replication. The fact that at least two of the enzymes whose changes in level have been examined are intimately concerned in DNA synthesis supports the suggestion of a connection between replication and transcription. If this proposal is accepted the effect of FUdR on the activity of certain enzymes may be taken as further evidence for the claim that the changes in enzyme level are a consequence of corresponding changes in transcription.

The evidence is not decisive as to whether the enzymes which control the course of division in G_1 and S are released serially. The data available for G_1 are at least not inconsistent with this interpretation, and this position may be accepted at least tentatively. If it is, then it may also be accepted that within each phase the serial release is the expression of a serial transcription in the genome. Further, it must be assumed that essentially the same situation operates within G_2 when it is present and within mitosis. Evidence has been presented for a succession of changes in catalytic state throughout the division cycle. If the changes in this state can be attributed to serial release of information from the genome in G_1 and S, then clearly a similar situation may be inferred for G_2 and mitosis. Below, this general position is considered further. It is recognized that the terms in which the general mechanism are developed can only be accepted as symbolic and indicative. Indeed certain aspects of the general proposal may seem naïve. For instance, it is suggested that each transition in the cytological sequence is mediated through the activation of a single gene. This clearly is a grossly inadequate assumption. On the other hand, while being such, it is possibly a convenient approximation since it provides an indication of the qualitative nature of the relevant events.

Clearly, the basic mechanism in the division cycle must incorporate a device which induces a transition from one transcription pattern to the next in a series. It may be suggested that appropriate genes are assembled in groups, each group being basic to and required in a particular phase of the cycle. Thus one group is basic to G_1, another to S, a third to G_2 and a fourth to mitosis. Each group, it is envisaged, is controlled by a master gene and within the group a set of operator genes is activated in a temporal sequence. The master gene itself requires activation. When this has been secured it has two effects: it sensitizes the group of genes which it dominates and it activates the first operator gene in the series. The gene that is first activated, activates the second, the second the third and so on. Activation, whether of the master gene or of the operator, generates a depressor which in due course deactivates the unit by which it is produced. Activation of the operator gene also leads to the release of an enzyme complex. The proposed mechanism is shown diagrammatically in Fig. 1.

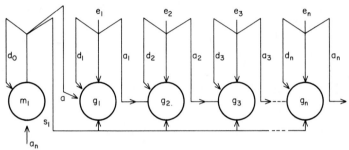

Fig. 1. Pattern for each section of the division cycle. For further explanation see text.

M_1 is the master gene which is activated by a_n. The impact of a_n leads to the generation of a sensitizer s_1, of a depressor d_0 and of an activator a_0. The depressor deactivates M_1 and the activator releases the first operator gene g_1. The promotion of activity in s_1 leads to the release of a depressor d_1, of an activator a_1 and of an enzyme complex e_1; g_1 is ultimately deactivated by d_1, and a_1 makes the next operator g_2 available, and so on until the operator gene g_n is reached.

A master gene which is linked to a particular group of operator genes is required since in any phase only a particular set of genes is transcribed. A selection mechanism is particularly necessary in S, since if replication promotes transcription, unless a device is provided which immobilizes all but a selected group, the whole gene complement should be activated. The master selects the particular group through the device of sensitization. This follows activation. The duration of any phase is limited and a deactivation mechanism is therefore required: this is provided by the depressor. Of the group of genes that are sensitized one is the first to be activated. This step is secured through the generation of an activator by the master. The release of any particular item of information, especially in G_1 and presumably in mitosis, is confined to a limited interval. Activation of an operator therefore requires the generation by the gene of a depressor for itself. The central consequence of activation is the discharge into the cell of a particular catalytic complex, but the restriction of this to a limited interval and the subsequent appearance of another complex implies that another consequence is the generation of an activator for the next gene in the sensitized series. Successive activation and deactivation through the series lead ultimately to the mobilization of the final operator gene g_n. This, after being activated in the normal pattern, generates a depressor, a quota of catalysts and a further activator, but this activator is not transferred to another operator. This component it may be supposed acts on the master for the next stage in the succession. Thus the master for S is activated by the activator from the final operator in the G_1 series; the master for G_2 by the activator from the final operator in S; and the master for mitosis by the final operator in G_2. Since the process is cyclical the situation requires that the master for G_1 be activated by the appropriate agent from the final operator in the mitotic series.

The complete cycle in diagrammatic form is shown in Fig. 2. This model is clearly inadequate and requires elaboration. It is inadequate since it does not provide for the temporal coincidence between replication and the enhancement of at least certain catalytic activities. It does not provide for the quantitative relation between DNA accumulation and the degree of enhancement. Further, it does not accommodate the situation that has recently been described, in which replication may begin in mitosis (Clowes, 1967). At the same time it is probable that a model based on the general pattern of that presented in Fig. 2 can be developed which will assimilate both the peculiar characteristics of the S-phase and the anomalous forms of the cycle. Further, it is significant that the considerations incorporated in the model can be invoked for the interpretation of a group of phenomena in which, although the cleavage of a parent cell into two progeny is not involved, the situation nevertheless may be considered as a modification of the normal division cycle. It is recognized that successful application of the model within this context does not provide evidence of validity. On the other hand it may be urged that it does endow it with feasibility (Brown and Dyer, 1972).

The phenomena all share the characteristic that through repeated replication within the same nucleus the DNA content may be very considerably enhanced. It seems that the normal division sequence is arrested at some point after the S-phase and that arrest is followed by a return to G_1. Three relevant intrinsic features of the division sequence

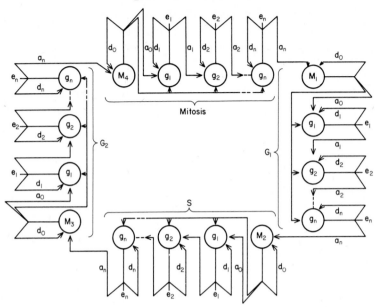

FIG. 2. Model for comprehensive division cycle. For explanation see text.

may be noticed in this connection: (1) the sequence is apparently not reversible at any stage—there is no stage at which a change comparable to a development from metaphase back into prophase can occur; (2) the elaboration of the division apparatus almost certainly involves a series of successive steps in which each depends on the earlier. The dependence is clearly such that a particular elaboration cannot occur until an earlier has been completed. (3) It is probable that every distinctive transformation is liable to arrest through a distinctive inhibitor. Evidence for an inhibitor that acts only in the earliest stage of G_1 has already been considered. Particular compounds that are each effective in S, in metaphase and in cytokinesis are known. FUdR has been shown to be effective only at the beginning of the S-phase, colchicine only at metaphase and ethoxy-caffeine only during cytokinesis.

In terms of the model, reversibility is excluded. It cannot occur since every activation, whether it be of a master or an operator, can only occur in a particular sequence, and in each case activation necessarily entails deactivation. Even if g_3 could transfer an activator to g_2 this must be ineffective since g_2 has already been immobilized. The model does not show that structural elaboration must occur in a particular sequence, but this is highly probable; nevertheless since the catalysts that produce components are only available in a particular sequence, it might be supposed that each component can only become effective if it is integrated with another which is already present. The general principle of inhibition which becomes dominant only in particularly restricted intervals is readily intelligible from the provisions of the model. The activation of each operator gene leads to the release of a particular catalytic complex. The complex is peculiar to the gene that releases it and is different from every other complex. In any phase the complex e_1 released from g_1 is different from any other complex generated during the same phase. Further, it is also different from a corresponding e_1 complex discharged in any other phase. Since the whole cycle is characterized by a sequential provision of distinctive catalytic states it is clear that a situation is available in which arrest can be imposed when any one operator has been activated. The principle that a specific inhibitor is conceivable for every enzyme is generally conceded. It is generally accepted that it is probable that an inhibitor may be available for each enzyme which acts only on that enzyme. If this is the case, then, since each operator produces a different complex, and since the components of each complex may be susceptible to particular inhibitors, it is clear that arrest may develop with the activation of any operator. The inhibitor may of course be supplied experimentally. It may on the other hand be generated spontaneously. A change in external circumstances or a change in metabolite supply may lead to the formation of a compound that can act as an inhibitor for some catalyst in the sequence. It is evident from the model that the inhibitor may not be formed at the stage at which it is effective. Since it is highly particular in its action it may persist through several stages and only become inhibitory when the appropriate enzyme has been released.

In the aberrant conditions that are being considered, after arrest in any one of a number of stages after the S-phase, the system returns to G_1. This is inferred from the fact that after return to an interphase replication may begin. A system such as that of the model allows precisely for this invariable development. When inhibition is imposed at any stage after S a catalyst or a particular group of catalysts is immobilized. This development does not, however, affect the release and transfer of an activator to the next operator in the sequence. Thus if e_1 is immobilized in G_2 this does not affect the action of a_1 or its transfer to g_2. Thus inhibition of the catalyst complex at one point does not affect the subsequent sequence of activations. Since these are not affected, the release of the corresponding enzyme complexes is also not affected. The basic mechanism ensures that inhibition at any point does not interfere with the development that returns the system to G_1. On the other hand, inhibition necessarily prevents the formation of the structural component that is normally formed at that stage, and it is undoubtedly this that imposes the apparent arrest. Since one structural transformation is not completed the whole apparatus cannot be elaborated and the system does not develop cytologically beyond the point of arrest.

An illustration of the pattern of events is provided by the action of colchicine. This substance interferes with the formation of the spindle. It does so possibly by inhibiting the action of a particular group of enzymes. Since the spindle is not formed the structural requirements for anaphase cannot be fulfilled and the cell remains apparently suspended in metaphase. Through activation of the gene for anaphase, however, the development through to the catalytic state of telophase continues and eventually the nucleus is re-formed with a return to G_1. The arrest in metaphase has not interfered with the activation sequence, but it has interfered with structural elaboration, and division therefore does not occur.

It is probable that inhibition may be imposed immediately after replication. It has been proposed that the effect of replication is the formation of so-called "chromonemata". Normally these are distributed between the two chromatids. In certain systems the chromonemata remain in a single bundle. It may be supposed that an inhibitor is formed which suppresses the activity of an enzyme system which secures the separation of the chromatids. This effectively prevents further structural elaboration, but it does not prevent serial activation and development continues through to G_1 and through to another phase of replication. The apparently curtailed cycle may be repeated several times, and eventually a giant chromosome is generated. The most notorious instance of this phenomenon is the formation of the giant chromosomes in *Drosophila*; in plants the abnormality has been observed in certain suspensor cells of the embryo sac. When chromatids have been formed in certain instances the centromere does not divide. Undoubtedly an inhibitor is generated which acts on the enzyme complex that promotes the cleavage of the centromere. Since the chromatids cannot separate, the structural requirements for anaphase cannot be fulfilled, but development continues and the normal activation sequence through telophase is completed. The cell reaches G_1 and another phase of replication begins in due course. The sequence continues and when metaphase is reached diplochromosomes emerge in which four chromatids are held by the same centromere: this condition is frequently observed in the root cortex. A situation similar to that induced by colchicine has been observed in certain tapetal cells. Chromatids are formed and the centromere divides, but a spindle is not organized. Anaphase cannot occur, but development continues and a nucleus emerges in G_1. Another phase of replication is traversed and again metaphase is reached, but now with the diploid number of chromosomes.

In the final stage of division in a number of tissues cytokinesis may be suppressed. This can be induced experimentally by a number of compounds, two of the most effective of these being caffeine and ethoxy-caffeine. When it occurs spontaneously another but similar inhibitor is no doubt generated by the cell itself. Wall formation is suppressed, but this does not depress further development and in each nucleus the master gene for G_1 is activated and another cycle begins. In this second cycle cytokinesis is again suppressed with the consequent generation of the multinucleate state. This elaboration has been observed in the tapetum and has frequently been described in callus cultures.

With most of the situations that may be considered as modifications of the normal cycle, inhibition is probably effective only at a single stage. With some, however, inhibition may be imposed at two stages. This is possibly the case when endopolyploidy is established. This phenomenon is relatively common in tissues such as those of the cortex in the root. In normal division, in the earliest stage of mitosis, after replication, a process of spiralization begins. This it may be supposed is the consequence of the activation of a gene from which an enzyme is derived that induces a particular form of bonding. This enzyme is inhibited, but the suppression of spiralization is not fatal to further development and the nucleus reaches the stage when the nuclear membrane is normally dispersed. At this stage a second catalytic system is inhibited and the nucleus remains intact. The development that carries with it the cleavage of the centromere continues and this ultimately brings the nucleus back to G_1 and back to another phase of replication. The process with double inhibition is repeated through several cycles and arrays of cells become established in which as the linear series is traversed, the DNA content changes and each change involves a doubling.

VI. DIVISION AND ENLARGEMENT

It has already been suggested that division is a primary requirement for growth in the whole organism. The discussion developed above provides the basis for examining the scope of this requirement further. The data of Fig. 3 indicate a close connection between the number of cells and enlargement. It is evident that change in each of the two variables follows a similar course. Since the rate of cell accumulation necessarily

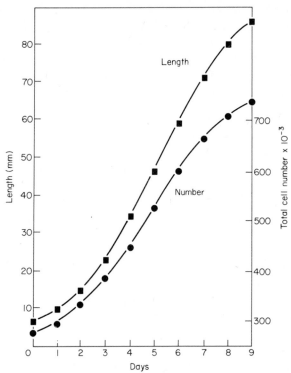

Fig. 3. Comparison of change in length and in cell number in cultured roots.

depends on the rate of division, the similarity suggests that the rate of division may be a rate-limiting step for the rate of enlargement of the system as a whole. The examination of the circumstances through which it can be such, requires some reference to certain decisive characteristics of growth in the intact plant, of which two may be emphasized. Most of the cells in the plant do not divide and division is normally linked to a process of cell expansion. The connection between division and the enlargement of a culture of unicellular organisms is relatively simple. In such a system it can be assumed that every cell divides and that after division the progeny return to the size of the parent cell. Further, it can be assumed that every cell before division is the same size. In this situation the rate of change in the mass of the culture must be directly related to the rate of division. In a higher plant active and repeated division is confined to primary and secondary meristems and to embryonic tissues derived from them. In these systems the cells are not all the same size before division, but they are structurally distinctive and different from the bulk of cells in the plant. Divisions do occur in mature tissues composed of fully expanded cells, but they are not frequent and when they occur they are certainly not followed by other divisions. Primary meristems tend to be at the extremities of the axial structures. Secondary meristems are those such as the cambium (Chapter 10) which are enveloped by mature tissues. Embryonic tissues are of two types: those immediately derived from the fertilized egg, and those that are incorporated in such systems as primordia which are generated from the terminal meristem of the shoot. Clearly, dividing cells in the higher plant are highly localized. Not only so but they tend to be relatively small. The cells in terminal meristems and in embryonic tissues are of the order of 15–25 μm in width. Cells in secondary meristems are approximately 100 μm in length, about 50 μm in breadth and 15 μm in width. The bulk of cells in the plant which are derived from these meristematic cells tend to be much larger. Meristematic cells are non-vacuolated and thin walled. The cells they donate to the plant develop vacuoles which enlarge spectacularly. The enlargement frequently involves a change in length from about 15 to about 200 μm and a change in volume of about 200-fold. Consequently the final size of the plant is primarily an expression of cell expansion. This being the case, and the meristems being highly localized, the circumstances which establish division as a rate-limiting step for the enlargement of the whole plant require definition. Clearly, in a peculiar situation such as that of the higher plant the relation between division and enlargement of the whole system must be considerably more intricate and subtle than it is in a culture of micro-organisms.

Undoubtedly division is decisive for overall growth in two respects. It provides the fundamental units from which the whole is organized and it provides the state from which expansion of the cell can proceed. The first contribution is self-evident and has already been referred to in an earlier connection. The higher the rate of division the greater is the number of cells donated to the system in unit time and the greater to that extent is likely to be the rate of enlargement. The nature of the second is perhaps not immediately apparent. The nature of the mechanism that terminates the expansion of the cell is relevant in this connection. It has been shown in at least three investigations that the arrest of cell growth is under nuclear control. The effect on cell growth of exposing the nucleus alone in *Micrasterias* to ultraviolet light has been examined (Selman, 1966). The treatment disrupts nucleic acid components and it was observed that the cell becomes abnormally large. It has been recorded that while the cells in cereal embryos that have been exposed to intense gamma irradiation do not divide, nevertheless they expand to an abnormally large volume (Haber, 1962). The irradiation disrupts the nucleus and it is no doubt as a result of this that division is arrested and expansion promoted. The third item of evidence is derived from observations on the effect of certain analogues of the nucleic acid bases on the growth of segments excised from the extending zone of the root (Heyes, 1963). It has been shown that the stimulator accumulates in the nucleus and therefore certainly interferes with the formation of the nucleic acids that are normally released from it. In each of the three connections the treatment has had the effect of depressing the release of information that normally arrests growth, and since this control from the genome is suppressed, growth can proceed at a higher rate over a longer time. The implications of this conclusion are of decisive importance. Clearly, the situation is such that when growth ceases a metabolic state is established in which further growth cannot occur, and however small the cell may be when growth has been arrested, in normal circumstances growth cannot be resumed. From this a further important inference may be drawn. If the capacity to grow is dissipated by growth then that capacity must be available only before growth begins. Evidently the cell is only in a state to grow when it has just been formed. The cell is formed in division, and division in a sense prepares the cell for growth and is a necessary condition for that growth.

It has been shown that the interphase necessarily requires the assembling of a growth component. This activity is independent of the elaboration of the division apparatus, but it proceeds in the environment of the interphase and consequently acquires particular qualitative characteristics. In the circumstances of the interphase particular aspects of the genome are mobilized and the information released becomes incorporated in the growth component. This, it may be supposed, endows the system with the capacity for further growth. It is of interest that this inference is also implied in the claim that division is necessary for redifferentiation. It is a fact of observation that the transformation of cells from one morphological form to another normally only proceeds when the cells divide. Mature cells may divide and the products may then develop into another form. The significant aspect here is the fact that the assumption of a particular form is only possible when growth is proceeding. A cell grows into a particular form and it cannot acquire that form through any other device. A cell is not constrained into another form while retaining the same volume. Thus for a transformation in form division is required. It is required since it is only through division that the growth can be secured through which another form can be established.

Since division establishes the condition for the expansion of the cell it must constitute a rate-limiting step in the determination of the enlargement of the integrated system, and this is evidently the basis for the observed correlation between rate of change in cell number and rate of change in size (however that size is expressed). Since this is the position, then the factors that influence the rate at which the cycle of Fig. 2 rotates are of basic importance for the analysis of growth. Since the cycle depends on several categories of metabolic reactions it is perhaps not surprising to find that the rate of rotation is influenced by the normal conditions of temperature, light and aeration that might be expected to influence such reactions. In roots all stages of the cycle are accelerated as the temperature is raised from 15 to 30°C, the least sensitive stages being anaphase and metaphase. As the temperature is raised from 15 to 25°C the overall cycle and the duration of each of the individual stages are reduced approximately by half. Evidently the transcription process and the assembling of the division mechanism are both accelerated with increasing temperature. Observations on artichoke tuber tissue have also shown an intimate effect of temperature on the different stages of the interphase. It is of particular significance that replication is apparently as sensitive to this factor as any other stage of the cycle.

The effect of anaerobic conditions is more complex (Amoore, 1961). It has already been stated in another connection that while interphase is arrested by anaerobic conditions mitosis is not. Evidently the synthetic processes that continue in interphase require a higher energy level than do the transformations of mitosis. The evidence suggests that mitosis can be maintained by the energy released in glycolysis. The difference in reaction suggests the possibility that the synthetic reactions of the cycle are particularly extensive in interphase. In mitosis it is probable that the changes that are enzyme mediated involve principally minor adjustments through bond formation or bond breaking in a framework that is acquired from the interphase. The difference in reaction has a peculiar effect on the conformation of meristems. Since in anaerobic conditions interphase is arrested but mitosis continues, in these conditions the mitoses that have begun at the time oxygen is withdrawn are completed, but they are not replaced by others from the interphase. As a result mitotic figures disappear from the meristem completely and it is a characteristic feature of meristems that have been exposed to anaerobic conditions for some hours that they only contain interphase nuclei. Normally, anaerobic conditions are established by introducing into the experimental vessels cylinder nitrogen. This always contains traces of oxygen. It has been reported that if these traces are removed, mitosis is then arrested. The traces are certainly not great enough to influence respiration, and it is therefore probable that some component in the system has a high affinity for oxygen, and that mitosis can only continue when this component is in an oxidized state.

The influence of light varies. It has been shown in cultured artichoke tuber tissue that the proportion of dividing cells participating in the first cycle after transfer of the explant to the culture medium is reduced drastically by exposure to light (Yeoman and Davidson, 1971). It has also been shown that sensitivity to light is restricted to G_1. A similar depression due to exposure to light has recently been recorded with other systems (Street, 1973). In leaves the reverse effect has been recorded (Dale and Murray, 1969). With certain species when plants are grown in darkness they become etiolated. One of the features of this state is a restriction in the growth of the leaf due to a failure of division. In certain instances an exposure to light for 10 min has an immediate effect. Division is induced and the number of cells increases abruptly. There is some evidence that this effect is phytochrome mediated.

The rate of division, while being influenced by environmental factors, is also controlled by internal states of the multicellular system: when external conditions are maintained at a constant level, variations in the rate of division still occur. This category of effects is clearly of profound importance for the interpretation of the course of enlargement in the whole plant. Of the factors internal to the plant two seem to be of primary significance: the supply of nutrients to the meristem from mature tissue, and the accumulation of inhibitors in the meristem. An effect of mature tissue on division in the meristem may be demonstrated by comparing the effects of root-tip inocula of different sizes on the subsequent course of change in the number of cells (Brown and Wightman, 1952). In one investigation inocula of three lengths were used: 3, 6 and 10 mm. Considerable differences were recorded in the times at which division began. With the shortest inoculum this stage was only reached after a lag phase of 3 days had elapsed. With the inoculum of intermediate length a lag phase was still traversed, but it was considerably shorter than that required with the smallest. With the longest tip no lag phase was observed. Division began immediately the inoculum was transferred to the culture medium. The significance that can be attached to these results is derived from the morphological differences between the different inocula. The shortest was composed only of non-vacuolated cells, the intermediate incorporated some cells that had begun expansion and the longest carried some cells which had reached the mature state. It may be suggested that division began immediately in the longest inocula since they carried mature cells. The long lag required with the shortest represented time required for the cells nearest to the cut surface to traverse a development into the fully expanded state. With the intermediate inoculum the shorter lag was a consequence of the inclusion in it of cells that had already begun expansion and which therefore required a shorter time to reach the mature state.

The influence of mature tissue is further shown by a consideration of the course of change in the number of cells when division has begun. The number of meristematic cells does not change, but the total number of cells in the root tends in the initial stages of culture to increase exponentially. Evidently the absolute increment increases with time. This, since the number of meristematic cells is remaining constant, necessarily implies that the rate of division is increasing correspondingly. The acceleration is being accompanied by an increase in the mass of mature cells. The pattern of events is certainly compatible with the suggestion that division in the meristem depends on metabolites supplied from mature regions and that as the mass of these increases so consequently does the rate of division. It is of some importance that further precision can be given to this generalization. The discussion developed earlier indicates that by the rate of division is meant the rate at which the normal cycle of division shown in Fig. 2 rotates. When intact cells are being discharged the cycle involves two phases: a primary transcription and a secondary structural elaboration. In terms of this formulation it is clear that the enhanced metabolite supply accelerates a transcription process and the assembling of structural components.

The necessity for mature tissue implies limited synthetic capacity in the meristem. It has been shown that the respiration rate in the meristem as a whole is relatively low. This is clearly a feature of some importance. It has been emphasized that respiration in the interphase is higher than it is in mitosis. But it is evident that the interphase rate is higher only in a relative sense. The general rate for the meristem indicates that it is lower than the expanding and mature tissue rates. In view of this the limited synthetic capacity is intelligible. The only form of synthesis which is unavoidable in either phase of the cycle is the formation of linkages in macromolecules. A situation can be envisaged in which the primary units such as amino acids and nucleotides are elaborated in mature tissue and supplied to the meristem, and these are elaborated into macromolecules in this zone.

The period during which the rate of cell accumulation increases is followed by another in which the rate declines. In the same sense, as the first is due to an accelerating rate of division, so the second is an expression of a progressive decline in this rate. The deceleration is intensified with time and a stage is reached when division is arrested completely. The decline may be due to one of two factors, or to a combination of both. It may be due to a change in the metabolite supply that is provided from mature tissue or to an accumulation of an inhibitor which is generated by the meristem itself. It is conceivable that while mature tissue may provide the primary requirements for meristematic activity, nevertheless senescent tissue may contribute either an inhibitor or a metabolite supply with an inadequate composition, and inevitably as growth proceeds senescent tissue becomes incorporated into the oldest regions. On the other hand several considerations suggest that the alternative interpretation is the more probable. It is a common experience that when sub-culturing from root cultures, the most effective inocula are those taken from comparatively young roots. Older meristems, even after isolation, do not give the most vigorous subsequent growth. This suggests that an inhibitor has accumulated in the dividing system with age, and that it must have been generated in the meristem. That such a succession does occur is shown more decisively by the change in the rate of division in different parts of the terminal meristem of the shoot.

Data have been assembled which show the number of cells and the rates of division in the dome in the first five primordia and in the first five internodes in vegetative plants of *Lupinus albus* in each plastochron from the tenth to the eighteenth (Brown and Sunderland, 1975). The data are shown in Fig. 4. The derivation of the situation in plastochron 10 is indicated by the additional extrapolated data that originate in plastochron 6. It is evident that in the dome, in each internode, and in each primordium, the rate of division decreases by a constant factor with the transition from one plastochron to the next. These changes must be interpreted in the context that the length of the plastochron is constant. Clearly, in all morphological segments the rate of division decreases with a constant rate of deceleration. It is difficult to reconcile a constant rate of deceleration with a changing metabolite supply from mature regions, since this might be expected to be subject to erratic fluctuation. The rigidity in the rate of deceleration is more closely compatible with the production, in each division cycle, of a constant quantity of a depressor which tends to decelerate proportionately. Since the rate is declining when the mass of the plant is increasing, the restriction is probably not being exerted through the assembling of the division apparatus. It is almost certainly being exerted through the rate at which the successive genes in the cycle are being activated. In other words it is probably being exerted through the process of sequential transcription.

The considerations presented with regard to internal regulation in terminal meristems are applicable also to embryonic systems such as primordia. In these also, it is almost certainly the case that the rate is determined by a metabolite supply and by the deposition of a depressor within each cycle of division. It is significant that both in the terminal meristem and in the embryonic system the final state is often one of cessation of division. At this stage it may be supposed that sufficient depressor has accumulated to arrest division completely. Frequently of course the terminal meristem in the shoot develops into a flower. In many species, on the other hand, it does not. In contrast no comparable development complicates the situation in the root. In fact both in terminal meristems and in embryonic systems the succession of developmental stages may be the same: an initial period of rapid division, another of declining rate and a final state of arrest. It is of some interest to notice that in view of the similarity between the sequence, in say the leaf and the root, the traditional distinction

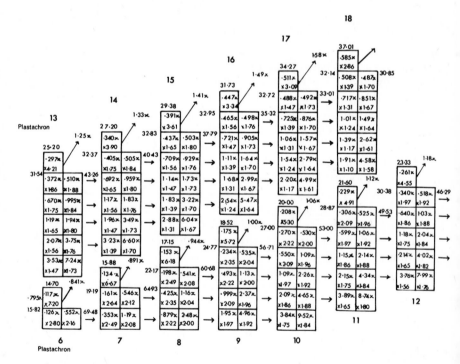

Fig. 4. Diagram showing the change in the number of cells in each internode and at the apex of the shoot of *Lupinus albus*. The diagram shows the change in the average rate of division in the progression from one plastochron to the next.

between organs of limited and organs of unlimited growth becomes difficult to maintain.

The mechanism that controls the decline in the rate of division in primary meristems must of course in certain circumstances be complemented by another which dispels the inhibition. If repeated division promotes the accumulation of a depressor in the cambium of the woody perennial, circumstances must be such that the depressor is not formed or is dissipated after formation. A similar condition must operate when the primary meristem or the embryo is first formed since they are both elaborated after a series of divisions. The case of the perennial cambium is instructive. This may persist for hundreds of years and activity in it may continue apparently indefinitely. Unfortunately little is known about activity in the cambium, but two characteristics of division within it are evident from simple observation: (1) activity is highest in the spring and gradually declines throughout the summer (Chapter 10); (2) within it activity is intermittent, since it is completely dormant throughout the winter. The first feature is compatible with a declining rate due to the accumulation of a depressor; the second suggests that with the cessation of activity a development may occur in the growth component which involves the release of a catalyst which degrades the depressor. It is significant that activity in the meristem in an annual plant is preceded by a period of dormancy in the seed. While the seed is being formed and water content is still high a depressor is no doubt degraded through the same mechanism as it is in the cambium. Immediately division begins the dissipating mechanism is dispersed since the growth component returns to a primitive state.

The discussion that has been developed above shows that in the process of enlargement division is a rate-limiting step and that variations in the rate of division, however induced, must therefore have a corresponding effect on the rate of enlargement of the whole plant. Various considerations indicate that division must also have a profound effect on development and differentiation.

VII. DIVISION AND DIFFERENTIATION

It is widely accepted that the development to the flowering state is accompanied by changes in the conformation of the vegetative meristem. The data of Fig. 4 show that the number of cells in the terminal meristem increases from about 300 to about 1600 over 12 plastochrons. In the determination of this increase a change in the rate of division is a contributory factor. To the extent, therefore, that increase in the size of the dome is an important developmental change, division is contributing to the progress of development.

The contribution to differentiation is more spectacular, although possibly not more decisive. The effect on differentiation is exerted through changes in rate, and through two aspects that have not been considered above—the plane and position of the partitioning wall, and the relation of this to the determination of the pattern of polarity.

The data of Fig. 4 show that a variety of effects on differentiation of the whole plant may follow from changes in the rate of division. With time, the rate of division in the dome decreases. This has the effect of inducing a progressive decrease in the size of the first primordium. This decrease is accompanied by a decrease in the rate of division in this primordium. These two changes must undoubtedly affect the final size of the leaf and thus through changes in rates of division a progressive change in the mature size of the leaf is induced. The rate of division decreases in each primordium and in each internode as it develops from one plastochron into the next. The change in the primordium ultimately leads to the cessation of division in the emergent leaf, in the internode it presumably has another, and for the plant as a whole, a highly significant effect. It will necessarily lead at a certain level from the apex to a generalized cessation of division except in a remnant of cells which constitute the cambium. This generalized cessation marks the limit of the apical meristem and the deceleration in the rate of division has therefore the morphological effect of confining the meristem to the apex.

The position and plane of the partitioning wall may affect differentiation by determining the size or the metabolic characteristics of the product cells (Chapter 6). The topic is one of some importance and the mechanism involved merits some attention, especially as this may clarify the nature of the significance that can be attached to it. In the earlier stages of the investigation of the subject various rules were enunciated which were supposed to describe the spatial restrictions to which additional walls conformed. Sachs proposed that the new wall tended to segment the cell into two equal progeny and that it tended to be aligned at right angles to parental walls. Later Errera (1888) formulated the general rule that the additional wall is always deposited in the plane of the shortest axis since this provided for the condition of minimum surface. These generalizations, even if valid, are not of course explanatory. But they are not in fact universally valid. It is not true to say that the product cells are always the same size. In a large number of instances the difference in size is considerable. Again it is not invariably the case that the additional wall is disposed at right angles to one of those of the parent cells. The additional wall in callus cultures, for instance, is often slanted obliquely across the cell. It is true that frequently the wall is deposited along the shortest axis but certainly not invariably. In a cambial cell it is formed in the plane of the greatest width and the greatest length.

Two factors have been identified as being decisive in the determination of the plane of division: mechanical stress and polarity. It is possible and indeed probable that dissimilar though they are both factors exert an effect through essentially the same mechanism. The plane of division can certainly be altered by the application of abnormal mechanical stress (Yeoman and Brown, 1971). The effect of this has been examined by culturing a long inoculum which has been bent through 180°. Samples were taken and after sectioning the distribution of walls examined. Bending has the effect of stretching the tissue over the external surface and of compressing it over the internal. Over both surfaces the plane of wall formation was disturbed and over both the nature of the disturbance was the same. Whereas in the normal situation the additional walls are deposited in a plane parallel to the surface, in the stressed tissue over both inner and outer edges they are disposed at random. Clearly, in the interpretation of this observation two aspects must be distinguished, the transmission of the stress and the reaction to it. It is inconceivable that mechanical stress can be propagated through the protoplast. For this at least, a semi-rigid structure such as the wall is required. At the same time it is inconceivable that the reaction could be due to the wall. This must arise in the cytoplasm. The general effect is intelligible if it is assumed that the reaction is due to a disturbance in the plasmalemma and that this is in some way anchored to the wall. It may be supposed that stress induces a stretching or compression in the wall, that this transmits a corresponding distortion to the plasmalemma, and that this is the immediate origin of the disturbance to the normal arrangement. It has been suggested that the plasmalemma carries two reactive groups A and B which are distributed in the following pattern:

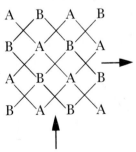

It has further been suggested that when the two reactive groups are brought together the structures that determine the plane of the wall are propagated from this position. It is evident that the two groups may be brought together either through stretching or compression. When a cell is subjected to stress it is highly improbable that the mechanical effect can be uniformly distributed. It is likely to be greater in certain positions than it is in others. It is likely to be least in a longitudinal wall where tensile strength is enhanced by conjunction with a transverse septum and greatest in the mid-position of an uninterrupted course. In the experimental situation in which bending is applied abruptly the lines along which association is established are necessarily random. In normal circumstances, the stress increases gradually along a single axis, and the two reactive groups tend to be brought together first at the point where the effect of the stress is greatest, which is midway between the points of attachment of transverse walls. In terms of mechanism it is possible that when association has occurred along a certain line the protein referred to earlier is propagated across the cell from this line. Microtubules become concentrated in the periphery of the baffle and the nucleus migrates into it. In due course the microtubules are displaced into the baffle and become organized into a spindle which is aligned at right angles to it. With this provision the wall is inevitably formed in the plane defined by the limit of associated reactive centres.

Mechanical stress undoubtedly determines the plane of division in the apical meristem, in the cambium and in tissues such as the cortex. Various considerations suggest that in the apical meristem of the root the tissue is compressed. It is a prominent feature of at least the basal region of this system that walls are deposited at right angles to the long axis of the root and this, it may be supposed, is a consequence of the compression stress exerted along this axis. The cambium is also undoubtedly compressed through a radial stress and it is this which certainly determines the formation of an additional wall in a plane which is parallel to the surface and in the long axis of the cell. Divisions in the cortex are frequently anticlinal. In this instance the plane of the additional wall is certainly being determined by a stretching tangential stress. As secondary thickening continues the circumference increases and the peripheral tissues become stretched. The component cells react by forming walls in a longitudinal plane aligned along the radius.

In a filament of cells stress exerted along a single axis cannot be the operative factor. In a system such as that of the protonema of the fern it has been suggested that transverse alignment is a consequence of a polar state in the direction of the long axis. Along this axis the cell is no doubt polar with respect to a variety of conditions, but in particular it is claimed that it is polar with respect to the concentration of auxin (Wettstein, 1965). Convincing evidence is available which shows that the polarity induced in the *Fucus* egg and in the spore of *Equisetum* by unilateral illumination depends on an auxin gradient. If a comparable concentration gradient is generated in the dividing cells of the protonema, and if this determines the transverse alignment, then it should be possible to disturb the normal pattern by supplying auxin since this should tend to establish uniformity. It has been shown experimentally that the application of auxin does in fact lead to random alignment. When auxin is supplied in the medium a callus tends to develop.

The auxin gradient cannot of course promote either a longitudinal stretching or a longitudinal compression. On the other hand the reaction may depend on a mechanism similar to that which is involved with stress. It is relevant in this connection that in the *Equisetum* spore it has been shown that the probable auxin gradient is carried in the superficial layers of the protoplast. A reaction from the plasmalemma is therefore conceivable. One of the few well established primary effects of auxin is that on the plasticity of the wall. It tends to enhance this, but probably only at particular concentrations. It is probable that the plasticity reaction changes in the same way as other responses to auxin. As the concentration is increased from a low level the effect increases until a maximum is reached. With further increase in concentration the effect is likely to decrease. It is possible that in the protonema cell a sub-optimal level is established at one end and a supra-optimal level at the other. The optimal concentration giving maximum plasticity would in that event be realized at an intermediate point. At this point maximum plasticity and therefore maximum stretchability is realized. The wall is under uniform stress through turgor pressure, but due to the auxin gradient, at one localized point along the axis the degree of stretching will be greater there than it is elsewhere. This localized maximum stretching which is induced from the plasmalemma necessarily imposes a distortion. The effect of this is the same as that due to mechanical stress. The two sets of reactive centres are brought together, and from a narrow region a wall is consequently generated across the long axis.

The plane of division is important in two connections, both of which are decisive for the pattern of differentiation. It may determine the subsequent direction of growth in the product cells and it may control the elaboration of different metabolic pattern in the progeny. It is likely that the direction of growth must be at right angles to the plane of the additional wall for two reasons. Firstly, while the partition is being deposited the walls of the parental cell at right angles to it are necessarily growing. The growth is controlled probably by the elaboration of microtubules within the plasmalemma. These tend to be disposed in what are virtually loops within the wall. From these, corresponding loops of cellulose are developed. When additional microtubules are contributed to the system they will tend to be deposited between those already present and thus the constraining pattern of cellulose microfibrils will tend to perpetuate itself. Secondly, a system of continuous microtubular loops in the direction at right angles to the plane of division is unlikely, since this probably requires a uniform plasmalemma. In a later paragraph it is shown that the membranes against the fresh wall are probably metabolically different from others. This being the case a uniform plasmalemma can only be secured against the walls at right angles to the additional member. Growth will therefore tend to be restricted to these walls and they will tend, through the deposition of cellulose microfibrils in loops, to grow in a direction at right angles to the plane of division. Casual observation shows that the expectation is justified. In the axial structures in which division, in the basal zone of the meristem, leads to the deposition of walls predominantly at right angles to the long axis, enlargement growth is in the direction of this axis. In the cambium the deposition of additional walls is in a tangential plane but subsequent growth is in the radial direction and therefore at right angles to the plane of wall formation. It may be noted that if the interpretation proposed is justified then the rule of Errera (1888) acquires a peculiar significance. In the mature cell the additional wall is often across the shortest axis, since this is precisely the plane in which growth has been restricted. Since growth is most intense in the plane at right angles to the plane of division this growth has in a sense established the additional wall along the shortest axis.

The plane of division has the further significance for differentiation that it may determine the emergence of two cells with different metabolic endowments. The additional wall may be deposited between the two extremities of a strongly polar cell. If it is, then the product cells immediately after division are likely to carry protoplasts with different constitutions and therefore with different developmental capacities.

VIII. INEQUALITY OF PRODUCT CELLS

The release of cells with different potentialities introduces a topic which merits more extensive treatment than it normally receives. It is commonly assumed and frequently asserted that because division leads to the formation of two identical nuclei it therefore leads to the establishment of two identical cells. The doctrine, it must be conceded, is certainly plausible. If it is accepted that the genome controls the characteristics of the organism as a whole then it is surely appropriate to infer that it determines those of the individual cell. And if this implication is recognized, then there can be no objection to the further inference that identical nuclei necessarily generate identical cells. But this argument entails an embarrassing paradox. If the cells that carry the identical nuclei are themselves identical how do they ultimately differentiate into different structural and dynamic entities? The conventional escape from this difficulty requires an appeal to the influence of a non-uniform environment. It is maintained, in principle, that although the initial position may be the same, the ultimate state is different since the product cells are exposed to different environmental conditions during development. The argument is specious and is belied by observation. The argument is specious since it relies on the assumption that the cells initially are in fact identical and it is belied by observation since it is difficult to envisage an environmental diversity that could match the intricacy of differentiation. That could account, for instance, for the alternation of two different types of cells in a longitudinal rank. The doctrine of product identity is manifestly untenable and a convincing argument can be elaborated to support the contention that, far from yielding identity, division ensures diversity. Appeal both to the implications of certain characteristics of division and to the evidence supplied by observation suggests that invariably the two product cells generated from a single initial are different and distinctive from each other and may subsequently traverse different developmental histories.

The nature of the difference may vary. One form is described in an earlier paragraph. Instances are common in which ranks are formed in meristems in which the two ends of each cell are visually different. One end is often denser than the other, and it is always the same extremity that is the denser. Such a system develops in the epidermal layer of the apical meristem of the roots of certain grasses. In due course these cells divide transversely, giving an alternation of cells with different configurations. The two types of cells incorporate different metabolic characteristics and therefore different developmental capacities. The cells with the less dense contents enlarge, but they retain their original form. These with the denser contents enlarge slightly and form a lateral protrusion which develops into a root hair. In the final state a rank is established in which cells carrying root hairs alternate with simple rectangular parenchymatous cells. Another instance of the same type may be seen in the epidermis of the intercalary meristem of the green leaf. Again, ranks are generated in which each cell has a dense

apical and a less dense basal end. The cells divide transversely and alternating clear and dense cells emerge. The dense units become stomatal
mother cells, and ultimately a system is established in which successive
stomata are separated by simple rectangular parenchymatous cells.

The two product cells may differ in size and this catagory of distinction, it has been claimed, may have important morphological consequences. The circumstances that promote the difference may vary.
It has been shown that with slow increase of stress, whether it be stretching or compression, association of the appropriate reactive centres is
likely to occur along a transverse line which is remote from zones of
stability associated with transverse walls already formed. In an isolated
cylindrical cell the line is likely to form midway between the two
extremities. In a tissue this tendency may be disturbed by the incidence
of transverse walls in adjacent cells. Transverse walls in different cells
are never at the same level. A transverse wall in one cell may be midway
between the transverse walls of an adjacent cell. The transverse wall
of one cell is likely to influence the mechanical properties of the longitudinal wall of another and thus when stress is applied the line of
association in any one cell is not likely to occur at the mid-point between
two extremities. It is likely to be nearer one end than the other, and
this must lead to one product cell being smaller than the other.

The difference may arise through another variant. The possible effect
of an auxin gradient has been discussed. It has been proposed that auxin
concentration may be too high at one end and too low at the other.
The optimal concentration is reached at an intermediate point, but
the precise distance of this point from one end or the other may vary.
The origin of this variation is shown by the diagram of Fig. 5. The
broken horizontal line indicates the optimal concentration. It is clear
that the point at which this is established will depend on the course
of change in concentration from one end to the other. In the diagram
the effect of variation in this is shown by continuous lines. If the change
with unit displacement is constant then the critical level may be
attained at a central point. If, on the other hand, the change is sharper,
with displacement at one end or the other, then the critical level is likely
to develop at the end which carries the sharpest rate of change. The
ultimate effect must be to promote deposition of the transverse wall
nearer one end than the other, with the emergence of cells that differ
markedly in size.

It has been claimed that size difference is the basis for all differences
in cellular differentiation between the two cells from the same parent.
Certainly the number of instances in which a size difference is associated with ultimate different states is impressive (Bünning, 1957). When

the sieve tube initially divides it gives two cells that differ in size. The ultimate state and form of each is different, one becomes the sieve tube and the other a companion cell. The *Equisetum* spore divides to give a smaller cell which becomes a rhizoid and a larger which becomes

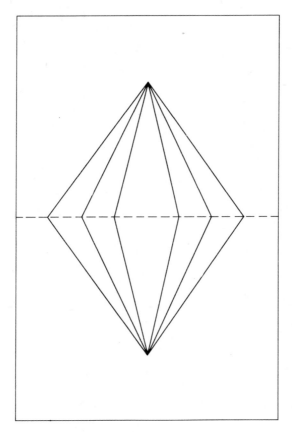

Fig. 5. Diagram illustrating polarizing effect of cell division. Broken line represents additional wall formed at the equator of the spindle.

the origin of the thallus. Embryonic cells in the leaf of *Sphagnum* divide to give cells that differ in size. The smaller develops peculiar wall thickenings and the protoplast that it acquires ultimately disintegrates. The larger cell divides again at right angles to the first division and again yields two cells that differ in size. The smaller cell changes in the same pattern as the smaller cell from the first division. The larger cell becomes a typical mesophyll cell.

IX. DIVISION AND POLARITY

While the association between size difference and difference in differentiation is frequent it must be recognized that difference in differentiation often follows from a situation in which no size difference is discernible, and when a size difference has been generated this is not invariably followed by different developmental histories. General observation suggests that the connection between size difference and differentiation difference is not causal. When difference in size is generated other distinctions are almost certainly involved. It is probable that the difference in size is incidental, and when this is accompanied by difference in development, it is these other distinctions that are decisive. The case of the piliferous layer is instructive. The root hair cell is segmented from a cell which is visibly polar. The plane of division separates two systems with different metabolic states. The division is not median and the two product cells are not the same size. In this instance there can be little doubt that the difference in size is irrelevant and that the decisive factor is the difference in metabolic state.

The evidence reviewed above shows that division depends on the polar state and that one aspect of diversity is also derived from it. But it is probable that, while exploiting incident polarity, division in a sense generates it; further, it is a necessary consequence that the product cells must be different. The implications of this for the progress of differentiation in the plant as a whole are extensive.

The diagram of Fig. 6 shows that division has a polarizing effect. When the wall is deposited across the equator of the spindle a plasmalemma and associated cytoplasm are elaborated on both sides. There

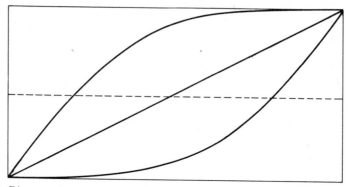

Fig. 6. Diagram showing possible effect of pattern of change in the oxygen level on the position of the wall. Broken line indicates a particular auxin concentration.

is little doubt that the metabolic properties of different components of the protoplast change with age. Thus the recently formed cytoplasmic system with its boundary membrane is likely to be opposed by an older corresponding system with different metabolic characteristics. Between the young and the opposed surfaces gradients are inevitably generated with respect to a variety of metabolic attributes. It is probable that many of the catalytic products generated in the cycle are deposited at the periphery. Many of these are released before cytokinesis and can therefore only be deposited in the older surface. If division has a polarizing effect, then since all cells are generated through division, all cells are necessarily polar. The suggestion that all cells are polar is not recent: it was first made by Vöchting in 1906. The suggestion is certainly compatible with observation. While a polar state need not have visible attributes, a surprisingly large variety of visible expressions of this state have been described (Vöchting, 1906).

While division induces polarity, through this it also necessarily imposes diversity, and with respect to two characteristics. Since the membranes against the additional wall are of the same age and since they are both between two older systems it is evident that the gradients must be disposed in opposite directions. Secondly, since opposite walls in any cell must be formed on different occasions the ages of the membranes associated with them must be different and their metabolic characteristics correspondingly different. This being the case, the steepness of the gradients in the product cells must differ. It is an important feature that if in any one cell repeated divisions occur, with the additional cell always being released on the same side of a parent cell, then the gradient within the terminal cell must increase with each division cycle. This must occur since with each cycle the cytoplasmic system within the outermost wall acquires an additional age increment with an additional endowment of catalytic division products.

The situation that is necessarily generated through division yields distinctive patterns of polar states in a tissue according to the distribution and the plane of additional walls. The simplest situation is that shown in the diagram of Fig. 7. In this, division always occurs in a single plane and only in a single terminal cell, the additional cell always being contributed on the same side of the terminal cell. The wall that is formed in the last division is designated by 0, the one formed in the division before that by 1, and the one formed in the division before that again by 2. In the diagram the two original walls are both marked 1 in the interests of convenience. With each division cycle the relative age of each wall increases by one unit. What was wall 1 becomes wall 2 in the next cycle, and what was 2 becomes 3, and so on. The changes

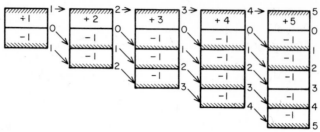

Fɪɢ. 7. Diagram showing consequence in terms of polarity of division, where this is restricted to a terminal cell.

in age of the walls are shown by the arrows connecting one rank with the next. The age of the wall shows the corresponding age of the cytoplasmic system on either side. But the cytoplasm on the upper side differs from that on the lower since it incorporates the products of one cycle. It is only the surface that incorporates these products, since when the cell is formed it is this surface that is undisturbed. This feature is emphasized by the difference between surfaces 0 and 1 in the same cell. The surface occupied by division products is indicated in the diagram by shading. It will be noted that with this pattern of division an array of cells is produced in which successive walls differ by one unit and the increment is always in the same direction. Further, the lower surface in any one cell always differs from the upper. In the system an array is generated in which the gradient is always the same, is always disposed in the same direction, and in each cell the basal surface is always the same and differs from the apical. The formation of a rank of cells in which the polar gradient is always the same is not incompatible with the claim that division promotes diversity. The products of the terminal dividing cell differ markedly with respect to the direction of the gradient of polarity, and with each cycle of division the gradient becomes steeper in the terminal cell.

Clearly, the hypothetical situation that develops when divisions are restricted to a single terminal cell may be considered as a model for the mechanism through which ranks of polar cells are established, in for instance the epidermis generated from the intercalary meristem of the grass leaf. It is probable that the ranks are each generated through some mechanism similar to that suggested in the model. It is evident from the diagram that when the array has been generated a further division in the same plane would separate different extremities and give cells which carry division products alternating with others that do not.

The situation becomes more complicated when all the cells in the array divide. The consequences of this situation are most sharply defined when these divisions are all in the same plane. The implications of this pattern are shown diagrammatically in Fig. 8. For convenience again an original cell is envisaged with opposite walls of the same age. This divides and a wall number 0 is deposited between the two original walls each marked 1 (stage I). The division products are arranged against the original walls. In the next cycle each cell divides and additional walls are laid down between the walls of the product cells of the first cycle (stage II). When the second division has been completed the walls of stage I have each aged through one unit. The walls marked 1 become 2 and the cytoplasm associated with them acquires two increments of division products. The additional wall of the first division acquires a relative age of 1 and it is significant that the cytoplasm on each side of it carries one increment of division products. In the diagram the four cells of stage 2 divide to give eight cells and these in due course divide again to give sixteen cells. When this fourth division has been completed a highly suggestive pattern emerges. The relative ages of the walls range from 4 to 0. At one end of each cell, division product accumulates and it is on both sides of all walls except the one

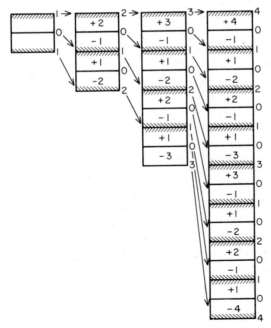

Fig. 8. Diagram showing consequence in terms of polarity of division where every cell in the lineage divides.

formed in the last division. Two features in the array are particularly significant. The direction of the gradient alternates and the steepness fluctuates. In the whole series the gradient oscillates and in the centre two cells emerge with steeper gradients than those in any other. The two cells with the gradient values of 3 are polar in opposite directions. One result of this particular pattern of division has therefore been to promote the development of cells which are unique.

   The intriguing phenomenon has been described of the incidence of a colourless cell in a filament in which all the other cells contain antho-cyanin (Blakely and Steward, 1961). The filament was developed in a callus culture and probably through a pattern of division similar to that of Fig. 8. The unique cell in the environment of the culture could not possibly have been generated by peculiar local conditions, and it is probable that it represents the effect of a unique polar state. The sequence of Fig. 8 may be taken as a model for situations in the intact plant, and it may be the basis for a variety of differentiations which require special pleading to attribute them to local conditions. It is diffi-cult for instance to invoke a non-uniform environment to account for the development of hairs from an epidermis or for the commitment of certain cells in the dome of the terminal meristem to the formation of leaf primordia. These emerge from an apparently uniform embryonic tissue and it is probable that they are derived from unique cells de-veloped in a sequence where every cell divides.

X. CONCLUDING COMMENT

Finally, while it must be recognized that division through ensuring genetic stability and cytoplasmic diversity has a profound influence on all aspects of growth, it must also be recognized that it does not deter-mine the whole content of enlargement, development and differentia-tion. Factors intrinsic to other processes make additional contributions to each of the three phases of growth. Moreover, it is probable that the consideration of division as a separate stage in the overall growth process is in itself a distortion. It is not improbable that the full impact of division is only exercised through an interaction with other processes. But this consideration introduces a topic which is beyond the scope of this chapter.

B. The Process of Cell Division

2. The Visible Events of Mitotic Cell Division

A. F. DYER

Department of Botany, University of Edinburgh, Scotland

I. INTRODUCTION

Many of the advances in our understanding of the mechanism of cell division are now coming from studies at the molecular level and many of the biochemical events recognized take place when the cell preparing for division shows little visible change (see Chapter 3). Nevertheless it is microscopy that still yields the basic framework for our attempts to assemble a complete description of the remarkable process by which cells replicate. Light microscopy has been extended by electron microscopy, which has both provided more detailed information on previously recognized structures such as centrioles and centromeres, and revealed unsuspected structures such as the synaptonemal complex and ribosomes.

The chronological sequence of visible changes in structure and distribution of cell components during the division cycle provides a means of correlating the detailed results of more sophisticated analyses. For practical reasons these analyses frequently involve a very limited number of cell types from a few experimentally amenable tissues and organisms. Comparative microscope studies of a wide range of material can help to put these observations into a broader biological context, drawing attention to the features which are common to many dividing cell types and those which are unique to one type.

The same microscope studies can remind us of neglected problems. Early cytological studies raised important questions concerning the division of the nucleus and the cytoplasm, but because the nucleus is conspicuous and was found to have a remarkably precise division process linked to its unique role in heredity and development, attention was soon concentrated on it in preference to other structures visible during division. Despite this concentration of effort and the acquisition of much detailed information, most of the basic questions about the nucleus remain with us. We still do not understand the structural organization of the chromosome, the mechanisms of chromosome movement and spindle action or the process which controls the sequence of linked events in division. However, now that more is known of the structure and activities of the cytoplasm, it is clear that the same attention must be paid to the replication and segregation of extra-nuclear cell components in order to achieve a balanced picture of cell division in relation to subsequent cell development (see Chapter 4). Here microscopy can still yield valuable new information.

It is not the intention in this article to recount in chronological sequence the co-ordinated events visible during cell division. Such accounts can be found elsewhere (Mazia, 1961; Rhoades, 1961; Brown

and Dyer, 1972; Lewis and John, 1964; Mitchison, 1971; Luykx, 1974).
Nor is it the intention to present a balanced review of the available
information on all the structures concerned with division, such as the
spindle, centriole and chromosome (see for example Luykx, 1970, 1974;
Henderson, 1969; Forer, 1969; Wettstein and Sotelo, 1971; Bajer and
Molè-Bajer, 1971, 1972; Newcomb, 1969; Jensen and Bajer, 1973;
Nicklas, 1971; John and Lewis, 1969; Holliday, 1970; Went, 1966;
Pickett-Heaps, 1974; Lafontaine, 1974; Hartmann and Zimmerman,
1974). Instead, it will highlight certain selected or neglected aspects
of cell division which deserve closer attention.

II. DIVIDING CELLS OF HIGHER PLANTS

What are the essential and characteristic features of a dividing cell in
a higher plant? Our answer to this question is frequently distorted by
the fact that most investigations of the division process have concentrated on mitosis in the root apical meristem and microspore mother
cells at meiosis. A brief survey of the range of dividing cells during the
angiosperm life cycle will restore the balance (Fig. 1).

A. The Vegetative Sporophyte

In the very young embryo, microscope studies indicate that most cells
divide, perhaps with some synchrony (Schulz and Jensen, 1968a, b;
Pollock and Jensen, 1964; Norstog, 1972), but as the sporophyte grows,
divisions though widely distributed through the plant become restricted
to particular regions or stages of development. In the young sporophyte
they are most numerous in the localized meristems of primary, lateral
and adventitious root apices, of shoot apices and leaf primordia and
in growing leaves. The dividing cells of these meristematic regions are
similar but by no means identical even within one organ. As in the
familiar root tip, the cells can be seen to vary in size, shape and degree
of vacuolation. Between successive mitotic divisions, the cells expand
(Prescott, 1961). Without cell expansion there is no organ growth
(Haber and Foard, 1964a, b). Higher plants in general have no tissue
where repeated cell divisions occur without cell growth, as occurs in
animal embryos in the cleavage stage, although in some angiosperms
there is little or no visible cell expansion during the first two or three
cycles of the embryo developing from the zygote (e.g. Schulz and Jensen,
1968a, b; Pollock and Jensen, 1964; Neubauer, 1972; Haskell and Post-
lethwaite, 1971; Norstog, 1972). Consequently, in most meristems the
cell volume is approximately constant in the initial cells and gradually

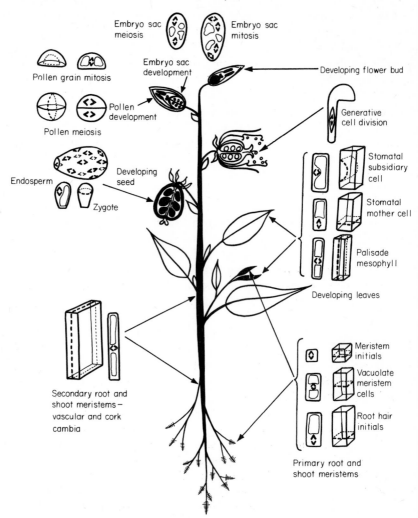

FIG. 1. A generalized flowering plant indicating the main regions of mitotic activity and the more distinctive types of dividing cell.

increases, as a result of progressive vacuolation over several mitotic cycles, as cells cut off from the initials progress towards differentiation (Jensen and Kavaljian, 1958; Kaufmann *et al.*, 1965). Cell division can take place in highly vacuolated cells (Roberts and Northcote, 1970; Esau and Gill, 1965; Sinnott and Bloch, 1941; Yeoman and Street, 1973). Cell division and expansion are thus not alternative processes, and indeed usually occur together, although either can proceed without the other. Only when division has ceased do changes in cell shape directly reflect the expansion of the entire organ (Haber and Foard, 1964a).

Expansion of a cell may be more or less equal in all directions or predominantly in one direction as in most cells of an elongating organ. Subsequently, the cell frequently but not always divides by a new wall perpendicular to the longest cell axis (Haber and Foard, 1964a) to form two daughter cells of approximately equal size. The rate of division varies within a meristem (e.g. Barlow, 1973; Phillips and Torrey, 1972). Differences in the rate and orientation of cell expansion relative to cell division contribute to the variation in cell size and shape seen even within one meristem (Jensen and Kavaljian, 1958). Sometimes, sister cells differ conspicuously in size when they are first formed by an asymmetrical division (e.g. Schulz and Jensen, 1968b; Heslop-Harrison, 1968; Kaufman et al., 1965; Avers, 1963; Stebbins and Jain, 1960; Bünning, 1957; Stange, 1965; Zeiger and Stebbins, 1972; Zepf, 1952; Corti and Cecchi, 1970; Resch, 1958; Pickett-Heaps and Northcote, 1966; Cutter and Feldman, 1970). No doubt other less obvious differences also occur in many cells (e.g. Czernik and Avers, 1964). Indeed, no two identical cells can exist within such a cell system, nor can they occupy an identical position in the cellular environment. Each wall of a cell, and probably of the plasmalemma membranes lining it, is of a different age and traceable back to a different cytokinetic event. There is no way of dividing such a cell so that the daughter cells are equal in respect of wall age. This may be of morphogenetic significance.

As the vegetative structure develops further, small localized regions of meristematic activity are associated with particular morphogenetic events. For example, there is a highly ordered sequence of equal and unequal divisions with precisely determined orientation in the formation from the epidermis of the stomatal complex (e.g. Stebbins and Shah, 1960; Stebbins and Jain, 1960; Zeiger and Stebbins, 1972; Pickett-Heaps and Northcote, 1966a) and multicellular hairs (e.g. Kaufmann et al., 1970; Cutter and Feldman, 1970) and glands. Characteristic divisions also occur in the later stages of development of certain tissues such as leaf palisade and phloem. Mitoses without subsequent cytokinesis are found in the later stages of development of certain cell types such as anther wall and tapetum (see Chapter 6, Section II D1), xylem vessels (List, 1963) and laticifers (Mahlberg and Sabharwal, 1967).

In most plants, during the first season of growth, a major revival of meristematic activity occurs in regions completely removed from the primary apical meristems, namely the intercalary cambia (Kaufmann *et al.*, 1965), vascular cambia (Evert and Deshpande, 1970; Catesson, 1974) and cork cambia. In large perennials, a large proportion of its dividing cells must be in the secondary or lateral meristems. These cells can be very elongated along the axis of the stem or root (up to 100 times their tangential width) but are probably not still elongating in that direction. They are expanding mainly radially so that unlike a cell of the apical meristem the direction of most active expansion coincides with the shortest cell axis. The new wall after division usually forms during a prolonged telophase (Catesson, 1974) parallel to the longest axis and in the plane of maximum area, as in palisade cells. If there is a correlation between the orientation of division and expansion in meristematic cells it must be that the long axis of the spindle is parallel to the direction of greatest active expansion, regardless of cell shape.

Primary and secondary meristems are similar in that the initial cells maintain a steady state by restoring the original cell condition after each mitosis while the derivative cells commence a sequence of changes related to their subsequent differentiation, even while they proceed through several further division cycles. The organization of the meristem results in an orderly sequence of progressively more differentiated cells extending away from the initials. The more distant cells are frequently referred to as "older". It is important to be precise in referring to cell age, as there are several possible reference points in development. At any given moment, all cells of one plant, whether meristematic or highly differentiated, are of the same chronological age in that all have developed from the same zygote. Their "developmental age" in relation to division, however, clearly differs throughout the plant, and can be referred either to the last division which occurred in their development or to the time when they were cut off from the apical initials of a primary meristem, which in a vegetatively propagating plant are potentially immortal. The cambial initials represent an intermediate point, being themselves the partially differentiated derivatives of the primary meristems. Whichever reference point is used, the concept of developmental age in relation to previous division activity carries with it the implication that cessation of division is a significant step in differentiation and that developmental potential is preserved in dividing cells (Brown and Dyer, 1972).

B. The Flowering Sporophyte

In the floral apex and the subsequent development of the unique structures of the flower such as the anther, ovary and ovule walls, divisions occur which are superficially like those of the vegetative apex (Lyndon, 1973). Even in the sporogenous cells only the increasing synchrony in the anther and increasing cell size in the ovule visibly distinguish the divisions from earlier ones. Only with meiosis, which can be considered as an extreme and unique modification of the mitotic cycle, and the subsequent development of the gametophytes within sporophyte structures, do strikingly different dividing cells appear and reappear in each succeeding generation.

Pollen mother cells become recognizable quite early in development as specialized wall formation cuts them off from their neighbours (Heslop-Harrison, 1972). Despite the important differences in chromosome and perhaps cytoplasmic activities, many features of meiotic cell division are reminiscent of mitosis and the subsequent cytokinesis (Bennett, 1973; Rhoades, 1961; John and Lewis, 1965; Stern and Hotta, 1969). The absence of DNA replication and sometimes of cytokinesis and even interphase between the two divisions of meiosis (Heslop-Harrison, 1968, 1971a) is one of the more striking differences. Typically the products of both divisions are held for a while, even up to anthesis, within the mother cell wall as a tetrad of daughter cells before they are released as pollen grains, the first cell of the gametophyte generation. The arrangement of cells in the tetrad varies with the species, depending on the relative orientation of the three meiotic spindles. Rarely, three microspores abort after each meiosis (e.g. Lewis and John, 1961).

The events leading up to and during meiosis in the ovule are in many ways similar, although the single and uniquely large mother cell within each nucellus is not released from its neighbours. Meiosis produces four nuclei, and in most angiosperms, cytokinesis produces a tetrad of cells before three of them abort, leaving one megaspore. Again spindle orientation determines the tetrad arrangement. In the remaining species, cytokinesis fails after the second or even both divisions and two or all four megaspores contribute towards the embryo sac (Fig. 2).

C. The Gametophyte

Both the male and female gametophytes are much reduced and have a definite growth pattern (Heslop-Harrison, 1972) unlike the indefinite development of the sporophyte. The mitotic divisions are unlike any others in the life cycle, particularly in respect of cytokinesis.

In the microspore an asymmetrical mitosis with a spindle perpendicular to the long axis of the cell divides the pollen grain very unequally into the larger vegetative cell and the generative cell (Mepham and

Lane, 1970; Burgess, 1970a; Brumfield, 1941). The curved wall at cyto-kinesis cuts off the generative nucleus and a minimal amount of cyto-plasm against the proximal wall of the spore (Sax and Husted, 1936). The generative cell, which may have few or no visible organelles (Lombardo and Gerola, 1968a, b), then detaches from the spore wall and moves into the cytoplasm of its sister generative cell (Sanger and Jackson, 1971). There is no comparable event elsewhere in the life cycle.

Fig. 2. The effect of the embryo sac development on gene segregation. The pattern of inheritance of alleles at a single heterozygous locus by the zygote via the egg and by the endosperm via the polar nuclei will be determined by the sequence of divisions during embryo sac development. In a diploid, each of the alternative alleles will be inherited by 50% of the eggs assuming equal viability, but the genotype of a polar nucleus in relation to the egg and the other polar nucleus will depend on the type of embryo sac development. In some, the polar nuclei, usually two, will contain the same allele ("homozygous"), while in others, some polar nuclei will contain the alternative allele ("heterozygous"). In some types of embryo sac development, the relative frequency of the first and second division segregation of the locus in question will affect the endosperm genotype. The frequency of first division segregation relative to second division segregation depends on the location of the gene relative to the centromere and on chiasma distribution. The closer the locus is to the centromere and the more distal is the nearest chiasma, the greater is the proportion of first division segregation.

(a) **Monosporic embryo sac development** (*Polygonum* and *Oenothera* types)

All polar nuclei are "homozygous" and of the same genotype as the egg. The products of first and second division segregation are the same. Heterozygosity of the endosperm depends entirely on the genotype contributed by the pollen.

(b) **Bisporic embryo sac development** (*Allium* type)

After first division segregation (locus between the centromere and the most proximal chiasma) the results are similar to those of monosporic embryo sacs. After second division segregation (locus distal to proximal chiasma) the polar nuclei are "heterozygous" and thus all endosperms are heterozygous whatever the male contribution.

(b–d) **Tetrasporic embryo sac development**

Adoxa type: After first division segregation and 50% of the cases of second division segregation, the polar nuclei are "heterozygous". The remainder are "homozygous" for one or other allele, and of the same genotype as the egg.

Drusa type: Nuclear migration and an extra mitosis results in similar products of first and second division segregation and probably a preponderance of "heterozygous" polar nuclei.

Fritillaria type: Nuclear migration and fusion ensures that all pairs of polar nuclei are "heterozygous" containing a total of two copies of each allele. Genetically similar "heterozygous" polar nuclei result in the 16-nucleate *Penaea* type and the 8-nucleate *Plumbago* type of embryo sacs. In the 16-nucleate *Peperomia* embryo sacs, the eight polar nuclei are likely to be "heterozygous".

NUMBER OF MEGASPORES DEVELOPING	TYPE OF EMBRYO SAC DEVELOPMENT	SEGREGATION	MEIOSIS	GAMETOPHYTIC MITOSIS	EMBRYO SAC	POLAR NUCLEI

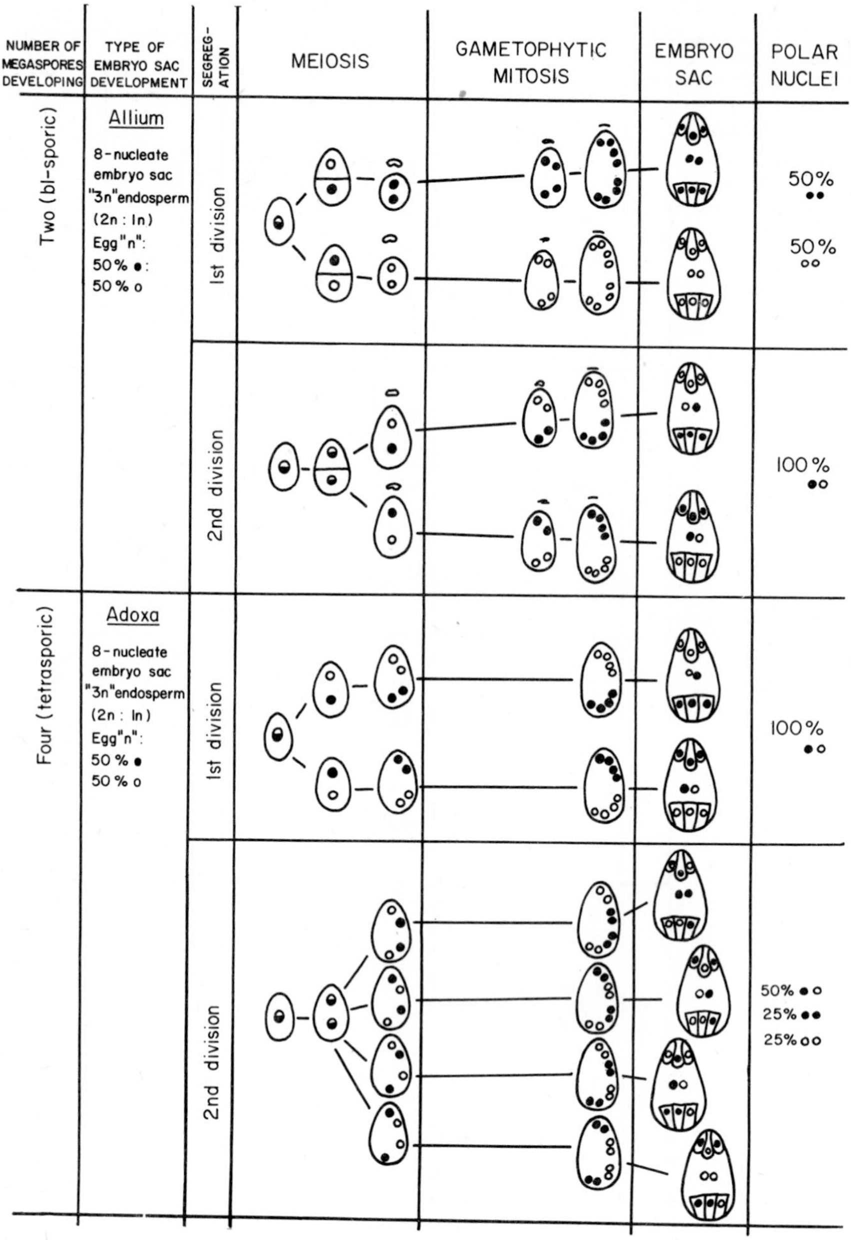
NUMBER OF MEGASPORES DEVELOPING
TYPE OF EMBRYO SAC DEVELOPMENT
SEGREG-ATION
MEIOSIS
GAMETOPHYTIC MITOSIS
EMBRYO SAC
POLAR NUCLEI
Two (bi-sporic)
Allium
8-nucleate embryo sac
"3n" endosperm
(2n : 1n)
Egg "n":
50% ●:
50% o
1st division
50% ●●
50% oo
2nd division
100% ●o
Four (tetrasporic)
Adoxa
8-nucleate embryo sac
"3n" endosperm
(2n : 1n)
Egg "n":
50% ●
50% o
1st division
100% ●o
2nd division
50% ●o
25% ●●
25% oo

NUMBER OF MEGASPORES DEVELOPING	TYPE OF EMBRYO SAC DEVELOPMENT	SEGREGATION	MEIOSIS	GAMETOPHYTIC MITOSIS	EMBRYO SAC	POLAR NUCLEI
Four (tetrasporic) cont.	Drusa 16-nucleate embryo sac "3n" endosperm (2n : 1n) Egg "n": 50% ● : 50% ○	1st division				Probably mostly ●○
		2nd division				Probably mostly ●○
Four (tetrasporic) cont.	Fritillaria 8-nucleate embryo sac "5n" endosperm (4n : 1n) Egg "n": 50% ● : 50% ○	1st division				100% ○○●●
		2nd division				100% ○○●●

NUMBER OF MEGASPORES DEVELOPING	TYPE OF EMBRYO SAC DEVELOPMENT	SEGREGATION	MEIOSIS	GAMETOPHYTIC MITOSIS	EMBRYO SAC	POLAR NUCLEI
Four (tetrasporic) cont.	Penaea 16-nucleate embryo sac "5n"endosperm (4n : In) Egg "n": 50%• : 50%o	1st and 2nd divisions	[diagram]	2 mitoses	[diagram]	100%oo••
	Plumbago 8-nucleate embryo sac "5n"endosperm (4n : In) Egg "n": 50% • : 50% o	1st and 2nd divisions	[diagram]	1 mitosis	[diagram]	100% o o••
	Peperomia 16-nucleate embryo sac "9n"endosperm (8n : In) Egg"n": 50% • : 50% o	1st and 2nd divisions	[diagram]	2 mitoses	e.g. [diagram]	Probably 100% "heterozygous"

The next division, that of the generative cell to form two male gametes, may occur within the developing anther so that trinucleate pollen grains are liberated at anthesis as in grasses, or within the developing pollen tube after successful pollination. The vegetative nucleus, which differs from the generative nuclei in its metabolism (La Fountain and Mascarenhas, 1972), does not normally divide again. In some species, the generative cells appear to divide by an equatorial constriction similar to cleavage of animal cells rather than by the expanding cell plate typical of plant cytokinesis. Where chromosomes are large the division in the narrow pollen tube may be considerably distorted, so that at metaphase the chromosomes are scattered from pole to pole instead of being aligned across the spindle equator (Burgess, 1970c). At anaphase, daughter chromosomes have to slide past each other, moving in opposite directions.

In most developing female gametophytes, an 8-nucleate coenocyte develops from the megaspore by three successive synchronized mitoses with no intervening cytokinesis. As a result of the controlled orientation of mitotic spindles and of nuclear migration, the characteristic arrangement of three nuclei at each end and two in the centre of the embryo sac is achieved before cell wall formation occurs simultaneously round six of the nuclei to form the egg cell, the two synergids and the three antipodals, leaving two polar nuclei free in the vacuolated cytoplasm of the sac (Jensen, 1965a; Schulz and Jensen, 1968c). In the minority of species where two or even four megaspores contribute to the embryo sac, the mitoses are usually correspondingly reduced in number to produce a similar gametophyte (Fig. 2). Only rarely are embryo sacs with more or less than eight nuclei found (Fryxell, 1957; Maheshwari, 1950). The significance of the widespread 8-nucleate embryo sac is not known. The modifications of development will in certain circumstances affect the heterozygosity and ploidy level of the endosperm (Fig. 2), but it is not known if this is the significant effect.

Another unusual feature of the female gametophyte is the fusion of non-sister nuclei which can occur. In most natural coenocytes (Erickson, 1964; Mahlberg and Sabharwal, 1967; Johnson and Rao, 1971), nuclei do not fuse despite the absence of intervening walls, and fusion of non-sister nuclei is almost entirely restricted to the unique event of fertilization. However, in the embryo sacs of many species, the polar nuclei fuse before fertilization, and in the lily, three of the genetically different megaspore nuclei fuse at the 4-nucleate stage of embryo sac development, producing ultimately triploid antipodal cells and one triploid polar nucleus.

D. The Endosperm

The endosperm is a tissue, neither gametophyte nor sporophyte, with no equivalent outside the angiosperms, resulting from the second fertilization. It has a unique genotype and chromosome complement, usually triploid, with female and male complements in the ratio 2 : 1. In some species, development of the endosperm commences, or continues after a single initial cytokinesis, with about ten consecutive synchronous mitoses without cell wall formation (Erickson, 1964; Dixon, 1946; see p. 216). The synchrony may be complete, or in the form of a wave moving along the endosperm (Ram, 1960). When several hundred or even thousand nuclei have been formed in a sac-like sheet of cytoplasm, cell walls form almost simultaneously between all nuclei (Newcombe and Fowke, 1973) and synchrony is then lost as cytokinesis follows each subsequent mitosis. There is here, as in the developing embryo sac and other natural and induced coenocytes (Johnson and Rao, 1971), a correlation between coenocytic development and a very high degree of synchrony. The asynchrony in binucleate pollen (Haque, 1953) probably has a genetic basis. However, both natural and induced synchrony can occur in cells separated by cell walls as shown by pollen mother cells (Erickson, 1964), in pollen grains (Heslop-Harrison, 1968), in root tips (Jensen and Kavaljian, 1958; Alfieri and Evert, 1968; Clowes, 1962; Jakob, 1972; Mattingly, 1966) and tissue cultures (e.g. Yeoman and Evans, 1967). In other species, there is no coenocytic stage in endosperm development.

E. The Dividing Angiosperm Cell

From this survey, some very general conclusions can be drawn about the dividing cell of an angiosperm:

(i) Mitotic division of the nucleus is frequently but not inevitably linked to cytokinesis. Either event can take place independently of the other.

(ii) Preparation for division does not exclude, and is normally accompanied by, cytoplasmic growth, vacuolation and a degree of differentiation. The differentiation accompanying division in the cell lineage which links generations through the stem apex, gametophytes and gametes is itself cyclic over the period of the plant's life history (see Chapter 6, Section I). The morphology of dividing cells thus varies widely. Division of the cytoplasm may be visibly unequal, and in no case will sister cells be truly identical at the sub-microscopic level.

(iii) The orientation of division is not determined by cell shape, but the spindle axis of an expanding cell is frequently parallel to the direction of maximum cell expansion.

(iv) The organization of a meristem depends on co-ordination of the timing and orientation of division and expansion of the cells within it. Division synchrony is not restricted to coenocytic tissues and independent control of spindle orientation of adjacent nuclei is not restricted to cellular tissues.

For practical reasons, much of our knowledge of the physiology, fine structure and biochemistry of cell division in the intact plant comes from the primary root meristem. A little is known about divisions in the development of the shoot apex, leaf and pollen (Heslop-Harrison, 1972; Lyndon, 1973), but for the female gametophyte, embryo, cambium and, for example, stomatal complexes, we rely almost entirely on the results of electron microscope studies. In all of these, the events of interphase must include stages in cell growth and morphogenesis as well as preparations for cell division, and they may not be easy to distinguish. It is therefore all the more surprising that the endosperm has not received more attention from those interested in the biochemistry of mitosis in higher plants (see Jackson, 1969; Hepler and Jackson, 1969). Although the tissue is not without its experimental limitations, it can be cultured in controlled conditions (e.g. Norstog *et al.*, 1969), it is the most acceptable material for correlating the fine structure (e.g. Bajer, 1968a, b) with the light microscopy of living cells, and cell wall formation and differentiation within the young tissue do not occur.

III. REPLICATION OF THE NUCLEUS

Growth and reproduction of eukaryotes involves replication of all protoplasmic components. In all unicellular organisms and most tissues of multicellular organisms including angiosperms, the culmination of the repeating cycle of replication is the partitioning of the cytoplasm by a cell membrane and, in plants, a cell wall. Cytokinesis or cleavage therefore provides a point of reference in most division cycles before which replication and segregation of all cell components must be completed and after which a new cycle of replication events can begin. Among these events, the division of the nucleus initially received most attention because of its conspicuous appearance and the early discovery of its role in heredity and development and of a specific staining technique for DNA. This concentration of effort to the detriment of our understanding of cytoplasmic events has continued using the more sophisticated techniques of the last 15–20 years. As division of the nucleus, mitosis, can take place independently of cell division and, at least in animal embryo cleavage, of cell growth, it is convenient to consider it separately.

Nuclear replication by the precise transmission of exact replicates of the chromosomal complement to separate daughter nuclei clearly requires the integration in a pre-determined sequence of a multitude of component events at the microscopic, macromolecular and molecular levels, and involves not only the chromosomes but also intimately related nuclear and extra-nuclear structures such as nuclear envelope, nucleolus, spindle and centriole. We are here only concerned with the visible events, and in order to determine the relationship of these to the fundamental steps in the mitotic cycle, it is useful to consider the cyclical changes in each of these structures separately before discussing their relationship with each other.

A. The Nuclear DNA Cycle

Only rarely (Nagl, 1970, 1972; Tokuyasu, 1972; Dewse, 1974; Lafontaine, 1974) can visible changes in the nucleus be related accurately to the period of chromosomal DNA synthesis, known as S, which is normally identified by cytophotometric or autoradiographic methods (Mendelsohn and Takahashi, 1971; Vendrelly, 1971; Van't Hof, 1968a, c; Nachtwey and Cameron, 1968). However, S is one of the commonly identified components of the mitotic cycle and must be considered briefly here.

Unlike prokaryotes, in eukaryotes S normally occupies only part, commonly about half, of the total cell cycle (Mitchison, 1971). This localized period of DNA replication usually does not overlap with the period (M) when the chromosomes are contracted by coiling (but see Clowes, 1967) and actively separating, nor does it occupy the whole of the rest of the cycle (I, interphase) when the chromosomes are dispersed within the nuclear membrane. The precise significance of this temporal localization of nuclear DNA synthesis is not immediately clear, but it is presumably related to the greater complexity of cell structure, division and differentiation in the eukaryotes. Chromosomal histones duplicate in step with DNA, but there must be other interphase events which take longer than S or cannot overlap with it.

The position of S in relation to M can differ between organisms, between cells of the same organism and between similar cells under different growth conditions (see Table I for examples) and even between nuclei of the same cell (Gonzales-Fernandez $et\ al.$, 1971). This variation in the duration of the interval between S and M (known as "G_2") relative to the duration of the interval between M and S ("G_1") is superimposed on the variation in total time taken for the complete cycle. In a few animal and plant cells, cycles have been recorded which entirely lacked either G_1 (e.g. in the short cycles of root cap columella cells (e.g. Barlow, 1973), in generative cell mitosis (Taylor, 1958; Woodard, 1958)) or G_2 (e.g. xylem parenchyma explants of $Helianthus$ $tuberosus$ in callus culture (Mitchell, 1967) and pollen grain mitosis (Moses and Taylor, 1955; Stefferson, 1966; see Heslop-Harrison, 1972)). Only in the early cleavage stages of some animal embryos, where there is no cell growth and the cycle is very short, are both G_1 and G_2 known to be omitted (Mitchison, 1971; Malamud, 1971) so that S occupies the whole of the brief interphase. It may be that future investigation will reveal similar timing of DNA synthesis in the first few cycles of the plant embryo before it begins to enlarge.

Meiosis shows several characteristic features in relation to DNA synthesis (see Chapter 5). There is no interval between S and the onset of the first division, that is, no G_2, and a very small amount of DNA synthesis, with possible significance in relation to chromosomes pairing, and crossing over is delayed until meiosis is well under way. Even when

TABLE I. The duration of the mitotic cell cycle, and the relative duration of individual identifiable stages, in root tip cells

Ref.	Organism	Cell type/treatment	Temp. °C	Duration of whole cycle, T (h)	I	G_1	S	G_2	M	Pro	Met	Ana	Telo
Barlow, 1973	Zea mays root	Cap columella		14	0·93	—	0·57	0·36	0·7				
		Quiescent centre (QC)		174	0·98	0·86	0·05	0·11	0·02				
		Near QC: stele (old root)		18	0·89	0·17	0·27	0·44	0·11				
		Near QC: stele (young root)		35	0·91	0·54	0·17	0·20	0·08				
		Near QC: cortex		42	0·90	0·52	0·23	0·14	0·09				
		200 µm from QC: cortex		30	0·90	0·60	0·16	0·13	0·10				
		400 µm from QC: cortex		25	0·88	0·52	0·16	0·20	0·12				
		700 µm from QC: cortex		20	0·85	0·45	0·15	0·25	0·15				
		1000 µm from QC: cortex		20	0·85	0·45	0·15	0·25	0·15				
Wimber, 1966	Tradescantia paludosa root		13	51	0·91	0·30	0·44	0·16	0·10	0·08	0·06	0·04	0·04
			21	21	0·91	0·28	0·51	0·12	0·08	0·06	0·06	0·04	0·08
			30	16	0·89	0·15	0·59	0·15	0·11	0·08	0·09	0·06	0·11
Brown and Klein, 1973	Pisum sativum excised roots	In dark		17·8	0·74	0·11	0·24	0·39	0·25				
		In white		17·0	0·91	0·26	0·29	0·36	0·09				
		In red		21·0	0·90	0·20	0·40	0·30	0·10				
		In green		24·6	0·75	0·42	0·23	0·09	0·25				
		In blue		19·8	0·88	0·30	0·34	0·23	0·12				

Relative duration of stages of cycle ($T = 1$)

there is an interphase and even cytokinesis between the two meiotic divisions, there is no S period as the undivided chromosomes from the first division proceed to the second division. Thus chromosome coiling, arrangement on a spindle and centromere division do not invariably require DNA synthesis during the immediately preceding interphase, nor does centromere division invariably occur on a spindle formed after an S period.

B. The Chromosome Cycle

Moving now to events at the level of organization visible under the microscope, the chromosomes usually follow a cyclical sequence of two conspicuous activities during mitotic division. Each forms two identical copies, and each contracts temporarily during M. These events can be recognized, but until the chromosome structure is better understood the underlying mechanism will remain obscure.

Replication of the chromosome to form two identical sub-units along most of its length occurs during interphase. The single unit of the daughter chromosome at anaphase becomes a replicated chromosome with two chromatids by the time it reappears at prophase. This reorganization of the chromosome structure appears to occur over a relatively short period of interphase overlapping with, or even just before, S, as judged by the consequences of the fusion of interphase cells with cells in mitosis (Johnson and Rao, 1971) and the damage caused by irradiation at different stages of interphase (Wolff, 1969; Monesi et al., 1967; Grant, 1965; Heddle and Trosko, 1966). As the chromosome enters M, only the centromere, usually a single localized and structurally distinct region (Luykx 1970, 1974; Bajer and Molè-Bajer, 1972), remains undivided, although at least part of this complex region is structurally and functionally double by the time it becomes attached by microtubules to opposite poles of the spindle. Replication of the chromosomes is not completed until the rest of the centromere region divides simultaneously with others in the cell at the end of anaphase. The sequence is less clear in organisms with dispersed centromeres.

In meiosis the timing is altered. Chromatids are not recognizable until the chromosomes have already partially contracted and their appearance at zygotene coincides with a small amount of late DNA synthesis. Also, centromeres remain intact on the spindle throughout the first division and the ensuing interphase. At first metaphase, each centromere only attaches to one pole, the sister centromere of the same bivalent attaching to the other. The centromeres sometimes appear as double structures, but if so, unlike mitosis, both halves are attached to the same pole (Luykx, 1974). Perhaps in meiosis the attachment region of the centromere only replicates after spindle attachment. Complete division of the centromere is delayed until anaphase of the second division. We may conclude that chromatid organization is not necessarily simultaneous with DNA replication and centromere division is not invariably linked with other events on a spindle.

It is still not certain whether any or all unreplicated chromosomes consist of more than one DNA double helix. The evidence is conflicting (Wolff, 1969; Whitehouse, 1973; Comings, 1971; Lafontaine, 1974; Luykx, 1974). If multistranded chromosomes exist, there is a unit, which we may call the chromonema, at a level of organization between the DNA helix replicated in S and the chromatid. New chromonemata, consisting of DNA and associated protein and perhaps other elements, could be formed in the period between DNA replication and the formation of chromatids if such a period exists. However, alternative models can be proposed for their organization before or after both events, or simultaneously with either. Because different regions of a chromosome replicate at different times within S, it is possible that DNA synthesis, chromonema formation and chromatid organization occur in a fixed sequence in relation to each other but can occur simultaneously at different points within the same chromosome. Many replicating chromosomes show semi-conservative distribution of new, labelled DNA. Such non-random distribution within a multistranded chromosome would require a high degree of structural organization of groups of chromonemata, particularly at the centromere. The suggestion that the distribution of newly synthesized DNA in separate, non-homologous chromosomes may also be non-random (Rosenberger and Kessel, 1968; Lark, 1967; but see Comings, 1970) implies that the two attachment regions of a mitotic centromere may be functionally differentiated so that for every chromosome in the complement, the half-centromere of the chromatid containing the newly synthesized DNA becomes attached by spindle fibres to the same spindle pole.

The coiling cycle which contributes to chromosome contraction involves the formation of a helix which can change direction along an arm (Ohnuki, 1968) but has a characteristic pitch and diameter at metaphase. This coil, and even a lower order of coiling, can be seen at mitosis after suitable pre-treatment and accounts for part of the enormous discrepancy in diameter between the smallest strands seen under the electron microscope and the chromatids as normally seen by light microscopy. Typically, this coiling reaches a maximum at late metaphase and disperses to a minimum during interphase. In all organisms there are specialized chromosome regions such as centromeres, telomeres and nucleolar organizers which are relatively over- or under-contracted at some stage of the cycle, while in many organisms there are other allocyclic regions or even whole chromosomes which show a different cycle of contraction (John and Lewis, 1968; see Chapter 5, Section IIA). They may for example appear under-contracted at metaphase or remain fully contracted in interphase. This behaviour may be related to regulation of gene activity, or specialized functions such as pairing (Stack and Clarke, 1973).

In meiosis, the coiling is more exaggerated but generally similar to mitosis and allocyly can occur. The large gyres of the "major coil" are often conspicuous at metaphase of the first meiotic division (Darlington and Vosa, 1963) and the lower order "minor coil" is not infrequently seen. Coiling at the second division is like mitosis except for the relic of the major coil and the absence of the relational coil of chromatids, even though no DNA replication occurred during the preceding interphase.

Observations of allocyly and meiotic division suggest that chromosome coiling normally accompanies chromosome replication and division, but is not a necessary integral part of the process. This is confirmed by the duplication of the chromosome complement while in the interphase condition which can occur in certain endopolyploid differentiating cells (see Chapter 6, Section IID2). The reverse situation of chromosomes remaining permanently visibly coiled throughout the mitotic cycle is known in some protozoa (Mazia, 1961).

C. The Nucleolar Cycle

It is usual for nucleolar material, mostly protein (Lafontaine, 1974), to detach from the organizers at late prophase and disperse. The fibrillar nucleolonema retracts into the organizer segment. No nucleoli are then seen until they reform at about the end of telophase, although some nucleolar RNA may be attached to the chromosomes. Some of the largely proteinaceous "pre-nucleolar material" which is produced in the cytoplasm and accumulated between the telophase chromosomes probably contributes to the new nucleolar organizer. Nucleoli reform, enlarge and even fuse by early interphase (Stockert et al., 1970; Dickinson and Heslop-Harrison, 1970). While final dispersal of nucleoli always appears to coincide with the transition from prophase to metaphase, the timing of the other stages of the nucleolar cycle varies in relation to the events of the chromosome cycle (De la Torre and Clowes, 1972) from one type of cell to another. In the very short cycles of some root cap and stele initials in the root meristem, the nucleolar cycle is brought forward, disorganization beginning in late interphase and reformation being completed within telophase. In the long cycles of the quiescent centre, the nucleolar cycle is relatively delayed, disorganization beginning in prophase and reformation not even commencing until early interphase. High temperature and other environmental treatments can disturb the cycle for example and cause nucleoli to persist through division. Occasionally, species or mutant individuals are found which regularly have persistent nucleoli (e.g. Frankel, 1937; Heslop-Harrison, 1971b; Brinkley, 1965; Brown and Emery, 1957; Mughaz and Godward, 1973). In some insects, fused persistent nucleoli enable sex bivalents to show regular co-orientation and disjunction in the absence of chiasmata. Whatever the significance of the abrupt dispersal of ribosomal RNA normally found during division, it is apparently not fundamental to successful mitotic division.

D. The Nuclear Envelope Cycle

In most if not all angiosperm cells, the nuclear envelope disintegrates and disperses at late prophase, at the time the nucleoli disperse, and becomes indistinguishable from the ER with which it frequently interconnects when intact (Stevens and Andre, 1969). In the generative cell division in the pollen tube, large pieces of envelope remain visible throughout division (Paolillo, 1974). It reforms closely around the chromosomes, starting near the spindle pole where the centromeres aggregate, at early telophase. During interphase, all the chromosomes are contained within one localized region of the cell by a membrane system which is at least intermittently continuous with the other membrane systems in the cell (see Section IV).

In many algae and fungi, nuclei which appear essentially similar to those of angiosperms follow an otherwise typical mitotic cycle entirely within an almost or totally intact nuclear envelope (e.g. Dodge, 1973; Du Praw, 1970; Pickett-Heaps, 1974). In some endopolyploid nuclei in higher plants, chromosomes replicate within an intact nuclear envelope. Although temporary dissolution of the nuclear envelope appears to occur invariably in higher plant cell divisions, it is difficult to believe it is a fundamental event in mitosis.

E. The Spindle Cycle

Until the centromeres divide, no visible extra-chromosomal structure has played a critical role in mitosis, and chromosome replication can take place within a fully dispersed nucleus. However, the orderly separation of daughter chromosomes to form two genetically equivalent nuclei requires the interaction of the chromosomes with another conspicuous structure, the spindle. In prokaryotes, where the whole gene complement is contained within a single structure or chromosome and regular disjunction is a simpler matter, this complex structure is absent. In many lower eukaryotes, the spindle is formed entirely within the persistent nuclear envelope (Heath, 1974), although material from the cytoplasm may be involved. In higher eukaryotes including angiosperms, the spindle only forms after the nuclear envelope has dispersed and nuclear and cytoplasmic material have merged. The question of whether the spindle has a nuclear or cytoplasmic origin is then a meaningless one, but there are visible features of the cytoplasm in late interphase and prophase which seem to be associated with spindle formation. In some animal cells at least, where the information is available, protein of the type which forms most of the spindle structure is synthesized in the latter half of interphase. Just before prophase in many dividing plant cells a group of microtubules similar to those of the spindle aggregate in a band running round the cell against the plasmalemma in a position which corresponds to that where the new wall is to be formed at the subsequent cytokinesis (Pickett-Heaps, 1969b, c; Burgess, 1970a; Evert and Deshpande, 1970; Cronshaw and Esau, 1968).

The tubules of the pre-prophase band disintegrate or disperse before the mitotic spindle appears (Burgess, 1970a; Pickett-Heaps, 1969d) and the relationship between the two structures is not clear. Perhaps the microtubules of the pre-prophase band contribute to the spindle (Pickett-Heaps, 1974). The pre-prophase band has not been reported prior to coenocyte mitosis. Just before the nuclear envelope breaks down, in some cells material accumulates immediately outside the nucleus, particularly in the two regions which will be close to the spindle poles. This probably represents an accumulation of as yet unorganized spindle protein. In some cells, microtubules are seen in this clear zone round the nucleus (Fowke et al., 1974; Bajer, 1968a), having moved from the cell margin.

Although the mechanism of spindle organization and of spindle–centromere interaction in chromosome movement is still not understood (Forer, 1969; Jensen and Bajer, 1973), the fine structure of the spindle has been studied in detail during mitosis (Bajer and Molè-Bajer, 1971, 1972; Luykx, 1970; Pickett-Heaps, 1974; Hartmann and Zimmerman, 1974). As anaphase progresses, spindle fibres become less conspicuous, only to reappear clearly in conjunction with cytokinesis where the microtubules may have a role in orientating the vesicles which form the cell plate. When the new wall is complete, the spindle fibres disappear, although presumably each daughter cell inherits the material of a half-spindle.

These observations reveal that a spindle is necessary for regular segregation of daughter chromosomes, but not for chromosome division. Spindles can form within intact nuclear envelopes in many lower organisms (Du Praw, 1970; Pickett-Heaps, 1974) but probably only in conjunction with chromosome coiling, although in some protozoan nuclei regular separation of chromosomes seems possible within an "interphase" nucleus. Finally, the orientation of the spindle is apparently determined late in the preceding interphase.

F. The Centriole Cycle

Although the centriole is generally thought of as an important feature of dividing cells of animals and algae, it requires some consideration here because it is found in plant groups as advanced as the cycads and because there has been no explanation of its absence or replacement in angiosperms which is satisfactory in the light of its supposed fundamental role in defining the position of spindle poles (e.g. Luykx, 1970), particularly in cells lacking rigid walls.

The structure of the centriole (Stubblefield and Brinkley, 1967) is remarkably uniform wherever it occurs. In at least some somatic cells of many species a pair of centrioles is inherited at each division and both begin to replicate at about mid-interphase (Pitelka, 1969; Went, 1966; Robbins et al., 1968), each forming an identical copy close to, but detached from, the existing centriole. In animal cells, the growing daughter centriole is frequently perpendicular to the parent centriole. In several bryophytes, fungi and ferns, end-to-end replication produces a daughter on the same axis as the parent (Paolillo, 1974). It has been suggested that the endoplasmic reticulum plays a role in centriole replication Hoage and Kessel, 1968). By late interphase, four centrioles exist close together, but prior to spindle organization the two mother–daughter pairs separate and migrate to opposite sides of the nucleus to take up positions at what will become the spindle poles. In a fully formed spindle, they are found at the focal point of the poles, but the converging tubules stop short of them, leaving a clear "centrosome".

The available evidence (e.g. Du Praw, 1970; Pickett-Heaps, 1969a) on the occurrence of visible centrioles reveals that they are only found in organisms with at least one flagellate or ciliate, and usually therefore motile, cell stage in the life cycle. Centrioles have not been seen in amoebae or myxomycetes. In angiosperms which have no centrioles, not even the male gametes have cilia or flagella. In some organisms, centrioles are not found in all dividing cells, but appear only in divisions preceding the formation of a motile cell (Moser and Kreitner, 1970; Pickett-Heaps, 1967c, 1968). In the cycads and some algae and bryophytes, centrioles are only seen during the division which produces the motile sperm, the only motile cell in the plant (Paolilo, 1974). This correlation brings to mind the extraordinary structural similarity of centrioles and the basal bodies of the cilia and flagella (see Mazia, 1961). Centrioles have been seen to develop into basal bodies (Renaud and Swift, 1964; Moser and Kreitner, 1970; Sorokin, 1962; Paolillo, 1974; Bajer and Molè-Bajer, 1972) and this transformation can be induced experimentally (Stubblefield and Brinkley, 1967). This introduces the possibility that centrioles are concerned with cilium formation instead of or as well as the supposed role in spindle organization or orientation which is frequently concluded from the circumstantial evidence of their location at spindle poles. Dietz (1969) reports that normal spindles formed after centrioles had been lost following experimental treatment. The centrioles are thought to contain DNA (Grannick and Gibor, 1967) and may be semi-autonomous, self-replicating, structurally independent organelles comparable to plastids and mitochondria, in which case the regular disjunction of one mother–daughter pair to each daughter cell is of vital importance. There is an interesting parallel here between this behaviour of centrioles and that of plastids in those plants, including many algae and *Anthoceros* and *Isoetes*, which have a single plastid in dividing cells (see Section IVA). In these the regular segregation of the photosynthetic organelle is critical. The plastids divide with, or just before, the nucleus and the daughter plastids move apparently without spindle attachment to opposite poles of the spindle, thus ensuring the inclusion of one in each daughter cell. Centrioles may be showing the same behaviour, thus revealing but not necessarily dictating spindle orientation (Bajer and Molè-Bajer, 1971, 1972; Pickett-Heaps, 1969a).

It appears that centrioles can develop from an invisible precursor, or in association with some structure other than centrioles, such as the nuclear envelope (Marchant and Pickett-Heaps, 1972). The replication at a distance seen under the electron microscope indicates this, as does the sudden appearance of centrioles in cycads and other organisms. In some fungi almost invisibly small centrioles can be found in most dividing cells, only enlarging and becoming conspicuous prior to the formation of a motile cell. Perhaps in some organisms, the structures in non-motile cells are further reduced. The invisible precursor could merely be a unit of self-replicating cytoplasmic DNA which directs the assembly of a visible centriole at appropriate stages of development.

If the function of this precursor is exclusively the formation of a centriole, it may well be totally absent from angiosperms. Alternatively, this unit may be present in all dividing cells because of a fundamental role in spindle formation, but its position is only clearly revealed when it also produces a centriole in preparation for eventual flagellum production (Pickett-Heaps, 1974). The various polar aggregations of ER (Wilson, 1970; Fowke et al., 1974) and other material (Pickett-Heaps, 1969a; Tai, 1970) and differentiated areas of permanent nuclear envelopes (Du Praw, 1970; McLaughlin, 1971; Robinow and Caten, 1969) which have been described as centriole equivalents could equally well mark the position of this spindle organizing body where centrioles are absent. The experimental induction of centriole formation in, for example, the dividing generative cell of angiosperms would be confirmation of this. A dual role in lower plants for this cytoplasmic unit in both spindle and centriole organization is not impossible as both structures are complex systems of similar protein microtubules. However, a role for these centriole precursors in determining the orientation of spindles seems less likely. The position of the poles is probably determined, presumably in response to extra-cellular stimuli, by late interphase and the diverging centrioles or their precursors presumably move to positions predetermined by some unknown mechanism (see Section VI).

G. The Integrated Mitotic Cycle

In the preceding sections, mitosis has been considered as several arbitrarily separated cycles, each related to a conspicuous structural component of the cell. The full cycle typical of many plant cells requires the integration of these events at the molecular and organelle level. This is indicated diagrammatically in Fig. 3 where the summarized cycles are arranged concentrically so that a radius cuts all cycles at points which occur simultaneously in a root meristem cell.

The control mechanism which provides the basis for this integration is quite unknown (see Mitchison, 1971), but any model proposed must take into account at least three remarkable features of the cell cycle: its complexity, repetition and flexibility.

The complexity at the molecular level must be considerable. In order that the visible events can occur in their proper sequence, molecular changes must be induced at an appropriate time before each event and in parallel with initially quite unrelated changes leading up to a

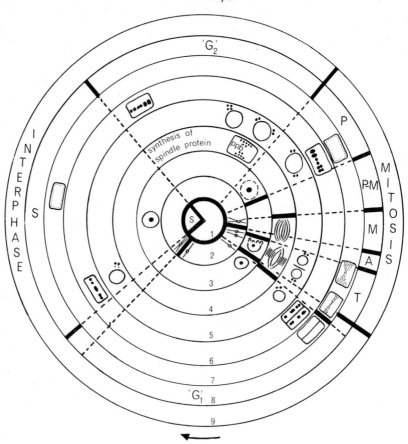

Fig. 3. The visible events of the generalized mitotic cell division cycle. These can be described in terms of component cycles of events affecting individual recognizable structural features of the dividing cell. Identifiable points in these cycles are indicated by solid lines. The corresponding points in other cycles are indicated by the dotted radii. The mitotic cycle is usually sub-divided arbitrarily, as indicated in the outer rings, 8 and 9, into phases defined by reference to events in several of the component cycles. The component cycles do not have a fixed relationship with each other. As a consequence, the limits of the normally recognized stages may be difficult to define. Cycle 7 is absent from certain cell types; Cycle 3 is absent from certain lower plants. The centriole (Cycle 5) is not recognizable in higher plant cells, and even where present is probably not directly involved in the mechanism of nuclear division and should be considered as a cytoplasmic organelle replicati g and segregating in close association with the nucleus. The timing of Cycle 1 can var considerably in relation to the other cycles, and "G_1" (Gap 1) or "G_2" (Gap 2) ph se may be absent from certain cell types. Variation in the relative duration of S (nuclear DNA synthesis) phase can in extreme cases result in the absence of G_1 or G_2.

Cycle 1, nuclear DNA cycle: The S period, recognized by micro-densitometry and separate event. Thus, for example, the self-contained sequence of bio-chemical changes leading to spindle formation will overlap within the cell with that sequence culminating in chromosome replication on that spindle.

The events of mitosis are repeated according to a pre-determined sequence. Thus, unlike many differentiating cells which follow a linear sequence of changes until eventual death, a mitotic cell follows a

---

specific staining or by autoradiography with specific precursors, represents the total period over which DNA replicates. Replication at a particular chromosome region occupies a shorter, probably much shorter, time.

Cycle 2, the chromosome cycle: Organization of the chromatids, as recognized by the effect on chromosomes at the next mitosis of irradiation at different stages of inter-phase or by the appearance of precociously condensed chromosomes after fusion of interphase cells with cells at metaphase, occurs at or just before the onset of the S phase. The centromere normally completes its replication when the chromosome is showing maximum contraction, but can occur in uncontracted chromosomes. Chromosome contraction (not indicated in the diagram), when it occurs, takes place over a period immediately preceding centromere division. An arbitrary point during this contraction period, when individual chromosome threads can be distinguished, is taken as the onset of prophase and the end of interphase. Dispersion of the chromosomes commences shortly after centromere division, usually when the daughter chromosomes have reached the spindle pole.

Cycle 3, nuclear envelope cycle: In many cells, the envelope disperses briefly as the spindle begins to form and the chromosomes approach maximum contraction. It re-forms shortly after centromere division. In several organisms, dispersion is in-complete or absent.

Nucleolus cycle: The nucleolus usually disperses and re-forms in step with the nuclear envelope.

Cycle 4, spindle cycle: Synthesis of spindle protein probably occurs in the second half of interphase. Organized microtubule arrangements appear briefly in many plant cells at the end of interphase as the "pre-prophase band" (PPB). The spindle proper organizes as the chromosomes reach maximum contraction and, where it occurs, as the nuclear envelope disperses. The distribution of microtubules changes during chro-mosome separation, particularly if a phragmoplast develops in conjunction with cell plate formation. Subsequently the spindle and its microtubules disperse.

Cycle 5, the centriole cycle: Both members of the pair of centrioles inherited at the spindle pole after the preceding division replicate during mid-interphase in most cell types investigated and by the time a new spindle is organized at the next division one mother–daughter pair has migrated to each polar position.

Cycle 6, plastid cycle: As far as has been determined, some or all the plastids (and probably mitochondria) usually replicate asynchronously throughout the mitotic cycle and approximately double by the next cell division. Segregation may not be regular.

Cycle 7, plasmalemma cycle: The cell membrane is extending throughout the cell cycle as the cell grows, but intensive and localized activity occurs at the equator of the spindle after the daughter chromosomes have separated, resulting in the rapid formation of a cell plate as vesicles aggregate, align and fuse.

cyclical sequence whereby a cell which completes the pattern of changes has returned to an earlier state and instead of dying can repeat the process almost indefinitely, each time returning to the same point in the cycle. To this extent, differentiation is the antithesis of division, and, while some cells show a degree of progressive differentiation at the same time as repeating the division cycle several times, total commitment to a linear sequence of normally irreversible differentiation can presumably only occur if the cyclical sequence of division is suppressed.

The flexibility of the control of mitosis is shown by the fact that the timing of one component cycle may vary relative to the others and that some cycles can be omitted altogether without disrupting other events. Thus the position of S in the cell cycle may differ even between cells of the same root tip meristem (Barlow, 1973), i.e. the DNA cycle in Fig. 3 is rotated relative to the rest. The timing of the nucleolar cycle can be similarly displaced (De la Torre and Clowes, 1972). Meiosis provides further examples. In the first division, the timing of S, chromatid organization and centromere division are delayed relative to the rest of the cycle, and in the second division, an essentially normal mitotic sequence follows an interphase in which there has been no DNA replication. Furthermore, in some endopolyploid cells, chromosome replication occurs in cells where the nuclear envelope, nucleolus, chromosome coiling and spindle cycles do not operate at all. It is unlikely that completion of these apparently dispensable events is critical for the completion of mitosis or plays an important role in the control mechanism. The only essential steps in the nuclear cycle can be summarized (Fig. 4) as DNA replication, chromonema replication (reorganization involving DNA and proteins), chromatid replication (=reorganization of chromonemata), chromosome replication (=centromere division) and nuclear replication (=daughter chromosome separation, probably always on a spindle), in their likely chronological sequence and order of ascending level of organizational complexity.

The mitotic cycle is more commonly defined, particularly with reference to timing, in terms of S, identified within interphase by a single molecular event; M, with its sub-divisions characterized by a complex of events at the organelle level involving nuclear envelope, nucleoli, chromosomes and spindle; and "G_1" and "G_2", the remaining intervals. As either G_1 or G_2 may be absent from a cycle, with S following on directly from telophase or with prophase following S, there is no specific event, fundamental or otherwise, by which these phases may be precisely defined, even though an identifiable biochemical step may regularly occur within one of them in a particular cell type.

It may be that all the interphase stages of chromosome replication, including chromatid organization, do or can occur within S. It is known that along a chromosome, replication can occur simultaneously at

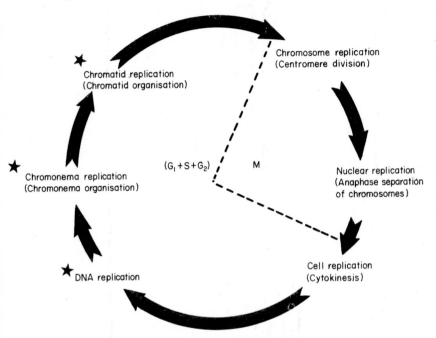

FIG. 4. The essential events of the mitotic cell division cycle.
★ These events could be almost synchronous within S, particularly if the chromosome were basically one DNA double helix, or in a different sequence, particularly if the chromosome contained two or more DNA helices. Chromosome replication implies the organization of the other structural components of the chromosome to be associated with the new DNA, while chromatid replication implies the organization of the chromosome into two visible and structurally distinct equal sub-units.

several points, and that the duration of S, the period necessary for all the chromosome DNA to replicate, is very much longer than the time needed for replication at any one point. Perhaps DNA replication, chromonema organization and chromatid formation can all take place simultaneously at different points within the same chromosome throughout S. If that is so, the rest of interphase is mostly concerned with cytoplasmic preparations for division, including cell enlargement, and the variation in duration and distribution of interphase time around S may reflect the qualitative and quantitative cytoplasmic differences between dividing cells even within one organism (Prescott,

1961). Some support for this suggestion comes from the observation that mitotic cycles lacking both "G_1" and "G_2" have only been found in those few cells which divide without cell enlargement, as in young animal embryos. However, events in a fertilized egg full of reserve materials may not be characteristic of dividing cells in general, and other synthetic events may be necessary to maintain the cycle throughout somatic development. If there are basic steps in the nuclear cycle which do not and cannot normally take place within S, the variable position of S relative to M can only be explained by differences in the order in which the events take place, the significance of which is not understood.

Whether or not discontinuous DNA synthesis precisely distinguishes eukaryotes, many of the other features of the angiosperm division cycle are absent from prokaryote cells. The feature which provides the basis for the terminology is the restriction of all the chromosomes to one part of the eukaryote cell by a bounding membrane to form the nucleus. Other conspicuous eukaryote features absent from prokaryotes are the spindle, centromeres and chromosome contraction. A less striking difference between the cell types which may, however, be of more fundamental significance is that in prokaryotes the entire genome is contained within one chromosome while in eukaryotes the gene complement is distributed over at least two, and usually considerably more, structurally distinct chromosomes. In conjunction with fertilization and meiosis, this creates increased opportunities for allele recombination with obvious evolutionary implications, but at the same time creates the necessity for extraordinarily precise mechanisms for chromosome distribution at division and for intimate communication between chromosomes during interphase. The structural features mentioned above no doubt evolved to meet these requirements.

IV. REPLICATION AND SEGREGATION OF CYTOPLASMIC ORGANELLES

Not only the nucleus but also all essential components of the cytoplasm must replicate and separate in preparation for cytokinesis if they are to be represented in the daughter cells (see Chapter 4). Very little is known about the processes involved and it is not even clear which of the visible structures are essential in this sense, because so little is known of the origin and formation of organelles during the life cycle. It is not unreasonable to suppose that under the control of a competent nucleus, daughter cells can produce new ribosomes and extend membranous components such as endoplasmic reticulum (ER), tonoplast, plasmalemma and perhaps Golgi bodies to augment whatever is inherited at cytokinesis. These elements are at least sometimes directly interconnected to form a continuous differentiated membrane system with linear continuity (Morré and Mollenhauer, 1974) (Fig. 5) and each may therefore be capable of giving rise to others in this group. Golgi bodies may in some cases replicate by fission (Lovlie and Bråten, 1970; Bisalputra et al., 1966).

The situation is less clear for the semi-autonomous organelles, the plastids and mitochondria, which are also occasionally seen to be interconnected with the other cell membranes (Fig. 5) but which are at least partly under the control of their own DNA "chromosome" (Tewari and Wildman, 1969; Herrmann and Kowallik, 1970, Kowallik and Herrmann, 1972) and also perhaps the centrioles. Mitochondria (Swift and Wolstenholme, 1969; Hanzely and Gilula, 1970; Mitchison, 1971;

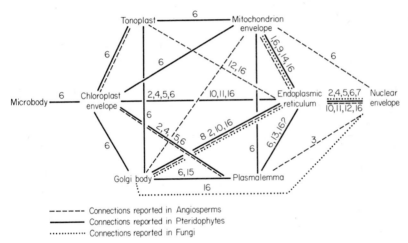

FIG. 5. The reported direct connections, with lumen continuity, of membrane systems within plant cells.

The infrequent appearance of some connections indicates that they are ephemeral and perhaps restricted to specific stages of cell or tissue development. The continuity via vesicles between the endoplasmic reticulum, golgi bodies and plasmalemma or tonoplast has been omitted.

FIG. 5a. The connections reported in the literature. The numbers relate to the following references:

1. Bracker and Grove (1971); 2. Brown and Dyer (1972); 3. Carothers (1972); 4. Cran and Dyer (1973); 5. Cran (1970); 6. Crotty and Ledbetter (1973); 7. Evert and Deshpande (1970); 8. Favard (1969); 9. Franke and Kartenbeck (1971); 10. Gullvag (1968); 11. Hanstein (1962); 12. Jensen (1965b); 13. Mepham and Lane (1970); 14. Morré et al. (1971); 15. Northcote (1971); 16. Morré and Mollenhauer (1974).

A. F. Dyer .

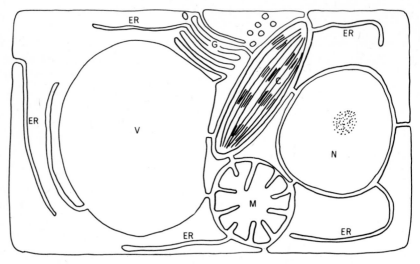

FIG. 5b. A generalized plant cell showing diagrammatically all the reported connections. C—chloroplast; ER—endoplasmic reticulum; G—golgi body; M—mitochondria; N—nucleus; V—vacuole.

Atkinson *et al.*, 1974) and plastids (Diers, 1970) are known to replicate in many different plant cells by division and perhaps by budding of pre-existing organelles, but it is less clear whether they can also arise *de novo* by the differentiation of new or existing membranes around a previously invisible precursor (Bell, 1970b; Gori *et al.*, 1971) in a way similar to that proposed by some authors for their evolutionary origin (Haldar *et al.*, 1966). Such an origin has been suggested in fern eggs (Bell *et al.*, 1966) and these organelles can apparently disperse or at least de-differentiate and re-form in some way during angiosperm meiosis (Heslop-Harrison, 1972; Diers, 1970; see Chapter 5). Centrioles, where present, do not replicate by division but only by the organization of a group of microtubules in the vicinity of an existing centriole, perhaps around a precursor, even its own DNA (see Section IIIF). If totipotency is to be retained through many cell generations, at least one of these organelles or its precursor must enter each daughter cell and their replication and separation prior to cytokinesis becomes critical. More is known in this respect about plastids than mitochondria because of their greater size, and as the available information for mitochondria is similar to that for plastids, only the latter will be considered further here.

A. The Regulation of Plastid Number

In those lower plants, including most unicellular algae, the multicellular alga *Chorda* and the bryophyte *Anthoceros* (Table II), which have only one or two, usually large, plastids in each cell, variation in plastid number is necessarily rare as failure of chloroplast segregation produces lethal albino cells which are incapable of regenerating a plastid. In these organisms, the plastids divide at about the same time as and in close association with the nucleus, and perhaps even making use of the spindle segregate to opposite poles prior to cytokinesis, thus ensuring regular transmission to daughter cells. The same organisms have evolved equally precise mechanisms of plastid fusion *(Chlamydomonas, Cosmarium)* or elimination (*Spirogyra, Zygnema, Anthoceros, Ulva, Vaucheria*) to regulate the plastid number after fertilization (Bråten, 1973).

Most higher plants have several or many small plastids in every cell (Clowes and Juniper, 1968), perhaps to provide greater flexibility of response to external factors like light intensity which induce rearrangement of the photosynthetic apparatus within the cell, particularly in land plants. In such plants, there are large differences in plastid number between cells of different tissue type (Bartels, 1964; Schröder, 1962; Butterfass, 1969, 1973) and of different age (Juniper and Clowes, 1965; Possingham and Saurer, 1969) and smaller but still considerable differences between neighbouring cells in the same tissue (e.g. Bartels, 1964). This suggests either an extremely precise control of plastid replication and segregation responding to subtle differences between cells such as increasing endopolyploidy (Butterfass, 1973) or cell volume (Kameya, 1972; Boasson *et al.*, 1972; Possingham and Smith, 1972) or an imprecise control permitting or even generating considerable random variation. The latter interpretation is supported by the variation in plastid number seen in identical sister guard cells of stomata where the number of chloroplasts is very low (Butterfass, 1969). There is certainly no evidence that chloroplasts divide with the nucleus, little evidence that they divide synchronously during interphase (Schröder, 1962; Mache *et al.*, 1973; Maruyama, 1968), and no visible arrangement of plastids in the cell which could ensure the distribution of a determined number of them to the daughter cells, let alone guarantee the segregation of sister plastids (see Section IVC).

The following indirect evidence from plastid numbers suggests that there is some relationship between plastid replication and cell division, but that the inclusion of plastids in the daughter cells is an accident of their cellular distribution at the time of cytokinesis. Different rates

TABLE II. The relationship between plastid and nuclear division in a variety of higher and lower plant genera

Timing of plastid division in dividing cells	A One or two plastids per cell. Plastid close to nucleus and divides once per mitosis		B One or two plastids per dividing cell. Plastid division continues in at least some differentiating cells		C One plastid in spore mother cells, young spores and/or sperm cells—several to many plastids in all other dividing and differentiating cells		D Three to many plastids in all cells—not closely associated with nucleus	
	UNICELLULAR/ FILAMENTOUS	MULTICELLULAR	UNICELLULAR/ FILAMENTOUS	MULTICELLULAR	UNICELLULAR/ FILAMENTOUS	MULTICELLULAR	UNICELLULAR/ FILAMENTOUS	MULTICELLULAR
Synchronous with mitosis or cell division	[a]Chlorella (1) [a]Spirogyra [a]Oedogonium [a]Chlamydomonas (2)	[a]Anthoceros (3, 4) [a]Notothylas (3) [a]Phaeoceros (3) [a]Ulva (5)	[a]Spirogyra?				[a]Spirogyra	
"Pre-mitosis"— mid to late interphase	[a]Mougeotia [a]Cosmarium (8) [a]Rhynconema (4) [a]Chlorella (1)	[a]Megaceros (3) [a]Chorda (9)		[a,b]Isoetes (4, 10) Megaceros (3) [c]Selaginella (4, 11)		Spores and mother cells of: Fissidens (6, 7) Physcomitrium (6) Funaria (6) Marchantia (6, 19) Polytrichum (19)		[a]Porphyridium

| | | *Fucas* (9)
Dictyota (15) [a]
Nitella (16) [b]
Dryopteris
   sporophyte +
   gametophyte
Equisetum (13, 19) [b]
Psilotum (17) [b]
Potato (4) [b]
Tobacco (4) [b]
Spinach (4) [b] |
| *Funaria* (6)
Marchantia (6) | | |

[a] Pyrenoids present.

[b] Proplastids present in light grown material.

[c] Only one plastid in differentiated epidermal cells in some species.

[d] Multinucleate cells with one reticulate or several small plastids.

Note: With further information, some examples in Column D may have to be transferred to Column C.

References:

(1) Atkinson *et al.*, 1974; Bertagnolli and Nadakavurkaren, 1970; Bisalputra *et al.*, 1966.
(2) Mihara and Hase, 1971; Warr, 1968; Ris and Plaut, 1962; Goodenough, 1970.
(3) Burr, 1969, 1970.
(4) Granick, 1961.
(5) Lovlie and Bråten, 1970.
(6) Sapehin, 1915.
(7) Mueller, 1974.
(8) Korn, 1969.
(9) Bouck, 1965.
(10) Stewart, 1948; Paolillo, 1962; Whatley, 1974.
(11) Jagels, 1970.
(12) Cook, 1973; Schiff and Epstein, 1965, 1966.
(13) Whatley, 1971; Jones and Hook, 1970.
(14) Giles and Sarafis, 1972.
(15) Chi, 1971.
(16) Green, 1964.
(17) Sun, 1961.
(18) Feldmann, 1969.
(19) Paolillo, 1974.

of plastid division relative to cell division could maintain, increase or even decrease the number of plastids present in each cell as tissues develop, and generate the wide differences for example between leaf epidermis, palisade and spongy mesophyll. Some of the variation between cells of the same tissue could be caused also by irregularities of plastid distribution at cytokinesis.

Presumably plastid replication occurs in mitotically active higher plant cells. Schröder's (1962) evidence suggests that in developing *Oenothera* leaves there may be a close relationship between cell and plastid division. However, there is evidence from several species, including tobacco and spinach (Boasson and Laetsch, 1969; Honda *et al.*, 1971; Rose *et al.*, 1974), which indicates that plastid replication also continues for a while after division has stopped in the expanding cells of developing leaves and cultured leaf disks, but is absent from mature cells. In the much simpler system of the gametophyte of *Dryopteris borreri* (Cran, 1970) growing as a filament by repeated divisions of the apical cell, a significant increase in plastid numbers occurs only in the dividing cell, although some plastid division figures are seen behind it in the differentiating cells. In at least these examples, plastid replication is restricted to cells which have recently divided. In some cases, including *Dryopteris* gametophytes, plastids become progressively larger as the cell differentiates. As it is known that the number of plastid genophores increases with plastid size (Tewari and Wildman, 1969; Herrmann, 1969, 1970; Hermann and Kowallik, 1970; Kowallik and Herrmann, 1972), it may be that plastid DNA continues to replicate even after plastid division has ceased in differentiating cells to produce large endopolyploid plastids comparable to the widely found endopolyploid nuclei.

In the *Dryopteris* protonema, the mean number of plastids per cell remains approximately constant during growth, indicating that on average, each plastid in the mitotic cell divides once per cell cycle. However, in a developing angiosperm leaf, the number of plastids in a palisade cell is clearly several times that in stem apex initial cells (Cran and Possingham, 1972; Bradbeer *et al.*, 1974; Possingham and Saurer, 1969), and in addition to chloroplast replication for a short period after cell division has ceased, this increase could result from an accelerated rate of plastid division compared with mitosis. Although no detailed analysis of chloroplast numbers throughout the period of cell division in the developing palisade has yet been achieved, Lyndon (unpublished) has evidence that in *Pisum sativum*, the number of plastids in the dividing cells of a new leaf primordium is larger, at about 14, than the 11 found in the nearby cells of the central zone from which the primordium developed. The plastids must have replicated slightly faster than the cells over the previous few cycles. A similar situation is known in bryophyte spores (Mueller, 1974). This accelerated plastid division may be tissue specific. In the pea plants, the stomatal guard cells each contain 8–13 plastids, that is about the same as a stem initial cell. It seems likely that in the development of the epidermis and possibly in the permanently meristematic stem initial cells, as also in the fern gametophyte, chloroplast replication is at the same rate as cell division and a constant plastid number is maintained, but in the palisade the number of chloroplasts is increased by accelerated or prolonged plastid replication.

It is not easy to analyse the cause of the wide variation in chloroplast numbers in the cells of a complex tissue like the palisade, but some conclusions can be drawn from the simpler fern gametophyte situation where the range of numbers is similar (Cran, 1970). There is no degeneration or *de novo* origin of plastids visible under the electron microscope, no differentiation of proplastids or significant increase in the average plastid number per cell after mitosis has ceased. The differences between the cells within a filament and of corresponding cells of similar filaments must have been present when the cells first formed. The mean number of cells and plastids per protonema in successive samples of a culture show that their average rate of increase is constant. Although fluctuations in plastid division rate in individual cells or filaments cannot be entirely ruled out, a more likely explanation of the differences is the uneven distribution of chloroplasts to daughter cells. Although in each dividing cell systrophy aggregates some chloroplasts more or less evenly round the nucleus, the scattered distribution of the remaining plastids would appear to result inevitably in uneven segregation even if the new wall were to halve exactly the cell volume. Thus variation in the number of plastids per protonema cell would appear during development even if every spore started with the same number and they replicated at the same rate. A similar situation may well occur in angiosperm cells where plastid number is increasing. Where a lower number is maintained, as perhaps in epidermis and primary meristems, distribution may be more even (Butterfass, 1969, 1973). However, it is difficult to see how, even in these cells, plastids can be maintained at a constant number through many cell generations of vegetative growth and through many plant generations. Presumably there is either strong selection against cells with deviant numbers, or the cumulative errors are "corrected" at some stage in each life cycle. The likely stage for this would be during a single cell phase such as the spore or egg. In several bryophytes the number of plastids per cell is reduced to one in the spore or sperm (Sapehin, 1915; Mueller, 1974; Paolillo, 1974). The reports of organelle degeneration and replacement in the fern (Bell *et al.*, 1966) and de-differentiation and re-differentiation in the bryophyte egg (Hanstein, 1962; Diers, 1966, 1970; but see Zinsmeister and Carothers, 1974) and angiosperm meiosis (Heslop-Harrison, 1971b, 1972; Maruyama, 1968; Bal and Deepesh, 1961; but see Willemse, 1972; Vasil and Aldrich, 1970) are of particular interest in this respect, as would be similar studies of the cells giving rise apogamously to sporophytes or aposporously to gametophytes. The gradual deterioration of old clonal material described by horticulturalists may also be relevant.

There are two further speculations to be made at this point about the regulation of plastid numbers. The first concerns the origin of organisms with multiplastid cells if they arose from more primitive forms with only one plastid. There is evidence of a progressive relaxation of the rigid association of plastid and nuclear replication found in most unicells and *Anthoceros* as the higher plant forms evolved (Table II). In unicellular organisms, where all cells are capable of division and reproduction and few have more than two plastids, any changes in the number of plastids per cell must be reversible. Thus in *Euglena* (Cook, 1973) cells in optimum growth conditions in new medium maintain about eight plastids which divide regularly once per cell cycle, but under low light this number increases to 15 and in the non-dividing cells of older cultures in the stationary phase up to 50 plastids form by plastid divisions continuing after the last cell division. However, when the cells are returned to optimum conditions, rapid cell division "dilutes" the plastids until there are once more about eight per cell. Thus, in this organism, the origin and maintenance of the multiplastid condition has been accompanied by a more flexible, though not uncontrolled, association between nuclear and plastid division. At the same time the plastids are scattered in the cell, and not situated close against the nucleus. In multicellular organisms, where some cells cease to divide further and become differentiated into somatic cells, irreversible changes in plastid number can occur during development and for example multiplastid vegetative cells can be produced even while maintaining the uniplastid condition in the meristem cell lines. In *Isoetes* (Paolillo, 1962; Stewart, 1948; Granick, 1961), some *Megaceros* species (Burr, 1970) and *Selaginella* (Jagels, 1970) a single plastid divides close to, and at the same time as, the nucleus of every meristematic cell as in *Anthoceros*, but as the cells cut off by the meristems begin to differentiate, the plastid usually moves away from the nucleus and continues to divide independently several times somewhere in the cytoplasm to produce the multiplastid vegetative cells. Only in epidermal cells of some *Selaginella* species (Jagels, 1970) is a single plastid found in a differentiated cell, but in all the examples listed the uniplastid condition is rigidly maintained throughout vegetative growth in the dividing cell lines perpetuated in the meristems and also probably in the spores (Sapehin, 1915). An even more flexible relationship between plastid replication and nuclear division which combines some of the features of *Euglena* and *Isoetes* is found in some bryophytes such as *Fissidens* and *Marchantia* (Sapehin, 1915; Mueller, 1974) where even the dividing cells of the meristem have several plastids and the uniplastid condition is only briefly restored during the sexual reproductive cycle in the spore mother cells, the young spores and perhaps the sperm. During development of the sporogenous cells, cell division continues without plastid division until the number of plastids per cell has been reduced to one in the spore mother cell. This plastid then divides twice before the next, meiotic, cell division, to produce four daughter plastids, one plastid then being included in each young spore of the tetrad. A similar sequence occurs at meiosis in *Anthoceros*. During subsequent spore development, plastid division continues without cell division, thus restoring the multiplastid condition which is then maintained, though not necessarily with a rigidly fixed plastid number, in every cell during gametophyte growth. Whether or not plastid replication also continues in differentiating cells, and it probably does at least to a limited extent, further opportunities have been created for changes in plastid number during development by the variable though precisely regulated relationship between plastid and nuclear division in dividing cells. The situations described in *Anthoceros, Euglena, Isoetes* and *Fissidens* may represent intermediate stages in the evolution of those angiosperms and ferns in which several plastids are found in all cells of the life cycle including meristems and spores. In these, at least some plastids remain scattered through the cytoplasm in all cells, and the timing of plastid division shows no clear correlation with cell division at any stage, although as argued earlier, some control of plastid number must operate, perhaps also in sporogenesis or gametogenesis but without reverting to the uniplastid state.

The second speculation concerns the pyrenoid, the presence of which appears to be as closely related to a precise correlation of plastid and nuclear replication (see Table II) or even to chloroplast size (Feldmann, 1969) as to the systematic affinites of the plant (Evans, 1966; Dodge, 1973). Pyrenoids of one type or another are almost always found in unicellular and multicellular algae with one or two chloroplasts in each cell. The bryophytes *Anthoceros, Notothylas, Phaeoceros* and *Megaceros* are unusual among other plant groups both in having only one plastid in every cell and in having pyrenoids (Manton, 1966a). In plants of the *Isoetes* and *Fissidens* types and in angiosperms, where there is a less rigid association between plastid and cell division, no pyrenoids are found. In the multi-plastid algae with pyrenoids, listed in Table II, there may also be a close relationship between cell and plastid division, maintaining a constant plastid number, at least under optimum growth conditions (Cook, 1973), but the data needed to confirm this are not readily available. Pyrenoids have been found near the nucleus in many species (e.g. Manton, 1966a) and occasionally near the plastid DNA (Ris and Plaut, 1962; Retallak and Butler, 1970) or near the centriole (Evans, 1966). The conspicuous pyrenoids of some algae appear to reproduce by budding (Evans, 1966) or fission (Manton, 1966b; Goodenough, 1970; Giles and Sarafis, 1972) before or during chloroplast division (Griffiths, 1970; Chi, 1971), although in other cases they arise *de novo* after plastid division (Atkinson *et al.*, 1974; Pickett-Heaps, 1970; Løvlie and Bråten, 1970). Their role in carbohydrate or protein storage or transport has been questioned (Manton, 1966a; Holdsworth, 1971; Bertagnolli and Nadakavurkaren, 1970). Could pyrenoids be involved in maintaining a rigid co-ordination of plastid and nuclear division and segregation?

B. The Visible Events of Chloroplast Division

The only confirmed mechanism for plastid replication is the binary fission of pre-existing organelles, either, where photosynthetic cells divide as in fern gametophyte or angiosperm leaves, as chloroplasts, or, where specialized meristematic areas exist distinct from the photosynthetic tissues as in the fern sporophyte, as undifferentiated proplastids, in which the process is presumably simpler and more economical. Plastid replication is dealt with in detail elsewhere (see Chapter 4). A brief description of the visible events in the chloroplasts of *Dryopteris borreri* gametophytes will serve as an example here. A comparison of light and electron microscope observations suggests that the chloroplasts follow a regular sequence of division stages (Fig. 6). The smallest, disk-shaped, plastids elongate along one diameter in the plane of the thylakoids until they are twice as long as wide. Then the plastid envelope constricts slightly around a circumference half way along the long axis to produce a "dumb-bell' plastid. The mechanism of constriction in plastids is not known, although microtubules have been seen (Sprey, 1968; Løvlie and Bråten, 1970) and contractile protein detected (Packer, 1966). Under the light microscope, the chlorophyll-containing

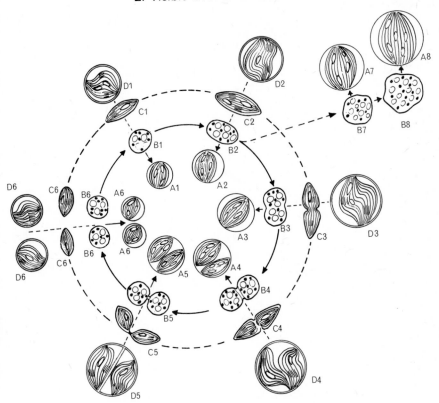

FIG. 6. Chloroplast division in *Dryopteris borreri*.

FIG. 6a. The chloroplast division cycle in *Dryopteris borreri*. Observed relationships are indicated by an arrow; suggested relationships by a broken line.

B_1–B_6 represent the stages in the division cycle as seen in surface view in living cells under the light microscope (see Fig. 6b(ii)). B_7 and B_8 represent enlarging plastids which have apparently ceased division (see plastids W, X, Y and Z, Fig. 6c(i)).

A_1–A_8 represent the appearance under the light microscope of plastids at stages corresponding to B_1–B_6 after osmotic shock following cell damage.

C_1–C_6 represent the electron microscope profiles of vertically sectioned chloroplasts which apparently correspond to the light microscope images. A plastid of the C_5 type, in which the inner envelope membrane, but not the outer one, has fused at the constriction has not yet been described in the literature.

D_1–D_6 represent profiles sometimes seen in EM sections in this and other material and variously attributed to an alternative division process (Cran, 1970; Brown and Dyer, 1972; Cran and Possingham, 1972) or plastid fusion (Esau, 1972), but here represented as fixation artefacts of stages C_1–C_6.

A. F. Dyer

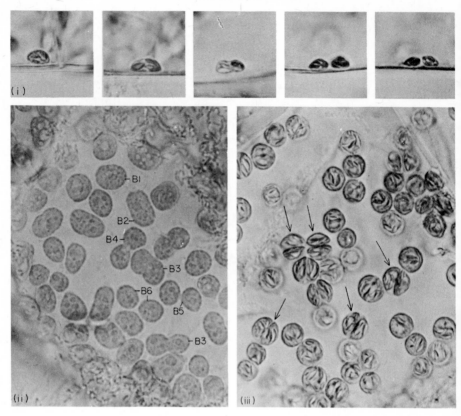

Fig. 6b. (i) Side view of stages in chloroplast division as seen in living cells under the light microscope. × 1100.

(ii) Surface view of stages in chloroplast divisions as seen in a living cell under the light microscope. The numbers refer to the stages identified in Fig. 6a. × 1100.

(iii) In a cell similar to that in (ii), the chloroplast envelopes have expanded and become spherical and the thylakoids in most of the larger plastids are in two distinct groups (arrows). × 800.

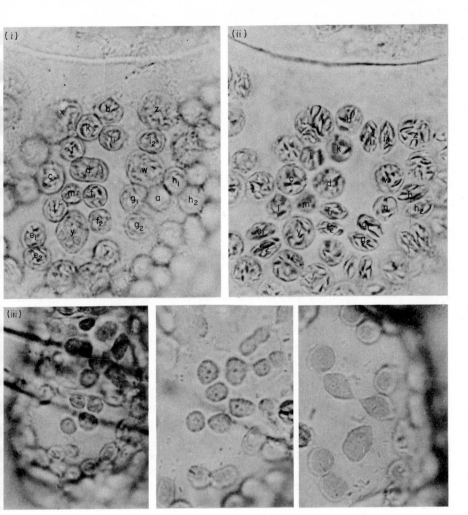

Fig. 6c. (i) Chloroplasts in a cell newly ruptured by the pressure of the cover slip. × 1100.

(ii) The same chloroplasts as in Fig. 6c(i) as they appeared three minutes later. The plastids have abruptly expanded and rounded off. The numbers refer to the corresponding plastids in Fig. 6c(i). × 1100. The smaller, unpaired chloroplasts (a, b, c ($=$Stage B_1, Fig. 6a) and j, k, l ($=$Stage B_6, Fig. 6a)) and the larger unconstricted plastids, which are either elongated (d ($=$Stage B_2, Fig. 6a)) or near spherical (w, x, y, z ($=$Stage B_7, ?Fig. 6a)) depending perhaps on whether or not they are preparing for another division, contain only one group of thylakoids, while constricted plastids (e, g, h ($=$Stage B_4, Fig. 6a)) and pairs of daughter plastids (f, i ($=$Stage B_5, Fig. 6a)) reform into one large plastid with two groups of thylakoids within a pellicle and colourless matrix, even when there is initially little or no evidence of a physical link between the sister plastids.

(iii) The chloroplasts of three cells of the same filament of a *D. borreri* gametophyte, showing the enlargement which occurs with cell age. The cells were cut off by the dividing apical cell approximately 2, 3 and 7 days respectively before the photographs were taken. × 1100.

material then appears to become separated into two equal and distinct disk-shaped areas separated by a narrow colourless region within the constriction of the dumb-bell. It is not easy to correlate accurately with these the plastid profiles seen in electron micrographs. In some sectioned across the thylakoids, the plastid shows a median constriction of the envelope, but the thylakoids are continuous across the constriction (Brown and Dyer, 1972; Cran, 1970). In other profiles, the thylakoids are in two equal groups, sometimes but not always separated by a partial or complete baffle across the plastid, derived from the inner membrane of the envelope, but these plastids are frequently almost circular in section rather than dumb-bell shaped or even elongated (Brown and Dyer, 1972; Cran, 1970). Similar profiles seen elsewhere (e.g. Cran and Possingham, 1972) have led to the suggestion that there are two types or mechanisms of division in chloroplasts, or that some profiles represent stages in chloroplast fusion (Esau, 1972). In the case of *Dryopteris*, the available evidence suggests that there is one division process but some profiles are affected by distortion during preparation of the material for electron microscopy. The likely sequence seems to be that after some constriction of the envelope, the thylakoids separate into two equal groups, with a gap containing only stroma appearing between them. The constriction is then accentuated so that the opposite sides of the envelope closely approach each other and the two daughter groups of thylakoids are further separated, eventually by a distance of several micrometres. At this stage fusion of the unit membranes of the envelope begins, fusion of the inner membranes preceding the completion of plastid fission by fusion of the outer membranes. This is the reverse of the sequence of membrane fusion when polar nuclei merge (Jensen, 1964). If the plastids are subjected during fission to osmotic shock following cell damage, it can be seen that chloroplasts at all stages tend to become spherical and in some cases the envelope swells, revealing one or two groups of thylakoids within a large volume of presumably diluted stroma. This process can be continuously followed over a period of several minutes after cell damage, and it becomes clear that chloroplasts which are initially spherical and those in the early stages of elongation and fission give rise to swollen plastids with one group of thylakoids, while later stages of dumb-bell plastids and even quite widely separated pairs of daughter plastids with no obvious connections between them yield larger spherical plastids in which there are two groups of thylakoids. Unfortunately, the inner membrane baffle, if present, cannot be distinguished under the light microscope. The fact that EM profiles showing the inner membrane baffle are also frequently circular with convoluted and often swollen thylakoids is consistent with the idea that they result from the "rounding off" of stages such as those described above following osmotic shock induced during preparation (Fig. 6).

Although there are no detailed descriptions of the complete chloroplast division cycle based on comparisons of light and electron microscope observations on the same material, stages have been frequently reported which indicate that a process very similar to that suggested for *Dryopteris* occurs in a wide range of plants (Kusunoki and Kawasaki, 1936; Gullvag, 1968; Gantt and Arnott, 1963; Bisalpultra and Bisalpultra, 1970; Whatley, 1971; Sprey, 1968; Possingham and Saurer, 1969; Vesk *et al.*, 1965; Stone, 1932; Diers, 1965; Green, 1964; Jones and Hook, 1970; Stetler and Laetsch, 1969; Giles and Taylor, 1971; Fasse-Fransisket, 1955; Schötz and Senser, 1964; Esau, 1972). In the few cases where the information is available, the chloroplast division cycle seems to last for 14–40 h, although the final stages of fission of a dumb-bell to give two daughter plastids probably takes only about 6 h Macnutt and von Maltzahn, 1960; Giles and Taylor, 1971; Kass and Paolillo, 1972; Ueda *et al.*, 1970; Green, 1964; Kusunoki and Kawasaki, 1936; Cran, 1970).

C. The Segretation of Chloroplasts at Division

Time lapse cine-micrography of dividing endosperm cells shows that at anaphase, cytoplasmic organelles surrounding the spindle are carried by cytoplasmic movement to whichever pole is nearest. If this is typical of dividing cells (e.g. Cronshaw and Esau, 1968), segregation of plastids at division is merely an accident of their previous distribution in the cell. Plastids are usually scattered throughout the cell, but they may not be distributed at random. Butterfass (1973) has shown that in haploid *Trifolium hybridum* the two chloroplasts in the guard mother cells separate into different daughter cells more often than would be expected from random distribution. Occasionally organelles including plastids are more obviously distributed in a non-random way. Organelle aggregation at spindle poles has been occasionally seen (Zirkle, 1927; Diers, 1965, 1966; Avers, 1963; Porter and Machado, 1960; Wilson, 1970; Pickett-Heaps, 1967c; Fowke and Pickett-Heaps, 1969; Tai, 1970; Fowke *et al.*, 1974; Roberts and Northcote, 1970). Though not an invariable precursor of division (Bui-Dang Ha and Mackenzie, 1973), systrophy is common in dividing photosynthetic and vacuolated cells of some plant groups, and the group of plastids around the nucleus may be approximately halved at division but the remaining plastids elsewhere in the cell may be unequally distributed in response to cytoplasmic gradients and light conditions. Prior to the unequal divisions during the development of, for example, root hairs (Avers, 1963; Cutter and Hung, 1972), stomata (Zeiger and Stebbins, 1972) and binucleate pollen grains (Heslop-Harrison, 1968; Burgess, 1970a), the cytoplasm, including plastids, concentrates towards the end which will produce the smaller daughter cell. However, because of the subsequent unequal cell division, this cytoplasmic gradient may merely maintain a more or less equal distribution of plastids to daughter cells.

If a constant number of plastids is maintained in angiosperm primary meristems by equal distribution at division it must be as a result of maintaining an approximately uniform distance between plastids in the cell followed by the formation of two cells of equal size. Even then it is inconceivable that sister chloroplasts of each plastid division could regularly separate to different cells. The necessary distribution could only be achieved if newly formed plastids detached and moved apart well before cytokinesis. But at least in *Nitella* and *Dryopteris* the pairs of new plastids remain attached and close together for a considerable time, and few if any interphase cells show the cytoplasmic movement necessary to separate them. It seems inevitable that sister plastids pass to the same cell for at least one, perhaps several, cell cycles.

This conclusion has implications for the "sorting out" of mutant plastids. Mixed cells with green and defective white plastids have been seen in several species (e.g. Kirk and Tilney-Bassett, 1967; Hagemann, 1965, Wettstein and Erikson, 1965; Röbbelen, 1966; Wildman *et al.*, 1973; Tilney-Basset, 1973; Anton-Lamprecht, 1966) and it is considered that these give rise to many cases of green–white variegation by chance sorting out of plastids at division to give pure green and pure white cells which in turn produce green or white tissues respectively (Neilson-Jones, 1969). The fact that sorting out occurs at all confirms that sister chloroplasts do not regularly segregate to opposite poles at the next division. The rate of sorting out will, however, depend not only on how often sister plastids go to the same cell at division, but also on how many plastids there are, and what proportion replicate, how many are green and how many white, how fast the green plastids replicate relative to white ones and how unequal the segregation is to the daughter cells.

It might be possible to investigate the effect of these different factors in the system provided by the seedlings of variegated *Pelargonium* (Tilney-Bassett, 1965, 1973), where, as a result of paternal plastid transmission, appropriate crosses produce zygotes with mixed green and white plastids. Sorting out then produces green and white sectors in variegated seedlings, or pure green or pure white seedlings if the sorting out is accomplished in an early division of the pro-embryo. Different varieties and reciprocal crosses give different rates of paternal transmission and subsequent sorting out, but within the products of a single cross, it should be possible to relate the rate of sorting out to observations on plastid numbers and on the cell division sequence during embryogenesis and draw conclusions about the segregation of plastids in this system.

V. CYTOKINESIS

A. The Significance of Cytokinesis

When the nuclear cycle and cytoplasmic replication is complete, cytokinesis frequently, but not always, follows to sub-divide the binucleate protoplast into two units with equal developmental potential. Except in uninucleate organisms where cell division is reproduction, the significance of cytokinesis seems to be in relation to differentiation of the sister cells to produce the complexity typical of multicellular organisms. As there is apparently a limit to the amount of cytoplasm which can be associated with a single nucleus, all the larger organisms are multinucleate. Many fungi and algae, and certain specialized tissues in angiosperms, are coenocytic, with large numbers of nuclei in a common mass of cytoplasm. Although no doubt the cytoplasm is functionally subdivided into areas of influence around each nucleus, very little structural complexity seems possible within a coenocyte. The greater complexity of most multinucleate higher plant tissues is associated with the formation and differentiation of small cytoplasmic units, each at least partially isolated from the rest by membranes and surrounded by rigid cell walls, containing a single nucleus. In such a system, even adjacent sister units can develop into structurally and functionally quite different cells, such as the sieve tube and companion cell of the phloem.

The role of the intervening membrane in the differentiation of sister cells is suggested by an experiment with caffeine on *Dryopteris* gametophytes (Fig. 7). Treatment with 10 mM caffeine for 4–6 h allows mitosis to proceed normally but prevents alignment of the vesicles which form the cell plate, and cytokinesis fails (Pickett-Heaps, 1969b; López-Sáez *et al.*, 1966). At predictable stages of normal development (Dyer and Cran, 1974) certain green protonema cells undergo an asymmetric mitosis followed by unequal cell division. The smaller of the two daughter cells rapidly elongates, losing both its chloroplasts and the ability to divide, and differentiates into a colourless rhizoid quite distinct from its sister cell which remains broad, green and potentially mitotically active. If cultures are treated with caffeine at an appropriate stage when many rhizoid-forming divisions are taking place and then returned to optimum culture conditions, a proportion of them, caught at telophase, will show inhibition of cytokinesis and form binucleate cells. In these binucleate cells, the cytoplasm which would have been included in the rhizoid fails to differentiate, even though rhizoid cells formed just before the treatment are capable of developing normally after treatment has ceased.

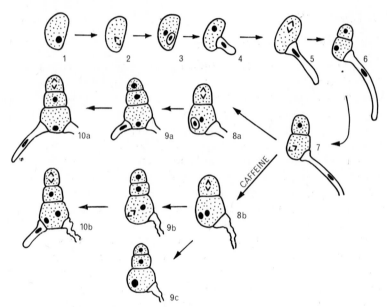

FIG. 7. The effect of caffeine treatment at mitosis on subsequent cell development in the young gametophyte of *Dryopteris borreri*.

FIG. 7a. The first four mitoses (stages 1–10a) produce a filament of three green cells with two elongated rhizoids extending from the basal cell. The first and third mitoses occur in the basal cell and are asymmetrical, cutting off a small daughter cell which then differentiates. The few chloroplasts degenerate and the cell elongates to form a colourless, unicellular rhizoid.

Treatment with 10 mM caffeine for 6 h during the third mitosis produces a filament with a binucleate basal cell (Stage 8b). Neither nucleus gives rise to a rhizoid when the gametophyte recovers on removal of caffeine, but the nuclei may fuse (Stage 9c) or one will divide again (Stage 9b) to produce a new and normal rhizoid and a still binucleate basal cell (Stage 10b). Differentiation of the rhizoid from the green cell does not occur after a mitosis after which cytokinesis fails, but the treatment does not affect development following subsequent divisions at which normal cytokinesis occurs. Caffeine treatment during interphase has little or no detectable effect on subsequent cell differentiation.

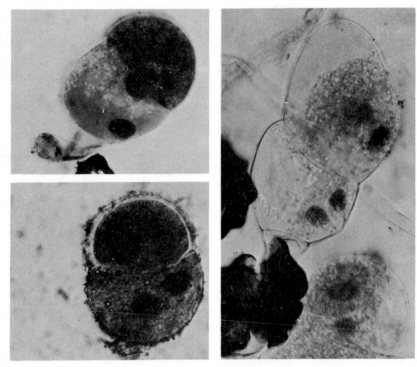

Fig. 7b. Gametophytes with binucleate basal cells but no second rhizoids after caffeine treatment during the third mitosis. × 400.

Cytokinesis therefore seems to achieve the degree of isolation necessary for one nucleus to initiate a pattern of differentiation in the cytoplasm which is quite different from that initiated or maintained by its neighbour. However, the isolation of cells can rarely be total in an intact developing organ because the development of one cell is integrated with that of the next according to an overall pattern and this integration must involve inter-cellular communication. While providing the isolation necessary for differentiation, the membrane produced at cytokinesis also allows or even provides the means for the communication necessary for integration.

B. The Events of Cytokinesis

Although cytokinesis is not inevitably linked with mitosis, the formation of a new cell wall across the equator of a spindle between newly formed daughter nuclei marks a definitive stage in the cell cycle when all component cycles, nuclear and cytoplasmic, are at an equivalent point: the point when a new protoplasmic unit is formed. However, the completion of the cell wall is itself the result of a cycle of events in the cytoplasm, although little is known as yet about any events prior to anaphase which are specifically in preparation for cytokinesis because at the biochemical level they cannot be distinguished from those associated with growth of the outer membrane and wall as the cell expands prior to division.

The first visible event leading to cytokinesis is the aggregation of ER or Golgi bodies or both near the equator of the spindle and formation from these of membrane bound vesicles (Brown and Dyer, 1972; Lehman and Schulz, 1969; Bajer, 1968b; Cronshaw and Esau, 1968). The vesicles aggregate and then align on the equator, perhaps under the influence of the spindle microtubules, and eventually fuse to form the cell plate (Roberts and Northcote, 1970). This process usually starts in the centre of the spindle and spreads gradually centrifugally until it meets the outer wall of the cell. In some cells, the plate may be curved as in the formation of stomatal subsidiary cells and pollen generative cells (see Section IIC). In cambial cells, the formation of a large cell plate can take up most of the time in mitosis (Catesson, 1974). The fused vesicle membranes form the two new plasmalemmas; the vesicle contents apparently form the first elements of the new wall. The wall later thickens and a middle lamella appears and unites with that of the parent cell wall.

As the cell plate forms, small interruptions occur at intervals and become plasmodesmata. Their origin is not clear, although they may represent elements of ER trapped as the plate forms (Evert and Deshpande, 1970; Burgess, 1971). However, as plasmodesmata can form in mature walls (Burgess, 1972) other origins must be possible. The structure of plasmodesmata is not fully understood (Robards, 1971; Burgess, 1971) but they seem to be lined by membrane inter-connecting the plasmalemmas of the two adjacent cells and to contain a central core which is frequently in contact with, and is possibly continuous with, elements of ER at each end (Frazer and Gunning, 1969; Evert and Deshpande, 1970; O'Brien and Thimann, 1967; Withers and Cocking, 1972).

C. The Role of Plasmodesmata

Protoplasmic continuity is demonstrably important in cell differentiation (Yoshida, 1962) and it is an obvious and by no means new suggestion that plasmodesmata provide the route for the selective transfer from cell to cell of the presumably chemical stimuli which integrate the differentiation of neighbouring cells (e.g. Arisz, 1969; Tyree, 1970; O'Brien and Thimann, 1967; Goldsmith, 1968; Kaufman *et al.*, 1970; Juniper and Barlow, 1969; Spanswick, 1972; Burgess, 1972; Northcote, 1971; Helder and Boerma, 1969) while the plasmalemma provides the barrier to other exchanges which in turn produces the partial isolation necessary for independent development. In conjunction with the plasmodesmata, the ER may provide a direct link between the perinuclear spaces of adjacent cells and the basis for nuclear interaction.

It is therefore significant that at least a few plasmodesmata exist between all vegetative cells of the multicellular angiosperm sporophyte (e.g. Kaufman *et al.*, 1970), even in graft chimaeras (Burgess, 1972) and also between cells within the embryo sac, but are absent from the two unicellular stages of the life cycle, the spore and the egg. Cytoplasmic connections are lost between the pollen mother cell and the surrounding tapetum by the end of the first meiosis (Vasil and Aldrich, 1970, 1971; Christensen and Horner, 1970; Horner and Lersten, 1971) and the spore is from then on isolated by a thick wall (Heslop-Harrison, 1972). Within the spore, the generative cell wall lacks plasmodesmata. There are few or no plasmodesmata penetrating the wall of the embryo sac as it develops from the megaspore (Rodkiewicz, 1970; Schulz and Jensen, 1971) and a cutin or callose deposition often provides further isolation within the embryo sac; there are no plasmodesmata through the outer wall of the egg, zygote, or developing embryo (Norstog, 1972). The egg is similarly isolated in the gametophyte of the bryophyte *Sphaerocarpus* (Diers, 1965) and *Fissidens* (Mueller, 1974).

It appears then that plasmodesmata traverse every wall within the sporophyte or gametophyte generations but either do not form or are blocked off in any wall which forms the boundary between generations. This raises interesting questions about the role of isolation in achieving the major developmental switch from one generation to the other through the spore or zygote (Rodkiewicz, 1970) and the effect of artificially induced isolation.

In ferns and bryophytes, surgical isolation of single gametophyte cells or tissue pieces from meristematic cells frequently induces regeneration without any other stimulus (Ootaki, 1967, 1968; Ito, 1962; Stange, 1964; Ootaki and Furuya, 1969; Stange and Kleinhauf, 1968; Giles, 1971; Giles and Taylor, 1971; Macnutt and von Maltzahn, 1960). Isolated cells of *Dryopteris borreri* gametophyte of all ages revert to a sporelike condition in that each one subsequently regenerates a complete and normal prothallus, having apparently returned to the earliest stage of gametophyte development. Other ferns and mosses show similar

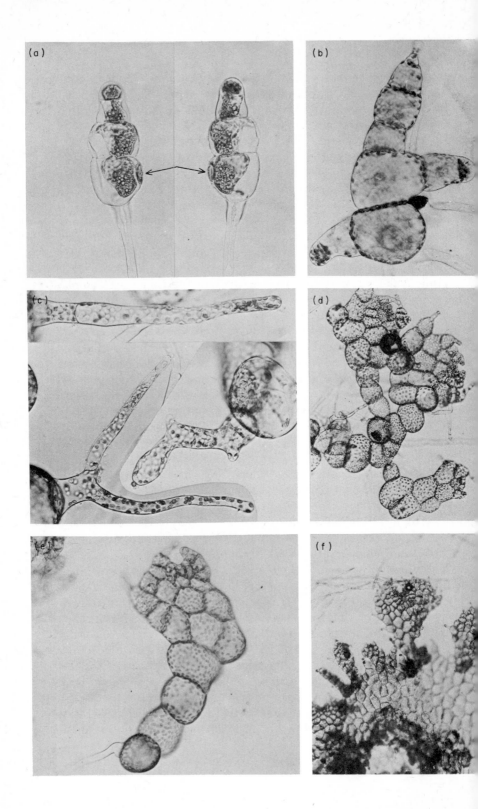

responses, but an exact parallel is not found in angiosperms. Except perhaps for the switch to embryo-sac formation in abnormal pollen grains (see Heslop-Harrison, 1972), the nearest equivalent is the stimulation of sporophyte cells or protoplasts isolated from tissue cultures or leaves to form free-cell cultures. Under appropriate culture conditions these cells may revert, especially if hormones are withdrawn (e.g. Reinert, 1959; McWilliam *et al.*, 1974), to a zygote-like state and give rise directly or indirectly to embryos (e.g. Nagata and Takebe, 1971; Davey *et al.*, 1974; Steward *et al.*, 1970; Backs-Husemann and Reinert, 1970). Induced cell aggregates without plasmodesmata do not form embryos (Sussex and Clutter, 1968).

In ferns (Nakazawa, 1963) and bryophytes (Zepf, 1952) plasmolysis has an effect similar to surgical isolation. In plasmolysed cells, thin strands of protoplasm can be seen running from the protoplast to the walls, and some of the strands may contain endoplasmic reticulum and even moving particles (Sitte, 1963; Falk and Sitte, 1962; Smith, 1972; Kiermayer and Jarosch, 1962) but some run to walls with no plasmodesmata and their significance is not clear. Nor is it clear what plasmolysis does to the plasmodesmata themselves (Burgess, 1970b), but it is probable that they are damaged or altered or even break and heal (Cocking, 1974) and no longer function normally in cell communications, thus at least partially isolating cells from their neighbours.

FIG. 8. The effects of temporary plasmolysis on the subsequent development of young gametophytes of *Dryopteris borreri*.

(a) Four-celled protonemata plasmolysed after 20 min in 0·5 M KCl. The small newly formed second rhizoid cell can be seen (arrows). This severe plasmolysis results in cell death in some gametophytes. At this stage of development, 5 min at incipient plasmolysis (i.e. in 0·15 M KCl) is sufficient to induce branching in 50% of protonemata with little or no cell death. A treatment of 5 min in 0·2 M KCl induces maximum branching (one or more branches in 80–90% of protonemata). Each branch represents regeneration of a new filament from a cell of the original protonema. × *c.175*.

(b) Branched protonema formed by regeneration from the two basal cells while the apical cell has continued to develop after 20 min treatment in 0·5 M KCl at the 4-cell stage. × *c.210*.

(c) Abnormal second rhizoids developed 2–3 days after plasmolysis of gametophytes with newly formed second rhizoid cells as shown in (a). The rhizoids are short, often broader than normal or forked, contain many replicating chloroplasts and in some cases have divided to become 2-celled. × *c.280*.

(d) Multiple gametophytes formed after further, normal development of branched protonemata following plasmolysis treatment as in (b). × *c.85*.

(e) Young unbranched prothallus in control culture. × *c.140*.

(f) Young gametophytes regenerated at the edge of a 3-month-old prothallus treated for 5 min with 0·5 M KCl. × *c.40*.

During culture under optimal conditions following only 5 min of incipient plasmolysis, the photosynthetic cells of *Dryopteris* protonemata regenerate normal new gametophytes as if they had been surgically isolated, and even young rhizoids fail to differentiate and instead remain green and capable of division (Fig. 8). Angiosperm cells also develop after plasmolysis as if induced to regenerate (Yeoman and Brown, 1971).

These observations suggest that isolation alone can cause a cell from an integrated multicellular system to revert to the immediately previous single, and therefore isolated, cell stage of the natural life cycle, that is the spore for a gametophyte cell and the zygote for a sporophyte cell (Stavitsky, 1970). Isolation apparently directly interrupts development and induces "de-differentiation", the immediate return to an undifferentiated state, and either induces or at least permits the resumption of division (Fig. 9). This may be accompanied by a change in balance of growth substances (Giles, 1971). It may be isolation from meristematic cells which has this effect on differentiated cells (Stange, 1964; Bunning, 1956). Adventitious embryos develop from the epidermis of the angiosperm *Torenia* if internode segments are excised and cultured (Chlyah, 1974). In *Ranunculus sceleratus* where embryos arise spontaneously from single hypocotyl epidermis cells of young intact plantlets developing on callus (Konar *et al.*, 1972), plasmodesmata are no longer visible in the outer cell walls as the embryo, presumably now isolated from surrounding tissues, develops. However, plasmodesmata can be seen in the initial cell at the time embryo development begins. If these plasmodesmata are fully functional at this time, isolation does not appear to be the cause of the reversion by the cell to a zygote-like state. Murashige (1974) has discussed evidence which suggests that the natural tendency for embryo initiation of certain "meristemoid" cells is normally suppressed by a diffusible substance. If this is generally true, and the substance is normally transmitted through plasmodesmata, either cell isolation or failure to produce the inhibitor would allow the embryogenic potential to be expressed.

While isolation alone may at least in some cases be sufficient to reverse development in this way, to induce a cell to cross the morphogenetic boundary from one generation to the other (Fig. 9) perhaps needs more than mere isolation. The number of chromosome complements is irrelevant to this, but the cellular environment including growth substances and carbohydrate source may be significant. For example, immature pollen grains of some species can be induced to produce embryos instead of male gametophytes if the culture medium contains sucrose and, in most cases, cytokinins and auxin (e.g. Sunderland and Wicks, 1971; Heslop-Harrison, 1972; Sunderland, 1973a; George and Narayanaswamy, 1973; Rashid and Street, 1974). Early stages of this embryo formation usually involve repeated "cleavage" divisions of the normally undivided vegetative cell within the intact exine. The comparable generation switch in ferns, the formation of apogamous sporophyte buds on prothalli, occurs naturally in a number of species (Steil, 1951), but has also been induced in several normally sexual species, usually by inducing a change from autotrophic to heterotrophic growth with a combination of darkness and sucrose in the medium (Whittier and Pratt, 1971; Whittier, 1974; White, 1971; Kato, 1970). In ferns, the complementary generation switch, from sporophyte to gametophyte, is also possible. Cultured, excised leaves, and even roots and stems, of young fern sporophytes regenerate aposporous gametophytes, sometimes from single cells in superficial layers, and frequently near necrotic tissues which are probably releasing growth substances (e.g. Lawton, 1932; Bell and Richards, 1958; Takahashi, 1962; White, 1971; Munroe and Bell, 1970). Older leaves lose the power to regenerate gametophyte tissue (Morel, 1963). As a final example of the way that the generation switch can be influenced by nutritional factors, undifferentiated fern gametophyte callus can be directed towards either gametophyte or sporophyte development merely by changing the sugar concentration in the medium (Bristow, 1962; Kato, 1965; White, 1971; Mehra and Sulklyan, 1969).

The generation switch of course occurs precisely and regularly in conjunction with the normal alternation of generations in cells at the appropriate stage of development without artificial cell isolation or exogenous carbohydrate and auxin, and sometimes, in apomicts, even without meiosis and fertilization (Fig. 9). Nevertheless, a degree of isolation, induced by the absence of plasmodesmata, and differences in nutritional and hormonal conditions, likely to exist between the sporangium and the fertilized female gametophyte, may still be critical in determining the developmental sequence to be followed. It has been suggested (e.g. Naf, 1962; see White, 1971) that the initial generation switch is determined within the cell although the subsequent characteristic development may only be maintained within a narrow range of conditions. Bell claims that in the fern egg this determination involves cytoplasmic reconstitution including organelle replacement (e.g. Bell, 1970a; see White, 1971), thus destroying the cytoplasmic components involved in gametophytic growth. Heslop-Harrison (1972) suggests that the organelle re-differentiation and ribosome depletion at angiosperm pollen meiosis (see Section IVA) has a similar role in removing the influence of previous sporophyte development and thus allowing

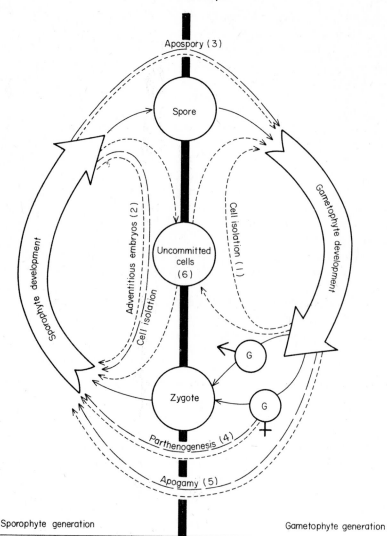

FIG. 9. The alternation of generations in higher plants.

In the normal sexual cycle (continuous lines) the boundary between the sporophyte and gametophyte generations is only crossed at the unique single-cell stages, the meiotic spore and the zygote. In natural apomictic life cycles (broken lines) *aposopory* can occur when the gametophyte develops directly from a sporophyte cell without intervening meiosis and spore formation, *parthenogenesis* occurs when an unfertilized egg gives rise to a sporophyte, and *apogamy* results from the development of a sporophytic embryo from a gametophyte cell directly by mitosis without gamete formation or fusion.

Other developmental changes can be induced in culture (dotted lines). These may the switch to gametophyte growth. Other evidence supports the conflicting "Lang" hypothesis that environmental factors alone determine the developmental path of the tissues derived from the zygote and the spore (De Maggio and Wetmore, 1961a, b). It is interesting that the switch back to sporophyte development induced by the conditions of culture in pollen grains also involves degeneration of cytoplasmic organelles (Sunderland, 1973a). Whatever may be the mechanism of determination of the next phase of development, isolation may be an essential precursor to the switch because it induces de-differentiation and division, and releases the cell from the consequences and influences of development in the previous, alternative generation. Much more information is needed on the biochemical and structural changes which follow artificial isolation during the early stages of regeneration and

---

involve a reversal of development within one generation (1, 2, 6) or a switch to the alternate generation (3, 4, 5, 6).

(1) *Cell isolation* in the fern gametophyte induces a repetition of the developmental sequence from the spore (see Fig. 8). It is almost impossible to test for the same response in angiosperm gametophytes, although the abnormal development of embryo sacs from pollen grains may involve the reversion to an uncommitted spore-like state.

(2) *Adventitious embryos* occur naturally in e.g. the nucellus of some apomictic angiosperms, and on the leaves of e.g. *Kalanchoe daigremontiana* and "viviparous" ferns such as *Asplenium viviparum*.

(3) *Apospory* occurs naturally in some apomictic life cycles, e.g. *Citrus*, and can be induced in culture in isolated excised fragments of young fern sporophytes.

(4) *Parthenogenetic* development of unfertilized angiosperm eggs occurs in some apomictic life cycles, and, in pseudogamous species, is a response to the stimulus of pollination. Where unfertilized eggs can be isolated and cultured as in *Fucus* and some animals including sea urchins and amphibia, parthenogenesis can be induced experimentally.

(5) *Apogamy* can occur naturally in angiosperms where e.g. a synergid develops into an embryo without fertilization, and is not uncommon in ferns where, as in *Dryopteris borreri*, a young sporophyte develops mitotically from one or more of the prothallus "cushion" cells. It can be induced in some other fern species when the gametophyte is grown heterotrophically in the dark and in some angiosperms when immature pollen grains are cultured in the presence of sugar and, in most cases, growth substances.

(6) Under experimental conditions, relatively undifferentiated cells can be derived from excised tissues, either gametophyte or sporophyte, and maintained in culture as callus, free cell cultures or protoplast cultures. Sugar and growth substances are usually required to initiate the cell divisions which produce these uncommitted cells. Further changes in the culture medium may induce the cells to revert to organized development of the same generation as the explant from which the cell culture arose. However, in some fern callus cultures originally derived from gametophytes, it is possible to initiate either gametophyte or sporophyte development by appropriate changes in the growth medium.

those which accompany a natural or induced switch from one generation to the other. A possible involvement of the plasmalemma may be
worth considering. If local differences in plasmalemma structure occur
during cell development (Northcote, 1971) at least in part as a response
to the influence of adjacent cells, then this membrane differentiation
may be important in restricting the range of possible alternatives for
development of the cell and thus in determining its future differentiation. Cell isolation, resulting in plasmalemma damage, may induce
membrane repair and replacement which erase the accumulated localized regions of differentiated structure and partially or completely restore a plasmalemma of a basic membrane structure, perhaps determined by the nucleus, which initially imposes no restrictions on future
cell development. This plasmalemma membrane would be similar to
that first laid down between daughter nuclei at cytokinesis. Perhaps
in permanently meristematic root and stem initial cells a plasmalemma
of this basic, undifferentiated structure surrounds the entire cell. A
similar model involving the plasmalemma could be used to explain
those situations, as frequently recorded in tissue explants in culture,
where re-differentiation of mature cells into a different cell type can
only occur if mitotic cell division is first induced, even when the cells
have not been isolated (Fosket, 1968, 1972; Torrey and Fosket, 1970).
In this situation, division itself will replace some, perhaps all, of the
plasmalemma surrounding the daughter cells with undifferentiated
membrane, thereby restoring a wider range of developmental possibilities. Thus, whether cell isolation or some other factor induces the
cell division, according to this model, the effects on the plasmalemma
could bring about de-differentiation.

    While all this is speculation, the possibility that the plasmalemma
is interacting with the nucleus in determining the way the cell develops
in response to external influences requires further investigation. It has
already been claimed that changes in cell membrane structure and
composition could be important in initiating and maintaining the
sequence of events in the dividing cell cycle (Mazia, 1974; Mueller,
1971). In the next section, the suggestion is made that it plays a role
in determining the orientation of cell division and expansion. If these
suggestions prove to be correct, the plasmalemma will be seen as a complex and important organelle, capable of differentiation, which is complementary to and interacting with the nucleus in providing a cellular
basis for morphogenesis. Then membrane connections within and
between cells will be seen to have even greater significance than has
already been suggested.

VI. THE ORIENTATION OF CELL DIVISION

Cell polarity, obviously of fundamental significance in differentiation and morphogenesis, is frequently revealed by visible structural asymmetry, but the physiological basis of this is beyond the scope of this article. However, an important and conspicuous aspect of polarity is the orientation of mitotic spindles. In plants, with cells fixed in position by rigid walls, division orientation is fundamental to morphogenesis. This orientation is defined by the positions of the spindle poles which in turn are influenced by surrounding cells during organized development. The mechanism is obscure but must involve points of reference within the cell. Polar "spindle organisers" if they exist (see Section III F) might be positioned in relation to cytoplasmic gradients, but differentiation of the plasmalemma and nuclear envelope could provide alternative points which remain fixed even when cytoplasmic streaming occurs.

Differentiation of the nuclear envelope is clear at the spindle poles in organisms which retain the membrane intact round the intranuclear spindle (e.g. Aldrich, 1968; Pickett-Heaps, 1969a; Heath, 1974). In organisms where the nuclear envelope disperses during division, there is much evidence that the interphase chromosomes are attached, particularly by the centromeres and telomeres, to the inner surface of the nuclear envelope and perhaps even to specific regions of the envelope (Comings, 1968; Moens, 1969; Wagenaar, 1969; Girbardt, 1971; Lafontaine, 1974; Luykx, 1974). Chromosomes apparently remain in a non-random arrangement during interphase (Rieger et al., 1973; Comings, 1968) and centromeres can be seen in early prophase to have retained the polarized arrangement produced at the preceding telophase (Wagenaar, 1969; Burgess, 1970b, Fig. 10). The region on the nuclear envelope on or near which all the centromeres are aggregated is likely to have unique properties, and in at least some cells, centrioles which are invariably close to the nuclear envelope (e.g. Tokuyasu, 1972; Marchant and Pickett-Heaps, 1972) seem to take up a position in relation to this point on the membrane (Moens, 1969). Non-random distribution of nuclear pores indicates locally differentiated regions of the envelope (Stevens and André, 1969).

The pre-prophase band (PPB) of microtubules, which reflects if not determines the subsequent position of the new cell wall at cytokinesis (Pickett-Heaps, 1969b, c; Evert and Deshpande, 1970; Burgess, 1970a), forms just inside the plasmalemma, which may have special properties at this point. Burgess (1970b) suggests that the PPB also interacts with the nuclear envelope in orientating the nucleus, as in

A. F. Dyer

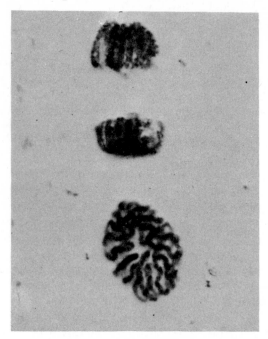

Fig. 10. Mitosis in *Allium cepa* root tips showing the polarized arrangement of the chromosomes within the nuclear envelope. × *c.600*.

At telophase (side view, upper cell) the centromeres aggregate in the polar regions of the two daughter nuclei. At prophase (polar view, lower cell) the same arrangement can be seen, with both chromosome arms of each chromosome extending away from the almost circular group of centromeres. Apparently the polarized chromosome arrangement in the nucleus is maintained throughout interphase, and in this case, either the orientation of the last division of the lower cell was perpendicular to the adjacent telophase, or the lower nucleus has rotated through 90° during interphase.

"closed" mitoses (Pickett-Heaps, 1969a). Northcote (1971) also suggests that microtubule/membrane associations are involved in the control of directed movements within the cell.

These facts suggest a further elaboration of the model discussed above involving a structurally differentiated plasmalemma (Fig. 11), in which the nuclear envelope shows a polarized differentiation which reflects the directions of the last division. The orientation of the interphase nucleus could then be determined in relation to differentiated regions of the plasmalemma through the mediation of microtubules such as those of the pre-prophase band. A change in division orientation from one mitosis to the next would involve first a change in the differentiation pattern of the plasmalemma, perhaps in response to external stimuli

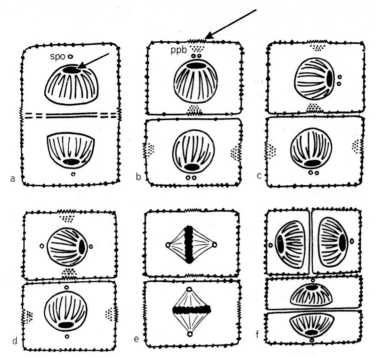

FIG. 11. A possible mechanism involved in the orientation of mitosis.

(a) In late telophase, a new undifferentiated membrane forms between the daughter nuclei. The locally differentiated region (arrow) of the nuclear membrane where the centromeres are aggregated is close to the spindle pole organizer (SPO). The older, outer cell membrane is no longer uniform, localized differentiated regions resulting from activities within the developing cells and from influences outside the cell, such as chemical gradients in the tissue.

(b) In interphase, the spindle pole organizer replicates and further differentiation of the outer cell membrane (arrow) establishes the position of the pre-prophase band (PPB) of microtubules close to the plasmalemma. The new regions of plasmalemma begin to differentiate.

(c) Interaction between the centromere region of the nuclear envelope and the microtubules establishes the proper orientation of the nucleus, if necessary rotating it until the centromeres and associated spindle pole organizers are facing away from the pre-prophase band.

(d) Spindle pole organizers separate, one remaining near the centromeres, and the other moving to a corresponding position on the other side of the nucleus.

(e) The nuclear envelope disperses and mitosis occurs on a spindle established between the polar organizers.

(f) Reformation of the nuclear envelope occurs, each daughter nucleus developing its own specialized region where centromeres are aggregated close to the spindle pole organizer. The new undifferentiated plasmalemmas extend between the daughter nuclei to separate the daughter cells and erase the differentiated regions of plasmalemma which determined the orientation of the division. Only if outside influences induce the appropriate localized differentiation of the plasmalemmas will the orientation of the next division be directed. If there are no external directional influences, random rotation of the interphase nucleus in the cytoplasm will result in random orientation of cell division.

(Yeoman and Brown, 1971; Nishi and Sugano, 1970) and then, through the influence of the microtubules, a rotation of the nucleus to restore the position of the centromere aggregation in relation to the differentiated plasmalemma structure. The differentiated nuclear envelope would then in turn establish the pre-determined positions for the centrioles or other spindle pole organizers. Consistent with this model is the effect of mercapto-ethanol on the formation of stomatal complexes (Stebbins *et al.*, 1967) which indicates that the orientation of division is established well before the spindle is organized. The same structural differentiation of the plasmalemma would no doubt influence any simultaneous cell expansion and thus provide the basis for the coordination between the direction of cell growth and the orientation of the division spindle (see e.g. Nishi and Sugano, 1970) discussed early in this chapter.

3. Molecular Events of the Cell Cycle: a Preparation for Division

M. M. YEOMAN and P. A. AITCHISON

Department of Botany, University of Edinburgh, Scotland

I. THE CONCEPT OF A CELL CYCLE

The majority of divisions in higher plants follow the pattern typical of most micro-organisms in that the mass of the cell doubles during each complete cycle. This situation applies to the component cells of the meristems of higher plants, both apical and lateral, and is a characteristic of their growth. However, all dividing cells in higher plants do not conform to this pattern. For example, in the developing fruit the average mass of a dividing cell increases markedly during a series of divisions (Sinnott, 1960). Conversely, in wounded tissues and during the establishment of a callus the average mass of the dividing cells decreases dramatically with successive divisions (Yeoman, 1970). However, in both these situations DNA is replicated at each division and it is the other cell components that fail to observe the rule of doubling.

We owe the concept of the cell cycle to Howard and Pelc (1953).

They showed that division is a cyclical process and sub-divided the period between mitoses into phases, G_1, S and G_2. The phases G_1 and G_2 are merely temporal gaps between the well-defined events of mitosis (M) and DNA synthesis (S). Since the publication of these results the general validity of this hypothesis has been confirmed and a considerable amount of information has been accumulated about the timing of these events in a variety of cells. Despite this we are still remarkably ignorant of the events which take place during the greater part of the cell cycle. Attempts have been made to investigate these largely unknown areas using asynchronously dividing cell populations of apical meristems, but such systems are not suited to investigations of this type for a variety of reasons. A characteristic of plant meristems is that they are made up of cells which divide out of step: they show a high degree of asynchrony. Indeed the use of [^{3}H]-thymidine to determine the timing of the phases of the cell cycle depends on this fact (Clowes and Juniper, 1968). Therefore, at any one time, the dividing population within an apical meristem consists of cells in many different stages of division. This makes a complete analysis of the population with respect to the position of every cell in the cycle impossible because a means of recognition of each stage is not at present available. Of course it is possible to define exactly the position of a cell within mitosis or to establish whether or not a cell is synthesizing or accumulating DNA, RNA or protein. However, the metabolic features of G_1 and G_2 are not readily determined with asynchronous systems and may not be characteristic of these phases. One ingenious approach to overcome this difficulty was published by Woodard *et al.* (1961). They showed that if the nuclear volumes of cells within the apical root meristem of *Vicia faba* were plotted against intermitotic time, a relationship was established in which the mass of the nucleus increased progressively during the cell cycle (Fig. 1). Using Feulgen microdensitometry they also determined which nuclei within the population were replicating DNA and were therefore able to delimit the "S" period. Once this was established it was possible to investigate changes in RNA and protein using histochemical procedures and to characterize the events of G_1 and G_2 in the root meristem of the bean. This technique has not been widely used by other workers, possibly because the small changes in nuclear volume which take place during the first half of the cycle make it difficult to position a cell with any accuracy during this period. Lyndon (1967) has shown that in dividing cells of pea roots, nuclear volume increases approximately five times during the cycle. Further analysis of this situation, bearing in mind the general paucity of techniques in this area, would prove fruitful.

A further disadvantage of the use of asynchronous meristems is that most chemical and biochemical procedures may not be employed to investigate the macromolecular events of division because it is impossible to collect large numbers of cells at a similar stage of the cell cycle. Even with the histochemical techniques available at present, it is impossible to study patterns of change in enzymes at the single cell level

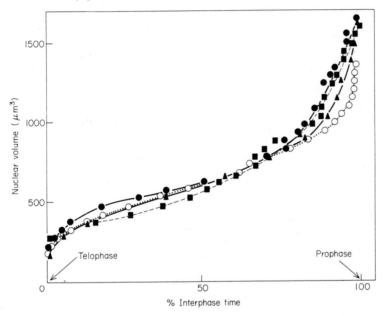

FIG. 1. Change in nuclear volume as a function of relative interphase time, in four root tips of *V. faba* (Woodard *et al.*, 1961).

or to examine the synthesis and degradation of various RNA species. This is a significant drawback because our present approach to an understanding of how cell division is controlled depends on a greater knowledge of how the synthesis and degradation of macromolecules is regulated.

II. SYNCHRONY

So far the most successful approach to a study of the cell cycle in a variety of organisms has been the use of synchrony and it is clear from the extent of published work that most is known about groups of organisms in which synchrony is most easily achieved. Natural synchrony is not a common phenomenon and therefore the majority of synchronously dividing cell populations have been produced by manipulation. Several methods of producing synchronous populations have been devised and these may be grouped together under two headings, selection synchrony and induction synchrony (Mitchison, 1971).

(a) Selection synchrony. This is the technique of selecting cells at a similar point in the cell cycle by physical separation. Two methods have been used extensively, the first with populations of micro-organisms and the second with animal tissue cultures.

(i) If a growing culture of unicellular micro-organisms is subjected to centrifugation on a sucrose gradient, the constituent cells are distributed along the gradient according to size (Mitchison and Vincent, 1965). As there is a relationship between cell size and the stage in the cell cycle reached by the individual, it is possible to collect cells of the same size and initiate a synchronous culture. Among the variety of micro-organisms with which this technique has been used, Mitchison and co-workers have achieved notable success with the fission yeast *Schizosaccharomyces pombe.*

(ii) Proliferating animal tissue cultures tend to adhere to a glass surface and this property facilitates the culture in liquid media of cells in monolayers on cover slips, or on the surface of roller bottles. During mitosis the cells round off and become less firmly attached to the glass and at this stage it is possible to remove the "mitotic" cells selectively by gentle washing of the culture (Petersen *et al.*, 1968). The cells present in the washing medium are therefore all at the same stage in their cell cycle, and can be used to initiate a synchronous culture. The low yields of cells produced by this technique limit its usefulness for biochemical investigations. So far, selection methods have not been used with higher plant cells, although the application of centrifugation on a suitable gradient for the separation of plant protoplasts remains an interesting possibility.

(b) Induction synchrony. Here the technique is to block the cell cycle at a particular point by adding a cytostatic agent, omitting an essential nutrient or altering the physical environment. These techniques include the use of inhibitors of DNA synthesis or mitosis, starvation and growth procedures, and multiple changes of temperature or light. After such treatment the cells may accumulate at one point in the cell cycle. Release of the populations from the effects of the inhibitor or addition of an essential nutrient then leads to a limited number of synchronous divisions.

Some attempts have been made to produce synchronously dividing populations of higher plant cells using induction techniques. The apex of the dicotyledonous seedling root has been used extensively. For instance one method employed has been to pre-treat primary or lateral roots of *Vicia faba* with the pyrimidine analogue 5-amino uracil (Clowes, 1965; Jakob and Trosko, 1965; Mattingly, 1966; Kovacs and Van't Hof, 1970) and then wash out the analogue. Other inhibitors of DNA synthesis, such as high concentrations of thymidine, fluordeoxyuridine or hydroxyurea, have been employed successfully with cell suspensions of *Haplopappus gracilis* (Eriksson, 1966); mitotic indices of 35% have been observed after treatment with hydroxyurea. Fluordeoxyuridine treatment together with starvation has been used to synchronize cell populations in the root apex of pea (Kovacs and Van 't Hof, 1970). Synchronous mitoses were observed in all cases, but no attempt was made to utilize the synchronous population so obtained for studies on the macromolecular changes during the cell cycle. Chemical methods of synchronization do of course produce side effects and may lead to a distortion of the cell cycle.

Cell suspension cultures of *Acer* (Roberts and Northcote, 1970; Wilson *et al.*, 1971) and *Nicotiana tabacum* (Jouanneau, 1971) have been induced to divide synchronously. The method employed by Roberts and Northcote (1970) and Jouanneau (1971) was to exclude kinetin from a culture dependent on that substance and then, after a suitable interval, add it to the resting culture. This results in partial synchronization of the cells and mitotic indices of 15% have been reported (Roberts and Northcote, 1970). Wilson *et al.* have used a starvation and growth technique with *Acer* cultures. They allowed a batch-propagated culture to grow out of the exponential phase and then subcultured the suspension at high dilution into fresh medium. This results in a cell population which, after a preliminary lag phase, divides synchronously for five or six divisions and is a system with great potential for examining the whole process of cell division in higher plants (King and Street, 1973).

There are some examples of naturally occurring synchrony in higher plants (Erickson, 1964). Two deserve special mention because they have already been used to study cell division. The developing endosperm of some angiosperms shows a high degree of natural synchrony with respect to nuclear divisions but is a difficult system to manipulate for physiological and biochemical investigations. It has been used extensively, however, for the study of mitosis in *Haemanthus katharinae* with time lapse cine-micrography (Bajer, 1958; Bajer and Molè-Bajer, 1954). The formation of the microspore in the anthers of higher plants is preceded by a meiotic and a mitotic division; both of these divisions exhibit a very high degree of natural synchrony. In two liliaceous genera *Trillium* and *Lilium* the synchrony of division has been clearly recognized and most of our knowledge of the biochemistry of meiosis and mitosis has been obtained with the use of these plants (Erickson, 1964; Stern, 1966). The approximate durations for the various stages of microsporogenesis at 4–6°C is more than 30 days for leptotene and zygotene, 40–60 days for the remaining meiotic stages, 28–40 days for the microspore interphase and 8–14 days for microspore mitosis. Even at normal laboratory temperatures (*c.* 20°C) the two divisions are extremely slow. Several authors have argued that this slow rate of division together with the natural synchrony of the system offers unparalleled advantages for investigations on cell division. However, it cannot be regarded as typical of higher plant cell cycles, being a meiotic division followed by a mitotic division in a highly specialized cell type.

In 1962 Adamson observed that pieces of parenchymatous tissue excised from the dormant tuber of the Jerusalem artichoke and brought into contact with 2,4-dichlorophenoxyacetic acid (2,4-D) were induced to grow, and that approximately 24 h after the commencement of the culture a distinct number of mitotic figures appeared in the outer cell layers of the disk. Similar observations were reported by Setterfield (1963), but no attempt was made to follow changes in cell number or mitotic index with time nor indeed was any mention made of the possibility that these divisions might be partially synchronous. At about the same time similar observations had been made by students in this laboratory (Yeoman *et al.*, 1965, 1966; Yeoman and Evans, 1967). When explants of tuber tissue, removed in the light, are brought into contact with a medium containing a mineral salts mixture with 2,4-D, coconut milk and sucrose, approximately 35% of the constituent cells divide synchronously (mitotic index reaching 22%). This first synchronous division is followed by a series of divisions during which synchrony is gradually lost. The system is completely asynchronous by the fifth division. Subsequent investigations (Fraser *et al.*, 1967; Yeoman and Davidson, 1971; Davidson and Yeoman, 1974) showed that if the explants were removed in low intensity green light and cultured in complete darkness, approximately 60% of the constituent cells divide synchronously (mitotic index reaching 45%). Of the cells which do not divide about half are damaged and eventually autolyse and collapse (Yeoman *et al.*, 1968) and the remainder which constitute an inert core in the explant exhibit little metabolic activity and do not accumulate RNA, DNA or protein, at least during the first 4 days of culture (Mitchell, 1967, 1968, 1969). Therefore the 60% of dividing cells constitute approximately 75% of the viable cells and dominate the synthetic activities of the explant. This system suffers from some disadvantages when it is used to study cell division: (a) while the synchrony is natural and is inherent to the system, the act of removal of the tissue from the parent tuber initiates a wound response which is involved in the induction of division and has an effect upon it; (b) the first division is longer than the subsequent divisions due to a prolonged G_1; (c) synchrony is only sufficient to study the first three interphases. There are, however, advantages: (a) it is freely available for most of the year and can be easily manipulated; (b) the synchronous divisions are within a multicellular explant and therefore the results obtained are likely to be directly applicable to the intact plant; (c) large amounts of material can be produced and therefore conventional biochemical approaches can be employed. This system has been used to study the molecular events during the cell division cycle in higher plants in this laboratory (Yeoman, 1970, 1974; Harland *et al.*, 1973; Yeoman and Aitchison, 1973a, b; Aitchison and Yeoman, 1973, 1974).

III. SYNTHESIS OF MACROMOLECULES

The timing and extent of DNA replication in higher plant cells has received considerable attention over the past 20 years (Howard and Pelc, 1951a, b, c, 1953; Wimber, 1960, 1961, 1966; Van 't Hof, 1963, 1965a; Evans, 1964; Clowes and Juniper, 1968), but little is known of the mechanism of DNA replication in these organisms. Most of the research has been performed on asynchronously dividing root meristems using the technique of pulse labelling with precursors of DNA, usually ^{14}C- or ^{3}H-labelled thymidine (TdR) in conjunction with autoradiography (see Chapter 7). A similar picture emerges for RNA and protein synthesis where the available information has come from autoradiographic and histochemical studies on root meristems. (Woodard *et al.*, 1961; Van 't Hof, 1963; McLeish, 1969; Barlow, 1970). It is within this general area that the real value of synchronous systems becomes apparent and allows a more detailed examination of the molecular events of the cell cycle.

A. DNA Replication

1. Measurement of DNA Synthesis

An indication of whether a particular cell is synthesizing DNA can be simply obtained by supplying a radioactive precursor (usually [^{3}H]-TdR) and visualizing the presence of the incorporated molecules in the nucleus by autoradiography. By making certain assumptions, an estimate of the rate of synthesis over any time interval can be made from the grain density figures. However, where attempts have been made to determine the course of DNA replication, Feulgen densitometry has been used in conjunction with measurements of the nuclear volume (Woodard *et al.*, 1961) in preference to the use of radioactive precursors where complications arise due to pool size and availability of the radioactive compound to the replicating sites. By using synchronized cultures the need for an independent estimate of the position of a cell in the cycle is avoided. Three basic methods have been employed: (a) histochemical procedures including Feulgen densitometry or the measurement of DNA content by gallocyanin after RNase treatment; (b) a chemical technique based on the method described by Burton (1956), for micro-organisms and animal tissues; (c) a pulse-chase technique in which the incorporated [³H]-TdR is measured in the extracted DNA (Harland *et al.*, 1973) by liquid scintillation counting. The first two procedures measure the amount of DNA accumulated and the third gives an estimate of the rate of incorporation of [³H]-TdR into DNA which may be used as a measure of the rate of DNA synthesis.

2. Time Course of DNA Replication

There are only a few accounts of the time course of DNA replication in higher plant cells (Woodard *et al.*, 1961; Mitchell, 1967, 1968; Harland *et al.*, 1973). So far the only pattern that has emerged is one in which DNA accumulates at a constant rate throughout the "S" phase (Mitchell, 1968; Harland *et al.*, 1973). The data presented in Fig. 2 are from Mitchell (1968) and were obtained using synchronously dividing artichoke tuber tissue (Yeoman and Evans, 1967) and Feulgen densitometry. These data provide evidence for a constant rate of accumulation of DNA in this tissue. This agrees with results published by Woodard *et al.* (1961), obtained using Feulgen densitometry with asynchronous cell populations in the root tip of *Vicia faba*. A similar linear increase in the amount of DNA per dividing cell has been shown in cultures of *Paramecium* (Woodard *et al.*, 1961), mouse fibroblasts (Zetterberg and Killander, 1965) and with *Euplotes* (Prescott, 1966).

Studies with mammalian tissue-culture cells (Huberman and Riggs, 1968), using the DNA-fibre autoradiography technique of Cairns (1962), suggested that DNA replication occurs simultaneously at large numbers of sites on the DNA template. These points are visualized as replication forks and probably represent sites of action of DNA polymerase as two daughter helices are formed from the parental template. This pattern has since been confirmed with a variety of mammalian cells (Callan, 1973). These studies also show that these forks occur in pairs, representing replication in opposite directions from a single initiation site. Furthermore, progression from these initiation sites seems to be linear. However, linearity of synthesis at the molecular level does not directly explain an overall linearity in the rate of DNA synthesis at the nuclear level, as fibre autoradiography shows that initiation does not occur simultaneously at all points. Because of the large number of initiation points involved, the net rate of synthesis will be controlled more by the rate of initiation than by variation in the rate of synthesis

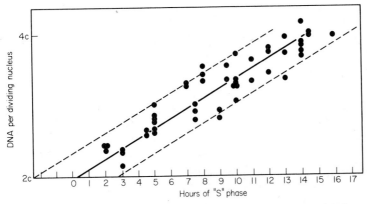

FIG. 2. DNA accumulation during the S period in synchronous artichoke explant cultures. Each point represents the mean Feulgen DNA value of DNA-synthesizing cells in one explant, expressed as a percentage of mean 2C value for that explant. Combined results of four experiments. 94% of the points fall within 2·5 h of the regression line, which agrees with a spread of *c.* 5 h in time of cell divisions (Mitchell, 1968).

from each initiation point. Linearity of net rate of synthesis throughout S suggests that there is some control over the number of sites at which DNA synthesis is occurring at any time. It is, however, worth noting that estimations of rates of synthesis based on chemical or cytochemical data (Woodard *et al.*, 1961a, b; Zetterberg and Killander, 1965; Mitchell, 1968) can only give rather crude average rates, which may well obscure short-term variations. It is also well known from chromosome autoradiography that different regions of the genome are labelled at different times during S. For instance Wimber (1961) found that the distal ends of chromosomes in *Tradescantia* root tips were more heavily labelled when the isotope was presented in late S than in early S. Conversely, in *Crepis* it seems that the region near the centromere incorporates more label late in S (Taylor, 1958; Kuriowa and Tanaka, 1970). A consistent observation from studies with a range of animal and plant material is that heterochromatic regions tend to replicate late in S (Lima-de-Faria, 1969). Barlow (1972a) has suggested that this may be explained if an initiating factor, which is produced throughout S, is required to make DNA accessible to polymerase action and that the number of initiator molecules required for initiation to occur varies with the degree of compactness of the DNA. Heterochromatin, requiring more initiators, will obviously start replication later than less compact chromatin. This means that an overall control in the rate of DNA synthesis may be exerted by the degree of condensation of different parts of the genome and the rate of production of initiator.

3. Timing and Position of S within the Cycle

An estimate of the timing and extent of DNA replication in the nucleus can be obtained with a high degree of accuracy using asynchronously dividing cell populations. The application of these techniques, first used by Howard and Pelc (1951a), has opened up a highly profitable field in the investigation of cell cycles in terminal meristems. The successful application of these methods, which include both continuous and pulse labelling with DNA precursors, depends on the asynchrony of the cell population under investigation. Unfortunately most biological systems develop rhythms and some degree of partial synchrony may be attained which will upset the basis for calculation of time parameters of the cell cycle. It is therefore important to choose tissues which are naturally asynchronous or in which a state of asynchrony has been induced (Clowes and Juniper, 1968). If the studies are located within the meristem then possible complications due to the occurrence of polyploidy or polyteny may be safely neglected. Methods of applying this technique are covered in the monograph by Clowes and Juniper (1968) and no further elaboration is necessary in this account.

Van't Hof has advanced the theory that under similar environmental conditions the average time taken for (diploid) cells of root meristems of different plants to pass through S depends on the amount of DNA to be replicated. This was illustrated with data for seven plants (Figs 3 and 4). He also showed that, for this group, the total average cycle time was determined by the duration of S. These data are not necessarily inconsistent with observations on large differences in cycle times in different regions of one root (Clowes, 1965b), as they refer to average times for whole roots, within each of which there will be a spectrum of cycle lengths. Sub-populations with very long cycles, e.g. cells of the quiescent centre, will contribute very little to the data.

Not all data, however, agree with Van't Hof's suggestion. Troy and Wimber (1968) compared diploid and autotetraploid pairs of *Lycopersicon esculentum*, *Tradescantia paludosa*, *Ornithogalum vivens*, a *Cymbidium*

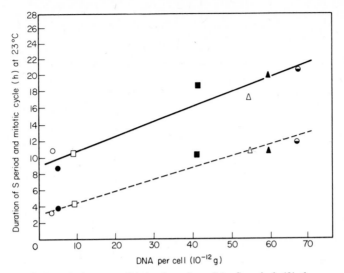

Fig. 3. The relationship between (1) the duration of the S period, (2) the average total cycle time, and 2C amount of DNA per root meristem cell of different species.

Crepis capillaris, ○; *Impatiens balsamina,* ●; *Lycopersicon esculentum,* □; *Allium fistulosum,* ■; *Allium cepa,* △; *Tradescantia paludosa,* ▲ ; *Allium tuberosum,* ◓ (Van't Hof, 1965a).

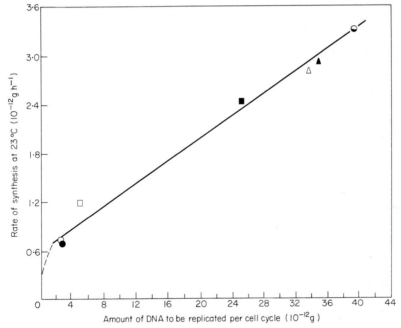

Fig. 4. The relationship between the average rate of DNA synthesis and amount of DNA to be replicated in root meristem cells of different species.
Key as for Fig. 3 (Van't Hof, 1965a).

culture and a series of several *Chrysanthemum* species. They found that the duration of the S-phase did not change between each pair and was therefore independent of DNA content. Also, within lateral roots of *Vicia faba*, Friedberg and Davidson (1970) found diploid and tetraploid cells to have similar mean cycle times. Of course, it may be argued that, in these cases, whatever genetic control element determines the rate of replication in the diploid will itself be present in double dose in the tetraploid, enabling a doubling in the net rate of synthesis. Alfert and Das (1969) found that both G_1 and S increased, but did not double, between 2n and 4n in *Antirrhinum majus* seedling roots. They found a closer correlation between the increased duration of S and increase in nuclear surface area (compared with volume) and suggested that this supported a role of the nuclear membrane in DNA replication. On the other hand Prasad and Godward (1965) reported that tetraploid *Phalaris* root cells do have an S-phase roughly double that of diploids. Also the presence of additional B-chromosomes can increase the length of the S-phase and total cycle time in rye (Ayonoadu and Rees, 1968).

The variation in total cycle time (usually achieved by extension of one of the G-phases) seen between non-differentiated cells, which may be closely related, suggests that there are internal controls on the progression of cells through their cycle. For instance the cycle time of the first tier of root-cap initials in *Convolvulus* was estimated as 13 h, while that of the second to fourth tiers increased to 155 h (Phillips and Torrey, 1972). They suggested that such differences might be explained by gradients of factors inhibiting or stimulating progression, originating from the quiescent centre, another sub-population of cells with a long cycle time in a region of generally high mitotic activity. Van't Hof has suggested that there are at least two control points in the cycle, one in G_1 which determines whether a root tip cell will progress to S, and one in G_2 which determines whether it will proceed to M (Rost and Van't Hof, 1973; Van't Hof *et al.*, 1973). Evans and Van't Hof (1973) also produced some evidence that a substance causing arrest of cells in G_2 in pea seedling roots could originate in the cotyledons.

It does seem established that between different cells in the same plant, the time taken to complete S is much less variable than G_1 and G_2. Within the root of *Zea mays* there are large differences in the length of G_1 from 0 to 151 h, with very little change in S (Clowes, 1965b). Wimber (1966) found that raising the temperature from 21 to 30°C decreased the total cycle time almost entirely at the expense of G_1 with S unaffected. A decrease to 13°C caused a general prolongation of all phases, but with S least affected. In developing antheridial filaments of *Chara* there is a decrease in cycle duration from the 2–16 cell stage,.

with S virtually unchanged and G_2 shortening (Godlewski and Ols-zewska, 1973). Lastly, in artichoke explants introduced into a 2,4-D medium, the length of time before the first division, which increases during storage of tubers, is almost entirely dependent on the duration of the pre-S-phase.

While DNA replication is a normal pre-requisite for mitosis, a cell may exceptionally undergo mitosis without a round of DNA synthesis. Rasch *et al.* (1959) have shown that in tumours produced by *Agrobacterium tumefaciens* on *Vicia faba* the cells can undergo a reduction in C value from 16 to 2 by successive mitotic cycles without intervening DNA synthesis. A treatment of abnormal cell cycles may be found in Brown and Dyer (1972).

B. Pattern of Synthesis and Accumulation of RNA

Inevitably, intense RNA synthesis and accumulation are associated with the cell cycle in all organisms (Mitchison, 1971). Once again, as with studies on DNA replication in plants, much use has been made of techniques in which labelled precursors of RNA, uridine, orotic acid or inorganic phosphate, are fed to asynchronously dividing populations of cells, usually in the root meristem, and their incorporation followed by autoradiography. Such an approach is extremely useful in deciding which cells are synthesizing RNA at a particular point in the cell cycle, and obtaining some measure of the rate of synthesis (Van't Hof, 1963; McLeish, 1969) but cannot be used effectively to establish the overall pattern of RNA accumulation during a cell cycle or to study the synthesis of particular RNA species. Here our basic knowledge of the pattern of RNA accumulation in dividing plant cells comes from the work of Woodard *et al.* (1961) with *Vicia faba* root tips. They were able to correlate DNA replication with the synthesis and accumulation of RNA and protein during a cell cycle by the use of cytochemical procedures coupled with autoradiography. They showed that cytoplasmic RNA accumulated essentially in three steps, one immediately post-telophase, one during the "S" phase and a third very sharp increase in the immediate pre-prophase period. These summated increases amounted to approximately a 3-fold rise in RNA during the course of one cell cycle. A similar pattern of RNA accumulation has been demonstrated by Mitchell (1969) with synchronously dividing cells from artichoke explants (Fig. 5). RNA synthesis is also variable in the cell cycle of *Physarum polycephalum*, showing minima in M and early G_2, and maxima during S and mid G_2 (Mittermayer *et al.*, 1964). An increase of just less than 3-fold has been reported in *Paramecium* with a sharp rise in the rate

M. M. Yeoman and P. A. Aitchison

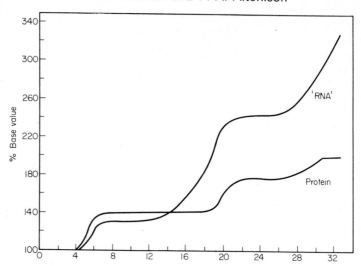

Fɪɢ. 5. Pattern of accumulation of "RNA" and protein in dividing cells of artichoke explants (Mitchell, 1969). S began at 17 hours, mitosis at 32 hours.

of accumulation in pre-prophase (Woodard *et al.*, 1961). It has not been satisfactorily explained why RNA, in these cases, shows more than the doubling in amount that would be expected in a normal cell cycle. Evidence for active RNA synthesis has also been obtained with yeast cells, Hela cells and mouse fibroblasts in culture (Mitchison and Walker, 1965; Terasima and Tolmach, 1963; Zetterberg and Killander, 1965), but in these systems the progress of RNA accumulation appears to be more or less exponential with an approximate doubling in amount during each cell cycle.

The synthesis and accumulation of RNA during the early synchronous divisions of artichoke explants have been examined in some detail by Evans (1967), Fraser (1968), Mitchell (1969) and Gore (1973). Fraser (1968) has fractionated the RNA present in cells at various times during the course of the cell cycle and has shown that the RNA seems to be accumulated in two major steps and not three as shown by Mitchell (1969). It is probable that the small initial step, very early in G_1, which is clearly evident from histochemical studies is not detectable when chemical techniques are employed because some non-dividing cells are lost from the periphery of the explant into the culture medium and the small initial increase on a per cell basis is negated by an overall loss of RNA (Yeoman *et al.*, 1968). The major constituent of the RNA accumulated is rRNA, this accounting for approximately 90% of the total (Mitchell, 1969). Apparently most of the tRNA is accumulated very late in the cycle just before mitosis. No estimate of the extent of mRNA production can be attempted from such an investigation. Jakob (1972), using a partially synchronous root meristem of *Vicia faba*, has demonstrated that the major proportion of RNA is synthesized during late G_1 and early S and that at least part of the RNA synthesized is essential for normal DNA replication. Webster and Van't Hof (1970) agreed that there was a requirement for RNA for pea root cells in G_1 to proceed to S, but that this was not necessarily furnished by recently synthesized RNA. The major point of interest from this and other investigations on RNA synthesis and accumulation in plants is the periodic nature of the major increases. This contrasts with the accumulation of RNA in most other organisms (Mitchison, 1971) which occurs throughout the cell cycle, though not necessarily at the same rate. Studies by Woods (1960), Woods and Taylor (1959) and Setterfield (1963) indicate that the nucleolus is the major site of RNA synthesis in meristematic cells and that the RNA made in the nucleolus is exported to the cytoplasm. A marked alteration to the structure of the nucleolus accompanies the progression of a cell through division. It becomes less compact and the fibrillar and granular regions become intermingled (Lafontaine and Chouinard, 1963; Brinkley, 1965; Birnstiel, 1967; Yeoman and Street, 1973). Frequently a very large, more electron-transparent central region appears in the nucleolus which contains and is surrounded by granular particles similar in size to ribosomes. Fibrillar material similar to the chromatin outside the nucleolus is also found within this body, and in addition smaller electron-transparent regions are present within the fibrillar zone surrounding the central area. Similar observations have been reported by Jordan and Chapman (1971) and Rose *et al.* (1972) in nucleoli of cells from ageing tuber discs of Jerusalem artichoke. King and Chapman (1972) have also demonstrated a 3-fold increase in nucleolar volume during 24 h incubation in water of similar tissue. Unpublished observations from this laboratory show a marked increase in nucleolar size during the first synchronous division in cultured cells.

C. Pattern of Synthesis and Accumulation of Proteins

1. General

In a meristem the growth of a cell between mitoses leads to a doubling of the cellular mass of which the two major constituents are polysaccharides of the wall and proteins of the cytoplasm and the nucleus. Some information on the pattern of protein synthesis has been obtained with asynchronously dividing cell populations in root meristems using pulse labelling techniques with ^{14}C-phenylalanine and [^{14}C]- and [^{3}H]-leucine (e.g. Van't Hof, 1963). These studies have demonstrated that rates of protein synthesis are high during interphase and mitosis with the highest degree of incorporation being recorded in G_2 and prophase. Data on protein accumulation obtained using cytochemical techniques both in the cytoplasm and in the nucleus of cells in the bean root tip have been published by Woodard et al. (1961). They demonstrated a doubling of protein in the nucleus coincident with the replication of DNA. Accumulation of cytoplasmic protein occurs at various times during interphase; Mitchell (1968), using the synchronously dividing artichoke explant system, has demonstrated that total cell protein increases in three steps, more or less coincident with the three steps in RNA accumulation (Fig. 5). The first periodic increase takes place early in G_1, the second at the commencement of S and a third, minor, rise immediately before mitosis. This pattern may be compared with that obtained in cultures of mouse fibroblasts (Zetterberg, 1966) in which protein per cell increased exponentially during interphase.

Within the context of a general increase in protein there is a marked change in the proportions and in the numbers of the many different proteins present. An indication of the pattern of proteins in the soluble fraction from synchronously dividing cells has been given by Yeoman and Aitchison (1973b). Clearly the pattern changes as the cells progress through a series of metabolic states which culminates in division. However, it is to changes in individual proteins and the mechanisms which regulate their synthesis and degradation that we must look to gain a better understanding of how the cell cycle is regulated.

2. Pattern of Enzyme Activities

Almost without exception, data on changes in enzyme levels during cell cycles have been based solely on the measurement of enzyme activity. It is always possible that the proportion of a given protein that is enzymatically active at any time is dependent on the presence of activators or inhibitors, and that changes of enzyme activity observed are a reflection of a change in amounts of such effectors. This possibility may be dismissed with some certainty by the use of mixed extract experiments in which enzyme extracts prepared at stages of the cycle characterized by different levels of activity give the arithmetic sum of their activities when mixed, only in the absence of effectors. In some cases an attempt has been made to show that a change in activity is prevented by the presence of an added inhibitor of protein synthesis and, if so, this is taken as suggestive of a requirement for new enzyme synthesis. However, even ignoring the alternative interpretations due to side effects of these agents, such an approach does no more than demonstrate the need for the synthesis of some species of protein which need not necessarily be that of the enzyme in question. To show that an increase in activity is entirely due to increased amounts of enzyme protein due to synthesis, it is necessary actually to measure the amount of enzyme protein present, to show that activity is proportional to the amount present during the period under investigation, and that the increased activity is entirely accounted for by additional protein synthesized *de novo* during that period. Even to demonstrate that the increased activity is at least partly due to increased specific synthesis, it is necessary to show that new enzyme protein appears during the period of activity increase, more rapidly than at other times. Such a demonstration is possible using immunological methods, or radioactive, or density labelling. The occurrence of *de novo* enzyme synthesis during periods of activity increase in plants has been reported by many authors (e.g. Johnson *et al.*, 1973; Shepard and Thurman, 1973) but these do not concern changes within the context of a cell cycle.

Early investigations in plants on the timing of the changes in activity of specific enzymes in relation to the cell cycle were carried out using the developing microspores of *Lilium longiflorum* which possess a high degree of natural synchrony (Hotta and Stern, 1961, 1963c, d; 1965). In a series of papers Hotta and Stern demonstrated a distinct periodicity with regard to levels of thymidine kinase (TdR-K), DNase and related enzymes. TdR-K appeared briefly, before DNA synthesis, coincidentally with the appearance of an increased pool of deoxyribosides, and evidence was presented (Hotta and Stern, 1965) that the periodicity in appearance of the enzyme was due to induction by thymidine (TdR). The sudden increase in deoxyribosides was in turn explained by an increase in DNase in the anther tissue surrounding the developing microspores. However, the increase in TdR-K activity was not solely due to variation in the supply of the inducer (TdR) for a permanently inducible system, as induction by added TdR could only be demonstrated at or shortly before the normal time of appearance of the enzyme in whole anthers (Hotta and Stern, 1965). McLeod (1971) has implicated high TdR-K and thymidylate synthetase in meristems of roots of *Vicia faba* using an autoradiographic procedure, and he also demonstrated TdR inducibility of TdR-kinase in this tissue, but it is not known whether changes occur during the cycle in the individual cells in meristems.

A study has also been made of the catalytic events which precede mitosis in cells from the Jerusalem artichoke tuber grown in culture

(Aitchison and Yeoman, 1973a, 1974; Harland *et al.*, 1973). It has been established that the first and second synchronous divisions are characterized by a changing pattern of enzyme activities. Characteristically, the activity of each of the enzymes investigated increases once per cell cycle and the period over which the increase occurs is fairly short (Aitchison and Yeoman, 1974).

It has also been established that the increase in level of particular enzymes is associated with different phases of the cell cycle. Perhaps the most striking example of this phenomenon is the relationship between the S period and the rise in level of DNA polymerase, TdR-K, and dTMP-kinase (Harland *et al.*, 1973; King and Street, 1973), the syntheses of which seem to be linked with DNA replication. A close coincidence between a rise in TdR-K and DNA synthesis has also been observed in synchronously dividing rat liver cells (Bollum and Potter, 1959), cultured cells from rabbit kidney (Lieberman *et al.*, 1963), and synchronously dividing Hela cells (Brent *et al.*, 1965; Stubblefield and Mueller, 1965). Similar coincidences have been demonstrated with dTMP-kinase (Brent *et al.*, 1965) and DNA polymerase (Lieberman *et al.*, 1963; Loeb and Agarwal, 1971; Hecht, 1972) with animal cells. In contrast, the increase in the levels of glucose-6-phosphate dehydrogenase (G6PDH) and ATP-glucokinase only occurs in the period preceding DNA replication (Yeoman and Aitchison, 1973a; Aitchison and Yeoman, 1973). Recent density labelling experiments have shown that, at least in the case of G6PDH, the increase in activity is associated with an increase in amount of catalytic protein, so that the pattern of activity probably represents a balance between synthesis and degradation of the enzyme protein, rather than a balance between activation and inactivation.

Within the limitations of the existing evidence the only pattern of change in enzyme activity encountered in higher plant cell cycles is periodic. In other organisms, although the majority of enzymes are synthesized periodically, during one or more stages of the cell cycle, continuous patterns of synthesis are known (Mitchison, 1969). It would not be surprising if further data showed this to be also the case in plants.

3. *Control of Synthesis*

From most of the evidence so far available, it appears that proteins and nucleic acids do not accumulate continuously throughout the cell cycle but exhibit patterns of periodicity. Changes in the levels of enzymes are abrupt and take place over a relatively brief interval, giving a stepped or peak pattern. The accumulation of DNA in most cases,

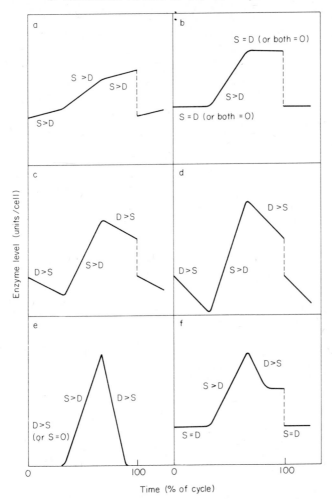

FIG. 6. Idealized cases of the relationship between turnover and patterns of enzyme levels for stepped (a, b), and peaked (c–f) enzymes (cf. Mitchison, 1971). S, synthesis; D, degradation.

and RNA in some cases at least, assumes a stepped configuration and the rates of incorporation of precursors into RNA or DNA exhibit a marked periodicity (Harland *et al.*, 1973; McLeod, 1974). It is important to understand both the relationships between the synthesis of a macromolecule and its accumulation, and the consequences of changing rates of synthesis and degradation. Figure 6 shows some aspects of such a relationship for both stepped and peaked increases in the level of an enzyme. A sudden increase in the amount of a particular molecule could be due to the commencement of synthesis (Fig. 6b–f) or an increase in the rate of synthesis (Fig. 6a–f). It could equally well be due in any of these cases to a decrease in the rate of degradation or its complete cessation. Conversely a sudden decrease may be caused by a decrease or cessation of synthesis, or an initiation or increase of degradation (Fig. 6b–f). The maintenance of a constant amount (Fig. 6b, f) depends on a balance between synthesis and degradation and does not necessarily imply that neither is occurring. It should be noted that in Fig. 6 each of the patterns a–e can result from only two adjustments in the ratio of synthesis to degradation during the cycle; the first being an increase in the ratio and the second a decrease (in the cases shown, back to the initial level). More complicated patterns may result from more than two changes in this ratio (Fig. 6f). The common feature is that for at least one period the rate of synthesis exceeds the rate of degradation, resulting in accumulation. The differences in these patterns are due to different balances between synthesis and degradation during the rest of the cycle, which may be so extreme that the enzyme accumulates throughout the cycle, albeit at different rates (Fig. 6a), or may fall below detectable levels (Fig. 6d, e) for a portion of the cycle. This interpretation implies that there is no qualitative causal difference between stepped and peaked patterns of accumulation, nor is it necessary to assume that synthesis is occurring only during periods of net accumulation (Mitchison, 1971). In any particular investigation the problems are these: (a) are the substances synthesized and degraded throughout the cell cycle, and is the pattern exhibited a reflection of changing rates of synthesis and degradation? or (b) are these substances, like DNA, only synthesized at a particular point during the cell cycle and the relatively constant level of substances merely due to the slow rate of degradation between rounds of synthesis? As a general rule most proteins, and enzymes in particular, are not indefinitely stable in the cell, but exhibit turnover, and indeed, the levels of some enzymes decrease quite rapidly when their synthesis is prevented by the presence of an added inhibitor. However, to date no reports have appeared concerning the turnover of enzymes during the cell cycle. We have made an attempt to obtain some data on turnover of G6PDH in the cell cycle of Jerusalem artichoke cells in culture. This enzyme exhibits a stepped increase in amount during G_1 (Aitchison and Yeoman, 1973). In the presence of cycloheximide or 6-methylpurine no increase in activity occurs, presumably because the enzyme is not synthesized, but neither does the level of G6PDH decrease significantly over a period of 24 h (Aitchison and Yeoman, 1973a). This was taken as showing that

G6PDH is stable within the cell. However, there is a difficulty in the interpretation of the data, because the effect of the inhibitor on the degradation of the enzyme is unknown. Indeed, subsequent studies on turnover using a density labelling technique with deuterium oxide suggest that these inhibitors may prevent or slow down the degradation of this enzyme (Aitchison, 1974). Therefore, the inhibitor approach to a study of turnover must be viewed with considerable caution. A better approach is offered by the density labelling method, for which deuterium oxide (D_2O) is the most convenient source of a heavy isotope. All proteins synthesized in the presence of excess D_2O exhibit higher buoyant densities than that of the corresponding pre-existing proteins. It is then possible to separate the "heavy" from the "light" proteins at equilibrium on a caesium chloride gradient formed by prolonged centrifugation when artichoke explants in synchronous culture were grown in the presence of 85% D_2O; after 5 days' culture (Fig. 7) only a small part of the G6PDH had a low density characteristic of the original enzyme. This indicated that most of the enzyme present (heavy) had been synthesized during culture and the majority of the original enzyme (light) had disappeared. By determining the density of G6PDH from explants cultured for various lengths of time during the cell cycle it has been possible to study the turnover of this enzyme within the context of the cell cycle, and the evidence so far suggests that G6PDH is much less stable than was formerly indicated by the inhibitor studies

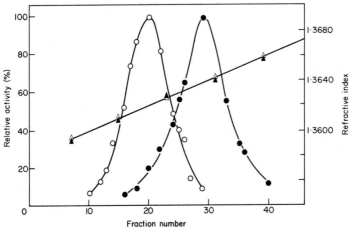

FIG. 7. The effect of growing artichoke explants in D_2O on the buoyant density of G6PDH. Batches of 100 explants were grown for 5 days in 85% D_2O ($\bullet$, $\blacktriangle$) or H_2O ($\bigcirc$, $\triangle$) and crude extracts were centrifuged for 3 days at 35 krpm on a MSE 50 SS centrifuge. Two drop fractions were collected by upward displacement and assayed for G6PDH ($\bullet$, $\bigcirc$) activity or refractive index ($\blacktriangle$, $\triangle$).

with cycloheximide and 6-MP (Aitchison, 1974). In this situation if the rate of turnover is appreciable then it follows that the maintenance of a relatively steady level between sharp increases is due to a balance between synthesis and degradation and not due to periods of synthesis separated by periods of non-synthesis coupled with a very slow rate of degradation. This presents a much more complicated picture than a simple "switching on and off" of genes. The evidence does not preclude the existence of a long-lived mRNA produced some time before the rise in the enzyme. However, it is unlikely that the rate of synthesis of G6PDH is controlled by changes in the rate of translation of preformed mRNA, because the addition of an inhibitor of RNA synthesis prevents the subsequent increase of the enzyme even when added immediately before the expected time of increase. It seems likely, therefore, that mRNA for G6PDH, at least, is synthesized throughout the cell cycle, but that the rate of synthesis increases at one point, and later reverts to the initial rate. An alternative explanation is that the effect of RNA inhibitors such as 6-MP and actinomycin-D block the synthesis of another RNA species essential for the synthesis of the protein and the mRNA for that species is present throughout the cell cycle. Possible candidates include tRNA or the mRNA for one of the many protein components involved in protein synthesis.

IV. CONCLUDING REMARKS

In writing this chapter, reference has been made to the majority of the publications available on this subject. The restricted literature serves to highlight how little is known about the molecular events which take place as cells divide and contrasts sharply with extensive information available about the visible events of division as exemplified by mitosis. Progress in the investigation of the biochemistry of cell division has been severely restricted by the lack of suitable experimental systems. With the techniques available at present, the study of asynchronously dividing populations is limited to investigations on the timing of division and its constituent phases and whether dividing cells are synthesizing particular classes of macromolecules, such as DNA, RNA and proteins. No estimate can be made of changes in the activity or amount of particular enzymes or the fluctuations in various RNA species. When there is a satisfactory method of deciding where a cell is placed with respect to the division cycle then histochemical procedures, including immunofluoresence, will allow a more thorough analysis of the cell cycle with regard to particular molecules. Until such techniques are available the major effort must be through the use of synchronously dividing cell populations. So far only a few systems have emerged, each having its own special advantages and disadvantages. None of these systems has been fully exploited. Clearly, much remains to be discovered with these, and other as yet undeveloped, systems. It is to be hoped that a stage will be reached when the information obtained with these systems will assist in the understanding of the process of cell division in the intact plant.

4. The Replication of Plastids in Higher Plants

RACHEL M. LEECH

Department of Biology, University of York, England

I. INTRODUCTION

The most constant feature distinguishing plants from animals is the presence of plastids within their cells. In addition, the entire process of
photosynthesis in green plants is mediated by the most complex type
of plastid, the chloroplast (Walker, 1970). It is perhaps not surprising,
therefore, that considerably more attention has been paid to the study
of the replication of plastids than to the study of division in plant mito-

chondria. Almost no information is currently available concerning mitochondrial division in higher plants (Leaver, 1974), so this present review will be limited to a consideration of division in the variety of plastids found in higher plants.

As electron microscopical studies of plastid structure have proliferated in recent years, it has become increasingly clear that the morphological diversity of mature chloroplasts found in leaves is immense. Yet, although functionally and morphologically the most

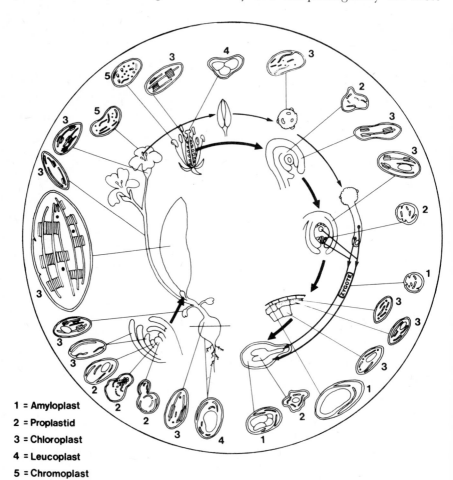

1 = Amyloplast
2 = Proplastid
3 = Chloroplast
4 = Leucoplast
5 = Chromoplast

Fig. 1. A diagram showing the type and distribution of plastids found in the organs of a higher plant during successive stages in the plant life cycle from seed development to fertilization. (The diagram is modified and redrawn from an article by Tageeva et al. (1971).)

complex plastid form, the chloroplast is only one of the many types of plastid, several of them interconvertible, found in higher plants (Kirk and Tilney-Bassett, 1967). In the different tissues of a single plant a wide variety of plastid types can be found ranging from morphologically simple forms such as amyloplasts and proplastids to the more complex

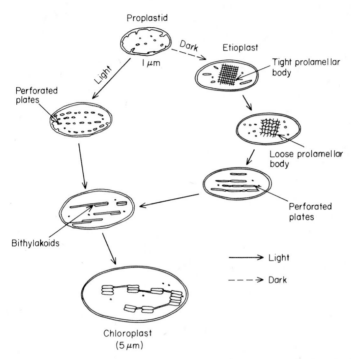

Fɪɢ. 2. Diagram illustrating the differentiation of proplastids into etioplasts in the dark (---) and into chloroplasts in the light (—). The diagram also shows the further development of an etioplast in an etiolated plant after illumination.

types such as the photosynthesizing chloroplasts, the plastids of dark-grown leaves (the etioplasts) and the pigmented chromoplasts of petals and senescing leaves. This range of structure for a single plant is illustrated diagrammatically in Figs 1 and 2. In common with the etioplast, the chloroplast develops from a proplastid and replication has so far only been studied in these three types of plastid.

II. THE RELATIONSHIP OF PLASTID DEVELOPMENT TO PLASTID REPLICATION

Questions concerning the timing and control of plastid replication cannot be realistically considered aside from the changes occurring during the sequential development of the mature chloroplast or etioplast. Chloroplast development takes place during cellular differentiation of leaf tissue and a brief outline of this development will be considered first before plastid replication itself is discussed. Further details can be obtained by reference to the accounts given by Clowes and Juniper (1968), by Kirk and Tilney-Bassett (1967) and by Possingham and Sauer (1969).

The leaves of angiosperms develop from the meristematic cells of leaf primordia which are initially recognizable as bulges on the external surfaces of stem apices (see Chapters 8 and 9). Within the cells of the leaf primordia, numerous small spherical organelles about 1 μm in diameter are present. One class of these organelles is characterized by a double bounding envelope, sparse internal membrane structure and frequently by the presence internally of osmiophilic droplets (the plastoglobuli) and small starch grains. These organelles are the proplastids and their origin is not definitely known although it is generally assumed they arise from even smaller organelles called initials which possess only a single bounding membrane. From the proplastids the chloroplasts of the mature leaf develop. Proplastids themselves can be formed in the dark in the enclosed meristems at the tip of the plant, but light is essential for their further development into chloroplasts (Cran and Possingham, 1972a; Srivastava and Paulson, 1968; Bain and Mercer, 1966). In young green leaves the chloroplasts are only partially differentiated and they increase further in size and in the complexity of their internal membrane structure to give the chloroplasts characteristic of fully expanded leaves. The general pattern of differentiation seems to be similar in the several angiosperms examined in detail. As the plastids increase in size, massive proliferation of the internal membranes takes place producing increasingly numerous membrane-bound flattened vesicles, the thylakoids, which in several areas of the chloroplast stroma become regularly stacked on top of each other, giving rise to grana. Concomitantly with the observable size and membrane changes, many additional biochemical syntheses take place. Pigments, Fraction I protein, nucleic acids and lipids are all rapidly synthesized.

III. THE REPLICATION OF PLASTIDS IN NORMAL GREEN LEAVES

The first question to be considered is whether plastid division is limited to one specific phase of plastid differentiation, i.e. is one, or more than one morphological form of plastid capable of self-replication? Several light microscopic and electron microscopical investigations have been addressed to the solution of this particular problem (Granick, 1938; Wildman, 1967; Fasse-Fransisket, 1955; Michaelis, 1962; Gyldenholm, 1968). Observations of proplastids and very young chloroplasts by light microscopy are rather difficult since these organelles are usually small, between 1 and 3 μm in diameter; but for the observation of larger differentiating chloroplasts a light microscope has considerable advantages if division is occurring, as the kinetics and morphological characteristics of the process and the splitting and separation of the daughter plastids can be followed visually.

A. Replication of Proplastids and Young Chloroplasts

It is now well established that both proplastids and "young" chloroplasts can divide. "Young" chloroplasts, in this context, can be defined as plastids which are still developing and have not yet reached their maximum size but already possess organized photosynthetic membranes and chlorophyll. The observations of proplastid division are limited to visual and electron microscopical observations of dumb-bell shaped configurations in still meristematic cells and to counts of profiles of proplastids in sections examined in the electron microscope. In the very young cells of the leaf meristem the number of proplastids per cell remains more or less constant since plastid division keeps pace with cell division. Fasse-Fransisket (1955) observed amoeboid proplastids in frequent division in the meristematic cells of the monocotyledon *Agapanthus umbellatus* and Michaelis (1962) also commented on the amoeboid nature of proplastids undergoing division in the meristematic cells of *Epilobium hirsutum*. At the later stages of development when young chloroplasts have already been formed it is easier to obtain much more direct visual evidence that replication is indeed occurring and has occurred by counts of the numbers of plastids per cell. Such quantitative studies are the only direct proof that the dumb-bell shaped profiles of young chloroplasts seen in electron micrographs are indeed in the process of replication (Lance, 1958; Frey-Wyssling and Mühlethaler, 1965; Possingham and Sauer, 1969; Boasson *et al.*, 1972a; Vesk *et al.*, 1965).

The first observations of leaf tissue cells suggesting that division could also occur in young chloroplasts, in addition to division at the proplastid stage, came from the work of Granick (1938) who found that during the later stages of growth of tomato leaves (from one-third to maximum size), the number of plastids per cell increased by an estimated 30%. More quantitative measurements supporting Granick's conclusions came from the work of Fasse-Fransisket (1955) who found in *Agapanthus* cells an increase in plastid number per cell from about 20 in the meristematic cells to 100–120 per cell in the mature palisade, providing clear evidence that during normal development in the light the plastid replicates not once but several times (Fasse-Fransisket, 1955). The important implication from this work was that young chloroplasts as well as proplastids could divide. Indeed in *Epilobium hirsutum* (Michaelis, 1962) and in *Spinacia oleracea* (spinach) (Possingham and Sauer, 1969) it is clear that the greatest increase in plastid number occurs by the division of young chloroplasts.

Once the young leaves of dicotyledons become green the development of the plastids has already passed beyond the proplastid stage and characteristically young chloroplasts are found in all the leaf cells. In order to obtain a full picture of proplastid development in these plants, therefore, in addition to young expanding leaves, the cells of the apical meristems and the expanding leaf primordia have also to be examined; this needs a very extensive study; and young developing strap-shaped leaves of some monocotyledons, in particular the grasses, provide rather better tissue in which to observe plastid development (Leech *et al.*, 1972; Leech *et al.*, 1973). Since these leaves grow largely by divisions of a basal intercalary meristem, the cells of the leaf are in sequential linear array with the youngest nearest the base and the oldest nearest the tip of the leaf. The plastids in these cells are also at successive stages of development (Leech *et al.*, 1973) and the representative micrographs from a 7-day-old maize leaf to illustrate this sequential development are shown in Fig. 3. The tissue of maize leaves is particularly useful for the study of plastid division as many dumb-bell shaped profiles of dividing chloroplasts can be seen in the cells in a band located between 2 and 2·5 cm from the leaf base. Plastids isolated from this region are

---

Fig. 3. Electron micrographs from different regions of a 7-day-old green maize leaf: **a,** cells 1 cm from the leaf base; **b** and **c,** cells 3 cm from the leaf base; **c,** shows a prolamellar body (PLB) in the same plastid in which grana are present; **d,** dividing young chloroplast in a cell 3 cm from the leaf base; **e,** the differentiated granal system of a well developed chloroplast 5 cm from the leaf base, still showing remnants of a PLB.

Magnification: **a,** $\times 36\,000$; **b,** $\times 54\,000$; **c,** $\times 45\,000$; **d,** $32\,000$; **e,** $\times 30\,000$.

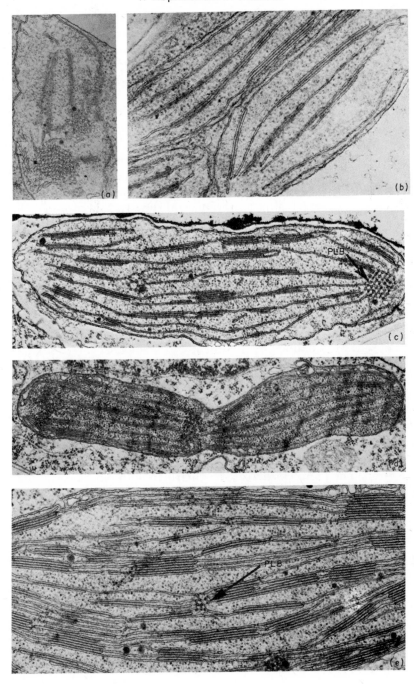

between 2 and 3 µm long and between 1 and 2 µm wide, have only a few rudimentary grana and show amoeboid movement. Isolated plastids of this type continue to replicate after isolation and the complete process of division can be followed visually under the light microscope. At this stage plastids in both mesophyll and bundle sheath cells look identical. No further division of the plastids has so far been observed at the later developmental stages when the plastids from the two types of cell become morphologically distinguishable (Laetsch, 1974).

Michaelis' observations in *Epilobium* led him to suggest that the division of young chloroplasts is synchronous within a single cell and also between several adjacent cells. In crab-grass (*Digitaria sanguinalis*) and also in maize (*Zea mays*) we have found a band of cells close to the leaf base in which almost every plastid shows a fission profile. The problem of synchrony will be discussed again later when the results of experiments using cultured leaf discs are considered.

The first extensive investigation of the division of young chloroplasts was carried out by Possingham and Sauer (1969). The success of their investigation owed much to their perfection of an elegant technique which made possible the unequivocal counting of plastid number per cell. Circular discs cut from leaves were pre-fixed in 3·5% glutaraldehyde, to stabilize the cell and organelle membranes, and then a suspension of cells was obtained by treatment with N HCl. The free mesophyll and palisade cells were firmly pressed under a coverslip to release the plastids as a discrete group. As the plastids lay slightly separated from each other but in the same plane they could be easily counted. Possingham and Sauer applied their technique to disks from young expanding spinach leaves and studied the change in plastid number as the cells of the intact leaf increased in size in the light. As the leaf grew over a period of 10 days, a 5-fold increase in plastid number per cell occurred. The numbers increased similarly in both palisade and mesophyll cells and during this time the leaf area increased from 1 to 50 cm². Later reports (Possingham and Smith, 1972) quoted a 10-fold increase in plastid number, i.e. from 50 to 500 per cell under similar conditions. No proplastids were found in the cells of these leaves and the divisions took place only in chloroplasts which already contained well defined grana. Electron micrographs showed plastids with grana containing up to five thylakoids at the time of division (Cran and Possingham, 1972b), but only rarely were the dividing chloroplasts more highly differentiated.

B. Replication of Young Isolated Chloroplasts *in vitro*

Division of young chloroplasts *in vitro* after isolation from growing leaves has also recently been observed (Ridley and Leech, 1970a; Kameya and Takahashi, 1971). Ridley and Leech isolated chloroplasts rapidly from either young spinach or broad bean leaves and suspended them in a defined buffered medium containing inorganic ions and an osmoticum but most essentially bovine serum albumin and Ficoll (Ridley and Leech, 1970b). The plastids were first seen to be in a state of division 9 h after isolation in an illuminated suspended drop culture on a coverslip. Division was much more frequently observed at times between 24 and 100 h after isolation when up to 60% of the plastids in any drop were simultaneously in the process of replication. Calculations from diameter measurements show that each plastid divides on average twice during the period of observation up to 100 h (Ridley and Leech, 1970a). The whole process of division can be followed visually and appears to be true binary fission. The timing and sequence of events during division are remarkably consistent and resemble in many respects the division in the alga *Nitella*, demonstrated elegantly using cinematography by Green (1964). The first indication that division is about to occur is the appearance of a central constriction in the plastid making it appear dumb-bell shaped (Fig. 4a). Generally the plastids remain in this configuration for several hours, but the final separation into two daughter plastids (usually of almost equal size) takes place fairly rapidly and is generally completed in about an hour. During this period the central constriction becomes narrower (Fig. 4b) and appears "tube-like" and eventually the two daughter plastids separate and move apart. After the division the plastids do not grow. The new plastids are green and therefore their formation can be easily distinguished from the simple separation of two overlapping plastids. Nor does division represent the separation of two previously fused chloroplasts as has been suggested can occur in *Mimosa* (Esau, 1972), since the daughter plastids are very much smaller than those originally isolated from the leaves. The average diameter of the plastids which divided in isolation is 5 μm as originally isolated and this becomes reduced to 4·6 μm 24 h after isolation and 3·3 μm 100 h after isolation. The change in size is too large to be accounted for by shrinkage.

Unfortunately, detailed electron microscopical examination of these isolated dividing chloroplasts has so far not been possible since the envelope membranes appear to be extremely labile at the time of division and are not retained during the fixation procedures; isolated lamellae systems from dividing plastids, however, do retain a dumb-bell shaped

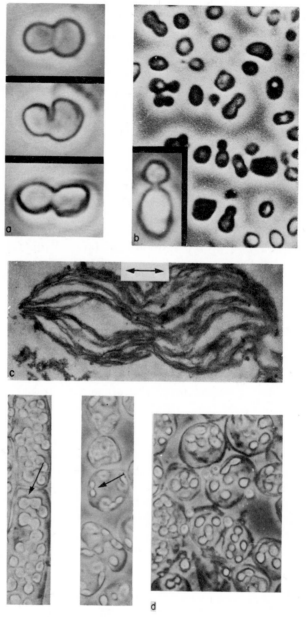

FIG. 4. **a–c,** division in isolated young chloroplasts (*Vicia faba*); **a–b,** light micrographs of dividing isolated chloroplasts 96 h after isolation; **c,** electron micrograph of a dumb-bell shaped lamellar system from b (Ridley and Leech, 1970a); **d,** dividing chloroplasts in leaf cells of *Digitaria sanguinalis* (Leech and Scarisbrick, unpublished).

configuration (Ridley and Leech, 1970b). Similar observations to those of Ridley and Leech have been obtained using tobacco plastids isolated from young expanding leaves and also from greening callus tissue cultures (Kameya and Takahashi, 1971; Kameya, 1972). Dextran 40 was added to the incubation medium used for these plastids and after 72 h they had increased in number by 50%, but their size had decreased.

The isolation of plastids capable of division *in vitro* has important implications in the search for a method of culturing plastids apart from the cell. However, the maintenance of a plastid suspension capable of growth as well as replication is as yet an unobtained objective.

The ultrastructure of plastids during replication, although not possible in the isolated system, has recently been examined in some detail in leaf disks cultured for several days (Cran and Possingham, 1972a, b; Boasson *et al.*, 1972). Both groups used leaf disks cultured on sterile agar supplemented with sucrose and salts (Murashige and Skoog, 1962). Possingham's group used disks from young spinach leaves and Laetsch's group material from tobacco leaves. Although the characteristics of the two tissue systems are not identical, the appearance of the profiles of dividing plastids is remarkably similar: electron micrographs from the two types of material are given together in Fig. 5 for comparison. The dividing young chloroplasts from an etiolated tobacco leaf disk after exposure to white light for 40 h (Fig. 5a) bear a striking resemblance to the young dividing plastids from the base of a green spinach leaf (Fig. 5b) (although Cran and Possingham showed no pictures of plastids from their disk cultures they stated that the photograph reproduced in Fig. 5b represents a typical profile seen in such cultures). Both chloroplasts could be described as "young chloroplasts" (see the definition on page 139) since they possess rather few grana with few thylakoids. Both plastids are small (*c.* 6–7 μm long), show a central constriction and appear to be undergoing binary fission. In a second paper (Cran and Possingham, 1972a) electron microscopical evidence is presented for the presence in spinach leaf disks of an additional form of dividing plastid exhibiting a central baffle profile. Chloroplasts with both dumb-bell shaped and central baffle profiles were found in different cells of the same leaf disk. Serial sections suggested that the membranes forming the baffle often extended across the entire lumen of the chloroplast. Mühlethaler (1960) had previously suggested that baffle formation may be the second stage in the division process after invagination and rupture of the thylakoids had already occurred. If this is the true sequence of events, baffle formation may also occur before thylakoid rupture is complete, since fission profiles with thylakoids present

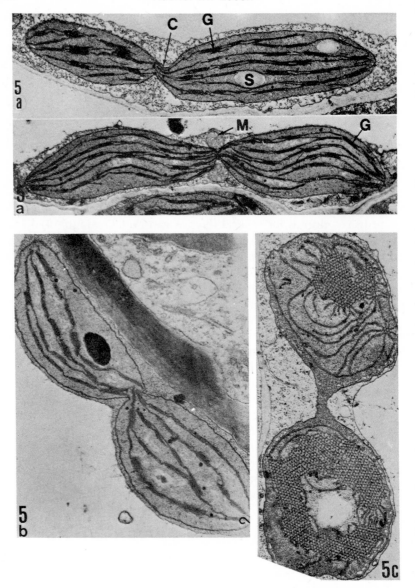

FIG. 5. Dividing young chloroplasts in (a) etiolated toacco leaf disks after exposure to white light for 40 h (upper) and 48 h (lower) (reproduced with permission from Boasson *et al.*, 1972); and in **b,** from the base of a 2 cm long spinach leaf (reproduced with permission from Cran and Possingham, 1972). **c.** A dividing etioplast from a 7-day-old maize (*Zea mays*) leaf (Leech and Crosby, unpublished results).

within the narrow central isthmus can be frequently found (e.g. Fig. 5b). A more detailed study of the membrane changes associated with plastid division is clearly required before the mechanism can be explained by which an already formed granum can fracture vertically.

C. Replication of Mature Chloroplasts

From the studies of leaf tissue and cultured leaf disks, it seems likely that most of the divisions of plastids responsible for increase in the plastid number per cell take place in the early stages of chloroplast development. Until recently much less was known of the ability of the "mature" higher plant chloroplast to undergo division, although pictures of well differentiated chloroplasts showing "hour-glass" profiles are occasionally seen and most electron microscopists who have examined leaf tissue have one or two such pictures hidden away in their files. On the other hand Wildman (1967) and Gyldenholm (1968), who both looked carefully at the cells of fully expanded leaves, could find no evidence of replication in "mature" chloroplasts. The term "mature" is frequently used but rarely defined and at present there are no criteria by which it is possible to judge whether a particular chloroplast has reached the state of maximum development before structural senescence begins. The definition to be adopted here recognizes a mature chloroplast as one in which ultrastructural development is as complex and granal development as extensive as normally found in fully expanded leaves of the species being examined. If we accept this description as a working definition it means we must necessarily accept the judgement of a particular observer about whether the plastids which he is examining are "characteristically mature". Indeed there is now clear evidence (Boasson et al., 1972b) that mature chloroplasts in leaf disks cut from fully expanded tobacco leaves can be induced to divide if the nutrient agar in which the disks are cultured is supplemented with 0·5 mg l⁻¹ kinetin (Boasson et al., 1972; Laetsch and Boasson, 1971). The disks were first cultured in the dark for 4 days prior to illumination, but almost immediately after transfer to the light a large increase in plastid number occurred and the increase continued steadily for a further 6 days until it reached a plateau: the results of these experiments are reproduced in Fig. 6. The plastids which undergo division have large and numerous grana, so clearly morphological complexity is no hindrance to division. Over the period of 10 days during which the disks were cultured they expanded to two or three times their original size and the numbers of chloroplasts per cell increased 5- or 6-fold. A three-way interaction

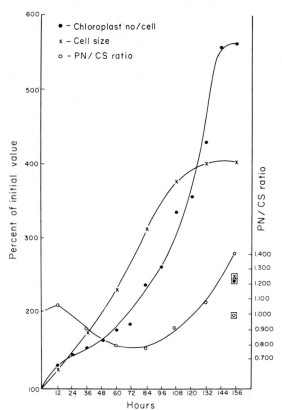

FIG. 6. Changes in chloroplast number per cell, cell size and the ratio between them
(P–N/C–S) in mature tobacco leaf disks. Cultured in the light on a medium supplemented with 0·5 mg l⁻¹ kinetin. (Diagram reproduced with permission from Boasson
et al., 1972.)

between light, cytokinin and cell expansion on the enhancement of
chloroplast replication was elegantly demonstrated. Light seemed to
be the critical factor since neither kinetin nor cell expansion induced
chloroplast replication in the dark.

D. The Effects of Physical Factors on Plastid Replication

The development of methods for the culture of leaf disks in which rapid
plastid replication is occurring has enabled the factors controlling plastid replication to be studied in greater detail in defined conditions.
Similar culture systems have been used by the two groups who have
so far examined plastid replication in leaf disks, but in one case the

disks were taken from the basal half of young green spinach leaves (e.g. Possingham and Smith, 1972) and in the other from fully expanded leaves of tobacco (Boasson et al., 1972b). Apart from the species difference, the disks differed in several other important respects. The spinach plastids were certainly at an earlier stage of maturity than the replicating tobacco plastids and whereas one of the most interesting characteristics of the tobacco system was the response of the plastids to kinetin treatment: kinetin applied in the same concentration to spinach leaf disks had no effect on plastid number per cell nor on cell area (Possingham and Smith, 1972). These authors suggested that lack of response to exogenously applied cytokinins might reflect a higher endogenous supply in their younger leaf disks. The disks used by Boasson et al. (1972b) were taken from fully expanded tobacco leaves in which cell division had long since ceased and which might well have a much lower level of endogenous cytokinin. An alternative explanation would be that plastid replication at different times during plastid development is under different types of regulatory control. The observed effects of temperature on chloroplast replication and chloroplast growth are interesting in this respect since sharp differences occur in the optima for these two processes (Possingham and Smith, 1972). Spinach disks grown at 12°C have larger plastids than those grown at higher temperatures (25°C day and 20°C night temperature) with higher plastid numbers per cell. Other growth substances, e.g. gibberellic acid (0.5 mg l^{-1}) or IAA (0.5 mg l^{-1}), had no effect on plastid replication in the tobacco system.

There are, however, two outstanding similarities between the two leaf disk systems: (1) the complete dependence of plastid replication on light and (2) a direct relationship between cell expansion and plastid replication. These two aspects of regulation will be considered in turn.

1. The Role of Light in Plastid Replication

Unlike cell enlargement which continues in the dark, chloroplast replication seems to be entirely dependent on light. A particularly high rate of division occurs immediately leaf disks are transferred from the dark to the light, so although some of the metabolic syntheses required for chloroplast replication may occur in the dark, the final steps are clearly light dependent. Most experiments have been conducted with white light only, but in a recent experiment Possingham (1973) showed that green light was totally ineffective in initiating division in spinach. The lack of synchrony often observed in plastid division in intact leaves grown under normal diurnal light regimes could be explained on the basis of the light stimulation of plastid division since cell division and enlargement occur continuously in both light and dark. The total daily illumination given to the disks also directly affects the number of plastids per cell (Possingham and Smith, 1972). Light measured as "quantity per day" was modulated by changing both the day length (5 min–24 h) and the intensity of illumination (2000–10 000 µW cm^{-2}) and a linear relationship was shown between chloroplast number per cell and the daily quantity of light up to a saturating value of 250 joules per day. Changes in day length alone do not affect chloroplast formation in spinach.

The nature of the light effect on chloroplast replication is unknown and the results of an experiment designed to determine whether the plastid photosystems were involved in the response were inconclusive, since the inhibitor used, 3-(3 :4-dichlorophenyl)-1 :1-dimethylurea (DCMU), a specific inhibitor of photosynthetic electron transport, inhibits plastid replication but also inhibits cell enlargement (Boasson et al., 1972b).

2. Cell Expansion and Plastid Replication

In addition to the light effect on plastid replication, Boasson et al. (1972b) noted a direct correlation between cell enlargement and plastid replication. Plastid replication was most frequent at the time when the rate of cell enlargement was decreasing but ceased abruptly at a point soon after cell expansion stopped. There is obviously a clear relationship between cell expansion and subsequent plastid division because it appears that replication does not occur in the absence of cell enlargement (Boasson et al., 1972b). Tobacco cells double in size in the dark and in a situation where space is not limiting, the light stimulus then triggers off plastid division.

Possingham and Smith (1972) and Kameya (1972) also observed a general relationship between cell size as measured by the area of a single cell face, and chloroplast number in spinach disks, but did not examine the relationship in detail.

In summary we can conclude that in monocotyledonous and dicotyledonous plants grown under diurnal light regimes, plastid division in the leaf primordia and growing leaves can occur at several different times during plastid differentiation. Increase in structural complexity is no impediment to replication. Whilst the main increase in plastid number is probably brought about by replication during the very early stages of development, division of "mature" chloroplasts can occur and can be induced in isolated chloroplasts and in chloroplasts in the cells of isolated leaf disks.

IV. REPLICATION OF ETIOPLASTS AND ETIO-CHLOROPLASTS

Another problem to which some attention has recently been paid is the behaviour of plastids in plants grown entirely in the dark and when subsequently illuminated. Angiosperms grown entirely in the dark have no chlorophyll and the proplastids in their leaf cells develop into etioplasts (Gunning, 1965). The internal membranes of etioplasts aggregate into a regular paracrystalline assembly known as a prolamellar body (Fig. 5c). On illumination the non-phytylated, non-reduced protochlorophyllide present in these bodies is photoreduced to chlorophyll and the membrane system disperses into a mass of tangled tubules. These tubules are transformed directly into the regular assembly of membranes characteristic of the thylakoid arrangement in a normal angiosperm chloroplast (Gunning and Jagoe, 1967). An excellent recent account describing the development of illuminated etioplasts into chloroplasts can be found in the paper by Weier et al. (1970). It now seems clear that early interpretations of electron micrographs suggesting vesicle formation (von Wettstein, 1958; Virgin et al., 1963; von Wettstein, 1967) were erroneous.

Many recent experiments point to the conclusion that etioplasts undergo considerable development in the dark but that this development is limited as the complete physiological differentiation of plastids depends on light. Etioplasts have been described as "grotesque abnormalities" and logically are therefore unworthy of further investigation. It is possible, however, that prolamellar body formation may actually be a more "normal" response than has previously been considered likely. Several observations support this suggestion since prolamellar bodies are routinely found in young chloroplasts of monocotyledonous leaf cells shielded from the light by the coleoptile or by the leaf sheath (Fig. 3c). In these cells prolamellar body formation may be a normal stage in plastid development. Such chloroplasts often have well developed grana in addition to one or two small prolamellar bodies. Another finding of interest is that prolamellar bodies can be induced to re-form in the dark, in periods as short as 45 min, in plastids from illuminated (but previously etiolated) leaves (Gunning and Jagoe, 1967). Prolamellar bodies also commonly occur in plastids in green tissue culture cells under conditions where chlorophyll synthesis is limited and may reflect a situation, as in dark-grown leaves, where membrane biogenesis occurs at a faster rate than chlorophyll synthesis (Laetsch and Stetler, 1967).

Despite reservations about its relevance to "normal" proplastid differentiation, illuminated etiolated tissue has frequently been used in studies of chloroplast development since the manipulation of differentiation in such tissue can be so precisely controlled externally by the onset of illumination. Plastid replication has also been studied in etiolated leaves before and after illumination and the leaf disk culture technique has been extended to etiolated leaves to study the effects of illumination on plastid division (Boasson et al., 1972a). In dark-grown leaves before illumination, dumb-bell shaped plastid profiles which probably represent dividing etioplasts can be seen, but it is not known whether this division can be completed in the dark. A picture of such a plastid from maize is shown in Fig. 6. Plastid numbers per cell do not appear to have been estimated for dark-grown tissue. Since the central lattice of the prolamellar body resembles a strutted pentagonal dodecahedron (Wehrmeyer, 1965) and consists of a regular array of tetrahedrally branched tubules, the mechanics of any division would be highly complex.

"Etio-chloroplasts" is a useful, if somewhat cumbersome, term which has been used to describe the plastids in cells of etiolated illuminated leaves. Many morphologically distinct types of etio-chloroplasts are formed sequentially as differentiation proceeds to the "mature" chloroplast. Plastid replication during these transformations has been studied in tobacco disks (Boasson et al., 1972a). In an elegant study it was shown that plastid replication had already begun 8 h after illumination commenced and that the number of etio-chloroplasts per cell had doubled after 32 h. Etio-chloroplast replication occurred during the period of maximum chlorophyll biosynthesis, although division ceased long before the chlorophyll synthesis began to decline. The dividing plastids had an elongated profile and many, but not all, divided almost equatorially by fission. Division was most frequently observed in plastids with rudimentary grana, characteristic of plastids in leaves which had been illuminated for 32–48 h and very reminiscent of the appearance of the plastids which divided in the spinach leaf disk system which had been continuously illuminated. The daughter plastids were at first smaller than the parent plastids but later grew in size. After 144 h in the light the chloroplast finally ceased to divide and grow and exhibited the distinct crescent-shaped profile characteristic of tobacco leaf chloroplasts.

It is clear then that plastids illuminated with white light after etiolation can continue replication and that in both spinach and tobacco a considerable size increase precedes plastid division.

V. ASYNCHRONY IN PLASTID DIVISION

One of the advantages of using the leaf disk system to study replication of the etio-chloroplast is the value of such a controllable system in the study of the synchrony of plastid division. However, no degree of synchrony has been observed either in the tobacco leaf disk system (Boasson *et al.*, 1972a) or in the spinach leaf disk system (Possingham, 1973; Possingham and Smith, 1972) when the tissues are transferred from the dark to the light. Using a different experimental approach Mache *et al.* (1973) have recently published some values for chloroplast numbers in leaves of etiolated greening maize which, they suggest, provide evidence for synchronized plastid replication in this tissue. The etio-chloroplast number doubles over the first 4 h of illumination and a further non-synchronous division of plastids occurs between 4 and 11 h after illumination starts. Unfortunately the authors give no evidence that their sample of isolated plastids is representative of the full complement of the leaves, nor do they consider the complications in interpretation following from the presence in maize leaves of two morphologically distinct types of plastid-containing cell (Laetsch, 1974). The experience of workers in many laboratories has been that it is impossible to obtain consistent yields of chloroplasts from leaves of the same species on consecutive occasions and the problem is particularly acute in plants such as maize, where the bundle sheath cells containing the "agranal" chloroplasts are much more resistant to fracture than the softer mesophyll cells containing the "granal" plastids (Woo *et al.*, 1970). The question of synchrony is an important one and it would be useful to have the ambiguities in this method resolved. Mache *et al.* (1973), suggested that the synchrony they had observed might be a reflection of a situation in which all the cells of the leaf have reached the same stage of physiological development in the dark. This may well be the case since they used only the top 10 cm of a 15 cm maize leaf. Etioplasts in such leaves have reached their maximum state of complexity in the middle region of the leaf and proplastid to complexity in the middle region of the leaf and proplastid to etioplast differentiation is completed in the lower few centimetres of the leaf (Mackender and Leech, in press). In view of the work of Mache *et al.* (1973), Possingham (1973) has re-examined replication in dark-grown illuminated spinach disks. He could find no immediate or synchronizing effect of light on plastid replication in spinach, but as previously observed showed that the non-synchronous light-stimulated replication occurred with a doubling time of 2–5 days. To establish whether or not synchrony in replication does

occur on illumination of etiolated tissue requires considerably more experimental investigation.

VI. THE RELATIONSHIP OF PLASTID NUCLEIC ACIDS TO PLASTID REPLICATION

The presence in chloroplasts of the biochemical elements necessary for classical genetical autonomy has been known for about 10 years. It was clear much earlier from genetical experiments using mutants that the plastid and nuclear genomes act as co-determinants in the regulation of plastid development. The genetic evidence is carefully reviewed by Kirk and Tilney-Bassett (1967) and the more recent biochemical information considered by Smillie and Steele-Scott (1969). In the case of plastid replication the interactions of plastid and nuclear genomes must also be considered. As in plastid development, progress has depended on the careful characterization of the nucleic acid components of the plastid in order that changes can be diagnostically recognized at different stages in the life cycle or under experimental manipulation. Chloroplasts possess DNA, RNA, polymerase systems for both DNA and RNA synthesis and also a protein synthesizing system (for a recent review see Boulter et al., 1972). Each chloroplast contains between 20 and 60 copies of a circular double-stranded molecule which is characterized by the absence of the base 5-methyl-cytosine. The contour length of the DNA molecules is about 40 μm and their molecular weight about 10^8. Calculations and kinetic complexity measurements suggest that this amount of double-stranded DNA of unique base sequence would be sufficient to code for about 125 proteins each of molecular weight 5×10^4 (Ellis, 1974). Most of the chloroplast RNA is found as ribosomal RNA in 70S chloroplast ribosomes of the leaf. Within mature plastids, the ribosomes are generally free, but polysomal configurations have been observed in etioplasts (Brown and Gunning, 1967) and a proportion of leaf ribosomes behave as polysomes in the ultracentrifuge. The high molecular weight RNA components of the chloroplast ribosomes are readily distinguishable on ultracentrifugation or polyacrylamide gel electrophoresis from their cytoplasmic ribosomal counterparts. The chloroplast ribosomal RNA species have molecular weights of $1·05 \times 10^6$ and $0·56 \times 10^6$ compared with their larger cytoplasmic counterparts of molecular weight $1·3 \times 10^6$ and $0·7 \times 10^6$. RNA synthesis by isolated chloroplasts has been demonstrated several times (see Woodcock and Bogorad, 1971, for references), but only recently has the nature of this newly synthesized RNA been examined. Isolated spinach chloroplasts will synthesize RNA species of molecular weight

1.2×10^6 and 0.65×10^6 and it has been suggested that these are in fact precursors for the normal chloroplast RNA (Hartley and Ellis, 1973). It would seem possible that the functional genes for chloroplast ribosomal RNA are therefore located in the plastid DNA. Several of the genes for chloroplast ribosomal protein, however, are coded for by nuclear genes (Bourque and Wildman, 1973). The established presence of these large amounts of DNA and ribosomal RNA in plastids has prompted recent preliminary attempts to examine the relationship between nucleic acid synthesis and the chloroplast replication process. The role of DNA in the regulation of this process is of key interest. Clearly if a "chloroplast cycle" exists in higher plants it might be anticipated that nucleic acid synthesis would occur between subsequent divisions. However, the presence of multiple copies of DNA in leaf chloroplasts may eliminate the necessity for such doubling. In the absence of an authenticated system in which synchronized plastid division is occurring, progress in the solution of this problem is difficult. Some evidence that replication of etio-chloroplasts can continue in the complete absence of DNA synthesis comes from experiments with greening tobacco disks treated with the inhibitor 5-fluordeoxyuridine (FUdR). This compound inhibits DNA synthesis in bacteria (Cohen et al., 1958) and in tobacco leaves (Flamm and Birnstiel, 1964) primarily by inhibiting thymidylate synthetase. Boasson and Laetsch (1969) showed that in illuminated etiolated tobacco disks, FUdR had no effect on normal plastid replication but inhibited chlorophyll synthesis by inhibiting chloroplast growth. Since the FUdR effects can be reversed by thymidine and since synthesis of nuclear DNA seems to occur during this recovery period, it would seem likely that it is this DNA synthesis which is inhibited by FUdR and is necessary for chloroplast growth. In contrast chloroplast replication itself can continue in the complete absence of DNA synthesis. The authors suggest that the reason why the plastids are able to replicate more than twice without synthesis of new DNA is that they in fact contain several copies of DNA. However, maintenance of the correct level of redundancy seems to be necessary for continuing chloroplast growth (lack of DNA synthesis is one explanation for the inability of isolated chloroplasts to grow in culture even though they are able to divide (Ridley and Leech, 1970b). However, it would appear that normally in the leaf DNA synthesis continues at the same time as replication. Indeed in chloroplasts from cut maize leaves (the system about which some reservations have already been expressed, p. 153), as the chloroplast number doubled, the DNA per plastid also increased during the first 4 h following illumination (Mache et al., 1973). It would be most interesting to determine whether the DNA per plastid also increases in the kinetin-treated tobacco leaf disks in which 100% increase in chloroplast numbers can be induced (Boasson and Laetsch, 1969).

In contrast to indications that the DNA per plastid may increase during plastid replication, the RNA per plastid seems to remain remarkably constant over the time period when plastids are dividing (Detchon and Possingham, 1972). During the time that green spinach grows most rapidly and the leaves increase from 2 to 7 cm in length, the chloroplasts are also increasing in size but the RNA content per plastid remains remarkably constant. It appears in this species that chloroplast RNA synthesis occurs when the leaves are very young prior to the development to the 2 cm length stage. The results of experiments on ^{32}P orthophosphate incorporation into chloroplast RNAs of wheat, swiss-chard and onion also suggest that chloroplast RNA is only synthesized over a very limited period when the leaves are young (Ingle et al., 1970).

Although the RNA per plastid may not change when etiolated tissue is illuminated, there is a large increase in the total RNA of the leaf (Ingle et al., 1970; Smith et al., 1970). The first investigations of RNA synthesis in the spinach disk system have recently been published (Detchon and Possingham, 1973). Over a period of 8 h, light stimulated the incorporation of ^{14}C- and ^{3}H-uridine into the RNAs of the leaf disk and these workers detected two rapidly labelled RNA species of 1.15×10^6 and 0.65×10^6 molecular weight which may possibly be precursors of chloroplast ribosomal RNA. At present no attempts have been made to correlate the RNA changes over this experimental period of 8 h with the plastid replications occurring over much longer periods of time to give a 10-fold increase in chloroplast per cell over a 7 day growth period. The spinach leaf disks have great potential as a system in which to investigate the relationship between nucleic acid turnover and plastid replication.

Fig. 7. Serial sections through a young dividing chloroplast of *Beta vulgaris* showing eight DNA regions differing in size and DNA content. Four of these regions [marked 1–4] are apparently separated in the first section (**a**), and seem to be connected in the succeeding sections (**b–d**) and to lie in a median position in the constriction area of the dividing plastid. Other DNA regions are outside the constriction area. (See text for further details.)

The leaf tissue was treated with trypsin between the glutaraldehyde and osmium fixations (Herrmann and Kowallik, 1970). × 1400.

I am most grateful to Dr Klaus Kowallik of the Botanisches Institut der Universität Düsseldorf for permission to use these electron micrographs from his own unpublished work.

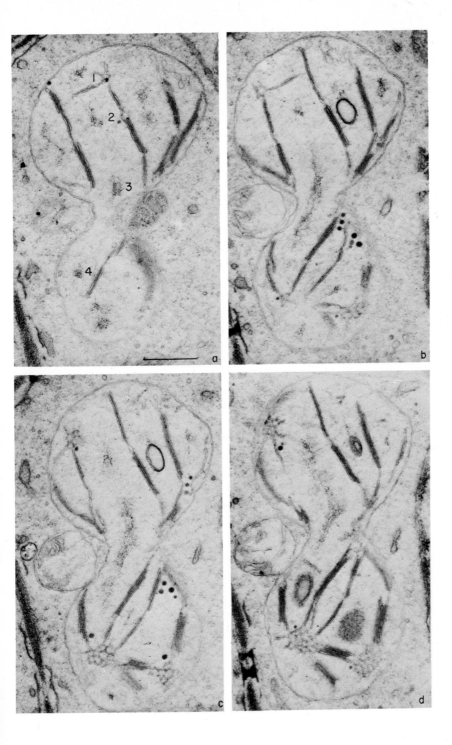

Electron microscopical examination of the behaviour of strands of DNA visualized in the chloroplast stroma in replicating chloroplasts have complemented biochemical studies on cultured leaf disks. Within the stroma of chloroplasts occur areas of low density which contain a network of fibrils about 250–300 n, in diameter and with characteristics resembling bacterial nucleoplasm (Ris and Plaut, 1962; Kislev *et al.*, 1965) and removed by DNase. Similar DNase-sensitive areas have been located within the stroma of several higher plant chloroplasts (for a summary of these findings see Woodcock and Bogorad, 1971). Autoradiographic studies following the administration of tritiated thymidine with adequate DNase controls confirm that in these fibrils active incorporation of thymidine is occurring. In swiss-chard (*Beta vulgaris*) the DNA is arranged in several regions (Herrmann and Kowallik, 1970) and there may be as many as 18 of these present in a single chloroplast (Kowallik and Herrmann, 1972). The number of such regions depends upon the size of the organelle and also upon its developmental state, and comparison with biochemical data giving the DNA content per plastid suggests that these regions may each be a single nucleoid. Kirk (1972) has discussed in detail the consequences of the presence of different amounts of plastid genetic material in separate regions or "chromosomes". The behaviour of these DNA-containing areas during plastid replication had until recently only been studied in barley (Sprey, 1968). Usually DNA areas, termed genophores, appear to be distributed evenly at either end of the plastid during division. However, recently, as part of a detailed investigation of chloroplast DNA, Kowallik and Hermann (1972) studied the behaviour of plastid DNA regions during chloroplast division. The development of an electron microscopical technique in which a proteinase digestion was used during fixation enabled the DNA-containing regions of the plastid to be more readily visualized. It was found that proplastids possess only a single DNA region but differentiated etioplasts and chloroplasts have several such regions. These observations suggest that the multiplication of the DNA regions may be independent of plastid division. Further support for the suggestion comes from the finding that in chloroplasts which only increase in size without further dividing a multiplication of DNA regions is regularly observed (Kowallik and Hermann, 1972). Occasionally a stretched DNA region can be observed in the constriction of a dividing chloroplast and electron micrographs of serial sections from such a dividing plastid are shown in Fig. 7. Sprey (1968) observed a similar arrangement in barley plastids and from light microscope observations suggested that splitting of the DNA region is caused by the division of the plastid itself. Similar separation of DNA areas in dividing plastids has been observed in algae (Bisalpultra and Bisalpultra, 1969). Cells in which synchronized plastid division is occurring and also plastids dividing in isolation may provide a valuable material in which to study the behaviour of DNA during plastid replication in greater detail.

VII. CONCLUSIONS

Plastids are intimately involved in the economy of the cells they inhabit, and it seems most likely that plastid replication, like plastid develop-ment, is essentially a cellular process and not a specific attribute which is under independent plastid control. The apparent dependence of chloroplast replication on a previous phase of cell enlargement and the response of plastid division to exogenous supplies of kinetin and the lack of response to gibberellic acid and IAA are examples of the way in which chloroplast behaviour reflects cellular and not merely plastid activity. Further investigation of the biochemical details of the in-tracellular interactions involved in plastid replication will depend very much on the availability of suitable experimental systems and the leaf disk systems currently in use seem to have several advantages, particu-larly in respect of ease of manipulation. The study of these systems and also of the characteristics of replicating chloroplasts after isolation are an essential pre-requisite to the study of plastid replication as it takes place normally in the leaves of the growing plant. Eventually, however, the characteristics of chloroplast division will have to be examined in the cells of these leaves themselves, still attached to the growing plant.

5. The Cell in Sporogenesis and Spore Development

M. D. BENNETT

Plant Breeding Institute, Cambridge, England

I. INTRODUCTION

There are several excellent reviews of the anatomy, morphology, physiology and biochemistry of sporogenesis and spore development in higher plants, including those by Maheshwari (1950), Stern and Hotta (1969) and Heslop-Harrison (1972). The object of the present paper is to review cell development in sporogenesis and spore development from several aspects which, hitherto, have received only passing consideration. First, attention will be paid to the premeiotic period of cell development in sporogenesis. Second, information will be presented concerning the rate and duration of cell development during the various stages of sporogenesis and spore development. Third, there is a considerable amount of evidence showing that non-genetic, physical nuclear characters, notably nuclear DNA content, play a major role in determining the rate of meiotic and gametophytic development in higher plants. The various factors controlling the rate of cell development in reproductive tissues will be discussed. Fourth, while the importance of meiosis in ovules has often been stressed on theoretical grounds (Darlington, 1971), nevertheless, the number of studies of female meiosis are too few to reflect the relative importance of the second component of "two-track heredity". Consequently, extra attention will be paid to describing the time and duration of meiosis in embryo sac mother cells (EMCs).

It should be noted that the data available concerning the duration of premeiotic, meiotic and post-meiotic stages are very limited and involve only a few higher plant species. Furthermore, in seeking to make meaningful comparisons both between species and between different cell types within a species, it is almost essential that the data for cell size and rate of development should come from plants grown under similar known environmental conditions and therefore this review will draw very heavily on data for four main species: *Lilium longiflorum* (Taylor and McMaster, 1954; Ito and Stern, 1967; Stern and Hotta, 1969; Walters, 1972), *Tradescantia paludosa* (Steinitz, 1944; Taylor, 1949, 1950; Bryan, 1951; Beatty and Beatty, 1953; Walker and Dietrich, 1961; Garot *et al.*, 1968), *Trillium erectum* (Sparrow and Sparrow, 1949; Hotta and Stern, 1963a, b; Kemp, 1964; Ito and Stern, 1967), and *Triticum aestivum* (Bennett *et al.*, 1971; Riley and Bennett, 1971; Bennett and Smith, 1972; Bennett *et al.*, 1973b), since only in these species are sufficient results available for most or all of the stages under consideration from suitably controlled experiments. Furthermore, the aspects of cell development during and around meiosis listed above have been the subject of the author's research (Bennett, 1971, 1972, 1973; Bennett and Smith, 1972, 1974; Bennett *et al.*, 1971, 1972, 1973a, b). Where no other data are available it will frequently be necessary to refer to the author's papers alone.

It seems useful to define some of the terms to be used in this review since some are subject to misuse and others may have several meanings. Sporogenesis is that period of development between the formation of the archespore and the production of spores at the end of meiosis. An archespore may either undergo meiosis itself, or will divide mitotically to produce two or more archesporial cells which will undergo meiosis. Sporogenesis consists of two developmental stages, namely premeiotic and meiotic. In species with many archesporial cells derived from a single archespore, for instance in most higher plant anthers, premeiotic development comprises several cell cycles. In many higher plants, however, where there is only one meiocyte per ovule and where the archesporial cell undergoes meiosis, both premeiotic and meiotic stages are accomplished within a single cell (as in unicellular micro-organisms such as yeast). Premeiotic development may be considered as ended when the first event unique to meiosis occurs. In practice this is a difficult moment to define and in most instances the start of meiosis has been defined in an arbitrary fashion. The term meiocyte is used to describe the cell which will undergo meiosis both during premeiotic interphase and while it is actually engaged in meiosis. As such, meiocyte is synonymous with the term sporocyte. In higher plants sporogenesis occurs in two types of cell line, one located within the anther (to produce microspores) and the other in the ovule (to produce megaspores). Meiocytes and spores are frequently described in accordance with the sex of the gametophytes which they will produce. Thus, a pollen mother cell (PMC) or microsporocyte, which gives rise to microspores, is sometimes called a male meiocyte, while an embryo sac mother cell or megasporocyte, which gives rise to megaspores, is sometimes called a female meiocyte. The terms PMC and EMC, or male and female meiocyte, are to be preferred since microspore and megaspore are misleading in many species where the size of the two types of meiocyte is indistinguishable (e.g. *Hordeum vulgare*) or where the PMC is bigger than the EMC. Sporogenesis ends at second telophase of meiosis with the production of four spores in a quartet or tetrad stage. Spore development begins when sporogenesis ends and includes the whole of pollen grain and embryo sac development until mature gametophytes capable of fertilizing or being fertilized are produced. A microspore is a spore which produces a gametophyte bearing only male gametes, and a megaspore is a spore giving gametophytes with only female gametes (Darlington, 1965). In fact, however, these latter definitions although generally true are only conveniences, since the production of an embryo sac can be readily induced in "microspores" derived from meiosis in PMCs of anthers in hyacinth (Stow, 1930, 1933).

II. PREMEIOTIC CELL DEVELOPMENT

A. Description of Premeiotic Cell Behaviour

1. The Size and Shape of Archesporial Cells in Anthers

The number of PMCs per anther loculus varies considerably both between species and between genotypes within a species. Thus, *Malva sylvestris* has only about 8 PMCs per loculus, *Secale cereale* has about 1500 and *Lilium longiflorum* about 7500 (Erickson, 1948). In *Triticum aestivum* the number of PMCs per loculus is reported to vary from about 48 to 322 (De Vries, 1971). Intraspecific variation in pollen mother cell number in *T. aestivum* and *Oryza sativa* (Beri and Anand, 1971; Oka and Morishima, 1967) is usually reflected in a proportional change in anther volume and does not involve large changes in PMC size.

The arrangement of PMCs within the loculus also varies. In some species, for instance *Gazania rigens* (Lima-de-Faria, 1964), *Crepis* and *Hieracium*, the PMCs within a loculus are arranged in a column of single cells. Consequently, the PMCs are in contact with tapetal cells over a high proportion of their surface area. In many other species, for instance *Triticum aestivum*, each loculus contains several rows of PMCs but each PMC has at least one cell wall surface in contact with the tapetal layer. In other species, for instance *Lilium longiflorum*, each loculus contains a column of meiocytes several cells thick so that many PMCs share no common interface with a tapetal cell.

The volume of PMCs and EMCs is frequently larger, often considerably so, than other diploid meristematic cells in the same species of higher plant. For instance, compare the size of the EMCs of *Solanum nigrum* and *Lilium* hybrid in Fig. 1a and b with the size of the nucellar cells surrounding them. In species with many PMCs per anther loculus the increase in cell volume apparently occurs gradually over several cell cycles prior to meiosis rather than suddenly prior to meiosis. For instance, in *Triticum aestivum* the mean volume of archesporial cells was estimated to be about $6.2 \times 10^3 \mu m^3$ when the loculus contained 12 archesporial cells, about $9.1 \times 10^3 \mu m^3$ when it contained 50 cells, about $13.6 \times 10^3 \mu m^3$ when it contained 100 cells and $31.1 \times 10^3 \mu m^3$ by first prophase of meiosis (Bennett *et al.*, 1973b). Similarly, analysis of data for anther development in *Zea mays* (Moss and Heslop-Harrison, 1967) shows that during the period of late archesporial cell development the number of archesporial cells per loculus increased about 3-fold and the mean volume per archesporial cell increased about 4-fold. During the same period the number of tapetal cells increased about 10-fold yet their volume remained almost constant. In species with only a single

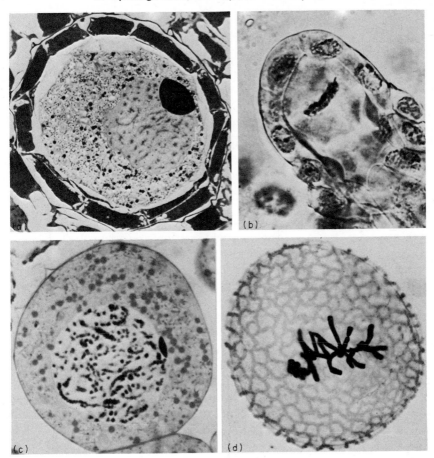

FIG. 1. **a.** Transverse section through an EMC of *Lilium* hybrid cv. 'Sonata' at pachytene (×675). **b.** Optical section through an EMC of *Solanum nigrum* at first metaphase (×1020). **c.** Median section through a PMC of *Lilium* hybrid cv. 'Black Beauty' at pachytene (×1350). **d.** Microspore at metaphase of pollen grain mitosis in *Lilium* hybrid cv. 'Sonata' (×750). Note the large size of the EMC compared with the surrounding somatic cells in **a** and **b**; also, the large amount of dense cytoplasm with numerous inclusions in **a** and **c**.

meiocyte per ovule, formed long before the initiation of meiosis (for instance *T. aestivum* and *Lilium* hybrid), this cell increases its volume slowly over a prolonged period. It seems likely that an increase in cell size may be an essential pre-requisite for meiosis to occur in many higher plant species rather than that it merely reflects a size change correlated with the growth in size of the anther or the ovule as a whole.

It is interesting to note in this context that even in the unicellular fungus *Saccharomyces cereviseae*, where premeiotic development need take no account of development in any other cell, a pronounced increase in cell size occurs immediately prior to the initiation of meiosis (Simchen *et al.*, 1972).

2. Cell Cycle Time

Studies of the rate of cell development in the male archesporium of *Triticum aestivum* cv. 'Chinese Spring' (Bennett *et al.*, 1973b) showed that the duration of successive premeiotic cell cycles was increased. Thus, the three cell cycles preceding meiosis, which commenced when the number of archesporial cells per loculus was 12, 25 and 50, were estimated to last 25, 35 h and 55 h respectively. Very few data are available for the durations of cell cycles in the archesporia of higher plant species; however, the data which are available for species other than *T. aestivum* indicate the existence of similar premeiotic cell behaviour. For instance, comparison of data for *Lilium longiflorum* shows that while the normal cell cycle time in root-tip meristem cells is about 24 h (Stern and Hotta, 1969) the duration of the cell cycle commencing at premeiotic mitosis and ending with the onset of meiosis is not less than 72 h (Erickson, 1948) and may be as long as 11 days. It should be noted that the developmental behaviour of cells prior to meiosis in higher plants appears to be the same in other groups of organisms, including fungi and some animals. Thus in the yeast *Saccharomyces cerevisiae* the normal cell cycle at 30°C was about 150 min but the cell cycle prior to meiosis was greatly prolonged and lasted up to 7–8 h (Simchen *et al.*, 1972). Monesi (1962) has shown that cell cycles in spermatogonial cells of the mouse are progressively prolonged as meiosis is approached.

In *Triticum aestivum* the duration of mitosis in root-tip meristem cells at 20°C was about 1·6 h when the duration of the cell cycle (T) was 12·5 h. In the male archesporium at the same temperature the duration of mitosis increased to 1·9 h, 2·3 h and 2·6 h in successive premeiotic cell cycles with durations of 25 h, 35 h and 55 h respectively. It should be noted that although the duration of mitosis increased with increasing cell cycle time, nevertheless it was by a smaller proportion (Bennett *et al.*, 1973b).

The author is unaware of any data giving the durations of stages of the cell cycle other than mitosis during successive premeiotic cell cycles in higher plants. However, data are available for the duration of DNA synthesis phase (S) in at least two animal species. Monesi (1962) noted that the duration of S-phase increased from 7 h to about

14 h during several successive spermatogonial cell cycles, and Callan (1972) reported that in male germ line cells of the newt *Triturus*, phases of DNA synthesis took progressively more time as meiosis was approached. It seems reasonable to expect that S becomes increasingly prolonged in higher plant archesporial cells as T increases during the approach to meiosis. Certainly the duration of the premeiotic DNA synthesis stage is greatly prolonged compared with S in root-tip meristem cells in several higher plant species (see Section B2).

3. Cytochemistry of Archesporial Cells

Little is known of the cytochemistry of cells during premeiotic development in higher plants. However, Moss and Heslop-Harrison (1967) have made a detailed study using the sporogenous tissue in *Zea mays* anthers. They showed that while the number of archesporial cells per loculus increased about 3-fold (from 92 to about 260) the total amount of RNA per loculus increased about 10-fold. Thus, the total RNA content per archesporial cell increased about 3-fold. However, because of the increase in archesporial cell volume which occurred during this period of development the concentration of RNA fell to about one-third of its initial concentration. Similarly, while the number of archesporial cells increased from 92 to 260 the total protein content of the sporogenous tissue increased about 10-fold, but due to the increase in cell size the concentration of protein fell to only about one-half of its initial value.

Cytological observations of archesporial cells during premeiotic development show that in many species (including *Triticum aestivum, Secale cereale, Hordeum vulgare* and *Zea mays*) the archesporial cells, like most meristematic cells, contain no vacuoles, or very small ones. Consequently, the large increase in archesporial cell size noted in wheat is accompanied by an increase in the amount of cytoplasm synthesized during successive cell cycles. If this occurs concurrently with a fall in the concentration of RNA in the cell then these two factors acting together may be largely responsible for the increase in cell cycle time noted in several species during premeiotic archesporial development.

B. Premeiotic Mitosis and Premeiotic Interphase

1. Synchronization of Meiocytes

Meiosis occurs more or less synchronously in all the meiocytes within a loculus in most higher plant species, yet premeiotic mitosis occurs asynchronously. Clearly then archesporial cells become synchronized in development just prior to the onset of meiosis. Premeiotic mitosis occurs asynchronously in archesporial cells of *Triticum aestivum*, *Lilium longiflorum* and *Trillium erectum*. In each of these species it has been shown that synchronization is achieved by a developmental hold which accumulates meiocytes in G_1 of premeiotic interphase (Taylor and McMaster, 1954; Bennett *et al.*, 1973b; H. Stern, personal communication). The mechanism of synchronization is unknown. It should be noted, however, that whenever cell populations are synchronized in natural situations it is almost always in G_1 or G_2 and usually the former (Avanzi *et al.*, 1970; Rost and Van't Hof, 1973).

It is not clear whether the developmental hold in G_1 in meiocytes is achieved by blocking the initiation of S-phase, or whether it is achieved by blocking cell development at some earlier stage of G_1. In *Triticum aestivum* grown at 20°C cells start accumulating in G_1 about 120 h before first metaphase of meiosis and it takes about 55 h for all the archesporial cells to assemble in G_1. Soon after PMCs must be released from the G_1 hold since they all enter DNA synthesis together about 33 h before first metaphase of meiosis. It appears as if the factor(s) which blocks PMC development in G_1 also acts upon the cells of the tapetal layer since they also become synchronized in G_1 by about 15 h after the PMCs; furthermore, they appear to be released from the G_1 hold synchronously with the PMCs since the tapetal cells enter DNA synthesis at the same time as PMCs in the same loculus (Bennett *et al.*, 1973b).

2. *The Stages of Premeiotic Interphase*

In terms of the normal cell cycle, premeiotic interphase consists of three possible stages, namely G_1, DNA synthesis stage (S), and G_2. In species such as *Lilium* where S precedes the initiation of meiosis all three of these stages are found, but they are subject to modification compared with corresponding stages in normal somatic meristematic cell cycles. As shown above the duration of G_1 may vary greatly when PMCs are synchronized by accumulation in this stage. Even the shortest duration of premeiotic G_1 stage in *T. aestivum* (20–30 h) is much longer than in root-tip meristem cells (4·7 h). The minimum duration of premeiotic G_1 in *Lilium* is about 60–70 h compared with a duration of less than 10 h in root-tip meristem cells. The long duration of G_1 is probably useful since it allows large amounts of molecules essential for subsequent DNA synthesis and meiosis to be accumulated. One of the last events during G_1 in *Lilium* is the sudden accumulation of the precursor pools of deoxyribosides necessary for DNA synthesis (Foster and Stern, 1959). It has also been suggested that long-lived messenger RNA required to control meiocyte and gametophyte development is transcribed during this period (Bennett and Hughes, 1972). Cummins and Day (1973) state that their experiments with *Ustilago violacea* indicate that during G_1 of premeiotic interphase all genes governing mating are transcribed and translated before the mating organelle is assembled and develops.

As with G_1 so with premeiotic DNA synthesis, its duration is much longer than S in root-tip meristem cell cycles. In *Triticum aestivum* the duration of DNA synthesis prior to first metaphase is about 12–15 h compared with an S-phase of only 3·8 h in root-tip meristem cells. In *Lilium longiflorum* premeiotic DNA synthesis was estimated by Taylor and McMaster (1954) to last between 24–48 h while the whole cell cycle only lasts about 24 h in *Lilium* root-tip meristem cells (Stern and Hotta, 1969), so presumably S only lasts 12 h or less. It seems likely that an extended DNA synthesis phase during the meiotic cell cycle is the normal situation in most organisms and may even be an essential prerequisite for normal meiosis to occur. A prolonged premeiotic DNA synthesis stage has been noted in the unicellular fungus *Saccharomyces cereviseae*. Thus premeiotic DNA synthesis lasted about 4 h while normal DNA synthesis only lasts about 40 min (Simchen *et al.*, 1972; Williamson, 1966). In animals Callan (1972) reported that in male germ line cells of *Triturus*, phases of DNA synthesis take progressively longer as meiosis is approached, and that the final S-phase takes 8–9 days at 18°C compared with a typical S-phase in somatic cells which lasts only about 24 h. Extended premeiotic S-phases have been noted in *Xenopus laevis* (Bird and Birnstiel, 1971; Coggins and Gall, 1972) and in several mammals including the mouse and the hamster (Ghosal and Mukherjee, 1971). In the mouse this observation has been made for both oocytes and spermatocytes (Monesi, 1962; Kofman-Alfaro and Chandley 1970; Crone *et al.*, 1965).

The reasons why premeiotic DNA synthesis is prolonged are unclear. Callan (1972) demonstrated that premeiotic chromosomes in *Triturus* have fewer DNA synthesis initiation sites than do normal somatic chromosomes; however, why this should be and how it is controlled is unknown. In *Triticum aestivum* the chromosomes are condensed and composed of distinct threads at the time of meiotic DNA synthesis and this physical state of the chromatin may prevent normal rates of DNA synthesis. It should be noted that some DNA synthesis occurs throughout meiosis in *T. aestivum* at stages when chromosomes are even more contracted (Riley and Bennett, 1971). According to Simchen *et al.* (1972), normal DNA synthesis and premeiotic DNA synthesis may be under separate genetic controls and may use different pools of precursors.

Higher plant species appear to vary with respect to whether or not they have a G_2 stage after DNA synthesis and before the initiation of meiosis. At one extreme there are species like *Trillium erectum* where premeiotic cell development is held in G_2 of the cycle until a cold requirement breaks the dormancy releasing cells into meiosis (Hotta and Stern, 1963; Stern and Hotta, 1969). In species which do not have a developmental hold in G_2 there is a wide range of cell behaviour. In *Lilium* hybrid cv. 'Black Beauty' grown at 20°C the duration of G_2 is at least 48 h. It seems certain that in *Triticum aestivum* there is no G_2 stage as normally defined in the meiotic cycle, and meiocytes pass directly from DNA synthesis into meiotic prophase. Some species (e.g. *Lilium longiflorum*) which do have a G_2 period after premeiotic DNA synthesis and before leptotene have a chromosome contraction stage followed by a phase of chromosome relaxation before the chromosomes commence normal meiotic behaviour (Walters, 1972). Other species of plant, for instance *Nicotiana* species (Burns, 1972), and animals, for instance *Homo sapiens* (Stahl, 1972), also have preleptotene contraction or prochromosome stages. These presumably also occur in G_2.

3. Effects on Meiotic and Post-meiotic Development Determined during Premeiotic Interphase

A considerable amount of evidence has accumulated which suggests that events occurring during premeiotic development, especially during premeiotic interphase, can affect cell development during subsequent stages of meiosis and spore development. Sparrow (1942) reported that treating seedlings of *Antirrhinum majus* with dilute colchicine solution 8 weeks before meiosis produced an increased univalent frequency at meiosis in anthers. Dover and Riley (1973) recently noted several developmental abnormalities in meiosis and pollen development after treating archesporial cells of *Triticum aestivum* at premeiotic mitosis and premeiotic interphase with colchicine. For instance, colchicine applied during early premeiotic interphase induced asynapsis in meiosis and also failure of germ pore formation in abnormal pollen. It was suggested that colchicine induced these effects by suppressing the formation of microtubules at premeiotic mitosis or some other stage, thereby disrupting the spatial arrangement of homologous chromosomes necessary for synapsis to occur. Although the effects are seen many hours after the end of the colchicine-sensitive stage, it is not difficult to understand the reasons for the induced aberrant meiocyte behaviour, and the hypothesis advanced seems highly plausible in view of the well established effect of colchicine on microtubule formation. There are other examples, however, of a treatment acting on a sensitive stage during premeiotic development which produces a developmental effect only after a prolonged period and where it is much harder to suggest a plausible or testable hypothesis to explain the cell behaviour.

It has been established that Ethrel (2-chloroethylphosphonic acid) can induce male sterility in several plant species, including wheat and cucumber. In experiments using *Triticum aestivum* (Bennett and Hughes, 1972) it was shown that to induce male sterility Ethrel must be applied during premeiotic anther development and no later than the start of meiosis. Meiotic and pollen development continue normally in treated plants for about 5 days after the end of the Ethrel-sensitive period before any visible cytological effect on pollen development was seen. When Ethrel was applied at the correct concentration to anthers at premeiotic interphase then extra pollen grain mitoses in the normally non-dividing vegetative nucleus were induced about 6 days after the treatment, at the time of second pollen grain mitosis in normal anthers. The mechanism of Ethrel's effect is unknown. However, Ethrel is known to break down in plant cells at pH values below 4·0 releasing ethylene, a molecule now widely acknowledged as having hormonal properties in plants. Perhaps Ethrel interferes with the synthesis of long-lived messenger RNA produced in premeiotic interphase but not active until much later in microspore development.

Another intriguing example of this type of phenomenon has been found using *Trillium erectum* meiocytes. Ito and Stern (1967) showed that it was possible to culture filaments of meiocytes extruded from *Trillium* anthers through meiosis provided they had already entered leptotene. Meiotic and pollen development is greatly prolonged in *Trillium* (Sparrow and Sparrow, 1949) even at warm temperatures (Ito and Stern, 1967). *Trillium* meiocytes placed in culture in G_2 do not enter meiosis, but under the appropriate conditions in culture each cell produced a pollen tube about the normal time after premeiotic G_2, despite their not having undergone meiosis or normal pollen development (H. Stern, personal communication) (see Fig. 4b).

III. MEIOSIS

A. The Approach to Meiosis

1. Developmental Switches

The history of the cell line which eventually produces a meiocyte as it progresses from one stage of development to another involves several distinct developmental switches. First, shoot apex meristem cells must switch their development from a vegetative to a flowering mode. This switch presumably involves changes in many cells of which very few contribute to the cell line from which meiocytes are later derived. It is not the purpose of this review to discuss the induction of flowering in higher plants. Nevertheless, it seems worth while to comment that in one species at least (*Sinapis alba*) cell behaviour during flowering evocation (after a single 22 h long day) was parallel in many respects to cell behaviour during premeiotic development of archesporial cells in wheat anthers described above. Thus, many cells first become synchronized in G_1, then undergo synchronous DNA synthesis followed by an increase in RNA and protein synthesis and an increase in cell size (Jacqmard *et al.*, 1972).

A later switch occurs when a cell within a flowering meristem divides to produce an archespore, that is a cell which will either produce a meiocyte itself or will divide mitotically to produce more archesporial cells. As described above (Section IIA2) and elsewhere (Simchen *et al.*, 1972; Bennett *et al.*, 1973b) the development of archesporial cells characteristically involves a progressive increase in cell cycle time, DNA synthesis time, and cell size. The preparation of the cell for meiosis may take place gradually over several cell cycles (McDermott, 1971). Eventually, within each floret, the first cell is produced which will itself undergo meiosis. That is, a cell is produced which, without first undergoing another mitotic division, either already has, or is capable of achieving, full meiotic readiness. This cell may be the archespore itself or it may be just one cell in a large population of archesporial cells. In the latter case production of meiocytes is correlated with the imposition of synchronous cell development induced by a developmental hold which accumulates cells in G_1 of the cell cycle. Superimposed upon the normal succession of stages of cell development (G_1, S and G_2) during premeiotic interphase is the progress of the cell through a series of states characteristic of the approach to meiosis irrespective of whether it is a single meiocyte in an ovule or one of a large population of meiocytes in an anther. During premeiotic interphase the cell must enter a state of meiotic readiness, that is it must be capable of undergoing meiosis without first undergoing another mitotic division. Just when this stage is reached in higher plant archesporial cells is unclear, although its start and duration have been clearly defined in some microorganisms, for instance in *Saccharomyces cerevisiae* (Simchen *et al.*, 1972). Stern and Hotta (1969) pointed out that premeiotic DNA synthesis is characteristic and necessary for meiotic cells in *Lilium*, but they noted that it does not commit the cell to meiosis. Throughout the period of meiotic readiness the meiocyte is still capable of reverting to a mitotic division should the remaining pre-conditions essential for initiation of a meiotic sequence not be fulfilled. It has been shown that cells of *Lilium* and *Trillium* cultured at G_2 of what would normally be premeiotic interphase can be induced to revert to mitosis so long as leptotene has not been initiated (Stern and Hotta, 1969). The existence in some *Lilium* genotypes of a premeiotic contraction stage after premeiotic DNA synthesis and before the start of leptotene indicates that the genes controlling a mitotic sequence of development are not yet suppressed, so that chromosomes reach prophase of mitosis before suddenly reverting within 5 or 6 h directly to a leptotene condition without any intervening anaphase condition (Walters, 1972; Bennett, unpublished data). Finally, readiness for meiosis is followed by commitment to meiosis itself.

Various photoperiod, temperature and chemical treatments have been demonstrated which will switch development from one stage to the next or redirect development back to a previous stage (Bernier, 1963; Heslop-Harrison, 1957; Ninneman and Epel, 1973; Bennett and Hughes, 1972). The nature of the switches involved remains totally unknown. However, the isolation of mutants may provide the means of investigating the nature of developmental switches in a less superficial manner.

2. Why does Meiosis not occur in some Somatic Cells in Higher Plants?

In many unicellular organisms (e.g. *Saccharomyces*) meiosis can be induced synchronously in every cell in a culture by appropriate environmental conditions. In such organisms loss of the ability to undergo meiosis would result in the extinction of the cell type, that is unless the function of meiosis were taken over by some other process. In fact this has apparently happened in organisms like *Penicillium notatum* in which meiosis is unknown but somatic cell crossing-over occurs (Herskowitz, 1969). In higher plants, however, only a very few cells initiate sporogenesis, although clearly many cells retain the ability to do so. Thus, many vegetatively propagated species still undergo normal meiosis, for example *Solanum tuberosum* and *Hyacinth* spp. Furthermore, single cells of *Daucus carota* from a suspension culture can regenerate normal whole plants which flower and produce fertile seed (see Chapter 12). Control of the occurrence of sporogenesis in specific cells is not, therefore, achieved by permanently removing the ability to undergo sporogenesis from all other cells but by a positive control determining that of all the cells capable of undergoing meiosis only a few shall do so. In view of this it might be questioned why meiosis does not occur sometimes in somatic cells of higher plants.

Meiosis is unique in always combining a reduction in nuclear DNA content to the 1C amount with recombination of DNA from different homologous chromosomes. Presumably the meiocyte is adapted to provide the optimum physical, physiological and biochemical environment for each of the component processes of meiosis to occur. It would be wrong to suppose, however, that the individual processes which together constitute normal meiosis can only occur in the meiocyte. Thus, somatic cell synapsis and recombination are known to occur in *Zea mays*, *Drosophila*, yeast and fungi (Herskowitz, 1969), and chiasmata have been claimed to occur between metaphase chromosomes in somatic cells in *Haplopappus gracilis* (Mitra and Steward, 1961). As far as the author is aware no example of the production of four cells with 1C DNA amounts is known other than in meiocytes. Providing the chromatid is uninene, then 1C type nuclei cannot be produced by subdivision of a 4C nucleus into four parts in two divisions lacking an intervening DNA synthesis stage. It could only occur if one division separated homologues (a reduction division) only, and the other separated chromatids, as in meiosis. The production of haploid somatic cells after the loss of *Hordeum bulbosum* chromosomes in embryos of hybrids made by crossing *H. vulgare* and *H. bulbosum* is known (Subrahmanyam and Kasha, 1973), but this is not a controlled reduction division; furthermore, it is not followed by a division involving chromatids alone without an intervening DNA synthesis. Thus, in higher plants the full sequence of meiotic behaviour resulting in the production of four 1C nuclei is not known to occur either naturally or as an induced response to any treatment in any cell type other than meiocytes in anthers or ovules. The reason for this may be that true meiosis can only occur after a whole series of pre-conditions have been met, in the correct sequence, spread over more than one premeiotic cell cycle. Such preconditions could conceivably include changes in the spatial arrangement of chromosomes, synthesis of a whole range of long-lived messenger RNAs, a special DNA synthesis, and changes in the physical environment of the cell. Presumably, therefore, the pre-conditions necessary for meiosis to occur are never met in somatic cells (Vasil, 1967; McDermott, 1971).

3. Developmental Holds during the Meiotic Cycle

In many species, even after meiocytes have been synchronized and have undergone premeiotic DNA synthesis, meiocyte development does not continue unchecked to the end of meiosis. Instead, development is blocked at some stage by some further requirement and only released when that requirement is met. In *Trillium erectum*, for instance, meiocytes become dormant after premeiotic DNA synthesis is complete in what is probably the G_2 stage of premeiotic interphase (Hotta and

Stern, 1963b). The dormancy is broken in the autumn by exposing to near freezing temperatures (less than 5°C). This break in dormancy is restricted to the reproductive tissues and meiosis is completed without further checks, taking about 90 days at 1°C. In the orchid *Dactylorhiza*, meiocytes in the ovule develop as far as leptotene. Without pollination, development is not resumed (Heslop-Harrison, 1957). In *Larix decidua* meiosis is initiated in the autumn and proceeds to diplotene, at which stage development is halted for several months. Here meiotic dormancy is broken by exposure to warm temperature (more than 5°C), in the spring, whereupon meiotic development proceeds to completion within 48 h (Ekberg and Erickson, 1967). It is seen that the same temperature treatment (exposure to temperatures below about 5°C) induces meiotic dormancy in one species, *L. decidua*, but breaks dormancy in another, *T. erectum*. Clearly, therefore, no great significance should be attached to the nature of the environmental "shock" required to induce or break meiotic dormancy other than to recognize that individual species are adapted so that stages of sporogenesis occur at the optimum time in the life cycle for plants growing in their normal environment. Thus, closely related species can differ widely in their behaviour. While *Trillium erectum* behaves as described above, *T. grandiflorum* undergoes meiosis during summer and has no developmental hold in G_2. Such variation in behaviour signifies adaptation of plants to their environments and involves optional controls of meiocyte behaviour which may be superimposed upon the mandatory controls of cell behaviour prerequisite for the initiation and completion of meiosis. Often variation in flowering time is consequent upon the adaptation of the life cycle of the organism as a whole to its normal environment, but sometimes it may indicate adaptation related to the peculiar needs of reproductive cell development *per se* (Section V; Bennett, 1972).

B. The Duration of Meiosis in Anthers

1. The Range of Meiotic Times in Higher Plants

The duration of meiosis has been accurately determined in less than 40 higher plant species. Reviews of the methods and techniques used for timing meiosis, and of the results obtained, have been made (Bennett, 1971, 1972). The duration of meiosis varies widely among higher plant species. Meiotic times for species completing meiosis without a developmental hold range from about 12 h in *Petunia* (Izhar and Frankel, 1973) to about 14 weeks in *Trillium erectum* (Hotta and Stern, 1963b). Of course, much longer meiotic times are known for species which have developmental holds at some stages during the division.

TABLE I. The duration of meiosis, nuclear DNA content, ploidy level and chromosome number in 37 species of higher plants

Taxon	Reference to meiotic timing	Duration of meiosis (h)	nuclear DNA content (pg)	chromosome number
DIPLOIDS				
1. *Petunia* hybrid	Izhar and Frankel, 1973	18	5·7	14
2. *Beta vulgaris*	Bennett, 1973	24	4·1	18
3. *Antirrhinum majus*	Ernst, 1938	24	5·5	16
4. *Haplopappus gracilis*	Marithamu and Threlkeld, 1966	24–36	5·5	4
5. *Vicia sativa*	This paper	24	8·2	12
6. *Lycopersicum esculentum*	Bennett, 1973	24–30	8·5	24
7. *Pisum sativum*	This paper	30	14·8	14
8. *Ornithogalum virens*	Church and Wimber, 1969	72	19·3	6
9. *Hordeum vulgare* cv. 'Sultan'	Bennett and Finch, 1971	39	20·3	14
H. vulgare cv. 'Ymer'	Finch and Bennett, 1972	39	20·3	14
10. *Triticum monococcum*	Bennett and Smith, 1972	42	21·0	14
11. *Rhoeo discolor*	Vasil, 1959	48	23·8	12
12. *Secale cereale*	Bennett *et al.*, 1971	51	28·4	14
13. *Vicia faba*	Maquardt, 1951	72	44·0	12
14. *Allium cepa*	Vasil, 1959	96	50·3	16
15. *Tradescantia paludosa*	Stenitz, 1944; Taylor, 1949, 1950; Beatty and Beatty, 1953	126	54·0	12
16. *Tulbaghia violacea*	Taylor, 1953	130	58·5	12
17. *Endymion nonscriptus*[a]	Wilson, 1959	48	69·9	16
18. *Convallaria majalis*[a]	Bennett, 1973	72	81·3	38
19. *Lilium henryi*	Pereira and Linskens, 1963	170	100·0	24
20. *Lilium longiflorum*	Taylor and McMaster, 1954; Ito and Stern, 1967	192	106·0	24
21. *Lilium candidum*	Sauerland, 1956	168	—	24
22. *Lilium* hybrid cv. 'Black Beauty'	This paper	264	—	24
23. *Trillium erectum*	Ito and Stern, 1967	274	120·0	10
24. *Fritillaria meleagris*	Barber, 1942	400	233·0	24
POLYPLOIDS				
25. *Lilium* hybrid cv. 'Sonata' (3x)	This paper	180	—	36
26. *Capsella bursa-pastoris*	(4x) Bennett, 1973	18	2·6	32
27. *Veronica chamaedrys*	(4x) Bennett, 1973	20	2·8	28
28. *Alliaria petiolata*	(4x) Bennett, 1973	24	7·1	36
29. *Triticum dicoccum*	(4x) Bennett and Smith, 1972	30	38·5	28
30. *Hordeum vulgare* cv. 'Ymer'	(4x) Finch and Bennett, 1972	31	40·6	28
31. *Triticale turgidum* var. *durum*	(4x) Bennett and Kaltsikes, 1974	31	37·9	28
32. *T. aestivum* × *S. cereale* (polyhaploid)	(4x) Bennett, 1973	35	41·8	28
33. *Secale cereale*	(4x) Bennett *et al.*, 1971	38	56·8	28
34. *Tradescantia reflexa*	(4x) Sax and Edmonds, 1933	144	144·9	24
35. *Triticum aestivum* cv. 'Chinese Spring'	(6x) Bennett *et al.*, 1971	24	54·3	42
36. *T. aestivum* cv. 'Holdfast'	(6x) Bennett *et al.*, 1972	24	54·3	42
Triticale cv. 'Rosner'	(6x) Bennett and Smith, 1972	35	66·3	42
37. *Triticale* genotype A	(8x) Bennett and Smith, 1972	21	82·7	56
genotype B	(8x) Bennett and Smith, 1972	22	82·7	56

The duration of meiosis was measured at 18° C in species 8, and at 20° C in species 2, 5, 6, 7, 9, 10, 12, 16, 17, 18, 20, 22, 25, 33, 35, 36 and 37. No temperature was given for species 3, 4, 11, 13, 14, 15, 19, 21 and 34. Species 23 was measured at 15°C; species 24 at 12–15°C; and species 1 at 15–17°C at night and 25–30°C during the day. The expected meiotic time at 20°C is given assuming a meiotic Q_{10} of 2·3. The values for species 8, 13 and 24 are only approximate although of the right order. Species marked [a] behave as tetraploids with respect to meiotic time (Bennett, 1973) and consequently they are not plotted as diploids in Fig. 2.

For instance, in *Larix decidua* meiosis can last for 4–5 months (Ekberg and Erickson, 1967). It seems probable that meiosis in higher plant species takes longer than in lower unicellular plants. Thus, while the shortest meiotic time for a higher plant is about 12 h (in *Petunia*) the duration of meiosis in *Saccharomyces cerevisiae* is only about 2 h (Simchen *et al.*, 1972).

Even when higher plant species are grown under constant environmental conditions a wide range of meiotic times have been recorded. Thus, estimates of meiotic time in higher plants grown at 20°C with continuous light range from about 18 h in *Capsella bursa-pastoris* to 8–10 days in *Lilium* genotypes (Table I). Because so few higher plants have been examined for meiotic time it seems certain that other species exist which complete meiosis either faster than *Capsella* or slower than *Lilium* when grown at 20°C with continuous light.

2. Factors affecting the Duration of Meiosis

Recent detailed studies of the rate and duration of meiotic development in higher plants has resulted in a greatly increased understanding of the factors which determine or affect meiotic time (Bennett, 1971, 1973; Bennett and Smith, 1972, 1973). Four main types of character are known which play major roles in determining meiotic time:

(*a*) *Nuclear DNA content.* A highly significant relationship ($P < 0.001$) between the duration of meiosis and nuclear DNA content exists in higher diploid plants (Bennett, 1971, 1973). This relationship is similar to the one between the duration of the cell cycle in root-tip meristem cells and nuclear DNA content in higher diploid plant species (Van't Hof and Sparrow, 1963; Evans and Rees, 1971; Chapter 3). Figure 2 shows that there is very little scatter of points from the regression line of DNA amount on meiotic time in diploids. Consequently it is clear that nuclear DNA content plays a major and precise role in determining meiotic time ($r = 0.98$).

(*b*) *Ploidy level.* When only tetraploid species were compared a highly significant ($P < 0.001$) positive relationship between nuclear DNA content and meiotic time was found (Bennett, 1973), similar to that for diploids. However, the slope of the regression line of nuclear DNA content on meiotic time for tetraploids ($b = 1.07 \pm 0.13$) was almost twice that calculated for diploids ($b = 0.50 \pm 0.03$). It was concluded that there is probably a positive correlation between nuclear DNA content and meiotic duration for species at each ploidy level, but that the slope of the regression line for nuclear DNA content on meiotic time is greater at successively increased ploidy levels.

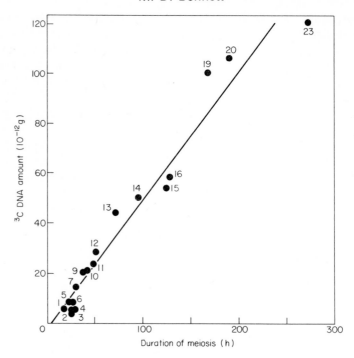

Fig. 2. The relationship between nuclear DNA content and the duration of meiosis in 18 diploid higher plant species (data plotted from Table I).

Key to points:

1. *Petunia* hybrid	11. *Rhoeo discolor*
2. *Beta vulgaris*	12. *Secale cereale*
3. *Antirrhinum majus*	13. *Vicia faba*
4. *Haplopappus gracilis*	14. *Allium cepa*
5. *Vicia sativa*	15. *Tradescantia paludosa*
6. *Lycopersicum esculentum*	16. *Tulbaghia violacea*
7. *Pisum sativum*	19. *Lilium henryi*
9. *Hordeum vulgare* cv. 'Sultan'	20. *Lilium longiflorum*
10. *Triticum monococcum*	23. *Trillium erectum*

Table I shows that the duration of meiosis in a polyploid species is much shorter than in a diploid species with an equivalent nuclear DNA content. Thus the duration of meiosis in hexploid *Triticum aestivum* (24 h) which has a 4C DNA content of 72 pg is much shorter than in diploid *Allium cepa* (96 h) with a 4C DNA content of 67 pg. Moreover, meiosis in polyploid species is usually shorter than in the diploid species from which the polyploid is derived. This has been demonstrated in autopolyploids and allopolyploids and also in a polyhaploid (Table I). For instance, meiosis lasts about 51 h in diploid *Secale cereale* but only 38 h in autotetraploid *S. cereale*. Similarly, meiosis lasts about 39 h and 31 h in diploid and tetraploid *Hordeum vulgare* cv. 'Ymer'. In the allopolyploid wheat series meiosis lasted 42 h in diploid *Triticum monococcum*; 30 h and 31 h in the tetraploid *T. dicoccum* and *T. turgidum* var. *durum*; and 24 h in hexaploid *T. aestivum*.

Faster meiotic development in polyploids occurs despite their increased nuclear DNA content compared with parent diploids. The reason for this behaviour remains unclear but must be associated with the changes in nuclear organization which accompany polyploidization. For instance, polyploids contain more gene doses per nucleus and more nucleolar organizers than do parent diploids. Increasing the number of genomes per nucleus not only results in an increased developmental rate during meiosis but also has a similar effect on pollen maturation time (Bennett and Smith, 1972).

(*c*) *Genotypic factors.* Experiments to screen for variation in meiotic time have been carried out using the wide range of genetic variation available in *Triticum aestivum* (Bennett and Smith, 1973). Meiosis was timed in nullisomic, tetrasomic, ditelocentric, substitution, addition and mutant lines of the variety 'Chinese Spring'. It was found that individual chromosomes differ in their effects on the rate of meiotic development. Thus, the absence of chromosome 5B had a large effect on slowing down development while the absence of chromosome 7B had no detectable effect. It was shown that the gene(s) responsible for the effect on meiotic time on chromosome 5B are located on the short arm; however, the mechanism responsible for the effect is unknown. It was noted that while the removal of chromosome 5B resulted in a decreased rate of meiotic development, the addition of extra "doses" of this chromosome did not produce any detectable increased rate of meiotic development above that found in euploid plants. These results might be expected since the removal of any chromosome-bearing genes controlling any meiotic event would upset the balance of the delicate meiotic control mechanism and result in a less efficient meiotic division. At the same time it might be expected that proportional changes in the gene dosage involving many or all of the chromosomes controlling meiosis would be required before an increased efficiency in meiotic behaviour would be found. Polyploids have undergone such a change involving every gene on every chromosome so that meiosis is completed more rapidly.

As far as the author is aware only one report of a mutant affecting the duration of meiosis in a higher plant has been published. Klein (1972) noted that two *Pisum sativum* mutants affected the duration on meiosis. He reported that first prophase of meiosis was prolonged and also that meiosis was less synchronous between PMCs in mutant plants than in normal plants. It remains to be seen whether mutants affecting the duration of meiosis are almost unknown because of their rarity or because insufficient screening for them has been carried out.

(*d*) *Environmental factors.* The environmental factor whose effects on the duration of meiosis are most widely known is temperature (Bennett, 1973; Wilson, 1959). Like the duration of mitosis and the somatic cell cycle (Brown, 1951; Burholt and Van't Hof, 1971) meiosis is very temperature sensitive. The most detailed investigation of the effect of temperature on meiotic time in a higher plant species was conducted by Wilson (1959) using *Endymion nonscriptus* grown at each 5°C interval over the temperature range 0–30°C. Over this range the duration of meiosis decreased with increasing temperature from 36 days at 0°C to only 0·83 days at 30°C. In general it appears that the duration of meiosis is inversely related to temperature for the range of temperatures normally encountered by each species of higher plant growing in its normal environment at the time when meiosis usually occurs. For instance, the duration of meiosis in *Trillium erectum* is about 90 days at 1°C (Hotta and Stern, 1963b), 70 days at 2°C (Kemp, 1964), 40 days at 3–5°C (Hotta and Stern, 1963a) and about 16 days at 15°C (Ito and Stern, 1967). In *E. nonscriptus* the Q_{10} for meiosis decreased with each increase in the mean temperature of the ten degree interval being compared: thus, for the range 0–10°C it was 5·2, for the range 10–20°C it was 3·5, and for the range 20–30°C it was only 2·4. It also appears that the Q_{10} for a constant ten degree interval is very similar in different higher plant species. Thus, for the interval 15–25°C the Q_{10} was about 2·4 in *Triticum aestivum* and *Secale cereale*, and 2·8 in *E. nonscriptus* (Bennett *et al.*, 1972; Wilson, 1959). In *T. aestivum* the Q_{10} for meiosis (2·4) over the range 15–25°C is greater than the Q_{10} for the somatic cell cycle (1·7). However, the Q_{10} values for meiosis listed above are mostly within the range of Q_{10} values for the somatic cell cycle quoted for higher plant species, namely 2·0 for *Pisum sativum* (Brown, 1951) to 3·0 in *Helianthus* (Burholt and Van't Hof, 1971).

The other factor about which some understanding has been gained is the molecular environment of the meiocyte necessary for meiotic development. There is one report that the growth factor kinetin can speed up late meiotic development in cultured anthers of *Tradescantia paludosa*

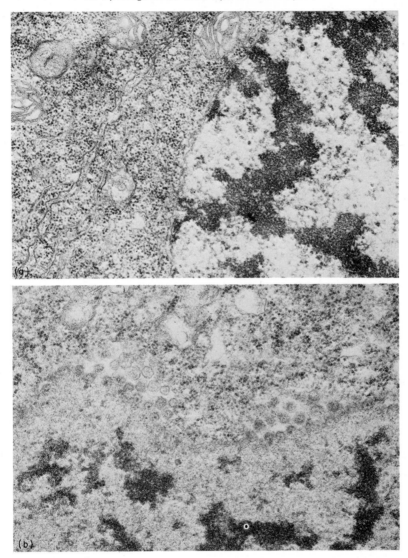

FIG. 3. Electron micrographs of PMCs in *Triticum aestivum* (× 35 900) at late premeiotic interphase showing (**a**) extremely large numbers of ribosomes in the cytoplasm (left-hand side); and (**b**) numerous pores in the nuclear membrane.

(Walker and Dietrich, 1961). *In vivo*, all the basic molecules necessary to sustain the growth and development of the meiocytes enter through the tapetum. Consequently it might be expected that any environmental factor acting to limit the supply of essential molecules, especially water and sugars, would affect the rate of meiotic development. It appears that meiocytes accumulate many necessary macromolecules prior to the initiation of meiosis or synthesize them from simple molecules during meiosis itself. Observations of meiocytes late in premeiotic interphase show that they are particularly well endowed with ribosomes (Fig. 3a). Ito and Stern (1967) showed that once meiosis had been initiated then *Lilium* meiocytes could be cultured through meiosis from leptotene to tetrad stage at the normal rate using only simple molecules in the culture solution. It is possible to culture *Triticum aestivum* anthers from just after DNA synthesis to tetrad stage using distilled water alone as a culture medium (Bennett, unpublished data). Observations of cultured anthers show that often development proceeds at the normal rate and then stops completely rather than slowing first and then stopping. Experiments using cultured anthers and anther sampling methods all show that dehydration of anthers causes almost immediate cessation of meiotic development. It seems reasonable to conclude that *in vivo* the meiocytes are protected as much as possible from stresses which would lead to meiotic failure. This seems true of EMCs even more than of PMCs in view of their more vital role in ensuring the survival of the species.

3. A Comparison of Meiocytes with Other Cells

When considered together the physical characters of meiocytes are strikingly different from those of other cell types. Meiocytes are very large compared with other diploid meristematic cells in most higher plants yet they usually contain little or no vacuolation (Fig. 1a and b). Furthermore, meiocyte nuclei are frequently much larger than nuclei in other meristematic cells (Fig. 1a). By leptotene the nuclear membrane in meiocytes contains very many more pores (Fig. 3b) than are seen in premeiotic archesporial cell nuclei or in the nuclei of most other meristematic cell nuclei. The PMCs in anthers come to form a syncytium during meiotic prophase, being linked by massive cytoplasmic connections which may form up to 20% of the interface between adjacent cells (Heslop-Harrison, 1972). Later these connections are broken completely and each meiocyte becomes isolated by a thick callose wall.

Despite the very large amount of cytoplasm in meiocytes their cytoplasm at the start of meiosis is extremely full of various organelles including ribosomes, and the cells have a high total protein and RNA content compared with other meristematic cells. In *Lilium* the PMC develops for 4–5 days during early prophase until pachytene stage before there is any detectable synthesis of ribosomal RNA in the nucleolus (Stern and Hotta, 1969). During this period the number of ribosomes per unit volume of cytoplasm steadily decreases (McKenzie *et al.*, 1967), and the kind of RNA that is synthesized differs from that synthesized at other stages: it may be mRNA (Hotta and Stern, 1963b).

The rate of cell development in meiocytes is always much slower than that found in other active meristematic cells (Bennett and Smith, 1972;

TABLE II. The duration (in hours) of the minimum cell cycle time in root-tip meristem cells and the minimum duration of the cell cycle from premeiotic mitosis to first telophase of meiosis in four higher plant species

Taxon	Cell type	
	Root-tip meristem cells	Meiocytes in anthers
Haplopappus gracilis	10·5 (Sparvoli *et al.*, 1966)	50 (Marithamu and Threlkeld, 1966)
Triticum aestivum	12·5 (Bayliss, 1974)	72 (Bennett *et al.*, 1973b)
Secale cereale	12·8 (Ayonoadu and Rees, 1968)	93 (Bennett *et al.*, 1971)
Lilium longiflorum	24·0 (Stern and Hotta, 1969)	288 (Erickson, 1948)

Bennett, 1973). The minimum duration of the cell cycle which commences after premeiotic mitosis and ends at first telophase of meiosis is always much longer than the minimum cell cycle time in root-tip cells of the same species at the same temperature (Table II). For instance, in *Haplopappus gracilis* development from premeiotic mitosis until first telophase of meiosis lasted about 50 h (Marithamu and Threlkeld, 1966) while the root-tip meristem cell cycle time was only 10·5 h (Sparvoli *et al.*, 1966). Slow development occurs in meiocytes at all stages of the cell cycle. In *Triticum aestivum* the duration of metaphase to telophase inclusive in root-tip meristem cells was about 35 min while in meiosis the duration of first metaphase to first telophase inclusive was about 160 min.

4. The Control of Meiosis

The existence of developmental gradients both along and across loculi in anthers in relation to the insertion point of the filament shows that there is an effect of the supply of some molecule(s) on the timing of meiocyte development (Carlson and Stuart, 1936; Gates and Rees, 1921; Rees and Naylor, 1960) and tapetal cell development. Such gradients might be produced in response to the supply of a stimulus which causes initiation of meiosis or, more probably, since the gradients affect the tapetal cells which do not undergo meiosis, they may represent the result of competition between cells for limited supplies of all the molecules required for cell development in the tapetum and sporogenous tissues.

It was once widely believed that the tapetal cells controlled the initiation of successive stages of meiotic development in PMCs and it was suggested that they might do this by producing macromolecules required by the PMCs at various stages (Cooper, 1952; Pereira and Linskens, 1963; Vasil, 1967). It now seems certain, however, that once a meiotic sequence has been initiated then the control of meiotic development is autonomous, certainly to syncytial groups of PMCs and probably to individual cells. This was demonstrated by Ito and Stern (1967) when they showed that filaments of *Lilium* and *Trillium* meiocytes extruded into a culture solution containing only simple molecules at leptotene can complete normal meiotic development to tetrad stage. The function of the tapetum must, therefore, be limited to providing an environment conducive to meiocyte development which includes providing simple molecules alone during meiosis. The autonomous control of meiocyte development once meiosis has begun is further illustrated by the fact that in cases where synchrony is destroyed, or where asynchrony is normal (e.g. in *Nicotiana* sp.: Burns, 1972), each individual meiocyte can still complete meiosis normally. Thus, although synchrony is the normal situation within a loculus it is not essential for normal meiocyte development.

Comparison of the durations of individual stages of meiosis in a wide range of higher plant species has shown that in general each meiotic stage occupies a constant proportion of the total time spent in meiosis. This holds true both for diploid and polyploid species; established species and newly synthesized genotypes; and for individual species grown at different temperatures (Bennett, 1971; Bennett and Smith, 1972; Bennett *et al.*, 1972). Such constancy during meiosis is striking compared with results obtained for other cell types (Webster and Davidson, 1968; Clowes, 1965b, 1970), at other stages of development. It has been concluded that the control of meiotic development appears to differ from the control of cell development seen elsewhere throughout most of the vegetative stages of the life cycle (Bennett, 1973). Premeiotic and meiotic cell behaviour indicates that development at these stages is highly canalized and lacks the normal feedback mechanisms which ensure the normal completion of one stage before a later stage is initiated (Fig. 4). Thus, it was noted (Riley and Bennett, 1971) that meiosis is a process which usually progresses to its termination irrespective of abnormalities of chromosomal or cell behaviour that may arise. For instance, cells in which chromosomes fail to pair at zygotene proceed to dyad and tetrad stages at the normal rate even though the failure of pairing inevitably means that sterile pollen will be produced. Similarly, if cross-walls are not formed to isolate the nuclei produced at one or both meiotic divisions and spores with unusual nuclear contents are produced these will still form a pollen wall and germ pore despite the previous failure of cytokinesis (Bennett et al., 1972; Finch and Bennett, 1972). The above behaviour together with that of cultured *Trillium* meiocytes (Section IIB3) suggests that meiocyte development is subject to pre-determined control, the operation of which continues without deviation irrespective of whether the development already completed was normal.

C. Meiosis in Embryo Sac Mother Cells

1. Time and Duration of Female Meiosis

The timing of meiosis in PMCs and EMCs within the floret varies both within and between monomorphic hermaphrodite higher plant species. In some species, for instance *Triticum aestivum*, *Hordeum vulgare* and *Secale cereale*, meiosis commences on average synchronously in PMCs and the EMC within individual florets (Bennett et al., 1973a; and Bennett and Smith, unpublished data). In other species, for instance *Lilium*, *Fritillaria* and *Tulipa* (Darlington and La Cour, 1941), meiosis in the embryo sac occurs after meiosis in PMCs, and may take place as long as 2–3 weeks later.

In some species the duration of meiosis in PMCs and EMCs is the same or nearly so, for instance (see Table III) in *Hordeum vulgare* and *Triticum aestivum* (Bennett et al., 1973a). In other species, however, the durations of meiosis in EMCs and PMCs differs. For instance, in *Secale cereale* where meiosis appears to commence almost synchronously in the two types of meiocyte, meiosis in the EMC lasts more than twice as long as meiosis in PMCs. In *Lilium* hybrid cv. 'Sonata', meiosis in the EMC begins at about the initiation of tetrad stage in anthers of the same floret and lasts about 10·5 days whereas in PMCs it only lasts about 7·5 days (Bennett and Stern, unpublished).

Limited though these data are it is possible to draw several conclusions regarding the time and duration of female meiosis in higher

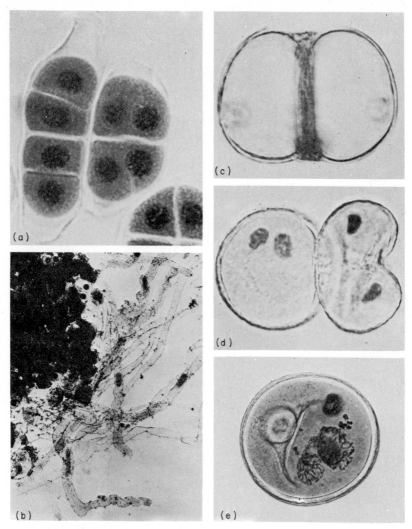

Fig. 4. Abnormal microspore development illustrating the lack of feedback mechanisms in the control of meiotic and pollen development. (a) Abnormal linear tetrad and normal iso-bilateral tetrad from *Hordeum vulgare* anther. The abnormal tetrad has formed a callose wall despite the abnormal meiotic division which it has undergone. (b) *Trillium erectum* meiocytes producing pollen tubes after a prolonged period in culture despite having not undergone meiotic or microspore development (see Section 11B3). (c) Pollen wall and germ pore formed around *Secale cereale* meiocytes in which meiosis was suspended at dyad stage after a 30°C heat shock. (d) Treatment as c except meiosis was completed but cross-walls were not formed in one cell after second telophase and the "spores" failed to separate. (e) Abnormal pollen development in *Triticum aestivum* treated with Ethrel before meiosis. Mitosis has been induced in the normally non-dividing vegetative nucleus.

(Fig. 4b is produced by permission of Professor H. Stern.)

plants. First, it has already been noted that DNA content and ploidy level, which are known to affect meiotic time in PMCs, also affect the duration of meiosis in EMCs (Bennett *et al.*, 1973a). Second, whether meiosis is synchronous in PMCs and the EMC or not appears to be related in part at least to the breeding system of the species. It is presumably important in in-breeding species (particularly in species in which

TABLE III. The duration (in hours) of meiosis in PMCs and the EMC in five higher plant species together with their type of breeding system and the timing of meiosis in PMCs and the EMC within the average floret, whether synchronous or asynchronous

Taxon	Duration of meiosis		Breeding system	Timing of meiosis in EMC and PMCs
	in EMCs	in PMCs		
Hordeum vulgare	c. 39	39	In-breeding	Synchronous
Secale cereale	c. 124	51	Out-breeding	Partly asynchronous
Triticum aestivum	c. 24	24	In-breeding	Synchronous
Tradescantia paludosa	c. 188	113	Out-breeding	Partly asynchronous
Lilium hybrid cv. 'Black Beauty'	c. 252	180	Out-breeding	Grossly asynchronous
Lilium hybrid cv. 'Sonata'	c. 396	252	Out-breeding	Grossly asynchronous

fertilization often involves fertilization of the egg with sperm nuclei from pollen produced in the same floret, as in *Hordeum vulgare*) that development of the male and female meiocytes and gametophytes should occur synchronously, thus facilitating self-pollination. At the same time, it would appear advantageous for out-breeding species to have male and female maturity occurring at different times within a floret, thus ensuring out-pollination. Table III shows that meiosis tends to occur asynchronously in PMCs and EMCs, and tends to have a longer duration in EMCs than PMCs in out-breeding species but not in in-breeding species.

2. A Comparison of EMCs with PMCs

In many essential respects meiosis in EMCs does not differ from meiosis in PMCs insofar as the sequence, behaviour and consequences of the various stages are concerned. Nevertheless, several important differences between meiosis in EMCs and PMCs have been noted. First, the synchrony and duration of meiosis can vary between EMCs and PMCs within a single floret as described above. Second, the physical characters of the two types of cell in which meiosis occurs may differ. For instance, in *Triticum aestivum* and *Hordeum vulgare* the shape of the PMC is sub-spherical while the EMC is elongated and pyriform (Bennett *et al.*, 1973a). In these two species the volume of the PMC and EMC at each meiotic stage is essentially similar; however, in *Lilium* hybrid cv. 'Sonata' the volume of the EMC at first metaphase (about $15 \times 10^6 \ \mu m^3$) is about 20 times that of the PMC at the same stage of meiosis (about $0.8 \times 10^6 \ \mu m^3$) (Bennett and Stern, unpublished data). Not only does meiocyte size vary in *Lilium* but so do nuclear and chromosome size. Thus, the degree of chromosome contraction is often much less in EMCs than in PMCs. For instance, Fogwill (1958) reported that in *Fritillaria* the size of first metaphase chromosomes was much larger in EMCs than in PMCs. This observation has been repeated in *Lilium* by the author, who also noted that first and second anaphase chromosomes were less contracted in EMCs than in PMCs. Fogwill (1958) also reported that the mean chiasma frequency per cell was much higher in EMCs of *Fritillaria* and *Lilium* than in their respective PMCs. This phenomenon has also been noted in several *Tulbaghia* species and in the orchid *Listera ovata* (Vosa, 1972; Vosa and Barlow, 1972). In *Hordeum vulgare* the mean chiasma frequency in EMCs and PMCs was not significantly different (Bennett *et al.*, 1973a). Fogwill (1958) suggested that chromosome size and chiasma frequency may be causally correlated. The evidence available so far is not inconsistent with this hypothesis. Another difference sometimes found between meiosis in PMCs and EMCs in the same species is in the type of tetrad produced. For instance, in *Triticum aestivum* the product of meiosis in PMCs is an iso-bilateral tetrad while the product of meiosis in the EMC is a T-tetrad.

Despite the differences listed above it is the similarities between PMCs and EMCs that are most striking. For instance, in *Lilium* hybrid cv. 'Black Beauty' there is a pronounced premeiotic contraction stage in PMCs in G_2 of premeiotic interphase. This stage is also found in the EMC even though there is only a single EMC per ovule, and even though premeiotic contraction stage in the EMC occurs about 10 days after the start of meiosis in PMCs in the same floret. Similarly, in *Triticum aestivum* which also has a single EMC per floret, the EMC forms a callose wall around itself during meiosis exactly as the PMCs do.

IV. POST-MEIOTIC DEVELOPMENT

A. Pollen Development

1. The Time and Duration of Pollen Development

The duration of pollen development varies greatly among higher plant species and ranges from a few days to several months. For instance, pollen development from the end of meiosis until anther dehiscence took about 7·5 days in *Triticum aestivum* grown at 20°C and about 16 days in *Secale cereale* at the same temperature. In *Trillium erectum* grown at about 5°C pollen development lasted at least 54 days (Sparrow and Sparrow, 1949) and in *Larix decidua* it took at least 57 days (Ekberg and Erickson, 1967), also mainly at cool temperatures.

A detailed study has been made of pollen development in *Triticum aestivum* grown at 20°C (Bennett *et al.*, 1973b). In this species, the tetrad stage lasts about 10 h and the first pollen grain mitosis occurs about 60 h after the end of meiosis. Second pollen grain mitosis takes place after a further 60 h period and anther dehiscence 60 h after that.

In *Tradescantia paludosa* several detailed studies of pollen development have been made (Beatty and Beatty, 1953; Bryan, 1951). In this species the tetrad stage persists for about 24 h. Pollen grain mitosis occurs about 6 days after the end of meiosis and lasts about 36 h. Anther dehiscence occurs about a further 7 days after the pollen grain mitosis. In *Lilium longiflorum* pollen development takes about 27 days (Erickson, 1948). Pollen grain mitosis occurs about 12 days after the end of meiosis and anther dehiscence about 13 days after that. In *Lilium* hybrid cv. 'Sonata' grown at 20°C pollen development took 15 days, and pollen grain mitosis (Fig. 1d) occurred about 8 days after the end of meiosis (Bennett and Stern, unpublished).

Like meiosis, pollen development is very temperature sensitive and has a similar Q_{10}. For instance, in *Triticum aestivum* grown at 15, 20 and 25°C the duration of pollen development from the end of meiosis until anther dehiscence was 5·8, 7·5 and 13·5 days respectively. Thus, over the temperature interval 15–25°C the Q_{10} for spore development was about 2·3. In *Secale cereale* the Q_{10} for the same temperature interval was about 2·2 (Bennett *et al.*, 1972). Some species (e.g. *Trillium erectum* and *Larix decidua*) normally complete pollen maturation at low temperatures. Because of the relationship between pollen development time and temperature such species always have long pollen maturation times.

Comparisons of several cereal genotypes have shown that the duration of pollen maturation shows a relatively constant relationship to the duration of meiosis in plants grown at 20°C (Bennett and Smith, 1972), and that this relationship is not altered in *T. aestivum* and *Secale cereale* by temperatures in the range 15–25°C (Bennett *et al.*, 1972). Furthermore, nuclear DNA content and ploidy level, both of which affect meiotic duration, also affect pollen maturation time in a similar manner (Bennett and Smith, 1972). Thus, pollen maturation time in diploid *Hordeum vulgare* (10·5 days) with a 4C DNA content of 27 pg

TABLE IV. A comparison of the duration (in hours) of the minimum cell cycle time in root-tip meristem cells with the duration of the cell cycle in developing microspores from second telophase of meiosis until the end of first pollen grain mitosis in six higher plant species

Taxon	Root-tip cell cycle time	Pollen cell cycle time
Hordeum vulgare	12·0 (Bennett and Finch, 1971)	108 (Finch and Bennett, unpublished)
Triticale cv. 'Rosner'	12·2 (Kaltsikes, 1972)	96 (Bennett and Smith, 1972)
Triticum aestivum	12·5 (Bayliss, unpublished)	60 (Bennett and Smith, 1972)
Secale cereale	12·7 Ayonoadu and Rees, 1968)	144 Bennett and Smith, 1972)
Tradescantia paludosa	21·0 (Wimber, 1966)	144 (Bennett and Smith, 1972)
Lilium longiflorum	24·0 (Stern and Hotta, 1969)	288 (Erickson, 1948)

is shorter than in diploid *Secale cereale* (16 days) with a 4C DNA value of 37·8 pg. The duration of pollen maturation in diploid *Triticum monococcum* (12 days), tetraploid *T. dicoccum* (9·3 days) and *T. turgidum* var. *durum* (9·5 days), and hexaploid *T. aestivum* (7·5 days) decreases with increasing ploidy level despite the increase in nuclear DNA content (see Section IVB2(b)).

The rate of cell development during microspore development is slow compared with that of other meristematic cells in the same species (Table IV). Thus, the period from second telophase of meiosis until first pollen grain mitosis, which may be considered as one cell cycle, lasted about 60 h in *Triticum aestivum* at 20°C compared with a somatic cell cycle of 12·5 h. In *Tradescantia paludosa* development from second telophase of meiosis to pollen grain mitosis took 168 h (Walker and Dietrich, 1961), but the cell cycle time in root-tip meristem cells only lasted 21 h (Wimber, 1966) at 21°C. Furthermore, the duration of mitosis in the developing *T. paludosa* microspore was estimated by Beatty and Beatty (1953) to last about 36 h and the inclusive durations of metaphase to telophase to be 1·6 h. The corresponding durations for root-tip cells at 21°C in *T. paludosa* were 1·7 h for mitosis and 0·4 h for metaphase to telophase inclusive (Wimber, 1966). Thus, the period of slow development, which, it was noted, begins during premeiotic development of archesporial cells and is found throughout meiosis, does not end with the conclusion of microsporogenesis but continues throughout microspore development in higher plants.

The reasons for slow development during microsporogenesis are unclear. One reason may be the haploid nature of the microspore nuclei. It is well known that haploid cell cultures often grow more slowly than do related diploid cell cultures in the same growth medium. The lower gene dosage in the haploid microspore nucleus compared with diploid nuclei is coupled in many species with a very large cell volume in developing microspores compared with most diploid cells. One thing is clear, rapid development at some stages can be induced by suitable treatments. Thus, Walker and Dietrich (1961) showed that kinetin given to cultured *Tradescantia paludosa* tetrads induces an immediate division in tetrad nuclei about 7 days before this would normally occur in developing microspores. It was noted (Bennett *et al.*, 1973b) that while cell cycle times for haploid nuclei in the microspore of *Triticum aestivum* were about 60 h long, in the embryo sac they were about 12 h long in some haploid nuclei. Clearly, therefore, slow development in the microspore is genotypically controlled and presumably has some adaptive significance. In in-breeding species (like *T. aestivum*), where meiosis occurs synchronously in PMCs and the EMC in each floret, slow microspore development would presumably facilitate synchrony of the time of anther maturity with that of ovule maturity in the same floret since embryo sac development involves more nuclear and cell cycles than does microspore development. This possible explanation is clearly not the whole truth concerning slow microspore development, since out-breeding species in which meiosis in PMCs is complete before meiosis in EMCs is initiated (such a *Lilium*) also show slow microspore development (Table IV). It seems possible, therefore, that slow development is an inherent property of the near synchronous development of numerous haploid cells within an anther.

2. Pollen Size

The volume of microspores increases markedly during microspore development. In *T. aestivum* it increases from about $1 \times 10^4 \mu m^3$ just after the break-up of tetrads, to about $11 \times 10^4 \mu m^3$ at the time of anther dehiscence. The increase in pollen volume was linear over the 7·5 day period between these two stages (Bennett *et al.*, 1973b). Within a species pollen size is influenced by the treatment given to the plant, in particular by the relative availability of substances essential for microspore development at the time of flowering (Natrovka, 1968). Thus, observations on four spring *Hordeum vulgare* varieties showed that early-sown plants had larger pollen grains than late-sown plants; also that pollen grains from the centre of the main tillers were larger than those from second and third tillers.

Pollen volume at maturity is a character widely adapted to the type of pollination found in individual species. Nevertheless, when several related species are compared a pronounced effect of species nuclear DNA value on pollen size is sometimes found. For instance, a highly significant ($P < 0.001$) relationship between pollen volume and 4C nuclear DNA content has been demonstrated for species in the Gramineae, and for species in the genera *Ranunculus* and *Vicia* (Bennett, 1972; Bennett, 1974). The correlation coefficient for 16 species in the Gramineae was 0·98. Clearly, therefore, nuclear DNA content can play a major role in determining pollen size in related species.

B. Embryo Sac Development

Very little is known concerning the duration of embryo sac development and the durations of the nuclear and cell cycles in its constituent stages. Certainly, the total duration of development in the female gametophyte varies considerably within and between species. In *Arabidopsis thaliana* female gametophyte development takes only a very few days, in *Lilium longiflorum* it took about 7–8 days (Erickson, 1948) and in *Secale cereale* it takes about 18 days at 20°C. It is logical to assume that the rate of embryo sac development is subject to some controls of the type shown to affect meiotic and microspore development. However, in species where each ovule contains only a single EMC and later only a single embryo sac, their development is more vital to the survival of the species than the development of any individual pollen grain. Consequently, special developmental behaviour designed to protect or promote embryo sac cell development not seen in microspore development might be expected.

In *Lilium longiflorum* the embryo sac containing the four haploid nuclei produced by meiosis persists for about 7 days when two conjugate mitoses occur. One mitosis involves just one haploid nucleus, and the other involves the formation of a common spindle for all the chromosomes of the other three haploid nuclei produced at meiosis. The result is a second 4-nucleate embryo sac containing two haploid and two triploid nuclei. This stage then persists for a further 6 days until the time of anthesis when each nucleus divides to form the 8-nucleate embryo sac (Erickson 1948).

In *Triticum aestivum*, embryo sac development is of the polygonum type (Maheshwari, 1950). Here only one haploid spore produced by meiosis in the EMC develops into an embryo sac while the other three nuclei degenerate. A detailed description of the timing of cell behaviour in the embryo sac of wheat grown at 20°C has been made (Bennett et al., 1973b). The functional megaspore does not divide until about 50 h after the end of meiosis in the EMC. After this time three successive synchronous mitoses occur within about 20 h producing an embryo sac containing eight haploid nuclei. The micropolar quartet differentiates into a 3-celled egg apparatus (consisting of an egg cell and two synergid cells) and the upper polar nucleus. The chalazal quartet differentiates into the lower polar nucleus and after repeated mitotic divisions of the remaining three nuclei a group of 20–30 antipodal cells which become highly polytene by the time of anthesis and pollination. Production of the 8-nucleate embryo sac occurs synchronous with the end of first pollen grain mitosis. Thereafter the remaining development of the female gametophyte to maturity takes about 86 h. At maturity the antipodal cell nuclei have between 64 and 256 times the haploid nuclear DNA amount. The volume of the largest antipodal cell nucleus is larger than that of a mature pollen grain, and the volume of the nucleolus equivalent to several diploid somatic cells. The function of the antipodal cells is not clearly understood, but they are thought to have a secretory role.

V. SPOROGENESIS AND THE LIFE CYCLE

Higher plants display a wide range of reproductive cell behaviour with respect to several aspects of the time of sporogenesis and spore development: (a) the durations of meiosis and gametophyte development; (b) how soon after germination these processes are first initiated; (c) the stage of life cycle and the prevailing environmental conditions under which meiosis and spore development are normally undertaken. The timing of meiotic and gametophytic development in individual genotypes is highly adaptive. It is interesting to consider, therefore, what determines the various types of reproductive development and what are the constraints that limit development, especially those acting at the cellular level.

It may be assumed that, in each genotype grown in its normal environment, initiation of sporogenesis usually occurs so that meiosis and spore development take place either under the most suitable environmental conditions or at the most suitable stage in the life history of the plant. In many species the initiation of flowering is correlated with the time of suitable environmental conditions for reproductive cell development by a requirement commensurate with the onset of such conditions. For instance, many annual temperate species have a long-day requirement for flowering which ensures that reproductive development occurs during summer when the substances necessary to sustain flowering are in plentiful supply.

There seem to be few constraints on flowering, meiosis, and spore development which also elongate the minimum generation time (MGT) of the plant. However, there are several constraints which limit how soon these events may occur after germination. In monocarpic herbaceous species the initiation of flowering primordia must be delayed until sufficient vegetative primordia have been made (Bennett, 1974). It is essential that meiotic and spore development should not occur before the vegetative tissues or the reserve food supply is sufficient to sustain these processes. Nucleotypic characters, such as nuclear DNA content, dictate limits to the maximum rate of development during meiosis, spore development and other stages in the life cycle (Bennett, 1971, 1972, 1974). The existence of a positive correlation between nuclear DNA content and the duration of meiosis and other reproductive processes allows several definite conclusions regarding limits to plant behaviour dictated by the DNA content of meiocytes. For instance, the minimum generation time (MGT) in *Arabidopsis thaliana* is 31 days at 23°C while in *Fritillaria meleagris* meiosis and microspore development at the same temperature together require a period not less than the MGT of *A. thaliana* (Barber, 1942). Plants which take a prolonged time completing meiosis at about 20°C do so because they have a high 4C nuclear DNA content. Consequently, it is concluded that no plant with a very high nuclear DNA content can behave like *A. thaliana* and have a very short MGT, that is, it cannot be an ephemeral. It is also concluded that all ephemeral species must have low 4C nuclear DNA amounts. Comparisons of the life cycle type and nuclear DNA values in higher plants have shown that these predictions are obeyed. For instance, in ten ephemeral species the highest DNA value was 10·1 pg and the mean 4·2 pg while DNA values for 261 non-ephemeral species ranged up to more than 200 pg with a mean of 55·9 pg (Bennett, 1972).

Many ephemeral species must complete their life cycle within a very few weeks if they are to survive. Effects of nuclear DNA content on the duration of meiotic and gametophytic development are important because any increase in the duration of these essential determinate stages of the life cycle will automatically increase the MGT by an equal amount. It has been noted that the effects of nuclear DNA content on the duration of developmental processes are best seen during phases of determinate development and are magnified if development is slow for any reason, for instance if it occurs at low temperature (Bennett, 1972). Sporogenesis and spore development from the start of archesporial cell development in the anther until the end of gametophyte development are all highly determinate and include stages when development is very slow compared with elsewhere. Consequently, the absolute duration of the sum of these stages plays a major part in determining minimum generation time and the relative proportion of the MGT which they occupy becomes increasingly important as MGT decreases. It appears that, not only does the switch to flowering occur relatively much sooner in ephemerals than in species with longer MGTs, but that the relative proportion of the total life cycle spent in reproductive development after floral initiation is much greater in ephemerals than in species with longer MGTs. For instance, in the ephemeral *Chenopodium rubrum* flower rudiments may be formed only 6 days after germination and in 36 days flowers appear. In *Hordeum vulgare* which has an MGT of about 70 days the switch to flowering occurs about 25 days after germination, while in *Fritillaria aurea* flowering does not commence until the end of the fourth year of the life cycle and the MGT is 4·5 years. The proportion of the MGT taken by reproductive development decreases with increasing MGT from about 80% in some ephemerals to only about 15% in long-lived herbaceous perennials.

Recently the existence of a positive correlation between nuclear DNA content and minimum generation time was demonstrated for higher herbaceous plants (Bennett, 1972). This relationship has important consequences for plant size, development and morphogenesis. It was shown that a short MGT defines a low nuclear DNA content, small cell size, and rapid development at every stage of the life cycle. Thus, ephemeral species are small in total plant size, and have small PMCs, pollen grains, flowers and fruits compared with species with longer MGTs (Salisbury, 1961; Stebbins, 1971; Bennett, 1972, 1974).

As shown above, a short total generation time can dictate limits to numerous cell, tissue and whole plant characters throughout the life of the plant, so too can the need for rapid cell development at only one critical stage in the life cycle. This phenomenon has been well demonstrated in Amphibia where it has been shown that the duration of the tadpole stage in the life cycle is positively correlated with species nuclear DNA content (Goin *et al.*, 1968). Some species undergo their tadpole stage in temporary water in desert areas. Such species must all have very short tadpole stages if they are to survive and, consequently, to facilitate this, they must have very low nuclear DNA values. However, the need for low DNA values at this one critical stage means that phenotypic characters at other stages in the life cycle are also limited by low DNA content. Similar examples in plant species probably exist where the requirements for rapid development at meiosis and sporogenesis set limits to plant development at other life cycle stages. It has been shown, for instance, that among annual species only those with very low nuclear DNA contents can flower at temperatures just above freezing point because of the interaction of the effects of nuclear DNA content and temperature (Bennett, 1972). The association of this ability with a very low nuclear DNA content sets limits to plant size and many other phenotypic characters.

Many perennial species with very high nuclear DNA contents are adapted to complete meiotic and/or spore development at low temperatures in their perennating organ outside the season of active vegetative growth, for instance *Endymion nonscriptus* (3C DNA=69·9 pg), *Tulipa kaufmanniana* (93·7 pg), *Trillium erectum* (120 pg) and *Fritillaria aurea* (270 pg). The interaction of the effects of high nuclear DNA content and low temperature require that such species take very long periods to complete meiosis and spore development in winter. *Trillium erectum*, for instance, takes 90 days to complete meiosis at 1°C (Hotta and Stern, 1963b). Completing meiosis and spore development outside the season of active growth is of obvious advantage to species such as *Fritillaria aurea* which grow in an extreme environment with a very short season where active growth is possible. The duration of meiosis and pollen development in *F. aurea* even at 15°C is longer than the season when active growth can occur in much of its normal environment. Consequently, it is essential that these processes should be largely complete so that flowering can occur within the strictly limited season of active growth.

The possession of a perennating organ has allowed species like *F. aurea* to escape from the constraint of a high DNA content limiting the development which can be completed in a limited growing season. However, *F. aurea* does not flower until several years after germination. This is because it takes several years of vegetative growth in restricted growing seasons before the plant accumulates a sufficient reserve of metabolites to sustain meiotic gametophytic and seed production. Here too, then, the behaviour of meiotic and spore development related to high nuclear DNA content dictates important plant phenotypic characters, including a greatly prolonged generation time.

VI. CONCLUSIONS

Several general conclusions regarding the nature and control of reproductive cell development during sporogenesis and spore development may be drawn from the data presented in the present review.

(1) Cells approaching meiosis undergo a variety of developmental changes including increased cell and nuclear size and a decreased rate of development. In some species these changes occur during premeiotic interphase while in other species they may be spread over more than one cell cycle. Each individual change may not be unique to cells in sporogenesis, but taken as a whole the changes may represent a mandatory series pre-requisite for the initiation and completion of normal meiosis. Whether the changes must occur in a set order remains an imponderable, as does the identification of the causal control of such behaviour. However, it does seem clear that the initiation of meiosis, recognized cytologically as the leptotene stage, cannot occur in cells programmed to undergo active somatic development without the interpolation of a transition phase during which normal somatic development is progressively suppressed or modified and while changes pre-requisite for, or consequent to, the activation of a meiotic sequence are completed.

(2) As cells are committed to sporogenesis they become subject to a very different mode of developmental control from that found in most other somatic meristems. Cell development during sporogenesis and spore development is highly canalized. Several examples illustrating the existence of this type of control have been described. In particular, examples have been cited in which cell development proceeds at the normal rate to later stages irrespective of whether a previous stage has been either initiated or completed normally, and irrespective of whether the consequences of proceeding include the death of the cell or the production of non-functional spores. In other words, the control of meiosis and spore development does not include feedback mechanisms which ensure that failure or faults in development at one stage are corrected before development proceeds to a later stage.

(3) The size of meiocytes and their nuclei, and the duration of meiotic development, varies widely between higher plant species. Such variation is not arbitrary, however, and these characters are all interrelated probably as a consequence of their common positive relationship with the species nuclear DNA content. It has been shown that in plants of a single ploidy level species DNA amount plays a major role in determining the duration of meiosis and its stages. Nuclear DNA content appears to determine the rate of meiotic development apparently by the physical and mechanical consequences of its mass, and indirectly and independently of its informational content. Conditions of the nucleus which influence the phenotype independently of

the informational content (the genotype) have been defined as nucleo-typic (Bennett, 1971, 1972, 1974). Nucleotypic characters play an important role in determining one component to the control of the development of cells in sporogenesis and spore development. The importance of the nucleotype lies in the fact that determines the minimum time limits for developmental stages. While genotypic control can prolong the duration of meiosis it cannot shorten the meiotic time to less than the nucleotypically determined minimum. Consequent upon such effects the developmental behaviour of cells in meiosis and spore development is often a major factor determining the types of life cycle which individual species may exhibit.

6. Modifications and Errors of Mitotic Cell Division in Relation to Differentiation

A. F. DYER

Department of Botany, University of Edinburgh, Scotland

I. INTRODUCTION

All living organisms, even unicellular prokaryotes, undergo cell division and enlargement, and most if not all show cell differentiation in the sense that cells originating from the same genotype may differ not only in size but also in recognizable features of structure and activity. Cell division, enlargement and differentiation are clear and distinct concepts and can be studied separately, but in many cells these processes are taking place concurrently and the relationships between them are complex. Cell division rarely occurs without cell enlargement at least sufficient to maintain cell size (see Chapter 2, Section IIA). Cell enlargement in turn is usually accompanied by cell differentiation, exceptions including any simple unicellular organisms where a uniform type is perpetuated by matching rates of cell division and growth and per-haps the meristematic initial cells of higher plants over restricted phases of the life cycle.

In unicellular organisms, limitations are imposed on the extent to which differentiation can modify the cell as it ages by the necessity for it to divide again. Except in those cells which die through mutation or environmental hazard, the life-time of one individual corresponds to interphase of cell division and continuation of the species depends on every cell retaining the ability to divide. Any structural or functional differences between cells therefore usually reflect differences in cell age relative to the duration of interphase, although reversible modifica-tions, such as resistant cyst formation, can occur in some species in re-sponse to environmental stimuli.

In cells of multicellular plants the relationships between cell division, expansion and differentiation are more complex and more varied (Stange, 1965). Instead of complete separation and isolation of cells following cytokinesis as in unicellular organisms, the products of cell division usually remain attached by their cell walls, although the proto-plasts are separated by membranes except at plasmodesmata. Only once or twice in the life cycle of a higher plant are sister cells separated physically: liberated spores including pollen grains, and motile gametes including perhaps the male gametes of angiosperms, do not remain linked in any way to other cells. These single-cell stages of the life cycle, themselves clearly differentiated from all other cells of the organism, represent the bridges between alternating generations across which must pass all that is necessary to maintain the full potential for develop-ment. At all other stages, cells remain attached and in communication. The co-ordination of cell activities within one organism creates the possibility of "division of labour" whereby no cell is required to carry out all the functions necessary for continued existence. Each cell is dele-gated a limited role integrated with the complementary activities of other cells. Probably the most significant aspect of this is that dif-ferentiation is released from the restrictions imposed by the necessity to divide again. While some cells perpetuate the species, in others division is eventually suppressed allowing the elaboration of a diversity of sometimes highly specialized cell types, even to the point of cell death (as in functional xylem vessels and ultimately in all somatic cells). If multicellular plants originated from unicellular ancestors, important initial evolutionary steps in achieving the complexity and specialization necessary to exploit a variety of demanding terrestrial habitats must have included the suppression of cell separation following cytokinesis, the maintenance of protoplasmic connections for cell co-ordination and the controlled suppression of division in certain cells (Stange, 1965). The first step alone, which can be induced experimentally (Tamiya, 1964), merely gives rise to a colony of individual unicells, filamentous or otherwise depending on the orientation of successive divisions. The first two steps together result in a simple multicellular system where cells within an individual can only diverge in structure and function to an extent which is compatible with the maintenance of mitotic activity in all of them. Only with the addition of the third step, with division restricted to specific cells, are the full possibilities for co-ordinated cell specialization realized. Simple examples of such systems are the 2-celled angiosperm micro-gametophytes in pollen grains, with a non-dividing vegetative cell and a mitotically active generative cell, and the young fern gametophyte with only two cell types, the green mitotically active cells and the colourless non-dividing rhizoids (Dyer and Cran, 1974). Most higher plant systems are much more complex than this, with many cell types organized into multicellular tissues, each closely specialized for a restricted range of activities.

It can be argued that the developmental sequence in a higher plant follows the evolutionary sequence. Single isolated cells express the fundamental cell activity, cell division (see Stange, 1965). In multicellular systems, maximum specialization only occurs when, following mutual interaction between cells, division in some of them is suppressed (Brown and Dyer, 1972). As has been discussed elsewhere (Chapter 2, Section VC), artificial isolation of a cell from the influence of its neighbours seems to eliminate the basis of this suppression and allow cell division to be resumed. At the same time, isolation from its differentiated neighbours, or the effect of resumed mitosis, eradicates those accumulated consequences of previous cell development which restrict the possibilities for further differentiation and thus restores totipotency in the division products and allows redifferentiation.

A simple model (Fig. 1) of the relationship between the cell division cycle and subsequent differentiation reflects the fact that multicellular plants consist of some cells which will divide again, and are therefore at some point in the cycle of events preparing for that division, and of other cells which will not divide again during normal development and are following a linear sequence of changes leading to specialization, maturity and eventual death. In the vegetative sporophyte at least, all the more highly specialized cells are of the latter type. The model also reflects the fact that in some differentiating cells there is no further DNA synthesis after the last full mitotic cell cycle, while in others, the telophase complement of DNA has been replicated before specialization is complete (e.g. Stange, 1965; Deeley et al., 1957; Adamson, 1962;

A. F. Dyer

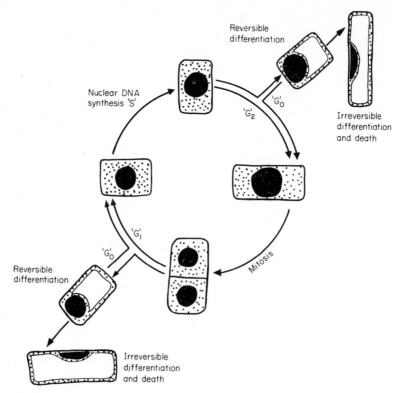

Fig. 1. A model for the relationship between the cell division cycle and differentiation.

Under natural conditions, some cells cease division activity and commence differentiation only after replication of nuclear DNA while others do so without any post-telophase DNA synthesis. In both cases, differentiation is sometimes reversible and the division cycle is resumed following, for example, wounding or hormone treatment. Ultimately, even if not until death of the protoplast, cell differentiation is irreversible.

Torrey, 1961; Patau *et al.*, 1957, 1961; Mitchell, 1967; Tobey *et al.*, 1971; Van't Hof, 1974). Where the telophase DNA level is maintained, and DNA replications must precede any induced resumption of mitosis (e.g. artichoke: Mitchell, 1967; Adamson, 1962; some cells in tobacco pith: Patau *et al.*, 1957, 1961), the cell is presumably initially diverted from the cycle after telophase and before S and thus in G_1, assuming that such a phase existed (see Chapter 2, Section IIIA). The model accurately indicates this but over-simplies the situation which occurs when DNA replication precedes or accompanies early stages in cell differentiation. Where the DNA is replicated once, and no further synthesis is required before division is resumed after treatment with growth substances (e.g. some cells of pea root segments: Torrey, 1961; tobacco pith: Patau *et al.*, 1957, 1961; Tobey *et al.*, 1971), it is reasonable to infer that the cell was diverted from division in G_2, as indicated in the model. However, the diagram does not represent those cells which have completed several rounds of DNA synthesis to become polyploid (e.g. D'Amato, 1952; Van't Hof, 1974) or those which, having already once replicated the telophase complement, still require further DNA synthesis before resuming division, as in many tobacco pith cells and pea root segment cells in culture (Patau *et al.*, 1961; Matthyse and Torrey, 1967). In those modified mitotic cycles producing endo-polyploid cells without chromosome contraction (see Section IID2), G_1 and G_2 are indistinguishable, and whether or not further DNA synthesis is necessary for the return to normal mitosis, the point at which the cell was originally diverted away from division preparations cannot be related to the model cycle in Fig. 1. In those other curtailed cycles in which chromosomes contract as they undergo some of the events of mitosis after the last DNA replication it could be equally well argued that the cycle was interrupted in M, mitosis. Similarly, where individual segments show differential duplication (Avanzi *et al.*, 1973) during cell specialization it could be claimed that the cycle was interrupted somewhere in S. Even where the interruption can be related satisfactorily to G_1 or G_2, different types of specialized cell may follow separate developmental sequences from an early stage, even from the point at which they "leave" the mitotic cycle. There may be many points within the cycle, even with G_1 and G_2, at which cells can be diverted away from division preparation, just as there are several points at which division can be "blocked" by experimental treatments (e.g. Mitchison, 1971).

According to this model, a distinction can be made between two types of cell in which preparation for division has at least temporarily ceased. Differentiated cells frequently show an unusually long interphase before the normal cycle is resumed after division induction (e.g. Yeoman and Evans, 1967; Mitchell, 1967). This has been interpreted as the result of an initial lag phase, sometimes called G_0 (e.g. Cooper, 1971), during which the events associated with differentiation are retraced to return the cell to the point on the cycle at which it was originally diverted. This contrasts with the situation considered to exist in dormant mitotic cells as found in embryos of dry seeds (Van't Hof, 1974) and quiescent lateral buds (Naylor, 1958) or induced in cultured roots (Van't Hof and Kovacs, 1972; Van't Hof and Rost, 1972; Van't Hof, 1974) where the cell is held in G_1 and G_2 by the inactivation of a critical step, as after the administration of certain mitotic poisons. In these, the sequence of division preparations is immediately resumed when the block is removed and protein synthesis restarts. In those species where dormant cells remain at the 2C level, there is little or no DNA replication associated with differentiation either (Van't Hof, 1974), indicating some overall genetic control over the blocking of the cycle.

Such a model leads to the concept of a control mechanism which can switch the cell between two mutually exclusive processes, cell division and cell differentiation. This may be helpful in certain restricted situations, such as continuously dividing meristem cells producing non-dividing differentiating derivatives or in tissue cultures where division and expansion are alternative responses to the culture conditions (e.g. Adamson, 1962). However, a major criticism of this model

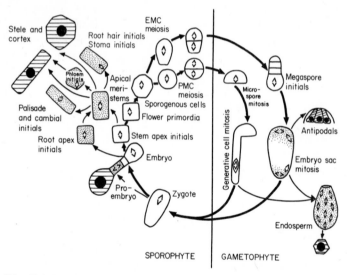

FIG. 2. The dividing cells of the angiosperm life cycle.
The cycle of broad arrows represents the direct cell lineage from one generation to the next, the "germ line". The dividing cells of this lineage (unshaded) belong to Type 1 as defined in the text and include a variety of differentiated cells differing in morphology and in detailed features of mitosis. The derived mitotic cells (stippled; Type 2 as defined in the text), linked by thin arrows, account for most of the divisions in a developing plant and produce the somatic tissues of the sporophyte, reduced gametophyte and endosperm. The differentiating cells (shown cross-hatched with large nuclei) represent some of the cells (Type 3 as defined in the text) in which the formation of endopolyploid nuclei replaces the full mitotic division cycle. Other differentiating and mature cells (not shown) in which no further chromosome replication normally occurs (Types 4 and 5 as defined in the text) usually develop from Type 2 and 3 cells but only occasionally (e.g. the vegetative cell of the pollen grain) form directly, without further mitosis, from Type 1 "germ line" cells.

is that it does not allow for the fact that differentiation and division can and often do occur simultaneously. Many dividing cells show considerable structural and functional specialization (see Chapter 2, Section II), while the division process itself shows modifications during the initial stages of the development of certain cell types. Consequently, rather than merely distinguishing between dividing and differentiating cells, it is probably more informative to group the cells of a higher plant into five types according to their division activity and degree of specialization (Fig. 2).

Type 1: These are the potentially immortal dividing "germ-line" cells in the direct lineage from one generation to the next through the shoot apex meristem initials, flower primordia, sporogenous tissue, spores, gametophytes, gametes and zygote. There is little or no visible specialization during vegetative growth, but considerable cell differentiation accompanies division through the reproductive stages (see Chapter 2, Section IIB–D). This differentiation is cyclic, returning to the same point once every life cycle, which may last many years, while the cellular events accompanying division are cyclic over the period of one cell generation, usually a matter of hours or a few days. These cells may enter periods of long or even permanent dormancy, as in lateral bud apices; this dormancy is important in maintaining the growth form of the plant. Once every life cycle, the cell cycle is modified to produce meiosis. Very rarely, there is an abbreviated cell cycle immediately preceding meiosis (see Section IID2).

Type 2: These are dividing cells, themselves derived from Type 1 cells, which undergo from one to many complete mitotic cycles to produce most of the somatic tissues of the plant. Division is often accompanied by net cell enlargement and even some specialization (see Chapter 2, Sections II, III) as their development becomes directed, in coordination with that of surrounding cells, towards a particular mature cell type. Specialization at this stage does not only involve structural differences; it includes differences in the duration of the mitotic cycle within a meristem (Barlow, 1973). Most dividing cells in a plant are of this type, which includes many mitotic cells of root and shoot apices and those of, for example, cambia, leaf and sporophyte primordia, developing leaves and ovary, ovule and anther walls. Occasionally, as in somatic segregation, the behaviour of certain chromosomes is modified during divisions at pre-determined stages of development. Sometimes the cells become dormant, as in the pericycle where only those which give rise to lateral roots or cambia will resume mitosis.

Type 3: These are cells, recently derived from Type 1 or 2, which undergo incomplete mitotic cycles in conjunction with further structural and functional specialization. In roots, for example, they are found in a short region just behind the apical meristem (Tschermak-Woess, 1956b). The result of the modified cycles is quantitative change in the nuclear material, usually if not always an increase, as in the later stages of development in the root cortex and anther wall. The products of Type 3 cell cycles include many of the cells of most plants (D'Amato, 1952; Tschermak-Woess, 1956b). Very rarely similar abbreviated cycles occur in Type 1 cells (see Section IID2).

Type 4: These are cells derived from Type 1, 2 or 3 in which all mitotic activity has been suppressed and under normal conditions of growth would not be resumed, but in which the potential for division has not been lost, as revealed by the resumption of mitosis after certain experimental treatments including excision, cell isolation and addition of growth substances. This potential has been demonstrated in a variety of tissues such as secondary xylem and phloem parenchyma, leaf mesophyll, leaf and stem epidermis and stem pith, and it may exist in other cell types, at least in young cells.

Type 5: These are cells derived from Type 4 in which the potential for division has been permanently lost during the later stages of cell specialization. While it is almost impossible to confirm experimentally, this stage may only be reached with senescence and death, as in later stages of xylem vessel development and ultimately in all non-dividing cells, while all other non-dividing cells are of Type 4, requiring only the necessary stimulus and culture conditions to resume division.

Most of the divisions seen will be in Type 2 cells, although the "germ line" in plants with indeterminate growth is ill defined and Type 1 cells are not always distinguishable. Most of the cells in all but the youngest plants will be of Type 4 or 5, derived after a relatively short period of Type 2 and perhaps Type 3 divisions from an almost infinite lineage of Type 1 cells extending back through development and reproduction to the early history of the species. Only rarely is this developmental sequence reversed with the induction of division in Type 3 or 4 cells by disturbances from outside the cell. For example, it has long been claimed (see D'Amato, 1964) that polyploid lateral buds formed after decapitation develop from endopolyploid differentiating cells which have been induced to resume Type 1 divisions.

Obviously the events of cell division are of fundamental significance to subsequent cell differentiation in the same or later plant generations. Some aspects of this relationship have been touched upon already

(Chapter 2, Section IIA, C). The segregation of cytoplasmic organelles in equal and unequal divisions is clearly important in determining the developmental potential of the products. Possible interpretations of the role of cytokinesis and plasmodesma formation (Chapter 2, Section VA, C), of the variation in position and duration of the S period within interphase (Chapter 2, Section IIIA, G), of the significance of cell membranes in orientating division and expansion (Chapter 2, Section VI) and of the basis and consequences of division induction in differentiated cells (Chapter 2, Section VC) have already been introduced. It remains to consider the changes in the nucleus occurring during division which may accompany or disturb normal development (Stern, 1958). These include the changes in chromosome behaviour and the abbreviated mitotic cycles which regularly occur at specific phases of development together with the errors of chromosome replication, spindle–centromere interaction and spindle organization which produce chromosome mutations and developmental anomalies. Chromosome mutations occurring in Type 1 cells may assume a further importance in addition to their possible effects on differentiation. Any such mutations which are perpetuated until meiosis may affect genetic recombination. One or two may be successful in later generations and contribute to karyotype evolution and speciation, but the cytogenetic implications of division errors are beyond the scope of this chapter.

II. MODIFICATIONS OF THE CELL CYCLE DURING NORMAL DEVELOPMENT

The chromosome complement seen in a dividing cell is not necessarily identical in appearance or in genetic content with all other cells or even all other dividing cells. Comparisons of cells undergoing a typical cell cycle may reveal differences in chromosome contraction or size. Controlled modifications of centromere behaviour resulting in non-disjunction produce cell lines in which specific chromosomes have been lost. Modified cell cycles in which certain fundamental steps have been omitted result in cells with duplicated or halved nuclear complements. The possibility that these differences are related to subsequent cell development needs further consideration.

A. Allocycly

Many examples are known of segments, whole chromosomes or even whole complements which undergo a cycle of contraction and dispersion which differs in timing during the mitotic cycle from the standard cycle which reaches maximum contraction at late metaphase and a minimum during interphase. Such "allocyclic" regions differ, at least at certain phases of division and development, from the rest of the nucleus in the intensity of staining in cytological preparations; they are "heterochromatic" or "heteropycnotic". Some authors refer to the chromosome material in these regions as "heterochromatin" (Lima-de-Faria, 1969), but to do so is probably undesirable because there is evidence to show that some facultatively allocyclic regions such as mammalian chromosomes do not differ fundamentally in chromatin structure or composition from the rest of the complement, although there are some other instances of heteropycnosity, attributed to "constitutive heterochromatin", where the chromatin may be of basically different organization (Flamm, 1972; Rees and Jones, 1972; John and Lewis, 1968).

A common pattern of allocycly in plants and animals is over-contraction during interphase and early prophase to form dense "chromocentres" with normal behaviour during mitosis. This frequently characterizes the regions of chromosome arms close to centromeres, particularly in species with small chromosomes, less than 3 μm (Lafontaine, 1974), which form interphase "pro-chromosomes" (Kurabayashi *et al.*, 1962). It also occurs in the region of telomeres and in sex chromosomes. Supernumerary B chromosomes are frequently similarly heteropycnotic, as in *Puschkinia* (Vosa, 1969; Barlow and Vosa, 1970), several grasses and cereals (Bosemark, 1956, 1957; Darlington and Thomas, 1941; Himes, 1967; Muntzing, 1966), *Plantago* (Paliwal and Hyde, 1959; Fröst, 1959), *Collinsia* (Dhillon and Garber, 1962), *Achillea* (Ehrendorfer, 1961), *Centaurea* (Fröst, 1948) and *Haplopappus* (Östegren and Fröst, 1962). This is not, however, an invariable character of B chromosomes, which are presumably derived from eucyclic autosomes, and they are still eucyclic with a normal contraction cycle in such species as *Trillium, Tradescantia, Listera ovata* (Barlow, 1972b), *Xanthisma* (Berger *et al.*, 1955), *Crepis* (Fröst and Ostegren, 1959), *Allium* (Vosa, 1966; Grun, 1959), *Godetia* (Håkansson, 1945), *Lilium* (Kayano, 1956, 1967; Ogihara, 1962), *Koeleria* (Larsen, 1960) and *Clarkia* (Mooring, 1960). In most of the species with eucyclic B chromosomes, the chromosome complement is constant throughout vegetative development. In some of the species with allocyclic supernumeraries, the B chromosomes are lost from certain cell types during plant development (see Section IIC). It is not known whether allocycly and somatic segregation are linked processes in these chromosomes.

In the large chromosomes of *Trillium* and several other unrelated plants and animals, allocyclic interphase chromocentres are always present and correspond to under-contracted "H"-segments which appear along the chromosome arms only when mitosis occurs at temperatures below about 9°C (Dyer, 1963) (Fig. 3). After conventional staining techniques the same chromosomes at higher temperatures are eucyclic from prophase to telophase and no segments can be distinguished, but new techniques involving quinacrine mustard fluorescence

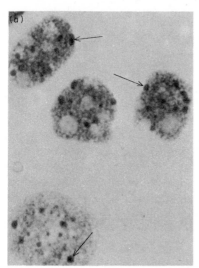
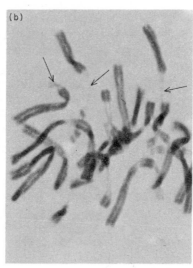

FIG. 3. Allocyclic segments in Feulgen stained nuclei of *Trillium grandiflorum*.
(a) Interphase root-tip nuclei revealing conspicuous positively heteropycnotic chromocentres (arrowed). × 2000.
(b) Metaphase chromosomes in the endosperm after 96 h at *c.* 3°C showing undercontracted negatively heteropycnotic segments (arrowed). The segments are constant in position and relative size throughout the plant and each corresponds to a chromocentre at interphase. The segments are not revealed in cells dividing at temperatures above 9°C, although the interphase chromocentres remain. × 2000.

or denaturation reannealing processes without low temperature treatment result in differential staining of regions which correspond to the H-segments (Vosa, 1973; Caspersson *et al.*, 1968; Schweizer, 1973). The same techniques also reveal otherwise undetected heteropycnotic segments in these and other species including man. These responses indicate an underlying difference in chromatin organization even when no chromocentres are detectable by the long-established staining techniques. Whether these segments are described as allocyclic or not is a semantic point.

Although chromocentres may vary in size and degree of fusion between cells, there are no clear cases among the plant examples where the pattern of allocycly is known to vary in a predictable way in conjunction with cell differentiation. The best known examples of changes in allocycly with development are found in animals. The coiling behaviour of X chromosomes in male orthopterans differs between comparable stages of mitosis and meiosis. Something similar may occur in dioecious plants such as *Humulus* and *Rumex* (Jacobsen, 1957; Shimizu, 1961). In mammals, one of the two X chromosomes in normal females becomes wholly allocyclic at an early stage in embryogenesis to form the heteropycnotic chromocentre known as the "Barr body". During subsequent somatic development, the heteropycnotic X chromosome is inactive as revealed by the phenotypic expression of heterozygous X-linked genes (e.g. Hamerton *et al.*, 1971). Taken in conjunction with the lack of synthetic activity which characterizes cells during division when the bulk of the chromosome complement is contracted (see Brown and Dyer, 1972; Lafontaine, 1974), this suggests that allocycly may be concerned with the regulation of gene activity in certain segments of the chromosome complement. As most cases of allocycly involve interphase chromocentres, the almost complete suppression of activity in specific blocks of genes is implied. In the relatively few cases where segments remain conspicuously under-contracted during mitosis or meiosis, the activity of certain genes may be extended into the otherwise biochemically inactive division stages.

Such a regulatory system would seem to be well suited to producing the differential gene activity which must underlie all cell specialization. However, the little information which is available for plants does not suggest that allocycly has an important role in differentiation. Many plants have no visible chromocentres at any of the stages examined, and in the remainder, no tissue-specific differences in allocycly have been reported. Some allocyclic segments contain repetitious DNA sequences, but the role of the latter is not known (Bostock, 1971; Rees and Jones, 1972; Flamm, 1972). The significance of allocycly remains to be discovered.

B. Changes in Chromosome Size

There is an enormous range in chromosome size between species. For example, each chromosome of *Trillium erectum* contains as much DNA as the whole complement of 48 chromosomes in *Nicotiana tabacum* (Bennett, 1972; Rees and Jones, 1972). Within a species, however, chromosome volume at comparable stages of division is remarkably constant, although a few striking exceptions have been recorded. Thomas (1936) described seedlings of *Lolium perenne* in which the chromosomes were of similar morphology but much reduced size compared with normal sister plants. There were no obvious effects on plant phenotype, but the frequency of plants with small chromosomes in the offspring of controlled crosses indicated that the chromosome change was produced by a single recessive gene mutation. Schwanitz (1967) illustrated two varieties of leek, the more robust of which had consistently larger chromosomes than the other. There seems little doubt that in these examples the changes in chromosome size are genetically determined, but it is not known whether the changes in volume involve changes in DNA content. The change is presumably reversible in *Lolium* because the small chromosomes inherited from the recessive mutant plants are restored to full size in the heterozygote.

Changes in chromosome size within a plant are also known. Chromosomes of root-tip mitoses can be larger than those in leaf cell and pollen grain divisions (John and Lewis, 1968), but smaller than those in endosperms (Kapoor and Tandon, 1963), although again it is not clear whether DNA differences are involved. Three-week-old lateral roots of *Vicia faba* have much smaller chromosomes than 1-week-old main roots (Bennett, 1970), but DNA differences are not involved. The change during development is again presumably reversible. A close and systematic comparison of a wide range of dividing cells from the same individual might reveal some consistent relationship with cell differentiation. In growing roots of rye and onion seedlings, chromosomes double in size over the first 3 weeks after germination, and then decrease (Bennett and Rees, 1969). Similar changes occur in the adventitious roots of germinating *Allium* bulbs. Chromosome size reaches a maximum at the time of highest relative growth rate. High phosphate levels, which stimulate growth, also induce increases of chromosome volume of up to 100%. The change in volume was shown to be accompanied by increases in protein content; the amount of DNA was unaffected.

A change in the amount of protein associated with the chromosomes, reflecting the overall level of activity in the cell, may also explain the observed differences in chromosome size between different tissues and even different genotypes, but even so its significance is obscure.

If some of the changes in chromosome volume are accompanied by changes in DNA content, a change in longitudinal repetition or in the number of component nucleo-protein chromonemata (Rees and Jones, 1972) is indicated. Many cytologists consider that at least some chromosomes are multi-stranded (Wolff, 1969) and a change in the number of chromonemata would result from an alteration in the relative rates of chromonema replication and separation. Gene amplification would result from differential gene replication (Rees and Jones, 1972). A different growth rate may then be a consequence rather than the cause of the nuclear change. Species with large chromosomes containing more DNA tend to have longer cell and life cycles than those with smaller chromosomes (Bennett and Smith, 1972; Bennett, 1972).

It remains to be determined whether differences in chromosome size between tissues, and between seedlings, are accompanied by differences in DNA content or in the level of overall activity in the cell concerned. We also need to know whether or not the chromosomal changes invariably accompany the development of certain Type 1 and Type 2 cells. If they do, further investigation will be necessary to ascertain whether the fluctuations in chromosome volume are the cause or consequence of cell differentiation.

C. Somatic Segregation

In the development of a number of plants and animals, certain divisions at pre-determined stages of development are modified so that one or both of the daughter cells inherit a chromosome complement which differs from that of the sister cell, the parent cell and the zygote from which it was derived. In most cases the change involves loss of chromosome segments, whole chromosomes or whole complements by controlled non-disjunction during division. In animals, where the germ line is clearly defined from an early stage, somatic segregation often occurs early in embryogenesis, leaving the germ line karyotype intact but producing an altered somatic complement. Presumably the genes lost are not required in somatic development. In higher plants, where the germ line is ill defined, somatic segregation has been found at all stages of sporophyte development from early embryogenesis to meiosis and in the male and female gametophytes. In nearly all the cases described in angiosperms, autosomes are not affected and only the supernumerary B chromosomes show somatic segregation. The only exceptions to this involve meiotic non-disjunction of whole genomes. In allopolyploid pentaploid *Rosa canina* (AABCD: Täckholm, 1922) and triploid *Leucopogon juniperum* (AAB: Smith-White, 1955), the unpaired genomes regularly form univalents which are entirely lost by non-disjunction from the microspores, but are all included in the megaspore following directed anaphase movement. The male gametophytes and gametes thus contain a single haploid genome, while the female gametophytes and gametes carry all the remaining genomes so that fertilization restores the full complement in the zygote. Thus in *Rosa*, for example, the haploid male gametophyte and tetraploid female gametophyte differ from each other and neither corresponds to half the sporophyte complement. This somatic segregation at spore formation is primarily concerned with the unusual sub-sexual genetic system rather than with cell development. Morphogenesis of the alternating generations is regular despite the quantitative and qualitative genetic differences between the gametophytes.

Somatic segregation of supernumerary B chromosomes can also occur at meiosis, as well as at gametophytic mitosis, and these modifications of division again affect few somatic cells but have important consequences for the inheritance of the chromosomes concerned. Non-disjunction of B chromosomes at "female" meiosis is directed towards the megaspore pole in *Trillium, Lilium* and *Plantago* (e.g. Kayano, 1956, 1957; Frost, 1959) and sister B chromosomes at anaphase move together towards the generative pole at pollen grain mitosis in *Collinsia solitaria, Secale cereale, Haplopappus gracilis* and several grasses (Bosemark, 1957; Dhillon and Garber, 1962; Håkansson, 1948; Pritchard, 1968), and also towards the egg cell pole in the first female gametophyte mitosis in *Secale* and *Collinsia*. This directed non-disjunction is apparently a response of a particular centromere type to a mechanically asymmetrical spindle. The effect of this at divisions where only one daughter cell continues towards gamete formation is to increase the number of B chromosomes passed on to the next sporophyte generation. This compensates for the random loss of supernumeraries which occurs during development in some species (Samejima, 1958; Muntzing, 1966; Pritchard, 1968) and, assuming that this compensation is an indication that these chromosomes can confer some advantage, allows an optimum frequency to be maintained. An alternative "boosting" mechanism is found in *Crepis capillaris* (Rutishauser and Rothlisberger, 1966) and in *Achillea* (Ehrendorfer, 1961). In these plants the zygotic complement is maintained throughout vegetative development, but, shortly after flowering has been induced by a change to the appropriate day length and before there are any visible morphogenetic changes in the apex, mitotic non-disjunction in the meristem produces cells with altered numbers of chromosomes. Those cells with increased numbers of supernumeraries contribute most to the developing flowers. For example, in a *Crepis* plant with a complement of $2n+1B$ during vegetative growth, cells with 0B to 4B appear at the stage of flowering induction. Of these, the $2n+2B$ cells become the most common type, indicating differential division rates in the apex. Eventually most or all the meiotic cells have two supernumerary chromosomes and as a result most or all gametes contain one B chromosome. Without this boosting mechanism no more than 50% of pollen grains would have inherited a B chromosome. The modification of mitotic division behaviour is associated with a particular physiological state, but the nuclear condition is the consequence rather than the cause of the developmental change. The limited evidence available supports that *Crepis pannonica* has a similar mechanism (Fröst, 1960).

These boosting mechanisms provide circumstantial evidence that B chromosomes can be beneficial, but they do not establish a direct relationship between somatic segregation of these chromosomes and subsequent cell development. Indeed, it has been claimed that B chromosomes, even those which are not allocyclic, are inert. However, in several of the 150 or so species in which B chromosomes have been observed, direct phenotypic effects of their presence have been recorded. There is, for example, an additive effect on pericarp pigment in *Haplopappus* (Jackson and Newmark, 1960) and a correlation with edaphic adaptation in *Festuca* and other grasses (Bosemark, 1956, 1957); altered vigour in *Centaurea* (Frost, 1958) and *Secale* (Muntzing, 1966; Jones and Rees, 1968) among others; modified pollen development in *Anthoxanthum* (Lima-de-Faria, 1947) and *Plantago* (Paliwal and Hyde, 1959); germination time in *Allium* (Vosa, 1966); changed chiasma frequency in *Puschkinia* and *Listera* (Barlow, 1972b; Vosa, 1969; Barlow and Vosa, 1970) and *Najas* (Vinikka, 1973); crossability with other species in *Achillea* (Ehrendorfer, 1961); nuclear phenotype in maize (Ayonoadu and Rees, 1971); variability, yield and chromosome size in rye (Moss, 1966; Jones and Rees, 1968; Frost, 1963); and increased chromosome breakage in *Trillium* (Rutishauser, 1956).

While in many cases the number of B chromosomes per cell is constant throughout a plant, the evidence that the presence of these chromosomes can be directly correlated with certain developmental features does focus particular interest on those examples where somatic segregation of supernumeraries occurs during vegetative development and results not in a "boosting" of B chromosomes in the germ line but in their total elimination from cells of certain somatic tissues (John and Lewis, 1968). In these plants, including several grasses and cereals (Baenziger, 1962; Milinkovic, 1957; Muntzing, 1966; Darlington and Thomas, 1941), *Haplopappus* (Östegren and Fröst, 1962; Pritchard, 1968) and *Xanthisma* (Berger and Witkus, 1954; Berger *et al.*, 1955), adventitious and secondary roots and sometimes leaves lack B chromosomes while most or all cells of the flowering shoot apex and all the spore mother cells retain them. At some stage in vegetative growth, the supernumeraries, most or all of which are heterochromatic, are lost, presumably during mitotic division, from cells giving rise to particular tissues removed from the "germ line". There are detectable differences between the cells of lateral root meristems and those of the primary meristems from which they derive (Davidson, 1972) which might provide the basis for a unique type of division at the stage when secondary roots arise, but the exact time and mechanism of chromosome loss has not been observed directly. It is possible that the stage at which they are lost is related to the stage at which the genes present on them exert their most significant effect. Where they are eliminated by somatic segregation at several independent points in the cycle from roots and leaves, as in some grasses, their influence may be restricted to the period of embryogenesis or subsequently to Type 1 cells. Where they are lost from roots only, they may exert a more general effect on shoot development throughout growth. Where they are present in dividing cells uniformly throughout all tissues, as in *Koeleria* (Larsen, 1960), *Lilium* (Ogihara, 1962), *Tradescantia*, *Trillium* and *Allium*, they may have a general and prolonged effect on development at all stages.

The limited evidence available indicates that B chromosomes which show somatic segregation are usually heteropycnotic, while those which maintain a uniform number throughout development are in general eucyclic, but the significance, if any, of this relationship is not immediately obvious. Detailed comparisons, including physiological and biochemical studies, are needed at different stages of development from the zygote between individuals of the same species with and without B chromosomes and in conjunction with a detailed survey of B chromosome distribution within the plant. Any direct relationship between somatic segregation and cell development might then be revealed. Meanwhile, it should be remembered that in a few species, even some with no B chromosomes in the root-tip divisions, a significant fraction of the nuclear DNA in shoot-tip cells represents that supernumerary chromosomes which originated as centric fragments duplicating segments of the A chromosomes.

D. Abbreviated Cell Cycles

Many of the nuclear changes which accompany normal development during the higher plant life cycle can be most easily interpreted as the products of abbreviated cell cycles in which one or more of the basic events of normal mitotic division (see Chapter 1, Section IIIG) are omitted as the cycle continues. In this way it is possible to relate all these conditions to each other and to the products of normal mitotic cell division. In all cases there is a controlled dislocation of the normal relationship between chromosome replication and cell division and, as all chromosomes in a cell are equally affected, the result is either a doubling or a halving of the nuclear complement.

Several cytological conditions including polyteny, diplochromosomes, polyploidy and multinucleate cells which arise in Type 3 cells embarking on their final sequence of developmental changes in plants and animals (Geitler, 1953; D'Amato, 1952; Tschermak-Woess, 1956b) can be interpreted as the related consequences of progressively more curtailed cycles in which chromosome DNA is replicated but not distributed in the usual way (Brown and Dyer, 1972). After each cycle, the cell re-enters the sequence prior to DNA synthesis and may repeat the process several times.

The two divisions of meiosis and the so-called "somatic reduction" also represent abbreviated cycles, but in these cell division is maintained while other steps, even including DNA replication, are omitted. The first division of meiosis halves the chromosome (i.e. centromere) complement, while the second achieves division of each chromosome. Confirmed cases of somatic reduction are confined to rare situations where the meristematic cell complement is restored by an appropriate number of modified cycles in cells which have previously undergone nuclear doubling.

1. Nuclear Doubling: Coenocytes

When cytokinesis is eliminated from a cell cycle but mitosis continues, a binucleate cell is formed. Further similar cycles will produce multinucleate coenocytes (Fig. 4). Coenocytic cells and tissues arise in tissue cultures (e.g. Cooper et al., 1964) and are not uncommon in intact higher plants (D'Amato, 1952; see Chapter 2, Section IIA, D). In the sporophyte, tapetal (Bennett and Smith, 1972; Mechelke, 1953) and other anther wall cells are frequently binucleate or quadrinucleate. Laticifers (Mahlberg and Sabharwal, 1967) and xylem vessels (List, 1963) can be multinucleate. In the female gametophyte of angiosperms, the early stages of development are coenocytic. Eventually all but the polar nuclei become individually enclosed by cell walls, but in Fritillaria-type embryo sacs, not until nuclear fusion in the coenocytic 4-nucleate stage has produced polyploid nuclei at the antipodal end. In endosperm development, the early stages are frequently coenocytic with synchronous mitoses (see Chapter 2, Section IID), although here too the nuclei are separated in the mature tissue by cell walls produced simultaneously between all nuclei after about ten abbreviated cell cycles. The coenocytic meiotic products of bisporic and tetrasporic embryo sac development result in genetic mosaics if the parent mother cell is heterozygous for gene or chromosome mutations because two

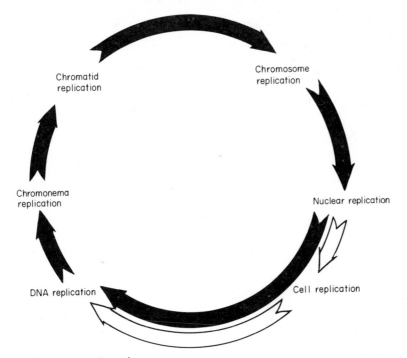

Fɪɢ. 4. The coenocytic cycle.

Binucleate cells result from an abbreviated mitotic cycle (solid arrows) in which cytokinesis (cell replication) has been suppressed while other events essential to nuclear divisions (see Chapter 2, Section IIIG) proceed. Repetition of this abbreviated cycle leads to the formation of a coenocyte. The same cycle can be induced in cells with a normally full cycle by treatment with caffeine and other chemicals (Section IID5).

or more of the recombinant nuclei are included. The significance of this and the effect, if any, of different genotypes within the one gametophyte tissue is not known.

2. Nuclear Doubling: Polyploidy

The commonest type of nuclear doubling in plants is polyploidy arising from the suppression of both cytokinesis and anaphase separation of daughter chromosomes (Fig. 5). Many plant cells can be shown to be polyploid if they are induced to resume normal mitosis by auxin treatment (D'Amato, 1952, 1965; Huskins and Steinitz, 1948; Sharma and Mookerjea, 1959; Therman, 1951; Chouinard, 1955; Geitler, 1953; Torrey, 1965a, 1961; Van't Hof and McMillan, 1969; Partanen, 1963; Fosket, 1968; Bennici et al., 1968). Without such treatment, the polyploid condition is usually only revealed by determinations of DNA

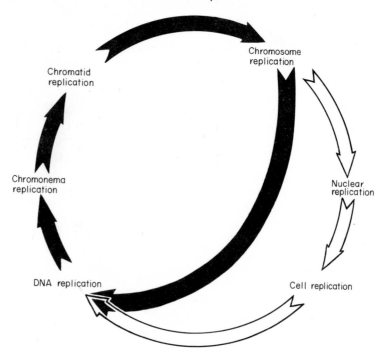

Fig. 5. The polyploidy cycle.

Polyploid cells result from an abbreviated mitotic cycle (solid arrows) in which cyto-kinesis (cell replication) and anaphase separation to poles of daughter centromeres (nuclear replication) have been suppressed while other events essential to nuclear division (see Chapter 2, Section IIIG) proceed. The same cycle can be induced in cells with a normally full cycle by treatment with colchicine and other chemicals. (see Section IID5). Treatment during successive cycles can lead to nuclei with eight or more times the original number of genomes.

content or indicated by increased nuclear volume or chromocentre number because in many tissues, such as root cortex and stele, the chromosomes replicate within an interphase nucleus without contracting and becoming distinguishable. More rarely, in, for example, tapetum (D'Amato, 1964) chromosomes do contract at the time of chromosome replication but they remain within an intact nuclear envelope, as in the insect "endomitosis" of Geitler (Geitler, 1939, 1953; Nur, 1968). In a few cases, mitosis is quite normal until metaphase, but subsequently, although the centromeres divide, spindle activity fails and the double chromosome complement is included within a single polyploid restitution nucleus. This has been described in the tapetum (Mechelke, 1953; Brown, 1949) of angiosperms, and in a variety of apomictic organisms, including insects (White and Webb, 1972), earthworms (Omodeo, 1952), planarians (Melander, 1948) and ferns such as allotriploid *Dryopteris borreri* (Manton, 1950), where the nucleus is duplicated in this way at the interphase before meiosis, which follows with regular bivalent formation and behaviour but results in no chromosome reduction or genetic recombination. In *Allium odorum* (Håkansson and Levan, 1957) a similar sequence involves diplochromosome formation during a curtailed mitotic cycle immediately before meiosis. The situation as in *Allium* and *Dryopteris* is unusual in that abbreviated cycles occur in Type 1 "germ line" cells, but they are immediately compensated by meiosis.

Polyploid nuclei rarely result from fusion of nuclei after mitosis has been completed in a natural coenocyte. An exception is the fusion of three megaspore nuclei at mitosis to form a single triploid metaphase in developing *Fritillaria*-type embryo sacs. Something similar occurs in the antipodal cells of *Caltha palustris* (Grafl, 1940). Nuclear fusion may be more common in tissue culture (Sunderland, 1973b; Partanen, 1963) and in animal cells (Hsu and Moorhead, 1956).

3. Nuclear Doubling: Diplochromosomes

Occasionally, usually in tissues mainly characterized by polyploidy, auxin-stimulated mitoses reveal diplochromosomes in which four identical chromatids are attached to each centromere (D'Amato, 1952, 1964; Dermen, 1941; Torrey, 1961; Stephen, 1974; Tschermak-Woess, 1956b). These result from a curtailed cycle in which centromere division as well as anaphase chromosome separation and cytokinesis are suppressed (Fig. 6). Diplochromosomes have also been reported occurring naturally without pre-treatment (e.g. Kato, 1955; Håkansson and Levan, 1957; Rasch *et al.*, 1959; Gatti *et al.*, 1973; D'Amato, 1952; Tschermak-Woess, 1956; Wipf and Cooper, 1940; Avanzani, 1950; Gentscheff and Gustafsson, 1939; Stange, 1965) and in cultured pollen grains (Dunwell and Perry, 1973). In these cells there is an extra centromere division, usually in late prophase, before entering a polyploid but otherwise normal metaphase and anaphase. Further curtailed cycles can produce chromosomes with 8, 16 or 32 chromatids attached to one centromere. Chromosomes with eight chromatids have been reported in crown gall divisions (Rasch *et al.*, 1959) and in pith of *Kniphofia* (Tschermak-Woess, 1956b). In *Allium odorum* (Håkansson and Levan, 1957) diplochromosomes are found in Type 1 cells immediately before mechanically normal but genetically ineffective meiosis in a diplosporous apomict (see Sections I and IID2).

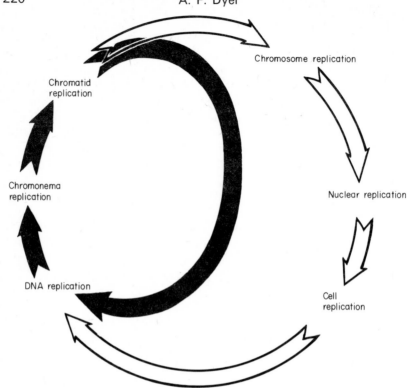

Fig. 6. The diplochromosome cycle.

Diplochromosomes result from an abbreviated mitotic cycle (solid arrows) in which cytokinesis (cell replication), anaphase separation to the poles of daughter contromeres (nuclear replication) and centromere division (chromosome replication) are all suppressed while the remaining events essential to nuclear division (see Chapter 2, Section IIIG) proceed. Repetition of this cycle produces up to 32 chromatids per chromosome and such cycles can be induced in cells previously following the full mitotic cycle (see Section 11D5).

4. Nuclear Doubling: Polyteny

The obviously multi-stranded condition of polytene chromosomes, traditionally described in *Drosophila* salivary gland cells, has been interpreted as the result of repeated replication of the basic structural unit of the chromosome, the nucleo-protein chromonema, without the organization of chromatids or any of the subsequent fundamental steps in the cycle (Fig. 7). The size and DNA content of these chromosomes frequently indicate many such rounds of replication. They have been widely found in certain specialized plant tissues (D'Amato, 1952;

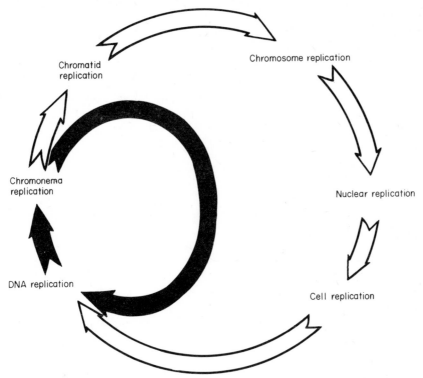

Fig. 7. The polyteny cycle.

A cycle (solid arrows) in which DNA and the basic structural unit of the chromosome, the chromonema, replicates without the subsequent organization of distinct chromatids, division of centromeres, anaphase separation of centromeres or formation of daughter cells results in polytene chromosomes. Repetition of this abbreviated cycle can produce identifiable chromosomes with several thousand times the initial DNA content.

Tschermak-Woess, 1956b), particularly those associated with embryo sac development, including antipodal (Tschermak-Woess, 1956a, 1957) and synergid cells (Håkansson, 1957), embryo suspensors (e.g. Avanzi *et al.*, 1970; Nagl, 1972, 1973) and endosperm (Geitler, 1955; Enzenberg, 1961) cells, and also tapetum cells in some species (Mechelke, 1953). They are frequently found in a prophase-like condition although they have been seen at metaphase (see John and Lewis, 1968, Fig. 20).

It is interesting to note in relation to the question of the order of fundamental events in the full cycle (see Chapter 2, Section IIIG) that in these modified cycles chromatid formation is an event distinct from, and presumably subsequent to, replication of DNA and chromone-mata.

5. Nuclear Doubling: the Induction of Abbreviated Cycles

It is possible to induce short-circuits of the mitotic cycle by chemical treatment and other environmental stimuli and exactly mimic some of the abbreviated cycles just described. The classic example is colchicine treatment (Rizzoni and Palitti, 1973; Eigsti and Dustin, 1955;

TABLE I. Some inhibitors of spindle activity

Inhibiting agent	Reference
PHYSICAL	
Hydrostatic pressure, heat shock	Risueno *et al.* (1968)
Low temperature	Clowes and Stewart (1967)
CHEMICAL	
Acenaphthene	Sharma and Sharma (1965)
Aesculin	Sharma and Sharma (1965)
4-Aminopyrazole (3,4-D) pyramidine	Schachtschabel *et al.* (1968)
α-Bromonaphthalene	Sharma and Sharma (1965)
Chloral hydrate	Sharma and Sharma (1965)
Chloramphenicol	Nasjileti and Spencer (1968)
Coumarin	Sharma and Sharma (1965)
para-Dichlorobenzene	Sharma and Sharma (1965)
Digitonin	Underbrink and Olah (1968)
8-Ethoxyquinoline	Sharma and Sharma (1965)
Maleic hydrazide	Grant and Harney (1960)
Mercaptoethanol	Nagl (1970b)
Mimosine	Pritchard and Court (1968)
Podophyllin	Cornman and Cornman (1948)

Pickett-Heaps, 1967a; Deysson, 1968) which induces a cycle exactly like that in *Dryopteris borreri* premeiotic mitosis. Polyploidy results from a cycle where the spindle fails to form and the daughter chromosomes after centromere division are all included within one restitution nucleus. By binding with microtubule protein, colchicine also affects the orientation of cell expansion. There are many other unrelated chemicals which also inhibit spindle activity (Sharma and Sharma, 1965) (Table I), but it is not known which if any achieve this in the same way as colchicine. There are many steps in the successful synthesis, organization and activity of the spindle which will be susceptible to interference by introduced chemicals. Nor is it clear that in all cases the cycle continues after spindle inactivation in the presence of the chemical. Many substances are used by cytologists to accumulate meta-phases with scattered, contracted chromosomes suitable for karyotype analysis, but many of these may cause a complete block of division with-out the eventual formation of restitution nuclei. A few treatments, such as chloral hydrate and cold shock, are known to produce polyploid cells, indicating that the result, if not the mechanism, is similar to that of colchicine treatment. However, colchicine treatment tells us little about the control mechanism regulating the sequence of events in normal mitosis or the abbreviated cell cycles which it initiates. It acts late in the preparation for chromosome separation rather than on any control system; it may not be totally specific as interphase events also appear to be sensitive to colchicine and centromere division can also be affected; and many other chemically unrelated substances have similar effects.

A treatment which may be of more interest in relation to mitotic control mechanisms involves caffeine and related substances. These in-terfere with the alignment of vesicles at the spindle equator during cell plate formation and prevent cytokinesis while having no permanent visible effect on other events of the cycle (Kihlmann, 1967; Lopez-Saez et al., 1966; Pickett-Heaps, 1969). There has been no detailed study of the effect of caffeine on cell division, but there are indications that its effect on cytokinesis is achieved during recovery. In the gameto-phytes of *Dryopteris borreri*, caffeine treatment induces an immediate total block of mitosis which is, however, reversible, even after several days, on removal of the chemical. Recovery is rapid with no evidence of induced synchrony, indicating that the cells are halted at all stages of division and not accumulated at specific points. If the treated cells are all in interphase, there is no visible effect in the cells after recovery. If some cells are in mitotic stages, binucleate cells are found after re-covery, presumably due to a specific effect on cell plate formation (see Chapter 2, Section VA).

As with spindle organization, there must be many steps in the forma-tion of a cell plate and it is not surprising that a variety of chemicals, not obviously related to each other, as well as physical agents inhibit cytokinesis (Table II). However, particular interest centres on the effect of caffeine and several related substances such as theophylline, theobro-mine, ethoxy-caffeine and ethyl trimethyl, allyl trimethyl and tetra-methyl uric acid (Deysson and Benbadis, 1966; Kihlmann, 1949; Kihl-mann and Levan, 1949; Gosselin, 1940; Grant, 1965), because of their chemical similarity to the purines in RNA and DNA, adenine and guanine. Taken together with the general, reversible, blocking effect on mitosis of the presence of caffeine and the apparently specific inhibition of vesicle alignment at cytokinesis during recovery from caffeine treatment, this suggests the possibility that caffeine is acting as a base analogue interfering with some fundamental RNA-mediated control

TABLE II. Some inhibitors of cytokinesis

Inhibiting agent	Reference
PHYSICAL	
Hydrostatic pressure, dehydration, pricking	Mazia (1961)
Heat shock	Ikeda (1965)
Low temperature	Clowes and Stewart (1967)
CHEMICAL	
Aminopyrine	Fourcade et al. (1963); Ostegren et al. (1953)
Colchicine	Risueno et al. (1968)
Diazouracil	Berter et al. (1966)
Ethylene glycol	Risueno et al. (1968)
	Giminez-Martin et al. (1964)
Lindane	Gonzales (1967)
Mercaptoethanol	Nagl (1970b)
2-Methyl-amino 1,3-diaza-azulene	Niitsu (1958)
Sulphur deficiency	Tamiya (1964)
Uric acid derivatives:	
Ethyl trimethyl uric acid	Grant (1965)
Allyl trimethyl uric acid	Gonzalez (1967)
Tetra methyl uric acid	Deysson and Benbadis (1966)
Xanthine derivatives:	Kihlmann (1949, 1967)
Dimethyl xanthine	Kihlmann and Leven (1949)
(theobromine, theophylline)	Gosselin (1940)
Trimethyl xanthine (caffeine)	Lopez-Saez et al. (1966)
Ethoxycaffeine	Pickett-Heaps (1969)

mechanism. Other effects of theophylline and theobromine which have been noted include those on histamine release and on the activity of $3',5'$-adenosine monophosphate phosphodiesterase and epinephosine (Lichtenstein and Margolis, 1968; Wood et al., 1972). Caffeine and 8-ethoxy-caffeine are known to inhibit DNA repair and they cause chromosome damage. As with theophylline and theobromine, they are mutagens. Their mutagenic effects are reduced by guanosine.

Another purine derivative with an effect on cytokinesis is kinetin. Naylor *et al.* (1954) and Patau *et al.* (1957, 1961) showed that there was frequent failure of cytokinesis in the 2,4-D or auxin stimulated mitoses of tobacco pith explants in culture. Only when kinetin was added to the culture medium did the full cell cycle with cytokinesis occur regularly. Kinetin and other cytokinins frequently differ from other growth substances in their effects on cell activities and even on the DNA itself (Spang and Platt, 1972; Kende, 1971). The observations on cytokinesis raise the interesting possibility that the chemically related substances caffeine and kinetin have opposite effects on cytokinesis, the former suppressing it where it occurs naturally and the latter stimulating it where absent. A mutual antagonism between certain substituted xanthines and cytokinins would create the basis for a control mechanism for cytokinesis. However, stimulation of cytokinesis in natural coenocytes by a cytokinin has not been demonstrated, nor has a direct antagonism between caffeine and kinetin when the substances are administered together to a system where one substance alone is known to affect cytokinesis. On the contrary, Wood *et al.* (1972) have shown that in the tobacco pith system of Patau *et al.* (1957), theophylline and theobromine, but not caffeine, promote cytokinesis in the same way as kinetin but at higher concentrations. Further investigation of the effects on division of kinetin and caffeine applied separately and together might yield valuable information on the control and mechanism of cytokinesis.

The chemical induction of diplochromosomes has rarely been reported, and then the effect is not specific as polyploid cells are also induced. Diplochromosomes have been induced in dividing human tissue culture cells after treatment with colchicine or colcemid (Rizzoni and Palitto, 1973; Herreros and Gianelli, 1967) and mercapto-ethanol (Schwarzacher and Schnedle, 1966). They appear in dividing plant cells after administering 8-azoguanine (Ronchi *et al.*, 1965) or mercapto-ethanol (Nagl, 1970b).

There are few examples of experimental treatments inducing abbreviated cycles equivalent to those which produce polytene nuclei, and none occurs in higher plants. In *Tetrahymena* (see Mitchison, 1971; Jeffrey *et al.*, 1970; Brewer and Rusch, 1968) heat shock can cause repeated DNA replication in the macronucleus without chromosome or cell division, and low light conditions in *Chlamydomonas* cultures (Kates *et al.*, 1968) can induce two successive S periods without an intervening division but it has not been confirmed cytologically that the chromosomes become polytene.

6. Nuclear Reduction: Meiosis

The normal meiotic sequence produces a tetrad of cells with half the initial chromosome number and differing from each other genetically as a result of recombination due to chiasmata and random assortment of chromosomes. The sequence consists of two modified cell cycles (Fig. 8). In the first division, chromosome replication (centromere division) is omitted from an otherwise normal sequence of fundamental steps.

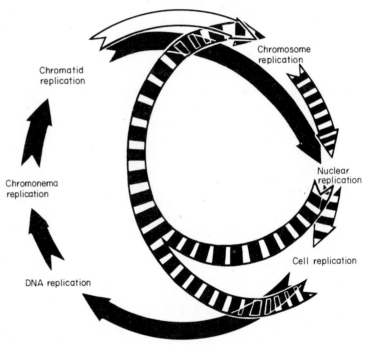

Chromosome replication

Chromatid replication

Nuclear replication

Chromonema replication

Cell replication

DNA replication

Fig. 8. The meiotic cycle.

Following in most cases a complete mitotic cycle, meiosis I (solid arrows) can be described as a modified cell cycle in which centromere division (chromosome replication) is omitted while the other essential events characterizing the mitotic cycle (Chapter 2, Section IIIG) continue, although their timing may be altered relative to other, non-essential events. There are also unique events in the meiotic cycle such as the formation of the synaptonemal complex and chiasmata. Meiosis II (broken arrows) follows with or without preceding cytokinesis and is characterized by the omission of DNA and chromonema replication and of chromatid formation. Centromere division (chromosome replication), anaphase separation of centromeres (nuclear replication) and usually cytokinesis (cell replication) are the essential events of meiosis II. The linked sequence of meiosis I followed by meiosis II provides only one replication of the DNA and subsequently of the chromosomes, associated with two replications of the nuclei. As a result, the chromosome number of the cells produced is half, and their nuclear DNA content is a quarter, of that in the cell entering meiosis.

There are other characteristic features of meiosis, involving the timing and extent of coiling and pairing and chiasma formation by homologous chromosomes. However, pairing is not unique to meiosis, being found widely in somatic mitoses in, for example, dipteran flies, and recombination between chromosomes is also known to occur in somatic tissues in, for example, fungi. As a result of the combination of events in meiosis, the first division does not result in the formation merely of two nuclei each containing some intact chromosomes, but of two equivalent nuclei each containing a balanced, matching set of chromosomes. In the second meiotic cycle, DNA replication (the S period), chromonema replication and chromatid replication are all omitted and the chromosomes of anaphase I enter prophase II essentially unchanged, except perhaps in the degree of coiling, and often without an intervening interphase. The subsequent events of meiosis II are typical of the full cycle, with centromere, nuclear and cell replication. Over the two cycles of meiosis, both nuclear and cell replication have occurred twice, while DNA, chromonema, chromatid and chromosome replication have only occurred once: hence the reduction in chromosome number.

Some of the modifications found in abbreviated mitotic cycles have also been imposed on the already modified cycles of meiosis. In some species, cytokinesis is omitted after the first or second or both divisions of meiosis in the pollen mother cell or embryo sac mother cell or both (see Chapter 2, Section IIB) to produce a two or four nucleate coenocyte. In the degenerate meiosis of some apomicts (Gustaffson, 1946) where the normal cytogenetical consequences of meiosis are disadvantageous, nuclear replication may fail at meiosis I or II, to give unreduced restitution nuclei. In other cases one division of meiosis may be omitted entirely or meiosis I may revert to an essentially mitotic type of division including centromere division.

7. Nuclear Reduction: Somatic Reduction

Several types of nuclear change can be included under the heading "somatic reduction". One is the halving of chromosome size with no change in number, most easily explained as the result of an extra division of the centromere in an unreplicated, multistranded chromosome (Fig. 9). This has been reported in the abnormal pollen grains of an infertile plant of *Alopecurus myosuroides* (Johnsson, 1944) and was apparently induced by 5-fluoro-uracil treatment of mouse ascites tumour cells (Lindner, 1959).

Usually, however, somatic reduction is taken to mean the separation of whole chromosomes at sporophytic mitosis, by suppression of

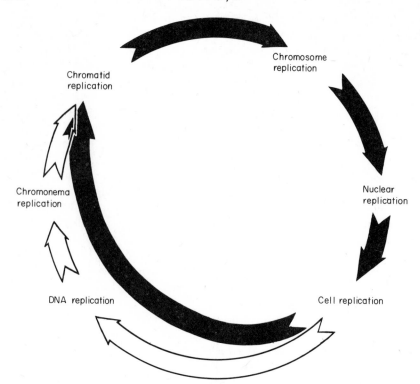

FIG. 9. Somatic reduction: Cycle 1.

An abbreviated cycle (solid arrows) in which chromatid formation, centromere division, nuclear replication and cytokinesis occur in the absence of the replication of DNA and other structural components of the chromosome would result in a reduction in size of a multistranded chromosome. Such a cycle may explain the occasional reports (see text) of chromosomes becoming smaller without a change in morphology or number.

centromere division, to give daughter nuclei with a reduced number of replicated chromosomes (Sinha, 1967; Sharma and Mookerjea, 1959). This does not necessarily imply the regular separation of balanced and equal genomes, but it has been deduced that this does happen in roots of cycads (Storey, 1968a) and angiosperms (e.g. Brown, 1947; Wilson *et al.*, 1951; Wilson and Chang, 1949) (Fig. 10). An exact somatic parallel with meiosis, "somatic meiosis", where homologue disjunction is accompanied by chromosome pairing and "chiasma formation", that is breakage and reunion at corresponding loci of homologous chromosomes, has been claimed to occur in *Haplopappus gracilis* tissue cultures (Mitra and Steward, 1961) and *Rhoeo* roots (Huskins, 1948b). However,

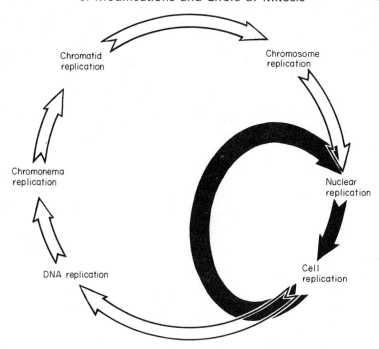

FIG. 10. Somatic reduction: Cycle 2.

An abbreviated cycle (solid arrows) in which separation of chromosomes to daughter nuclei without preceding replication would result in nuclei with a reduced chromosome number. The formation of nuclei with balanced and equal genomes in somatic tissues as has been occasionally claimed (see text) would presumably require in addition some mechanism such as chromosome pairing to ensure the regular disjunction of homologues.

pairing and chromatid translocations are not unique to meiosis and neither mitotic metaphases with reductional groupings of chromosomes nor the presence of haploid cells (Huskins, 1948a, b; Storey, 1968a) provides conclusive evidence of somatic reduction. For example, haploid cells may eventually predominate where mitotic instability (see Section IIID) gives rise to a range of hypodiploid aneuploid cells among which balanced haploids have a competitive advantage (Kafer, 1961). Reductional groupings at metaphase have not been followed through anaphase to confirm the formation of haploid telophase nuclei. If this occurs at all, it is very rare (Srinivasachar and Patau, 1958). Glass (1961), having statistically analysed the distribution of metaphase chromosomes in squashed cells of *Bellevalia romana* $(2x=8)$, concluded that in this species reductional groupings occurred no more frequently than expected on the basis of random distribution. Perhaps regular separation of intact genomes can only be expected in cells with previous somatic pairing, and a similar statistical analysis should be carried out on material in which this has been reported, such as *Haplopappus* cultures and *Rhoeo* roots. Meanwhile, until more critical evidence is available, the claims that somatic reduction occurs naturally in cells with a zygotic chromosome complement should perhaps be accepted with caution, as should the claims that it can be experimentally induced (Gallinsky, 1949; Chen and Ross, 1963; Kodani, 1948; Sharma and Bhattacharyya, 1956; Huskins, 1948a, b; Huskins and Chen, 1950).

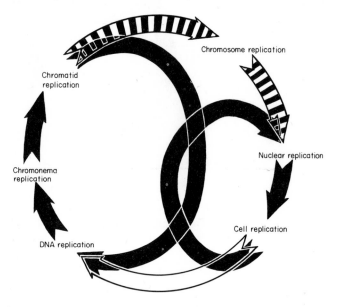

Fig. 11. Somatic reduction: Cycle 3.

Regular reduction in chromosome number over several cycles, with an exact halving of the chromosome complement as a result of chromosome pairing at each anaphase, occurs in the mosquito *Culex* in iliac epithelial cells which have previously become endopolyploid (Grell, 1946). Several successive cycles of DNA and chromatic replication without chromosome, nucleus or cell replication (solid arrows to left of diagram) produce chromosomes with up to 32 chromatids. Following a transition sequence (broken arrows) during which the centromeres divide within one cycle to release all the daughter chromosomes from each other, the cells undergo several successive abbreviated cycles consisting of only two essential events, nuclear replication by ana-phase separation of centromeres and cell replication by cytokinesis. Abbreviated cycles cease when the original complement existing prior to endopolyploidy has been restored by a compensating number of reduction divisions.

A similar sequence may explain the observations on *Vicia faba* crown gall tissue (Rasch *et al.*, 1959).

There is no doubt that precise somatic reduction can occur, if rarely, in cells which have previously undergone nuclear doubling. The clearest example concerns the endopolyploid iliac epithelium cells of the mosquito, *Culex* (Grell, 1946). Cells with up to 32 chromatids per chromosome pass through several modified cycles following complete separation of chromatids at the first division. Homologous unreplicated daughter chromosomes associate in pairs as they move to the poles at each anaphase, to reappear as pairs at metaphase after a short interphase with no DNA, chromonema or chromatid replication. Disjunction of each pair of unreplicated chromosomes transmits half the number of chromosomes to each daughter nucleus. This is repeated until the zygotic complement is restored. Thus a series of cycles lacking chromosome, nuclear and cell division has been followed by a compensating series lacking DNA, chromonema and chromatid replication (Fig. 11), homologue pairing at each anaphase providing the mechanism to ensure regular disjunction of balanced genomes. A very similar situation is indicated by the observations of Rasch *et al.* (1959) on crown gall tissue of *Vicia faba*. In *Chlamydomonas* cells grown under low light, the nuclear DNA content is reduced by two successive mitoses without any intervening DNA synthesis, compensating for an immediately preceding sequence of two successive S periods (Kates *et al.*, 1968).

8. Abbreviated Cycles and Cell Differentiation

Among the unanswered questions concerning the relationship of modified cycles to subsequent cell differentiation remain those relating to the origin and establishment of the cycles, the nature of the relationship, the effect on cell development and the control mechanism.

(a) *The origin and establishment of abbreviated cycles.* While there are differences between mitotic divisions in different organisms, particularly in the protozoa, algae and fungi (see e.g. Du Praw, 1970; Leedale, 1970), in all eukaryotes the same essential events occur (Chapter 2, Section IIIG). Presumably, the first fundamentally modified cycle to become established during evolution was meiosis. Involving a number of important modifications of the mitotic cycle, it arose with the development of sexual reproduction and is found in most eukaryotes with only minor variations. Other modifications of mitosis, resulting in nuclear doubling and reduction, could have been tolerated in a sexual species and become established as part of normal development only when multicellular organisms with somatic cells distinct from the germ line arose. As many of the modifications result in suppression of the cell division necessary for diverging specialization of the protoplasts, it is understandable that many of the established mitotic modifications are restricted to a stage of tissue development when growth and increased activity are required rather than the formation of new cells for further functional diversification. The resulting nuclear changes are presumably of physiological significance (see below), but similar changes in most Type 2 cell divisions would interfere with early stages of tissue development and in Type 1 "germ line" cells would disturb the normal cycle of nuclear changes associated with the alternating sporophyte and gametophyte generations. Many recent mutant deviations including "short-circuits" in mitosis and meiosis, clearly demonstrating the specific control of events during division, have been described in Type 1 and Type 2 cells (e.g. Rees, 1961; Warr and Durber, 1971; John and Lewis, 1965; Haque, 1953), but as the effects on development or reproduction are almost always deleterious, they rarely become established. The rare exceptions include the suppression of cell division at meiosis during certain types of embryo sac development (see Chapter 2, Section IIB) and in some diplosporous apomicts the suppression of nuclear and cell division at pre-meiotic mitosis or meiosis (see Sections I, IID2, 3) or the reversion of meiosis to mitosis (Gustafsson, 1946) to maintain fertility in the absence of recombination. No doubt new forms of modified cycles, particularly those leading to nuclear doubling, are still arising in Type 2 or Type 3 cells by mutation and becoming established in normal development, but in the absence of extensive and systematic comparisons of these relatively inaccessible and inconspicuous division cycles between individuals of the same species, this cannot be conclusively demonstrated. The fact that equivalent cells of related genera sometimes show different types of nuclear doubling (see below) provides circumstantial evidence of such changes in the recent past.

(b) *The relationship between abbreviated cycles and differentiation.* It is necessary to establish whether abbreviated cycles initiate differentiation, inevitably occur as part of differentiation or merely sometimes accompany differentiation.

The nuclear events of meiosis are not the cause of the accompanying morphogenetic events in the life cycle. Preparations for sporogenesis extend back many cell generations prior to meiosis and normal gametophyte development is maintained even if nuclear reduction fails. In the case of apospory, the female gametophyte can develop normally without any preceding meiosis. The modified nuclear cycle is, like the accompanying cytoplasmic changes, an expression of differentiation in the germ line mediated by the immediate environment of the cell.

Somatic reduction of the zygotic complement, if it occurs at all, has no apparent place in normal development. As a sporadic event, it may have important cytogenetical implications in reversing polyploidy (Huskins, 1948a, b; Sharma, 1956). Somatic reduction, or reversal, of previous nuclear duplication may be more widespread than the few examples already recognized and it is presumably of significance in cell development. Perhaps it is associated with, and even necessary for, resumption of mitosis and redirection of cell activities in endopolyploid cells at a particular stage of development.

Several observations indicate that nuclear doubling frequently accompanies but does not initiate cell specialization (Partanen, 1963; D'Amato, 1964). Although diplochromosomes and polyploidy occur widely in root and stem cortex and stele (D'Amato, 1952; Tschermak-Woess, 1956b) even in plants from haploid zygotes (Bennici *et al.*, 1968) there is a significant minority of plants, including *Crinum* $(2n=2x=22)$, *Helianthus tuberosus* $(2n=6x=102)$, *H. annuus* $(2n=2x=34)$, *Crepis capillaris* $(2n=2x=6)$, *Medicago sativa* $(2n=16, 32, 64)$ and *Lactuca sativa* $(2n=2x=18)$ (D'Amato, 1952; Street, 1966b; Sunderland, 1973b; Van't Hof, 1974; Brunori, 1971; John and Lewis, 1968; Partanen, 1963; Tschermak-Woess, 1956b), in which the corresponding nuclei remain at the zygotic telophase level. It is not known whether nuclear doubling is absent from all cells in such plants. Information on the nuclear condition of antipodal, suspensor and tapetal cells would be of interest. Clearly, although widespread, nuclear doubling does not invariably accompany cell differentiation, except possibly in very specialized anther and ovule cells. Nor does nuclear doubling initiate differentiation. Induced or spontaneous polyploidy in meristems of intact plants does not seriously disturb development. The polyploid cells perpetuate themselves by regular mitoses and apart from their greater volume, which is largely compensated in a chimaera by a reduced cell number, they and their polyploid derivatives appear no different from unaffected cells (Satina, 1959). There is no evidence of precocious or misdirected differentiation in the polyploid cells. In tissue and callus cultures, polyploid cells are frequently present (e.g. Torrey and Fosket, 1970; see Section IID2) and although they may have different nutritional requirements their ability to continue to divide does not appear to be restricted or their differentiation promoted or directed by their nuclear condition.

There is no evidence that nuclear doubling or reduction is a causal factor in cell differentiation (Hsu, 1961).

(c) The effect of abbreviated cycles on subsequent cell development. Critical

observations on the effect of nuclear changes brought about by abbreviated cell cycles are not easily obtained because the "control" cell, a similar cell in similar circumstances but with the nucleus unchanged, is usually unavailable, particularly within a multicellular system. However, it is possible to speculate on the basis of other known consequences of polyploidy or haploidy.

Regular meiosis will produce haploid or polyhaploid cells. Haploidy normally allows the expression of recessive alleles which would be masked by dominant alleles in a heterozygous diploid cell. However, at meiosis, and independently of the nuclear reduction, the sporophyte developmental sequence is suppressed and replaced by the gametophytic sequence. The whole complement of genes controlling sporophytic growth must be simultaneously inactivated and consequently selection will not operate against them in the haploid generation. Only those recessive alleles of the activated gametophyte gene complement which have not been eliminated by selection in previous generations will be exposed as a consequence of meiosis. Thus the direct effect of nuclear reduction in its normal context is unlikely to be deleterious.

Where further cycle modifications are superimposed on meiosis or the preceding mitosis, or indeed where meiosis reverts to mitosis (Section IID8(a)), and unreduced spores are formed, the effects on gametophyte development may even be in some ways beneficial. The switch from sporophyte to gametophyte development still takes place, but the gametophytes will inherit and express any heterozygosity for gametophyte-controlling genes which was present but latent in the parent sporophyte. This heterozygosity, not normally possible in a diploid species, may confer flexibility and vigour to the gametophyte over and above that derived merely from the increase in nuclear size. Increased gametophytic vigour produced in this way may be a neglected factor in the successful establishment of some polyploids and apomictic diploids.

However, if nuclear reduction occurs in somatic tissues, the effects are likely to be deleterious. The cells will probably be smaller (Storey, 1968a). The same set of genes, whether in the gametophyte or sporophyte, will be operating as in the closely neighbouring cells; there will be no generation switch and no isolation from the surrounding tissues. After somatic reduction in the gametophyte, genetically unbalanced nuclei are inevitable in all but a few balanced autopolyploids. In the sporophyte, nuclear reduction will create a cell with a reduced maximum size and in most cases one in which recessive alleles are operating. The resulting cells are likely to be at a competitive disadvantage in a developing tissue. Little is known about competition between cells of different genotype in an organized but chimerical tissue, but it is possible, even likely, that any cell type at a significant disadvantage will be eliminated.

The widespread occurrence of nuclear doubling in animals as well as plants (Geitler, 1953), and the high levels of ploidy reached, up to the equivalent of 32 768n though most cells are 4n, 8n or 16n, indicate that increases in nuclear DNA can be advantageous, at least in certain cells. The cells concerned are usually large, and while some are relatively simple parenchyma cells, others have clear secretory, storage and/or absorptive functions (e.g. Tschermak-Woess, 1956b). The more extreme examples of nuclear replication occur in highly specialized and unusually large cells thought to have high levels of synthetic activity, such as antipodal and tapetal cells and epidermal hairs. Increased cell size regularly accompanies nuclear doubling (see e.g. Tschermak-Woess, 1956b, Figs 4 and 6), even in Type 1 cells (Satina, 1959). It is likely that the increased number of identical DNA templates allows a higher rate of transcription, leading to increased metabolic activity. This in turn will support a larger volume of cytoplasm. Where a highly specialized tissue has a restricted range of functions, the potential for functional diversification resulting from division into many small cellular units (see Chapter 2, Section VA) is of no advantage, and tissue growth by cell enlargement alone achieves the necessary increase in activity while avoiding the unnecessary expenditure of resources or disruption of synthesis which cell division would cause (Nagl, 1973). Nuclear duplication permits this growth without division as indicated by the comparison of protophloem and metaphloem companion cells in *Vicia* (Resch, 1958). The more uniform the tissue, the greater is the extent to which growth can occur without cell division. In the minority of species where, at least in root and stem tissues, growth has remained closely linked to cell division, it may be that the abbreviated cycles necessary for nuclear doubling have never arisen, or for some reason smaller protoplast units are desirable. Alternatively, there may be sufficient numbers of replicates of the DNA templates, due to polyploidy, gene amplification or multistranded chromosome structure, to support extensive tissue growth of cell enlargement without additional nuclear doubling. Further information is needed on the relationship between the levels of somatic nuclear doubling reached and the number and DNA content of the genomes, and the effect of environmental conditions on the degree of nuclear doubling within a tissue (Witsch and Flügel, 1951).

Having considered the possible role in development of nuclear doubling, it remains to be considered whether there are significant

differences in effect between the four distinct types of duplication, polyteny, diplochromosomes, polyploidy and coenocytes. Certain tissues do
contain predominantly one type of nucleus. Cortex and stele cells are
predominantly polyploid. Tapetum is usually binucleate and may be
polyploid or polytene as well. Embryo suspensor cell nuclei are frequently polytene, antipodal cells are frequently polytene and sometimes
multinucleate, and endosperm often at least temporarily coenocytic
and later polytene or polyploid (D'Amato, 1952; Maheshwari, 1950;
Stephen, 1974). However, in many cells two or more types of nuclear
doubling are superimposed in the same nucleus (e.g. Hasitschka, 1956)
and in other cases a tissue characterized by one type in one plant is
characterized by a different one in another species. For example, the
premeiotic mitoses in apomictic *Allium odorum* produce diplochromosomes while in *Dryopteris borreri* they produce polyploid nuclei, and
xylem-conducting elements become coenocytic in some ferns (List,
1963) but polyploid in most angiosperms. Thus, it may be chance that
determines which of the alternative types of nuclear doubling become
established in a particular cell type. It is difficult to see that polyploidy
and diplochromosomes of the equivalent level of replication would
differ markedly in their effect. However, polytene nuclei may be different, as they frequently appear in a prophase-like condition rather
than fully dispersed, and they include the nuclei with the greatest DNA
content. It may be no coincidence that the most curtailed cycle, with
the minimum disruption of cell activities, is the one which produces
the highest levels of nuclear duplication. Coenocytes also differ in one
obvious respect from other types: the replicated nuclear DNA is dispersed in separate units throughout the cytoplasm. This may aid in-
tracellular communication, particularly important in very large cells,
and allow for greater cell volumes than could be supported by an equivalent complement localized within a single nucleus at one point in
the cell.

(*d*) *The mechanism of control of abbreviated cycles.* A conspicuous feature
of many of the known abbreviated cycles is that the cell cycle reverts
from some point between DNA synthesis and the onset of interphase
to a point prior to DNA synthesis where a new cycle can be initiated.
The cells are not held permanently at the point where the cycle is
blocked. Chemically induced mimics of these cycles, produced when
a single stage is specifically blocked, suggest that in natural cycles, control of each fundamental step in the cycle by means of specific inhibitors
is the basis of the regulation of the mitotic cycle and its modifications.
In view of the suggestion that division is the basic activity of a cell and

that its suppression is necessary for full differentiation (see Section I and Chapter 2, Section VC; Brown and Dyer, 1972), the curtailed cycles in differentiating cells could be interpreted as the step-wise suppression of mitosis over two or more cycles instead of an abrupt cessation of all division preparations. There are examples where more than one type of short cycle follow each other during the development of a cell in the sequence of progressive curtailment of the cycle. In the tapetum of tomato (Brown, 1949; Mechelke, 1953) the full cycle is first replaced by one in which cytokinesis is omitted, producing binucleate cells, and then by one which is further curtailed to produced polyploid nuclei, before nuclear doubling ceases altogether.

It is tempting, on the basis of these facts, to produce models to explain the control of mitosis and the abbreviated cycles during differentiation (Brown and Dyer, 1972). One possible model incorporates an underlying regulatory "clock" which progresses unaffected by the blockage of one of the steps it controls or by the consequent failure of subsequent steps. Another model does not incorporate a superimposed regulatory system, but postulates a sequential process in which the completion of one step initiates the next event in the cycle. Blocking one step thus inevitably eliminates all the later ones but does not entirely explain why the cell reverts to a stage just prior to DNA synthesis. One suggestion is that as at each stage an activator of the next stage is produced, so also is a repressor of DNA synthesis renewed. Another suggestion is that disruption of the mitotic apparatus prematurely releases enzymes which can break the chemical bonding which is responsible for chromosome contractions and consequent gene inactivation. The released enzymes would cause relaxation of the chromosomes and exposure and reactivation of the genes, including those responsible for initiating DNA replication (Brown and Dyer, 1972). In animal cells, concurrent spindle degeneration and chromosome despiralization have been observed as cells which reached a normal metaphase develop tetraploid restitution nuclei (Hsu and Moorhead, 1956).

However, it must be remembered that the mitotic cycle is more versatile than indicated by the examples used for these models. The predominance of nuclear doubling cycles among modified mitoses may be more a result of the adaptive advantage of the consequences than of any underlying control mechanism. Other "short-circuits" are known which involve omission of DNA synthesis, or even centromere division alone, as described previously. Thus not all cycles include the S period, or revert to G_1 after blockage. Furthermore, where more than one type of abbreviated cycle occurs within one differentiating cell, they do not always occur in a sequence corresponding to progressive contraction of the cycle. In some cases the sequence is reversed as in *Chlorophytum* roots (Storey, 1968b) and *Antirrhinum* tapetum (Mechelke, 1953) where the cells become first polytene and then binucleate. In other examples, short cycles alternate with normal mitosis. In endosperm development of the Helobial type, a single full mitotic cell cycle is followed by several cycles of mitosis without cytokinesis to form a coenocyte, before reverting to normal cell division again and in some cases finally to further modified cycles producing polyploid or polytene nuclei. In *Phaseolus* suspensor cells (Nagl, 1970a) and hybrid *Tulbaghia* endosperm (John and Lewis, 1968, Fig. 20) nuclei which have become polytene once more undergo mitotic division. In old fern roots, the apical initial cell becomes polyploid before resuming full mitoses (Avanzi and D'Amato,

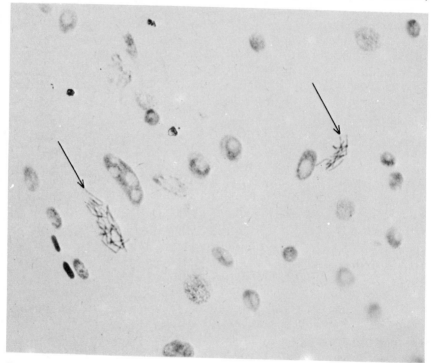

Fig. 12. Diploid and tetraploid mitoses in untreated root tips of *Tulbaghia violacea*, 2x = 12. × 400.

The diploid complement (arrow, right) predominates in root-tip squash preparations and occurs exclusively in pollen mother cells. The zygote was presumably diploid. The tetraploid mitoses (arrow, left) probably represent cells which have undergone one abbreviated cycle to become polyploid before resuming mitosis. Several of the larger interphase nuclei associated with the tetraploid mitoses are also probably endopolyploid.

1967). The tetraploid divisions found at the back of root-tip meristems in *Tulbaghia* species (Fig. 12) and the polyploid divisions in endosperm (e.g. Stephen, 1974) and several other tissues (Tschermak-Woess, 1956b) probably arise in a similar way. Thus, longer or even full cycles can be restored.

These additional facts must be accommodated in any satisfactory model for the control of mitosis and its suppression during differentiation. Such a model is not yet available.

III. ERRORS OF MITOTIC CELL DIVISION IN RELATION TO DEVELOPMENT

Estimates of the number of cells in a single mature leaf vary from 5×10^6 to 100×10^6 (Maksymowych, 1963; Sunderland, 1960). Clowes (1972a) has calculated that a maize root produces about 150 000 cells per day. Dittmer (1937, 1948) reports that fibrous-rooted species have more than 1000 roots (one rye plant had nearly 14×10^6), and even small species with a single main root have more than 100 subsidiaries. On the basis of these figures, a conservative estimate of the number of cells in a small annual would be 10^8 and most plants would contain a number several orders higher. As every cell formed requires a mitotic cell division, if there was a serious error in the complex mechanism of mitosis only once in every million divisions then every plant must contain many mutant karyotypes, some of them present in many cells. Mitosis would have to be almost unbelievably faultless for the majority of plants to preserve the zygotic complement unchanged throughout vegetative growth. Yet the level of detectable spontaneous karyotype mutation in somatic tissues, when it is reported at all, is very low, except in the rare cases of somatic instability (see Section IIID), compared with the frequency after physical or chemical treatments or irradiation (Satina, 1959, Dermen and Stewart, 1973; Davidson, 1959, 1960, 1965b; Brumfield, 1943). Davidson (1960) reports one spontaneous mutant karyotype for every 2000–4000 dividing cells in *Vicia faba*. Rutishauser (1956) reported no mutant karyotypes in over 2000 *Trillium* root-tip cells.

There are several reasons why the products of mitotic errors may be more frequent than the traditional root-tip preparations would seem to indicate. First, at any moment only a very small proportion of the cells in the plant are of Type 1 or 2 and still proceeding through the mitotic cycle, and if asynchronous, only about 2% of these will be in metaphase or anaphase, the only stages at which many of the errors and their consequences can be detected in cytological preparations. Only the more extreme euploid and aneuploid changes in chromosome number can be recognized with any certainty in interphase or non-dividing cells. Second, many of the tissues which contain dividing cells such as cambia, leaf and shoot meristems, and developing flowers, fruits and embryos, are rarely examined at all. Karyotype changes may be more common in ovules and endosperm than in roots (Rutishauser, 1956). Even in the familiar root-tip meristems, cytological analysis of a few thousand cells at metaphase is a major task, but there may be that many at any given moment in every root tip. Third, there may be strong selection pressures operating against some or all the products of aberrant divisions; cells with altered gene complements may divide and grow more slowly or not at all and be replaced by cells derived from unaffected initials (Davidson, 1959). More intense selection, rather than greater stability, may explain the lower frequency of aberrant karyotypes in root tips compared with other tissues (Rutishauser, 1956). Fourth, Type 1 and early Type 2 divisions might be less prone to error than mitoses taking place shortly before division ceases; there would then be fewer errors detected at the apex of the root than in the region at the back of the meristem which is less frequently examined, at least in squash preparations. Fifth, many of the more common structural changes such as paracentric inversions and small deletions and translocations are difficult to detect at mitosis and even pericentric inversions and major interchanges will be missed if the chromosomes are small or numerous. Thus the nuclear changes most likely to be seen are those which are accurately reproduced at mitosis early in root growth to produce a large number of derivatives with altered karyotype without seriously disturbing the genotype. Presumably many other karyotype mutations take place undetected.

These observations indicate that it is possible, even probable, that in most if not all plants the karyotype is not as stable as frequently assumed. Changes in the chromosome complement occur which are not an integral part of differentiation, as were those described earlier in this chapter. They are superimposed upon, and may interfere with, normal development. The nature of these spontaneous changes, and their possible effects, must be considered further.

Mitotic errors (Fig. 13) fall into five main groups:

(1) Failure to maintain the structural integrity of the chromosomes, probably due to errors at replication. These result in structural rearrangements within the complement. Breakage alone produces centric and acentric chromosome fragments and fragments with incomplete centromeres. Breakage followed by reunion also produces inversions, interchanges, deletions, duplications and dicentrics. Initially none of

① Phase of the cell cycle at which mitotic error occurs

② Division activity initially modified by error

③ Initial effect of mitotic error

④ Initial product of mitotic error

⑤ Ultimate effect on development and fertility

FIG. 13. The origins and consequences of chromosome mutations (see text).

these will seriously interfere with cell development except in instances of significant gene position effects on cell activity, and probably then only in haploids where there is no unchanged homologue present. However, during the succeeding mitosis, fragments with no or defective centromeres will fail to move to the poles, and daughter nuclei deficient in one or more chromosome segments, segmental aneuploids (Dyer *et al.*, 1970), will be formed. Dicentrics will also usually give rise to duplications and deficiencies (Davidson, 1959; but see Sunderland, 1973b). Inversions, interchanges, deletions and duplications, and centric fragments will undergo regular replication and division at subsequent mitoses which therefore do not result in further qualitative changes in the genotype (e.g. Rutishauser, 1956). However, deletions will expose to selection recessive alleles on the unaltered homologue and are less likely to be maintained (Davidson, 1959, 1960).

(2) Errors of centromere function at metaphase and anaphase. These will lead to failure of chromosome congression and disjunction, usually of only one or a few chromosomes of the basic genome, producing nuclei lacking or gaining one or more whole chromosomes: chromosomal aneuploids. The modified complement is genically unbalanced and may expose recessive alleles, but can be accurately reproduced at subsequent mitoses.

(3) Errors of spindle organization leading to "split" or multipolar spindles. This causes one or both sets of daughter chromosomes to be distributed between two or more daughter nuclei, which are therefore deficient in one or usually several chromosomes. These chromosomal aneuploid nuclei can reproduce themselves accurately if they are able to survive in competition with other genically balanced cells.

(4) Errors of spindle activity. These will usually result in failure of disjunction of the whole complement and result in a single, polyploid but genically balanced nucleus, which can be reproduced at later normal mitoses.

(5) Errors of spindle orientation. These will not alter the chromosome or gene complement of the daughter nuclei, but by altering the position of daughter cells relative to their neighbours, may affect the interaction of the genotype with its cellular environment, and thus disturb subsequent development.

Those mutant cells which survive through mitosis and development (Davidson, 1959) will later be subject to new selection pressures during reproduction. Structural rearrangements will affect pairing and bivalent formation at meiosis, and aneuploidy and polyploidy will result in univalents and multivalents. Some will result in genically unbalanced, inviable meiotic products; all will affect recombination. Those which survive in haploid gametophytes or sporophytes homogygous for the nuclear change will lose the protecting effect of heterozygosity, and gene deficiencies will put more of them at a disadvantage. Those which survive this further challenge make up the pool of altered complements which forms the basis of karyotype evolution. The cytogenetical implications of chromosome mutations are widely discussed elsewhere (e.g. John and Lewis, 1965, 1968; Lewis and John, 1963) and are beyond the scope of this chapter. Here we return to consideration of the implications of these changes to development in the heterozygous chimaera in which they first arose.

Apart from those in which the chromosome segments are merely rearranged, the cells produced following the completion of the mitosis in which the errors described above occurred will be aneuploid, polyploid or displaced. There is little direct information on the effect of the first two of these conditions on the development of individual cells within an otherwise normal intact plant, but it is possible to make some general statements on the likely consequences based on their effects on plants, developed from mutant zygotes, in which all the cells are affected.

A. Aneuploidy

In this context, it is of no significance whether the aneuploidy is segmental or chromosomal. Species differ in their tolerance of chromosomal imbalance (see John and Lewis, 1968). Cultivated hyacinths occur with all chromosome numbers from $2x=16$ to $4x=32$, and the aneuploids are viable and fertile. Hypodiploid aneuploids, however, are not represented, although 'Tubergen's Scarlet' and 'Scarlet Perfection' are diploids heterozygous for a large deletion in which the affected segment is larger than the smallest chromosomes of the set. It is generally true that gain of chromosomes is more easily tolerated than loss. Similar tolerance of aneuploidy is reported in several wild species, including *Narcissus bulbocodium*, *Erophila verna* and *Claytonia virginica*. The basis of this tolerance is not known, but past gene amplification in the diploid may have resulted in tolerance to additional genetic material but not to loss. In many other genera, for example *Tradescantia*, there is little tolerance of aneuploid variation and the triploid is almost sterile due to the inviability of the aneuploid gametophytes resulting from unequal meiotic disjunction.

In addition to the inherent differences between species, factors which affect the tolerance of aneuploidy, particularly in the gametophytes, include the degree of polyploidy, the proportion of the genome gained or lost, and the specific phenotypic effects of the chromosome concerned. Loss of chromosomes is rarely tolerated in diploids, where recessive alleles would be exposed, or in haploids, where total absence of part of the genome would result. In polyploids, loss of a particular chromosome cell can be correlated with a specific effect on the plant phenotype. Similarly in diploids with an additional chromosome, the particular chromosome concerned can be identified with the altered appearance of the plant.

It is likely that the same factors operate when an aneuploid cell first arises within a plant, and they will affect the survival of the mutant cell and its derivatives. Many aneuploid cells are probably eliminated. Aneuploid cells are most likely to prosper when, for example, a small chromosome or segment is added to a polyploid complement in a species tolerant of chromosome imbalance. No doubt all aneuploid cells are affected at the biochemical or fine structural level, but this has rarely been investigated. If the mutant cell gives rise to a significant amount

FIG. 14. Complementary adjacent mutant sectors in the epidermis of a Cox's Orange apple such as could form from the reciprocal products of chromosomal non-disjunction.

The sector to the right is narrower, paler and has several infected necrotic spots. The other sector is broader, more intensely pigmented and healthy. Such a chimaera could result from non-disjunction in the epidermis early in the development of the ovary primordium as a result of which one daughter cell, and the lineage derived from it, inherited an extra chromosome and along with it the ability to produce more pigment. The other daughter cell and its derivatives could then have inherited a complement lacking this chromosome and, apparently as a consequence, exhibited reduced pigment production, reduced growth rate and increased susceptibility to infection.

of tissue or a whole organ, then the plant phenotype may be visibly affected, depending on the chromosome concerned. Some "sports" may be caused by aneuploidy rather than gene mutations. Occasionally, both reciprocal products of non-disjunction may give rise to reciprocal twin "mutant" sectors (Fig. 14). Where large numbers of chromosomes are involved as a result of split spindles, cell size and division rate may be affected as for polyploidy. Balanced complements at lower ploidy levels, including haploidy, could also arise rarely in this way (see Section IIID).

B. Polyploidy

Increased levels of polyploidy, in which the gene balance is maintained, are more easily tolerated than aneuploidy. Polyploidy affects such cellular characteristics as cell size and the duration of the division cycle. This is true of chimaeras as well as wholly polyploid plants. Polyploid cells are maintained in meristems after colchicine treatment (Davidson, 1965a, b). The slower division rate of the polyploid cells largely compensates for the increased cell size at the shoot apex (Satina, 1959) and little distortion of the meristem results. As tissues develop, however, the effect of polyploidy may become more obvious. Colchicine-treated seedlings can develop mis-shapen or buckled leaves, presumably due to differential growth of tissues of different ploidy levels. Purely polyploid shoots and roots will show the typical "gigas" qualities and altered physiological characteristics (Hall, 1972).

C. Cell Displacement

The control of spindle orientation is clearly an important part of morphogenesis in an organism where cells are fixed in position by rigid walls.* At every stage of organized development, the orientation of the spindle of a cell is determined in relation to the development of the structure as a whole. Frequently it can be demonstrated that the fate of a cell is decided by its position in relation to other cells and to cytoplasmic gradients. Mis-orientation of the spindle will displace the daughter cells in relation to these influences and affect their subsequent development. This is particularly obvious in situations where, following an unequal cell division, the daughter cells normally follow strikingly different developmental paths. Thus errors of division orientation induced in the pattern of divisions resulting in formation of a stomatal complex cause abnormal development of the misplaced cells (Stebbins et al., 1967; Stebbins and Shah, 1960). Similarly, if the pollen grain

* See Chapter 2, Section VI.

of angiosperms divides in a direction perpendicular to the normal situation, where the generative nucleus forms against the so-called dorsal wall, the two daughter cells fail to differentiate into either vegetative or generative cells and the pollen grain is sterile (La Cour, 1949; Barber, 1941).

Another situation where the usual orientation of mitosis can be disturbed, but in this case usually with no visible effect on development, is in the layered shoot apex. Commonly, layer 1, the outer layer of the tunica, maintains its integrity and gives rise in the leaf only to the epidermis. When all the cells are genetically identical it would be very difficult to detect if this were not so, except by detailed examination of cell wall orientation during development. However, in some GWG and GWW chimaeras (Kirk and Tilney-Bassett, 1967; Nielsen-Jones, 1969), where the inner tunica layer, layer II, is marked by a mutation preventing chlorophyll formation, it can be seen that small islands or sectors of green tissue have intruded from the epidermis into the otherwise white margin of the leaves. White sectors in the otherwise green leaves of WGG plants are also known (Stewart et al., 1974). These can be attributed to mis-orientated periclinal divisions which have caused cells derived from layer I, which in the epidermis develop few chloroplasts even when genetically "green", to be displaced into the mesophyll tissue normally derived from layer II where any potential for chlorophyll production is revealed. In GWG *Saxifraga sarmentosa* (Fig. 15) the islands of green mesophyll cells are all small but frequent and scattered throughout the white tissue. Evidently the mis-orientated divisions occurred frequently and all at about the same time, late in leaf development. In *Abutilon striatum*, *Pelargonium zonale* (Kirk and Tilney-Bassett, 1967; Stewart et al., 1974) and *Pelargonium peltatum* cv. 'L'Elegante' (Tilney-Bassett, 1963) the "islands" are few and sometimes large, even affecting whole leaves or shoots. Here the mis-orientated divisions must have been rare and at all stages of leaf growth, and even in the shoot apex itself. A correlation between the presence of these "islands" of green in variegated leaves, of occasional green seedlings amongst albinos, and of periclinal divisions at the apex, has been demonstrated in variegated tobacco plants by Stewart and Burk (1970). These displacements of cells in the layered apex are only of developmental significance in chimaeras where the layers differ genetically. However, this may be more often the case than usually recognized (Neilsen-Jones, 1969). Some cases of frequent "sporting" of flower colour in, for example, *Chrysanthemum* and in roses may be due to cell displacement in plants where layers of cells not normally involved in petal formation carry mutations affecting flower colour. Following mis-orientation of

Fig. 15. Green "islands" in the white leaf margins of *Saxifraga stoloniferum* var. *tricolor* (1½ × natural size).

This cultivar has defective chloroplasts in tissues derived from layer II of a 3-layered stem apex (i.e. the inner layer of a 2-layered tunica). As a result, the leaf lamina margins are white, while the core of the leaf, derived from layer III, is green. The epidermis, derived from layer I, has, as is normal for that tissue, only a few, small chloroplasts in each cell and even the chloroplasts of the stomatal guard cells do not give the epidermis sufficient pigment to mask the underlying white tissues. The plant can be designated as a GWG chimaera.

The many, similar, small green flecks in the otherwise white margin probably arise as a result of frequent "mis-orientation" of cell division in the developing epidermis at a stage late in leaf growth. These divisions will insert individual cells capable of developing normal large chloroplasts in mesophyll tissues into the otherwise defective, white tissues derived from layer II. The few remaining mitoses before cell division ceases produce from each of these inserted cells a small island of green cells visible to the eye. Such "mis-orientations" of epidermal divisions may well occur during development of the normal all-green form, but there is no way of detecting them unless a chimaera for an alternative phenotypic marker can be recognized. In GWG chimaeras of other species, these mis-orientated divisions occur more rarely (and often earlier in development) or not at all (Tilney-Bassett, 1963; Kirk and Tilney-Bassett, 1967; Neilsen-Jones, 1969; Stewart and Burk, 1970).

division, some of these cells become included in developing corolla tissue, and the latent genotype is expressed.

D. Somatic Instability

Occasionally, plants show many and superimposed nuclear mutations resulting from frequent errors of mitosis. In extreme cases, the original zygotic complement may be totally obscured in the growing root or shoot apex, and all cells may inherit different chromosome complements. This has been found in a wide range of angiosperms, as well as in ferns (Sharma and Majundar, 1955). Mitotic instability is frequently associated with recently induced or high levels of polyploidy (Britton and Hull, 1957; Ehrendorfer, 1959; Levan, 1948; Thompson, 1962; Rajhathy, 1963; Vaarama, 1949; Snoad, 1955; Sharma, 1956) or recent hybridity, often with ancestral polyploidy (Håkansson, 1950; Hollingshead, 1932; Jones and Banford, 1941; Kattermann, 1933; Love, 1938; Menzel and Brown, 1952; Neilsen and Nath, 1961; Olden, 1953; Sachs, 1952; Alexander, 1968; Brown, 1949; Takehisa, 1961; Walters, 1958; Yang, 1965, Sharma, 1956; Heinz et al., 1969; Heinz and Mee, 1971). Only rarely does it occur in plants apparently of diploid an non-hybrid origin (Lewis, 1962; Hegwood and Hough, 1958; Sharma, 1956; Duncan, 1945).

The chromosome numbers in the affected nuclei are almost always below the number presumed to have occurred in the zygote and may include haploid or polyhaploid levels (Rajhathy, 1963; Sharma, 1956; Menzel and Brown, 1952). In several cases, this aneuploid variation in chromosome number within a tissue is known to be associated with split spindles at mitosis (Walters, 1958; Snoad, 1955; Vaarma, 1949; Thompson, 1962; Neilsen and Nath, 1961; Ehrendorfer, 1959; Britton and Hull, 1957; Alexander, 1968).

Peak frequencies are seen of cells with euploid numbers in *Rubus* (Britton and Hull, 1957), *Agroelymus* (Neilsen and Nath, 1961), *Malus* (Olden, 1953), *Claytonia* (Lewis, 1962) and *Ribes* (Vaarama, 1949). As individual chromosome sets are not easily distinguishable in these species, it is not certain that euploid numbers represent balanced chromosome complements, but it is difficult to explain their disproportionate representation in any other way. There are two possible explanations of their origin. The physiological disturbance, often a result of polyploidy or hybridity, may result in a random population of aneuploid cells among which those that have by chance inherited one or more complete sets are at an advantage over the rest due to the absence of genetic imbalance, and as a consequence grow and divide more rapidly to produce a greater proportion of the cell population than the aneuploids. Alternatively, genomes with different division characteristics tend to act independently when in the same cell (Gupta, 1969), so that if, by chance, random chromosomal distribution at late prophase results in the chromosomes of a complete genome being aggregated in one part of the nucleus, they tend to form an independent spindle and give rise to separate near-euploid products.

There is some doubt whether somatic instability has much significance for development or evolution. Affected plants are not strikingly abnormal in phenotype although they may show clonal variation (Heinz and Mee, 1971). While cells of lower euploid levels including haploidy may arise, instability seems to be inheritable (Heinz and Mee, 1971; De Torrock and White, 1959) and only if mitotic stability returns with the lower, balanced number will the new karyotype be of any significance in reproduction and evolution. It has, however, been suggested that this is an important mechanism for reversing polyploidy in evolution (Sharma, 1956).

Somatic instability for chromosome breakage is also known, particularly in hybrids (Rutishauser, 1956). In other cases, cultural conditions may be responsible (Rieger and Michaelis, 1958). In tissue, callus and tumour cultures all forms of nuclear change are frequently found (Hsu, 1961; Venketswaran, 1963; Sacristan and Wendt-Gallitelli, 1973; Mitra and Steward, 1961; Cooper et al., 1964; Sunderland, 1973; Norstog et al., 1969; Partanen, 1963; D'Amato, 1964). This is also true of animal tissue cultures (e.g. Hsu and Moorhead, 1956). No doubt the unnatural conditions cause physiological disturbances similar to those resulting from hybridity or polyploidy. The nuclear changes have been associated with changes in growth characteristics (e.g. Fox, 1963; De Torrock and Roderick, 1962; Blakely and Steward, 1964; Torrey, 1967), and the heterogeneity of cell cultures and the loss of regeneration and differentiation potential which frequently develops with prolonged culture have been attributed at least partially to the accumulation of deviant chromosome complements as a result of frequent mitotic errors. A full discussion of nuclear changes in tissue culture requires a separate chapter (see Sunderland, 1973). It is sufficient to note here that the regularity with which these cytological changes occur in culture and the potential importance of stable and uniform tissue cultures in, for example, cytological research and plant propagation (Reinert, 1973) make it necessary that the cause and cure for mitotic instability in plant cultures be discovered. This will no doubt require a better understanding of the division process in the intact plant. This book represents a progress report on the efforts being made in that direction.

C. Cell Division and Generation of Form

7. The Root Apex

F. A. L. CLOWES

Botany School, University of Oxford, England

I. APICAL MERISTEM

A. Structure in Relation to Root Anatomy

All the growth in length of a root occurs at the apex and, apart from secondary thickening, all the root cells are generated there in the apical meristem. The meristem has distinct regions corresponding to the tissues of the mature root. The centre of the root is differentiated from the rest as a stele. The stele is surrounded by a cortex and this by an epidermis. The cells of the stele and cortex lie in longitudinal files and the expansion of the cross-sectional area of root that occurs near the apex results from both an increase in the number of files and enlargement of the cells. The epidermis is, in some plants, the outermost layer of the cortex, while in others it is ontogenetically part of the root cap. In a few species it is independent of both cap and cortex in ontogeny. In angiosperms its cells differentiate differently from all other tissues, and in some it is ephemeral. At the tip of the root is the cap, the distal cells of which are sloughed off and replaced from the distal region of the meristem.

The vast majority of roots are ephemeral, lasting a few days or weeks before collapsing and being replaced by new roots exploiting new regions of the soil.

The differentiation of the root into stele, cortex, cap and, in some plants, epidermis is reflected in the structure of the meristem, but root meristems are more diverse than is necessary to account for differences in root anatomy. The diversity stems essentially from the degree of discreteness of the meristematic regions or histogens. This degree of discreteness of the histogens varies during the development of the root from the embryo or lateral primordium and reaches different levels in the fully developed apices of different groups of plants. Species of the Gramineae and Cyperaceae, for example, all have a separate meristem to produce the cap whereas species of the Leguminosae do not have an independent cap meristem. However, roots of grasses do not start with a discrete cap meristem and, in the Leguminosae, the cap almost becomes discrete in the fully developed apex. This will be explained more fully later, but the difference in the development of root apices in the two groups of plants results in meristems of quite different appear-

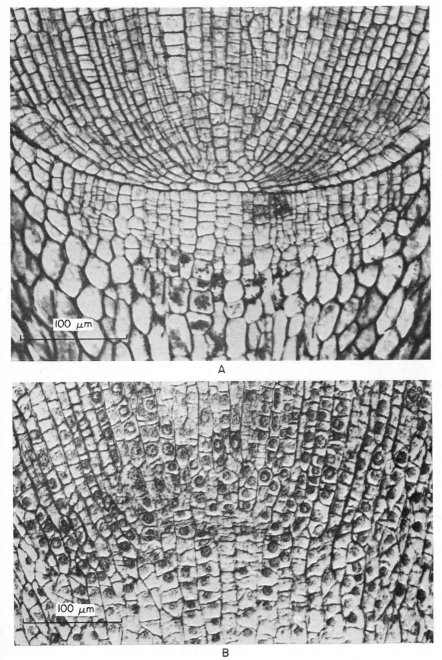

Fig. 1. Curvature of the files of cells in the region of the quiescent centre in root meristems of the "closed" type, A (*Zea mays*), and "open" type, B (*Vicia faba*).

ance even though they ultimately behave in much the same way. The conspicuous difference in structure is in the curvature of the files of cells in the apex (Fig. 1). In large roots of the grass type the files of outer cortical cells radiate horizontally from a pole on the axis turning upwards and so never run into the files of cap cells (Fig. 1A). In *Vicia faba*, on the other hand, files of cortical and stelar cells run into the files of cap cells (Fig. 1B) and may even appear continuous. This led many people to think that the meristem at the junction of the cap and the rest of the root was common to the three major regions of the root, i.e. that there were common initial cells for stele, cortex and cap. We shall see that this view can be shown to be wrong and it demonstrates how careful we must be in interpreting cell patterns in meristems.

Another example of a misleading cell pattern occurs in those crypto-gams whose roots have apical cells. The pattern here leads us to expect that the tetrahedral apical cell is the perennial source of all future cells. We know from the work of Avanzi and D'Amato (1967) that this is not so. The apical cell stops dividing quite early in the development of the meristem. The pattern of cells thus reflects past behaviour of the meristem as well as current behaviour. So strongly are people in-fluenced by pattern that Guttenberg (1947, 1960) always interpreted even angiosperm roots in terms of a "central cell" (or cells) as if this behaved like an apical cell during the normal growth of a root.

B. Extent and Productivity of the Meristem

Within the meristem the cells grow and divide, and the number of cells doing this may be determined by simple observation. In *Zea mays* the meristems of primary roots, which are about 1 mm across when alive, extend for some 2 mm above the tip of the root, which is occupied by non-meristematic cells of the cap. They contain 130 000 cells of which 110 000 are in one or other of the phases of the mitotic cycle. Most of the 20 000 cells not in cycle are scattered through the meristem and are probably out of cycle temporarily. There is, however, a compact group of 600 cells (the quiescent centre) at the pole of the stele and cortex, 60% of which are out of cycle, some of them temporarily and some of them permanently.

Such a meristem produces 180 000 cells a day at 23°C. Of these, 170 000 contribute to the growth in length of the root, but 10 000 are formed by the meristem of the cap, which in this plant is discrete from the rest of the meristem. Counting shows that 10 000 is also the number of cells in a root cap of *Zea* so the whole cap can be produced and sloughed off in a day (Clowes, 1971). This fact has surprised some people and it has been questioned by Harkes (1973), but only as a result of a misunderstanding of the data supporting it. Other species and other conditions of growth may support cell production at quite different levels.

Thicker roots have more extensive meristems than this and thinner roots have smaller meristems. The average rate of cell division for the whole meristem also varies with conditions of growth from zero to about two cells per cell per day.

II. PLANES OF DIVISION

A. Evidence from Pattern

Analysis of cell division in root meristems may be simplified by considering planes and rates of division separately. Planes of division can be discovered by examining sections for orientation of mitotic spindles, thickness of cell walls and, most importantly, by analysis of cell patterns. The task of analysis has been made simpler by the Körper-Kappe theory. In this Schüepp (1917) drew attention to those patterns of cells seen in median sections that indicate proliferation of files of cells. Most of the divisions in a root apex are transverse, adding to the number of cells in a longitudinal file. Divisions which do not add to the length of a file increase the number of files, producing the pattern of "T"s of the Körper-Kappe theory.

Most of the Körper-type divisions occur in the outer stele and inner cortex, widening the root near the tip. In some plants these periclinal divisions are frequent enough to produce regular radial rows of cells in the inner cortex seen in transverse sections. In the outer stele the radial seriation is nearly always obscured by longitudinal divisions in other planes centred on the protophloem and producing the cells of the conducting system. Similarly periclinal divisions in the outer cortex are compensated for by anticlinal divisions. The relative rate of increase in the number of files of cells reaches a maximum of 0·01 file per file per hour and, in Zea, this occurs at 0·5 mm from the tip of the root (Erickson and Sax, 1956). The Kappe-type divisions lead to an increase in the number of initial cells and so are not common except in the root cap where the increase is compensated for by the sloughing off of the peripheral cells. Elsewhere in the meristem Kappe-type divisions cause a permanent enlargement in the root as sometimes happens in the development of a thick root. But such increases in the numbers of initials can be compensated for by the loss of an initial or other meristematic cell by its failure to grow and divide or by conversion of such a cell in the cortex into a cap cell.

In the caps of wide roots the central initials often divide only transversely, producing a columella of unbranched files of cells. The peripheral initials, on the other hand, also divide longitudinally, producing the Kappe-type pattern of "T"'s. There is a gradual transition from the central cells with no longitudinal divisions at all to the peripheral initials with longitudinal divisions frequent in relation to the number of transverse divisions. In thick roots of grasses this gives rise to three different patterns of cells in the cap which catch the eye—the central columella, a surrounding cylinder with T divisions facing downwards towards the tip and an outer cylinder with T divisions facing obliquely upwards and outwards (Fig. 2).

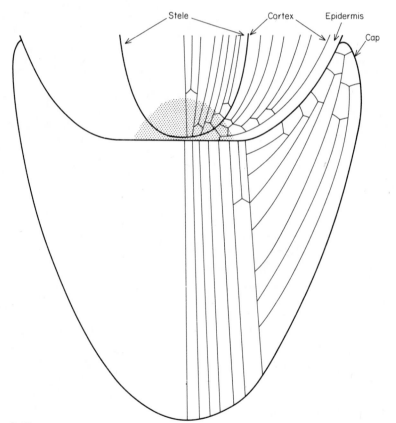

FIG. 2. Patterns of proliferation of files of cells in the root apex of *Zea mays*. The quiescent centre is shown as shading superimposed upon the cell pattern.

B. Discreteness of Histogens

In all roots of seed plants the stele is normally produced by a meristem separate from the other regions. In some species the cortex and the cap clearly have their own meristems and the epidermis is either produced by the meristem that produces the cortex or that which produces the cap. There is a small number of plants, all aquatic floating plants, which appear to have a discrete meristem for the epidermis from an early stage. This is a sheet one cell thick covering the surface of the cortical meristem and separating it from the discrete cap initials. In other species the cortex and cap appear to be continuous and to share a common meristem. The name cap is then given to the tissue, distal to the meristem, which is eventually sloughed off, and the name epidermis to the layer which survives the abrasure of the soil for a while. In such roots the property of extending the surface area by the outgrowth of root hairs is not confined to any particular layer except by the accident of being at the surface. Roots of many gymnosperms appear to be like this. Among angiosperms those with the type of meristem described as "open" (Fig. 1B) are rather like the gymnosperm type, though it is easier to demonstrate the boundaries of the histogens in them.

The fact that it is possible to associate a region of the root meristem with a tissue of the root is the basis of the histogen theory of Hanstein. The division of the meristem according to the destiny of the cells produced is still a useful way of regarding root meristems. It is not so useful for shoot apices merely because the vascular tissue of the stele has to cross the cortex to enter the leaves, but, in those shoots that have long apices, narrow steles and whorled leaves, the histogens can be seen as clearly as in root apices.

Although the histogens are fairly distinct in root meristems there is no mechanism such as there is in many shoots to maintain their discreteness in the face of accidents of growth. For this reason roots never maintain stable chimaeras. Genetically different cells introduced naturally or artificially into a root meristem by accidents of mitosis or by endoreduplication have sufficiently different rates of division and growth from the rest of the cells for them to intrude across tissue boundaries. There are no forces sufficient to maintain the boundaries intact, and produce "germ layers", as there are in shoot apices. When genetically abnormal cells exist in root tips, as after colchicine treatment, they are usually quickly eliminated from the meristem by the superior performance of the normal meristematic cells. When more drastic means of introducing genetic abnormality are used as in the X-ray experiments of Brumfield (1943) the longer survival of the meristem as a chimaera is due to the loss of reproductive integrity of the majority of meristematic cells and the formation of a virtually new meristem from a small group of survivors (Clowes, 1959a). Even in such roots the meristem will eventually lose its chimerical nature by the elimination of all but one kind of cell. This reorganization involves breaching the integrity of histogen boundaries.

C. Renewal of Initials

A meristem obviously varies in structure during its development from the embryo or primordium. Structural changes of this sort have been studied by Guttenberg *et al.* (1955), but they have not been fully analysed in terms of changes of planes and rates of division, though such an analysis would be possible with the techniques now available. However, changes also occur in mature roots and these are better known. One change already mentioned briefly has been detected as a change in the Körper-Kappe pattern in *Fagus* (Clowes, 1950) and may well occur in other thick roots of similar construction.

Many changes must occur that are undetectable. One kind that can be detected is the renewal of cap initials. In this process an existing set of cap initials for some reason ceases to divide and is then replaced by the proliferation of proximal cells. The old initials differentiate as cap cells: the new initials divide and push the old ones towards the tip. This internal reorganization of the apex can be detected because in the course of proliferation of the old cap initials these cells become slightly displaced in relation to their proximal neighbours, which are normally quiescent. The cells produced by the new initials are not in the same files as those produced by the old initials. This discontinuity may be seen in roots with conspicuous columellas and a single cap may show that its initials have been renewed once, or twice or three times during the time it takes for the progeny of the initials to reach the tip of the cap.

Replacement of initials in this way may occur in parts of the meristem other than the quiescent centre, but it is not so easy to detect as in the cap. In normal growth such replacements occur most often in developing apices such as in the radicles of *Helianthus* (Guttenberg, 1955; Clowes, 1959b). They are confined to meristems of the "open" type.

D. Polarization

Cell division in higher plants is polarized in two senses. It quickly becomes confined to the shoot and root poles in the developing embryo and this is enhanced by the later essential confinement of cell divisions to the apices. This mode of development is related to the extensive form of higher plant sporophytes, and is therefore absent from some lower plants. Cell division is also polarized in the sense that its plane is predominantly at right angles to the axis of the plant. This follows from the direction of cell growth, but planes of division become complexly organized in a manner that is genetically determined. The development of a lateral root primordium follows a pattern that produces an apical meristem constructed in the same way as that of a primary root of the same species whose initial development in the embryo is quite different. During this development the planes of division in the cells destined to become cap are not those predictable from the shape of the cells and the law attributed to Errera, for the cells become partitioned by walls of maximal area. This presumably indicates that these cells are not under equilateral stress, but the difference between these cells and those elsewhere in the primordium that do obey Errera's law is not clear.

There is a certain amount of local synchrony in division which is seen as groups of contiguous cells in the same phase of the mitotic cycle. This is soon lost and Ivanov (1971) has shown for *Zea* roots that, of two sister cells, the larger tends to divide first and this is usually the apical one.

III. RATES OF CELL DIVISION

A. Measurement

Root tips have for long been a favourite source of dividing cells for people interested in cell kinetics. Consequently many methods of measuring rates of mitosis within them have been devised (Clowes and Juniper, 1968). Some of these are laborious and today the field is dominated by methods that involve measuring cell flow after labelling the DNA or blocking the mitotic cycle. The early methods involved the measurement of root elongation and counting the number of cells in the meristem either by inspection of sections (Gray and Scholes, 1951) or by macerating and sampling in a haemocytometer (Brown and Rickless, 1949). These methods gave useful information on average rates of mitosis in whole apices.

Methods of measuring cell flow involve either marking cells at a particular phase of the mitotic cycle and letting them proceed through one cycle into the next or watching them accumulate at a recognizable phase. In practice a large number of roots is used and they are sampled at intervals. The most popular method today is pulse-labelling with tritiated thymidine. This was devised by Quastler and Sherman (1959) and, when used on roots, involves placing them in a solution of tritiated thymidine for a short time—the pulse—then placing them in a label-free solution and sampling the roots at intervals. Those cells that were in the DNA-synthetic phase (S) of the mitotic cycle at the time of the pulse are labelled permanently by the tritium incorporated into their DNA. They proceed through the next phase of the cycle, G_2, and then reach mitosis. The observer counts what fraction of cells in mitosis is labelled at each sampling period in autoradiographs made from sections or squashes. At the start of sampling almost no mitotic cells are labelled, but a little later the cohort of labelled cells reaches mitosis and the fraction of labelled mitoses may reach unity. It then falls to near zero and rises again as the progeny of the first labelled mitoses, themselves labelled, also reach mitosis. A graph of these events then shows two peaks of labelled mitoses and the distance separating them is the duration of a complete mitotic cycle, the reciprocal of the rate of mitosis. In practice the curve is damped by the variability of cell cycles and the finite length of the pulse, and there are devices used to allow for this. The graph also enables us to measure the durations of all the phases of the mitotic cycle.

The method assumes that the tritiated thymidine pulse has no effect on the behaviour of the meristem and that the labelled cells proceed as they would do if not labelled. The first assumption is justifiable, but not the second. Evidence is amassing to show that cells with labelled DNA do not behave like normal cells (De la Torre and Clowes, 1974). However, many people are willing to ignore this and elaborate statistical procedures are available for extracting information from pulse-labelled mitosis curves (e.g. Macdonald, 1970, for root tips).

A practical disadvantage of the pulse-labelled mitosis method is that meristems have to be sampled over at least one and a half average cell cycles and, partly for this reason, double labelling methods have been devised (Wimber and Quastler, 1963); sometimes two different isotopes are used to label the DNA, giving β-particles of different energies with two autoradiographic films, one to detect the isotope with high energy radiation and the other to detect both isotopes. The two isotopes are supplied in pulses separated by an interval of time. Two pulses of the same isotope and a single autoradiographic film may also be used (De la Torre and Clowes, 1974).

Methods that block the cell cycle take a shorter time than the pulse-labelled mitosis method. For plants the best cycle-blocker is colchicine. This drug is applied continuously and the observer watches the accumulation of metaphases in sections or squashes of sampled roots. The method is valid only if the drug does not alter the rate of entry into mitosis and only if escape from metaphase is negligible. In many kinds of roots it is possible to use a concentration of colchicine and a suitable accumulation period for these conditions to be fulfilled. Long accumulation periods are usually not acceptable (MacLeod, 1972). The rate of mitosis may be calculated directly from the measurement of cell flow or by the more sophisticated method of Evans et al. (1957).

There is an important difference between the pulse-labelled mitosis method and the other methods in what is measured. In the former only cells in mitosis are counted and therefore the rate of mitosis calculated is the average for cycling cells only. Moreover those cycling cells with low rates of mitosis are also effectively excluded from the calculation for they do not contribute to the peaks of labelled mitoses. So this method gives a rate of mitosis which is the average for the fast-cycling cells of the meristem.

The other methods count all cells, either fractions of all cells that are labelled or fractions of all cells in metaphase. The rate of mitosis calculated here, then, is the average for all cells and the duration of the mitotic cycle is the cell-doubling time. The difference between the two calculated rates provides an estimate of the number of cells in the meristem that are not dividing. This point is discussed further on p. 271.

B. Values and Variation

Of all the meristems of plants the root apices have the highest rates of mitosis. Shoot apices and cambia seem always to have lower rates in the few plants that have been investigated. Average rates of mitosis of between once and twice a day are common in root tips. The rates actually measured vary with temperature and with how much of the root tip is included in the meristem. Gray and Scholes (1951), for example, showed that the region of the apex of *Vicia faba* between 0·19 and 3·04 mm from the tip had a cell-doubling time of 24·5 h, whereas between 0·19 and 1·52 mm the cell-doubling time was only 19·4 h. The level at which cells stop dividing depends on their position in the root. The central stelar cells stop dividing first (i.e. nearer to the tip), making these cells the largest. In the cortex the middle cells stop dividing first and cessation of division spreads inwards and outwards so that the outer stele, inner and outer cortex continue dividing after other cells have stopped.

Probably the cell-doubling time for any region increases as the level at which it is measured approaches the zone of differentiation.

However, it is likely that this is not due to a gradual slowing of the cycle, as we shall see later. The effect of temperature upon rates of mitosis is complicated by the different reactions of different tissues and will also be described later. In several species the higher average rates occur at about 30°C. The duration of cycles in fast cycling cells in *Trade-scantia* varies from 16 h at 30°C to 21 h at 21°C and 51 h at 13°C (Wimber, 1966). These figures are averages for whole meristems selected from the root by the intensity of Feulgen staining.

In *Zea mays* growing at 23°C the cell doubling time of the meristem is about 17 h. This time corresponds to the production of 180 000 cells per day already mentioned.

The influence of levels of ploidy on rates of mitosis has interested people because they may affect rates of synthesis. Skult (1969) showed that in *Hordeum vulgare* autotetraploid plants had average rates of mitosis in their root apices slightly lower than in the diploids, which have a similar growth fraction but a slightly higher mitotic index. The cycle durations were 10·4 and 11·4 h for the diploid and tetraploid respectively.

A survey of root meristems suggests that the greater the DNA content per diploid nucleus the longer the mitotic cycle (Van't Hof, 1965a; Van't Hof and Sparrow, 1963). Three anomalies are known in this generalization. Dicotyledons appear to have longer cycle durations than mono-cotyledons of the same DNA content over a range up to 30 pg per nucleus, though both show a similar trend of increasing cycle length with increasing DNA content (Evans and Rees, 1971). Plants with B chromosomes show a disproportionate increase in cycle duration compared with plants without and the more B chromosomes the greater the increase (Evans and Rees, 1971; Barlow, 1973). The third anomaly concerns the large and small genotrophs of a plastic cultivar of *Linum*. Data of Timmis used by Barlow (1973) show that the large genotrophs, which have 16% more DNA, have shorter cycles (12·4 h) than the small genotroph (13·6 h).

IV. THE QUIESCENT CENTRE

A. Existence

One of the things that has interested people about meristems in recent years is the discovery that their cells do not divide all at the same time. Rates of mitosis are controlled within the meristem in an organized way just as planes of division are organized. The fact that root meristems display widely different rates of cell division could be exploited in research on control of cycling in cells, a subject that has medical applications as well as biological importance.

The first meristems where regions of differing rates of mitosis were shown to exist were the root apices of *Zea mays*. This plant has root tips where an analysis of cell patterns leads to the unambiguous conclusion that a region exists at the pole of the stele and cortex where rates of cell growth and division must be small compared with those in surrounding cells (Clowes, 1954). Until this discovery it had always been assumed that this region, now called the quiescent centre, had rates of mitosis similar to the rest of the meristem and, in fact, contained the initials of the root. The only assumption made in the geometrical argument for the existence of a quiescent centre is that the cells do not slip in relation to each other in the apex. The pattern of cells in the cap has already been described and shows that the initials at the head of the central rows of the cap do not expand transversely, and do not divide longitudinally. The contiguous cells of the cortex–epidermis complex are subject to the same restraint and, at the pole, lie one cell deep and therefore they do not divide transversely either. Thus they do not divide at all and the pattern of cells shows that this applies to a hemisphere of cells above the central cap initials. In the primary roots of *Zea* this hemisphere comprises some 600 cells.

The validity of the argument from the pattern of cells was demonstrated by autoradiographs of roots fed with labelled precursors of DNA for a day or more (Clowes, 1956). There is a region where none of the cells synthesize DNA in preparation for mitosis and this corresponds exactly with the quiescent centre postulated above; it is surrounded by cells most of whose nuclei are labelled. The DNA precursor used originally was ^{32}P-phosphate, but, after tritiated thymidine of high specific activity became available, this has been the precursor of choice. It allows quantitative work to be done to compare rates of mitosis in the regions of the meristem and provides a popular class exercise for undergraduates.

The pattern of cells cannot be interpreted unambiguously in root apices of other types, but autoradiography shows that there is a quiescent centre in all normally growing roots. In roots of the Leguminosae it separates the initials of the central cap from those of the stele, showing that the initials cannot be common. It was formerly thought that pteridophyte roots with apical cells were exceptions because the apical cell, which corresponds in position to the quiescent centre of larger roots, usually does become labelled by procedures that leave the quiescent centre cells unlabelled. However, D'Amato and Avanzi (1965) and Avanzi and D'Amato (1970) have shown that the synthesis of DNA that occurs in apical cells prepares the cell for endore-duplication rather than mitosis, and that the apical cell and its imme-diate neighbours in the stele and cortex are in fact quiescent except in the early stages of root development. Reports of roots without quiescent centres exist (Wimber, 1960; Raju *et al.*, 1964; Alfieri and Evert, 1968), but all of them can be discounted. In thin roots, where the quiescent centre consists of 50 cells or less, it may be difficult to pick it out in autoradiographs either because the section is not truly median and has therefore missed the quiescent centre, or because not a large enough proportion of the cycling cells round about has been labelled. Short exposures to labelling solutions are inadequate here. Quiescent centre cells are not inherently quiescent and we shall see that the various treatments and accidents can relieve the inhibition; this has led some people to find no quiescent centre in roots that have one in normal growth. Even excised root tips in culture have quiescent centres (Thomas, 1967; Phillips and Torrey, 1971a, b).

B. Development

Neither primary roots nor laterals have quiescent centres when the root is first formed. In primary roots of *Sinapis* a quiescent centre forms at the pole of the stele and cortex after germination of the seed when the root is a few millimetres long. At first it consists of few cells, but quite quickly neighbouring cells lose their ability to grow and divide, and the quiescent centre reaches its maximum size of 500–600 cells (Clowes, 1958a). In *Malva* the quiescent centre develops more slowly. When a root is 30 mm long it is non-existent, at 60 mm it contains 180 cells, at 160 mm 190 cells, at 230 mm 454 cells and at 330 mm 699 cells (Byrne and Heimsch, 1970).

The development of the quiescent centre in lateral roots has been investigated in *Pistia* and *Eichhornia* (Clowes, 1958a). In the lateral pri-mordia all the cells are at first meristematic, including those derived from the endodermis of the mother root. By the time the primordium is fully organized, but before it emerges from the mother root, qui-escence sets in at the pole of the stele and cortex, and spreads to cells occupying half the width of the root at this level. *Pistia* has a discrete dermatogen derived from the endodermis of the mother root and the quiescent centre here includes the axial cells of this layer. As in many thin roots, mitosis in the root apex ceases completely soon after emer-gence from the mother tissue and the root grows solely by cell elonga-tion. The laterals of these two floating plants are initiated within the apical meristem of the mother root and the formation of the quiescent centre may consequently be precocious. Byrne (1973) has shown that in *Malva*, which has lateral primordia initiated at the normal level, outside the meristem, the quiescent centre is not formed until the lateral has emerged from the mother root and has broken through the endodermal cover.

Another point of interest found by the autoradiographic investigation of the laterals of *Pistia* and *Eichhornia* is that the distal cells of the cap stop their mitotic cycles when the primordium is only about as long as wide. These plants may be precocious in this respect too.

Some root primordia have proved difficult to investigate experimentally (MacLeod and Davidson, 1968, 1970; MacLeod, 1971, 1973a; Davidson, 1965b, 1969; Davidson *et al.*, 1968; D'Amato and Ronchi, 1968). Barriers to the penetration of solutions appear to exist at some stages but not at others in *Vicia faba*, but there also appear to be changes in the response of cells during the development of primordia (MacLeod, 1973b).

C. Concept of Initials

The discovery of the quiescent centre makes us re-examine the classical concept of initial cells. It used to be thought that in each meristem there were one or more cells from which all cells were ultimately derived, a promeristem in one sense of the word. In some pteridophytes this promeristem would consist of a single tetrahedral apical cell: in others the tetrahedral cell is the initial for the root proper and there is another independent initial for the cap. In seed plants most people considered that there were several initials for each of the histogens. For the stele, for example, there might be three initial cells and these were usually considered to be those at its pole, close to the axis of the root. Similarly for the cortex and cap there would be small groups of initials near the axis. Guttenberg (1947) rejected this view in favour of either a single central cell or a small group of initials which was common for all tissues. The evidence presented for all these theories was based on patterns of cells and this, as we have already seen, can be misleading. However, Guttenberg's concept appeared to be given experimental backing by the X-ray-induced chimaeras studied by Brumfield (1943). These showed that a complete sector of an angiosperm root crossing all the tissues could be derived from a single cell and that there were about three such sectors. Brumfield interpreted this as indicating the existence of three initials.

The quiescent centre theory resolved all these problems. For many purposes the cells lying over the surface of the quiescent centre can be regarded as the initials, but these initials are not permanent for, as we shall see, the quiescent centre is not completely inert and is capable of giving cells to the meristem which will head the files of dividing cells. The fact that the active initials are those cells lying over the surface of the quiescent centre implies that there is a large number of them—about 800 in primary roots of *Zea*. This accounts for the way in which roots recover when part of the meristem is cut away (Clowes, 1953).

Brumfield's experiment indeed shows that roots can be derived from three cells, but the reason for this is not that roots have three initials, but that after X-irradiation the root is regenerated from a few cells of the quiescent centre after the reproductive integrity of the normally meristematic cells has been impaired. This is shown by watching which cells synthesize DNA after irradiation (Clowes, 1959a, 1961a). The reason why the quiescent centre is apparently less sensitive to irradiation than the rest of the meristem will be discussed later, but the difference in susceptibility between the cells of the quiescent centre and elsewhere need only be quite small because the effect is amplified by the necessity of having a group of contiguous, viable cells to regenerate the meristem.

Some people have felt that there is dilemma in choosing between the quiescent centre and the pattern of cells in helping to understand how root meristems behave. The quiescence of cells at the pole of the stele and cortex can be demonstrated to everyone's satisfaction and yet the pattern of cells leads us to suppose that the initials lie here. This dilemma has been presented by Kadej (1963) for the large roots of *Cyperus gracilis* which show clear segmentation patterns such as exist in fern roots. The pattern suggests that a single axial cell must be the origin of the cortical segments, yet this cell ought to be in the quiescent centre whose existence can be demonstrated by a geometrical analysis in this root as well as experimentally. How does the segmentation pattern arise? It is not the fact that there is a slow rate of mitosis in the quiescent centre that accounts for the pattern by giving the status of initials to some of its cells. Rather it is the fact that in some previous era of the root's existence there was an axial initial which imposed the cell pattern on the root for ever. The onset of quiescence merely stabilizes the pattern. The anatomical stages through which the root primordium passes in this process have been described for *Elodea* by Kadej (1966).

Thus Guttenberg's central cell reflects the mode of development of the meristem and not its current behaviour as he thought. Some plants, including some varieties of *Zea mays* (Stallard, 1962), do have a conspicuous central cell, but the temptation to ascribe a special morphogenetic significance to it must be resisted. In the same way, as we have already seen, the existence of an apical cell in a fern root must not make us believe, without proper investigation, that it has a current morphogenetic role.

V. MITOTIC CYCLES

A. Diversity within the Meristem

1. Time Parameters

The first indication that there is a diversity in the cell cycles of root meristems was the discovery of the quiescent centre. There followed measurements of the duration of the cycle and its phases in different regions of the meristem (Clowes, 1961b, 1965, 1971; Thompson and Clowes, 1968; Barlow, 1973; Barlow and Macdonald, 1973). Barlow

TABLE I. Durations[a] of the mitotic cycle (T) and its phases (G_1, S, G_2 and M) in hours in four regions of the root meristem of *Zea mays* growing in water at 23°C

	T	G_1	S	G_2	M	Tcd
Cap initials	10·4	−0·4	4·3	5·6	0·9	15·5
Quiescent centre	39·6	21·4	11·7	5·6	0·9	231·0
Stele just above QC	14·2	2·7	6·1	3·9	1·5	24·5
Stele 200 μm above QC	14·4	3·6	5·3	3·7	1·8	17·5

[a] The durations given are for cycling cells only except that the cell-doubling time, Tcd (the average duration of the whole cycle for all cells), is given as well as the duration for cycling cells, T. (Data from Clowes, 1971.)

and Macdonald (1973) have the most complete account of a meristem in terms of mitotic cycles, but their data do not allow them to provide a cycle duration for the cycling cells of the quiescent centre, the most interesting region. Table I provides data for roots of *Zea* growing at 23°C from curves showing the fraction of labelled mitoses at intervals after a pulse of ³H-thymidine. The negative duration of G_1 in the fast-dividing cap initials indicates that this phase is eliminated, and that DNA synthesis in preparation for the next mitosis is advanced into the telophase of the previous mitosis. This is confirmed by the immediate appearance of labelled telophases in this region after a brief exposure to ³H-thymidine (Clowes, 1967; Barlow, 1973). G_1 is the phase that varies most in the different regions of the meristem—in *Zea* from zero to over half the cycle. This appears to be true of animal cells too. The values given for G_1 in Table I are for the cycling cells displayed by the pulse-labelled mitosis curve. Measurement of the DNA content of nuclei also shows that over half of the cells of the quiescent centre are at the 2C level (Clowes, 1968). This method does not separate cycling cells in G_1 from non-cycling cells because cells of this region always drop out of cycle at G_1 in *Zea* though apparently not in *Marsilea* (D'Amato and Avanzi, 1965).

The rate of DNA synthesis during S also varies in the regions. In root meristems considered as a whole, MacLeod (1968) found that kinetin affects the duration of S, though not the time of its onset.

Barlow and Macdonald (1973) separated the cap initials in *Zea* into two groups, the central ones which produce the columella and the peripheral ones which produce the rest of the cap. They found that the central initials divided much faster than the peripheral ones with cycle times of 14·0 h compared with 22·5 h at 21°C. Only the peripheral initials had a G_1-phase. This lasted 6·3 h and accounted for most of the difference in the rates of mitosis.

Barlow and Macdonald also took their measurements on *Zea* to 1000 μm above the boundary between the cap and the quiescent centre and measured the cycles in cortex and epidermis as well as the stele. In the stele there is not much change in the cycle durations over this 1000 μm which covers most of the meristem. In the cortex the rate of mitosis actually increases with distance from the tip of the root. Near the quiescent centre the cortical cells divide more slowly than the stelar cells at the same level, but the rates become similar by 700 μm. Where the rates differ, it is due to the difference in the duration of G_1. Just above the quiescent centre G_1 is 3·0 h in the stele and 21·8 h in the cortex, and even at 200 μm, G_1 durations are 7·5 h and 17·6 h in the stele and cortex respectively. Balodis and Ivanov (1970) also believe that the rate of mitosis remains the same along the root and that the decrease in the relative rate of cell formation is due only to a drop in the fraction of proliferating cells.

The only other root meristem for which data on mitotic cycles are available with this detail is that of *Allium sativum* (Thompson and Clowes, 1968). But here no rate for the cycling cells of the quiescent centre could be measured, though a cell-doubling time of 165 h was found from the pulse-label data. This compares with about 33 h in the cap initials and stele at 21°C. The cycle duration for cycling cells is about 26 h in these regions. In these roots the cap initials were not the fastest dividing cells of the meristem as they usually are in *Zea*, but Taylor (unpublished) has subsequently found that in roots growing at between 23°C and 30°C the cap initials do have the highest rate of division in *Allium sativum*.

2. Growth Fractions

Not only is there diversity in the cycles of cycling cells within a root meristem, but there is also diversity in the fraction of cells that are in cycle. This fraction, called the growth fraction, is important in medicine in understanding the behaviour of tumours and in recent years methods of calculating it have been devised from pulse-labelling (Mendelsohn, 1962; Lala, 1968; Lamerton and Steel, 1968; Brown and Berry, 1968, 1969). In the calculation it is assumed that, after pulse-labelling, both labelled cells and cells in mitosis are cycling cells whereas others can be cycling or non-cycling. The growth fraction equals the ratio of the fraction of all cells labelled to the fraction of cycling cells labelled. Cells that come into mitosis during the sampling of pulse-labelled meristems are labelled and unlabelled in the same proportion as the numbers of labelled and unlabelled cells in any other phase of the mitotic cycle. Therefore the fraction of cycling cells labelled equals the fraction of mitoses labelled and so the growth fraction equals the percentage of all cells labelled divided by the percentage of all mitoses labelled. The numerator remains constant throughout a sampling period, but the denominator varies cyclically as the cells in S during the pulse come into and out of mitosis. Its value, therefore, is conveniently integrated over one mitotic cycle. Other measurements may be used in the calculation, and the fraction of the cells contributing to the fast cycle shown in the labelled mitosis curve may also be estimated (Clowes, 1971).

The only root meristem for which figures are available for different regions is that of *Zea* growing at 23°C. Of the regions measured only the stele at 200 µm had a growth fraction of one. Even the cap initials, with their high average rate of mitosis, have a growth fraction of less than 0·9. The quiescent centre's growth fraction is under 0·5. Thus there are some 300 non-cycling cells in the quiescent centre and some 20 000 elsewhere in the meristem. The importance of those in the quiescent centre is that they form a compact group whereas the others are scattered and mixed with cycling cells.

We do not know if the non-cycling cells are permanently out of cycle or only temporarily so, but there is no evidence to suggest permanent loss of ability to divide. In the quiescent centre, however, most obviously cycling cells (labelled or in mitosis) occur near the proximal face of the region. Here, therefore, there may be a core of cells permanently out of cycle so long as the root is growing normally. This core would be in the position of the classical initials of the root. These cells, as we shall see, are not incapable of growth and division, for they are readily stimulated into cycle by processes that damage the dividing cells.

B. Interactions between Cells

It is instructive to measure the rates of mitosis in the different regions of a meristem functioning in different conditions of growth. Environmental factors that affect rates of mitosis include temperature, the medium in which the roots are grown and the age of the roots. In *Zea* there is a general correlation between average rates of mitosis (the reciprocal of cell-doubling time) in the cap initials and the meristematic stelar regions. High rates in one region are accompanied by high rates in another, but this is not true of the quiescent centre. Here there appears to be an inverse relationship within certain limits. A rise in the rate in any other region caused, for example, by a rise in temperature is accompanied by a fall in the quiescent centre and vice versa so long as the temperature range is not extended too far (Fig. 3). Interference with the meristem, as when it is subjected to ionizing radiation or many chemical treatments, also produces this diversity of reaction between the quiescent centre and the rest of the meristem. This will be described later. There must be some fine control operating on the meristem to adjust rates of cell growth and division in one region in relation to another.

Any root meristem displays a certain amount of local synchrony in cell cycles in that it is common to find several closely related cells in the same phase of the mitotic cycle; it was this observation that led to the idea of a mitotic hormone, but it can equally well be ascribed to the general asynchrony of cell growth in the meristem. A cell and its daughters will expand and divide to fill a space, for, although cells in adjacent files grow conformably, they need not grow equally.

C. Nucleolar Cycles

Meristems with such a diversity in mitotic cycles and growth fractions as root tips display would be expected to show morphological differences in the constituent cells. There are, in fact, differences in the RNA content of the cytoplasm and nucleoli easily seen by suitable staining, in numbers of mitochondria and ribosomes and in the size of the golgi bodies and nucleoli (Clowes and Juniper, 1964; Pilet and Lance-Nougarède, 1965). There are also differences in the rates of synthesis of RNA and protein (Clowes, 1958b; Barlow, 1970).

Nucleoli show a cycle of formation, disorganization and reformation

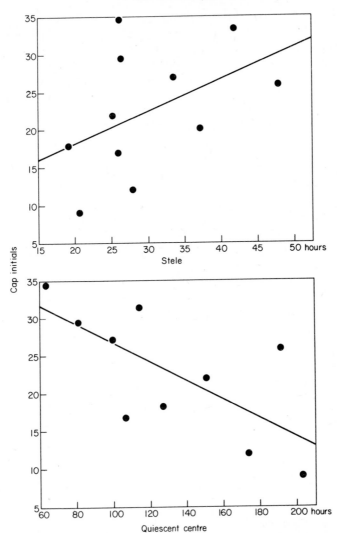

Fig. 3. Cell-doubling times in regions of the root meristem of *Zea mays* growing under different conditions. The times of the cap initials are plotted against the times for the quiescent centre and the stele just above the quiescent centre in the same roots. The linear relationships have been fitted by computer.

which normally fits the mitotic cycle in the following manner. Nucleoli begin to be formed during the last half of telophase and complete their reorganization during G_1. They remain fully organized during S and G_2 and start to disorganize after the onset of prophase. They become completely dispersed as the nucleolar membrane breaks down at the prophase–metaphase boundary and remain dispersed until the middle of telophase. This is what happens in the stelar meristem in *Zea*, the period spent without fully organized nucleoli lasting 2 h, the same as the period spent in mitosis, but in the cap initials the nucleolar and mitotic cycles fit differently (De la Torre and Clowes, 1972). The rapid cell cycle of the cap initials with no G_1-phase has nucleolar disorganization advanced into G_2 instead of prophase and reorganization completed half-way through telophase when DNA synthesis starts, the whole process lasting 0·8 h compared with 0·9 h for mitosis. In the quiescent centre nucleoli remain dispersed throughout metaphase, anaphase and telophase, and reorganization starts and finishes in G_1. An interesting point about the quiescent centre in *Zea* is that telophase in its cells lasts a shorter time (0·4 h) than in any other region of the meristem. Fernández-Gómez *et al.* (1972) have shown that slight artificial inhibition of protein synthesis accelerates telophase and nucleolar reorganization. Protein synthesis is lower in the quiescent centre than elsewhere and the nucleoli are smaller. It may be that, with the demand for ribosomal RNA minimal, the nucleolar organizer genes are able to reorganize the nucleolus faster than the protracted cell cycle would suggest.

D. Cessation of Cycling

Odd as it may seem, the nature of the response of cells as they approach the proximal margin of the meristem towards the zone of elongation and differentiation is not known. Some people expect there to be a gradual lengthening of cell cycle, but, if Barlow and Macdonald (1973) are right in thinking that the meristem ends at about 1000 μm above the cap–quiescent centre boundary in their roots, this view must be wrong. The duration of the cycles measured on the pulse-labelled mitosis curve by Barlow and Macdonald have already been mentioned and they suggest that cycling either proceeds at normal rates or not at all as the end of the meristem is approached. Erickson and Sax (1956) showed, also for *Zea*, that the relative elemental rate of cell formation reaches a maximum of 0·16 cell per cell per hour at 1250 μm from the tip and declines to zero at 2500 μm. It is possible to reconcile these two sets of data, apart from differences in cultivars and ways of measuring the root, if the growth fraction declines with distance from the tip. But this has not been ascertained.

Barlow (1973) explains the sudden end of cycling at the proximal margin of the meristem that his experiments suggest by pointing out that, if cells of successive generations become bigger, there may come a time when the concentration of some substance in the cytoplasm essential for mitosis becomes reduced below a critical level. Erickson and Sax (1956) have shown that the relative rate of change of mean cell length reaches a maximum at the proximal boundary of the meristem, but it is negative for the first 1000–1500 μm. Börner and Ramshorn (1968) found changes in the purine : pyrimidine ratio in RNA at the onset of elongation and suppose that there is a change in the ribosomal RNA as cells switch from proliferation to differentiation.

There is a similar lack of information on what happens as a meristem ages prior to stopping cell production. Barlow (1973) suggests on the basis of an experiment on *Zea* roots 250 mm long that rates of mitosis fall in all regions except the quiescent centre where the rate increases. But the experiment was not designed to compare young and old roots, and so may not provide the necessary data. Street (1968a) suggests that a changing hormonal balance may bring a meristem to the end of its life or to dormancy.

The phase at which the cycle stops is of some interest. In the quiescent centre it always stops at G_1 except in ferns and, in many plants, this is where the cycle stops in cells about to differentiate. But, in many species, DNA synthesis continues after the last mitosis and the cells become polyploid or polytene. In some species this happens generally, but in others it occurs only in certain tissues. In *Zea* most cells double or quadruple their DNA content after the final mitosis, but the cells that become vessel elements go further to reach the 32C level of DNA (Swift, 1950). Phillips and Torrey (1971a) found that in cultured roots of *Convolvulus* no DNA synthesis occurs after the cap initials have stopped dividing. In *Zea* caps there appears to be a break between the ending of mitosis and the resumption of DNA synthesis where this occurs (Clowes, 1968).

In maturing embryos we know how the DNA level of the cells is regulated as they stop dividing from the investigations of Avanzi *et al.* (1963, 1969), Brunori (1967), Brunori and D'Amato (1967) and D'Amato (1972). In *Vicia faba* embryos DNA synthesis stops when the water content drops to 75%, but mitosis continues until the water content drops to 65%. This depletes the population of cells at the 4C level of DNA. In some species including *Lactuca sativa* this population is completely depleted in the seeds. In other plants a few cells are in G_2 in the ripe seed and occasionally cells in the middle of S occur as well. The water content is here merely a convenient measure of the maturity of the embryo and is not necessarily the cause of the operation of the control points of the cycle. In growing roots we have less idea how the control points operate at the edge of the meristem, but it appears to be more complex than in seeds in that tissues may differ as well as species. In *Vicia faba* and *Zea* DNA synthesis goes on after mitosis has stopped, but in *Lactuca sativa* DNA synthesis occurs only in the meristem and mitosis depletes the G_2 population ás in *Lactuca* embryos (Brunori, 1971).

E. Generations of Cells

Erickson and Sax (1956) considered that a root initial cell would go through six mitotic cycles on average before stopping division. This view depends upon a particular conception of the nature of the initials and the authors themselves believed that there was evidence for variation. Luxova and Murin (1973) came to a similar conclusion for *Vicia faba* roots, as did González-Fernández et al. (1968) for *Allium* roots. Although 6 is a small number, there is no reason to think that it imposes a limit to the life of a meristem in practice. There are many quiescent cells which could repopulate the meristem from time to time, and we know that they do this when the meristematic cells are perturbed (Clowes, 1959a, 1961a; Clowes and Stewart, 1967).

Is there some limit to the number of cell generations in a plant as a whole? There does appear to be such a limit of about 50 in mammalian cells *in vitro* and *in vivo* (Hayflick and Moorhead, 1964). Large plants have about the same total number of cells as large mammals like ourselves—6×10^{13}. Plants lose cells in large numbers just as mammals do, though mostly by abscission and decay of organs. Vegetative reproduction, though, makes it possible that plants are different from mammals in having no limit to the number of cell generations or in having a higher limit. People who grow cultures of *Lemna* fronds do not consider that there is a limit to the number of generations of fronds which reproduce themselves by budding. But in this and some other examples of apparent vegetative reproduction it would be easy to fail to see that sexual reproduction had intervened.

VI. PERTURBATION OF THE MERISTEM

A. Differential Response

The first indication that stress causes different reactions in different parts of the meristem came from the study of X-rayed roots (Clowes, 1959a). Acute doses of a wide range of severity of X-rays cause roots to reduce their rate of growth, and after an interval the growth rate is restored to normal. What happens within the meristem is that the cell cycles are stopped in the normally dividing cells and started in the quiescent centre. This is seen as a reversal of the usual pattern in autoradiographs with the cells of the quiescent centre labelled and many of those outside not labelled at a suitable interval after irradiation. The result of this is that a new meristem is reconstituted from a group of cells in the quiescent centre. The number of cells involved may be quite small, as we have seen. There is no external sign of this reorganization except for the changes in the growth rate of the root. Internally the reorganization can be seen in those species with discrete root caps, for here the proliferating cells of the old quiescent centre burst through the cap boundary to form new cap initials which behave like the previous ones before they stopped dividing. The cells that come to lie at the pole of the stele and cortex in the reorganized meristem become quiescent and the apex acquires its normal internal structure.

A dose that produces cellular injury causes the average rate of mitosis in the meristem to fall, but the rate increases in the quiescent centre.

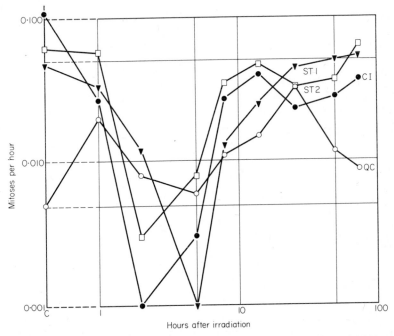

Fig. 4. Rates of mitosis before (C) and at intervals after an acute dose of 1700 rad of X-rays given to root meristems of *Zea mays*. Rates for four regions are given separately (QC, quiescent centre; CI, cap initials; ST1, stele just above quiescent centre; ST2, stele 200 μm above quiescent centre). Log–log plot.

In *Zea* and *Allium sativum* the cap initials are more sensitive to the radiation than any other region (Clowes, 1963a, b; Thompson and Clowes, 1968). This is shown by their loss of ability to incorporate thymidine and by measurement of rates of mitosis after irradiation. In the cap initials of *Zea* the rate drops to less than one-hundredth of normal 2 h after an acute dose of 1700 rad. In no other regions is the fall as great as this nor does recovery take so long (Clowes, 1972b). In the quiescent centre there is an immediate rise in the rate of mitosis to four times its normal value by 1 h after irradiation. It then drops slightly and rises to a peak of six times normal at 24 h, but remains above normal for at least 4 days, by which time the stelar cells have resumed their normal rates. The data above and in Fig. 4 refer to average rates for all cells in the region because there is no way of measuring a growth fraction over a short interval of time. The increase in the rate of mitosis in the quiescent centre, though, is certainly due at least partially to an increase in the fraction of cells in cycle, for a lot of prophases can be seen within a few minutes of irradiation, and this is followed by a wave of metaphases so that the mitotic index rises from its normal level of 3·3 to 13·4% after 30 min (Clowes, 1970a). In other parts of the meristem the mitotic index falls steadily over 120 h from 13·8% to 2·3% in the cap initials and from 14·8% to 4·9% in the stele.

Stimulation of cycling in the quiescent centre and inhibition elsewhere is confirmed by autoradiography which shows that DNA synthesis is stimulated in the quiescent centre immediately after irradiation and inhibited elsewhere. In some cells of the quiescent centre the cycle is so accelerated that they reach metaphase from S within 30 min. Moreover, in an experiment where roots were labelled with tritiated thymidine for 48 h prior to X-irradiation most of the quiescent centre's prophases were unlabelled after 30 min. This suggests that some of the cells reach mitosis from G_1 without synthesizing DNA at all and this probably accounts for some prophases and metaphases that are found to have the 2C level of DNA instead of 4C. There is also the possibility that mitosis without DNA synthesis occurs in non-irradiated roots in the quiescent centre to account for the 1C nuclei that every investigation reveals in root tips (McLeish and Sunderland, 1961).

Most of the work on the diversity of reaction in root tips after perturbation has been done with X-rays, but other agents produce a similar stimulation of the quiescent centre and reduction of proliferation outside. Various chemical treatments including those with cyclic AMP have this effect. A somewhat similar reaction occurs when the temperature at which plants are grown is lowered beyond the level at which mitosis is sustained in any region and then raised to the usual level. The recovery of growth of the roots is then due to the mitotic activity of the quiescent centre in repopulating the meristem with new initials. The cells of the other regions do not recover their normal rates of division until replaced from the quiescent centre and, in *Zea*, the cap initials are again the most susceptible to damage by the cold regime (Clowes and Stewart, 1967). Webster and Langenauer (1973) induced dormancy in cultured excised roots of *Zea* by carbohydrate starvation and, when released, all cells, including those of the quiescent centre, entered mitosis. After a while a new quiescent centre was formed.

In looking for a reason for the differential response of the quiescent centre and the rest of the meristem, the factor most likely to be responsible seems to be the predominant phase of the mitotic cycle in the

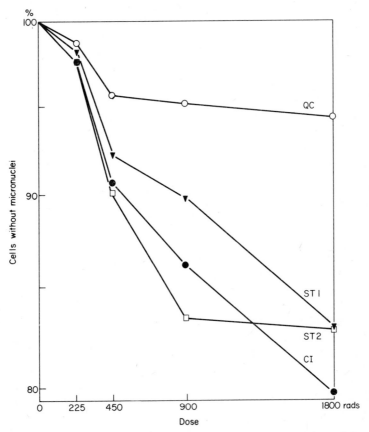

FIG. 5. Dose–response curves for chromosomal damage in the four regions of the root meristem of *Zea mays* given in Fig. 4. Damage was scored as the fraction of cells with micronuclei and integrated over 6 days. The survival scale is logarithmic.

constituent cells at any time. In the quiescent centre most cells are in the G_0 or G_1 condition. Outside it most cells are in some other phase. In the quiescent centre 50–60% of the cells are in G_0 and, of the cycling cells, 54% are in G_1. In the cap initials none of the cells is in G_1 and the percentage out of cycle is about 14. Probably not all of these are in G_0 because some cells exceed the 4C level of DNA. The different response to X-rays could be due to the difference in DNA content—the size of the target. It is not likely to be due to the differences in growth fraction, for the stele 200 μm above the cap boundary ought then to be more sensitive than the cap initials.

It might be thought that the response of the quiescent centre is due to relief from the competing claims of the surrounding cells, which are more severely injured by irradiation. But the rapidity of the response suggests that this cannot be the only explanation, although it must contribute to the complete reorganization of the meristem that occurs after irradiation and other treatments. There is no doubt that the cells of the quiescent centre are less severely damaged by X-rays than the rest of the meristem, for, using micronuclei as a measure of chromosomal damage, the dose-response curves are found to differ with the cap initials showing the greatest response and the quiescent centre the least in accordance with the law of Bergonié and Tribondeau (Fig. 5). Root tips are extensively used to investigate radiation effects and hitherto it has been assumed that their cells showed a uniform response. Other work has now shown that this assumption is untenable (Savage and Wigglesworth, 1972).

B. Regeneration and Homeostasis

The regeneration of a virtually new meristem from cells of the quiescent centre leads to a clonal situation within the root, and where the quiescent cells have mutated as after irradiation this results, as we have seen, in a chimaera which can be stable enough to exist some weeks after the start of the cellular reorganization (Brumfield, 1943). Although the severe irradiation used to produce regeneration from the quiescent centre is not likely to be met in nature, the same kind of regeneration is known to occur during recovery from cold- and starvation-induced dormancy and probably occurs in nature whenever a root stops growing and restarts. The advantage to a root of the stimulation of the quiescent centre rather than of the rest of the meristem is probably that the cells of the quiescent centre have been held at the 2C level of DNA and they therefore cannot have become polyploid or polytene during the period of dormancy. If a root meristem becomes completely inactive the growth of the root is usually by a distal lateral primordium and so it is not clear whether there is an advantage to the plant in having regeneration from the quiescent centre rather than from the pericycle in recovery from dormancy. However, there is no doubt that the quiescent centre serves as a reservoir of cells relatively insensitive to accidents of the environment. A similar reservoir exists in the pericycle, which is the site of lateral root production. Corsi and Avanzi (1970) have shown for *Allium cepa* that DNA content falls in the pericycle between 500 and 1500 μm from the tip, although it increases in the rest of the stele.

One of the most remarkable things about roots is the homeostasis exhibited in their arrangements of cells. There is considerable diversity in cell pattern in root tips of different species. Yet roots derived from embryos, pericycles, quiescent centres and other meristems all have similar cell patterns within a species. If a root meristem is damaged surgically, the regeneration which then occurs by the formation of a callus also quickly restores the original cell pattern.

VII. CONTROL OF CYCLING

A. Nature of Inhibition in the Quiescent Centre

The differences in cell cycles and growth fractions between the quiescent centre and the surrounding cells is so great in so wide a range of root meristems that there must be some control exerted over the apex as a whole. We have seen that RNA synthesis is minimal in the quiescent centre and that nucleolar cycles are modified in the different regions of the meristem in relation to the mitotic cycles. It could be that the synthesis of ribosomal RNA determines the rates of cell growth and division, but, while this may be the proximate cause, the ultimate cause of cell cycle diversity must be sought elsewhere.

Experiments on cell cycles have led to the suggestion that there are some principal control points along the cycle which have to be passed by some metabolic act. Sugar starvation of roots of *Pisum* suggests that these controls operate at G_1 and G_2 for all cells (Van't Hof and Kovacs, 1972). Other species may have different controls.

A meristem-wide control could exist if the quiescent centre was the site of production of some hormone affecting cell division and this has been suggested several times. A substance could stimulate division at some distance from the site of production where its concentration is low and inhibit division at the site of production where the concentration is high. An auxin (Gonzales-Fernandez et al., 1968) and a cytokinin

(Torrey, 1972) have been mentioned in this connection, but assay methods are not yet adequate to test this theory. There seems to be no doubt, however, that overall rates of mitosis in root tips depend upon the relative levels of auxins, kinins and gibberellins (Street, 1968a).

Another theory is that there is competition between cells for supplies of nutrients or hormones arriving at the apex from the base of the root and that the quiescent centre is starved, especially by the demands of the proximal meristematic cells. This can be tested and is shown to be an unlikely reason for the existence of the quiescent centre (Clowes, 1970b), although there is some evidence in its favour. Van't Hof (1968b) used carbohydrate starvation in excised roots to stop cycling and in *Pisum* 90% of the cells proceed to G_1 and stop there. Barlow (1969) showed that the duration of S increased from 3·3 h to 6·3 h towards the tip of the root meristem and then rose abruptly to 20·7 h in the quiescent centre. On the other hand, Barlow also showed that the grain count over nuclei in S in roots fed with tritiated thymidine for 30 min is higher in the quiescent centre than in the proximal cells of the stele. These facts can be reconciled by assuming that, in the proximal regions, the radioactive thymidine is diluted by a larger pool of endogenous DNA precursors than in the more distal quiescent centre. The supply of thymidine from the shoot or endosperm does indeed appear to be attenuated by passage down the root, but the cap initials, on the distal side, receive more than the quiescent centre (Clowes, 1970b). It is the problem of the cap initials with their high rate of cell production that makes a starvation theory difficult. The differentiated cells of the cap are peculiar in their metabolism and are eventually sloughed off. There is some evidence provided by exchanging caps on roots that some of the contents of the moribund cells can be returned to the meristem (Clowes, 1970b) and this could explain the high activity of the cap initials if the quiescent centre is starved. But the experiments that show this also show that material from the cap is equally available to the stele on the far (proximal) side of the quiescent centre. It is also used by the quiescent centre itself because the exchange of caps stimulates its cells to proliferate. Thus it seems unlikely that a quiescent centre is maintained by starvation and more likely that demand determines supply.

Another possibility is that the quiescent centre is somehow isolated from the surrounding cells. It is true that its cell walls are usually thicker than those outside, but this is due to its cells being older. Plasmodesmata exist between the quiescent centre cells and their active neighbours as well as between themselves. Even the thick wall at the cap boundary is traversed by plasmodesmata shared by quiescent cells and cap initials (Barlow, 1971). Also those cells that are in cycle within the quiescent centre can use precursors from outside in their syntheses and so it looks as if isolation is not important.

B. Regulation by the Root Cap

A different light is shed on the control of cycling by some experiments on root caps. It was known that when caps are prized off roots of *Zea* the quiescent centre is stimulated into division and regenerates a complete cap in a few days and, with it, regains the ability to respond to gravity which the root loses with its cap (Juniper *et al.*, 1966). The rate of mitosis in the quiescent centre rises from 0·12 to 0·46 cell per cell per day within 5 h of the loss of the cap while in the stele the rate drops from 1·15 to 0·72 per day (Clowes, 1972). If the distal half of the cap, containing no meristematic cells, is cut off, the change in cell division is even more dramatic in the quiescent centre where the rate of mitosis increases to 0·7 cell per cell per day at 5 h. The change in rate in the stele is slight, but in the cap initials the rate drops over 24 h. Since removing the cap from *Zea* roots does not alter the growth of the root, it is not clear how removing the cap effects the change in cell division, but in view of the ability of the cap to control growth rates in a geotropic reaction the influence of a hormone must be considered. Both Gibbons and Wilkins (1970) and Pilet (1972, 1973) have argued that an inhibitor of growth is produced by the cap, but the effects on rates of division show that the control exerted by the cap is complex.

C. Pressure

A survey of quiescent centres in root apices of different construction shows that the shape of the quiescent centre varies with the cell pattern. In *Zea* it is almost hemispherical with the flat surface facing the tip; in *Sinapis* it is hemispherical with the flat surface facing the base; in *Pisum* it is a flat disc. The way the files of cells run towards the quiescent centre suggests that it is the expansion of the initials and their derivatives near the surface of the quiescent centre that prevents the quiescent centre from expanding and therefore its cells from dividing. In other words it is the pressure exerted by the expanding cells and the arrangement of cells that make a quiescent region inevitable. Some support for this view comes from some preliminary experiments in which the pressure exerted by the cap is replaced artificially.

The experiment of Webster and Langenauer (1973) in which roots were released from starvation-induced dormancy and all cells divided before a new quiescent centre was formed also suggests that it is the dividing cells that are responsible for the existence of the quiescent centre rather than its being the site of production of an inhibitor.

8. The Shoot Apex

R. F. LYNDON

Department of Botany, University of Edinburgh, Scotland

I. INTRODUCTION

The outstanding feature of the shoot apex* is the initiation of new organs—leaves when vegetative and floral parts when reproductive. These processes depend on continued growth and in so far as cell division is a component of growth it is therefore an essential component

---

* Sometimes the term shoot apex is used to include the whole of the terminal bud. At the other extreme is its restriction to mean only that part of the meristem distal to the youngest leaf primordium. Here it will be used in a looser sense to include not only the apical dome (equivalent to the shoot "apex" as defined by Abbé and Phinney, 1951), but also the youngest leaf primordia, and the associated axial tissue, with no specific lower limit.

of morphogenesis at the shoot apex. The fact that new organs grow out in directions which are different from the direction of growth of the axis from which they have sprung implies that the direction, or plane, of growth alters and presumably the plane of cell division alters too when the leaves and floral parts are initiated and begin to grow.

The positioning of new leaves on the apex and the consequent leaf arrangement, or phyllotaxis, depends on the interrelations of the relative growth rate, the shape of the apex, and the size of the leaf primordia on initiation (Richards, 1951). At least the first two of these, if not all three, are a function of the rates or directions of cell growth and division. An understanding of the processes of leaf initiation and the siting of new leaves therefore requires an exact knowledge of the rates and planes of division in the shoot apex and how these may change with time in relation to the processes of morphogenesis and development with which they are inextricably bound up.

II. RATES OF CELL DIVISION IN THE VEGETATIVE SHOOT APEX

A. The Apex as a Whole

The shoot meristem originates in the embryo at the base of the cleft between the two cotyledons, in a dicotyledon, or at the side of the embryo in a monocotyledon. The changes in the growth rate of the shoot meristem in the developing embryo have been measured in the maize (Abbé and Stein, 1954). Since average cell size remained constant the changes in the rate of cell division could be measured as changes in the relative growth rate, which slowed down exponentially as successive leaves were formed and eventually became zero when five or six leaf primordia had been formed, and the embryo was mature.

On germination of a seedling, cell division is resumed, and the developing primordia and leaves continue division so that the size of the meristem increases to include not only the shoot apex but all the young unfolded leaves (Sunderland, 1961; Sunderland and Brown, 1956). The meristematic region therefore expands until the plant reaches a steady state of growth and at this stage the shoot meristem in its broadest sense encompasses most of the terminal bud, and consists of tens of millions of cells, all probably dividing at very much the same rate. In contrast to the root, in which the meristem extends back only 1–2 mm from the tip, in a shoot it may extend for 1 or even 2 cm down from the apex. The immense size of the shoot meristem compared with the root meristem means that the shoot can increase in size relatively rapidly even if the relative growth rate is low, because of the large number of dividing cells which it contains.

The mean cell generation time (MCGT), or mean cell doubling time for the whole apex, is a measurement of limited value as it is an average which includes all cells however different their division rates may be. It can, however, be useful in allowing a general comparison of shoot apices of different species and a comparison with the MCGTs of roots of the same species. While it is difficult to make exact comparisons, the MCGTs of shoots seem to be on the whole longer (and cell division is therefore slower) than in the roots of the same species under comparable conditions (Lyndon, 1973). A comparison of the shoot apices of different species shows that the MCGTs differ considerably but are often of the order of 1–3 days. The longest recorded is the oil palm, with a cell dividing on average only once every 7 weeks.

The MCGT of young leaf primordia is in most cases probably about the same as that of the cells in the apex itself (see Chapter 9). The apex, primordia and associated axial tissues can be regarded as all parts of a single system growing exponentially (Schuepp, 1938), the growth of which can be expressed by the equation

$$y = ae^{rt}$$

where y is the final number of cells, a the original number, r the relative growth rate and t is time. When this equation is applied to the growth of the apical dome then t can never be greater than one plastochron because the apical dome is re-defined when each new primordium is initiated. However, the growth of the primordia may continue for as many as six or more plastochrons, so that even when the relative growth rate (r) for the primordia is the same as, or even lower than, that for the apical dome, the primordia soon become much larger than the dome and overtop it (Lyndon, 1972a). The apparently rapid growth of the leaf primordia compared to the apical dome is therefore a consequence of following the growth of the primordia as distinct entities for a much longer time than is done for the dome, the growth of which is of necessity followed only for a single plastochron.

Some plants may reach a steady state of growth so that the MCGT of the apex remains constant for long periods of time. This may well be the case with plants having an absolute photoperiodic requirement and which are kept in non-inductive conditions. There is evidence that in some plants the apex does not reach a steady state. In maize the MCGT shortens, so that the relative growth rate of the apical dome steadily accelerates, during growth of the shoot (Abbé et al., 1951). In the lupin and the rye, the opposite occurs, the MCGT in the apex steadily lengthening during vegetative growth and shortening transiently only during the transition to flowering (Sunderland, 1961).

Detailed measurements of the growth of the apical region of the lupin during vegetative growth reveal that the relative growth rates of the apical dome, the primordia and the axial tissue are all different and all decrease, but at different rates. The data are consistent with the rates of decrease themselves being exponential (Brown, Chapter 1).

B. Cell Division in Different Regions of the Apex

1. Rates of Cell Division

Until about 10 years ago, the conclusions drawn about the relative rates of division within the shoot apex depended on indirect and circumstantial evidence. The cells at the summit of the vegetative shoot apex were believed to divide less rapidly than the cells on the flanks of the apex (where the leaves are initiated) because the summit cells characteristically had a lower frequency of mitotic figures than elsewhere in the apex and the cytological appearance of the summit cells was thought to resemble that of cells with low metabolic activity (Nougarède, 1967). Another point of view was that all the cells of the apex were dividing relatively rapidly. Some of the principal evidence advanced in support of this view depended on the measurement of rates of displacement of cells from the summit of the apex to the flanks (Ball, 1960; Soma and Ball, 1963; Ball, 1972). By analogy with the root meristem, in which rapidly dividing cells are rapidly labelled with ^{3}H-thymidine and slowly dividing cells are slowly labelled, it was expected that the rapidity and extent of labelling with DNA precursors would give an estimate of relative division rates. In the apex of some species the summit cells became labelled, but in other plants they did not (see review by Nougarède, 1967).

Much of the discussion has centred around the validity of the concept of the "méristème d'attente" propounded by Buvat (1952) and other workers, which requires that the cells at the summit of the apex do not divide rapidly enough during vegetative growth to allow cells to be displaced into the flanks of the apex, which is occupied by the "anneau initial", a self-perpetuating ring of rapidly dividing cells in which the leaves are initiated. Only at the onset of flowering would the cells of the "méristème d'attente" begin to divide rapidly and result in the formation of the flower or inflorescence.

The controversy has continued mainly because of a lack of reliable measurements of the rates of division within the shoot apex. Measurements of the frequency of mitotic figures cannot, in the absence of other information, be used as indications of the rates of cell division. It has been repeatedly pointed out (e.g. Brown, 1951) that the mitotic index is a measure only of the proportion of cells in the act of mitosis at the time of observation, and in a completely asynchronous meristem indicates the proportion of the whole cell cycle which is spent in mitosis. If, and only if, the absolute length of time spent in mitosis remains constant, and changes in the length of the cell cycle are due entirely to changes in the length of interphase, is the mitotic index proportional to the rate of division. There are, however, many instances in which this is not so. In *Zea* roots (Clowes, 1961) the mitotic index was shown not to be proportional to the length of the cell cycle. In *Helianthus* roots the mitotic index remained constant despite differences in the length of the cell cycle of more than 7-fold (Burholt and Van't Hof, 1971), and conversely, in *Pisum* roots, large variations in mitotic index occurred without changes in the length of the cell cycle (Van't Hof, 1965b). This was because, in both cases, the length of time spent in mitosis altered.

Labelling indices, based on the proportion of nuclei which become labelled after the application of radioactive DNA precursors (usually ^{3}H-thymidine), can sometimes be even more misleading. The percentage of the cells which becomes labelled after supplying ^{3}H-thymidine is proportional to the number of cells in S which, in an asynchronously dividing tissue, is a function of the proportion of the whole cell cycle spent in S. Like the mitotic index, the labelling index would be a function of the rate of division only if the length of S remained constant, irrespective of the length of the cell cycle. It has been shown that this is not the case for the shoot apices where it has been measured. The length of S varied with the length of the cell cycle in *Rudbeckia* (Jacqmard, 1970) and *Pisum* (Lyndon, 1973) so that the proportion of cells in S remained relatively constant in the different parts of the apex. It is quite clear in these plants that the labelling index does not give an estimate of division rates in the shoot apex. This can be demonstrated diagrammatically by comparing a map of the rates of division with a map of the labelling index (Fig. 1). Cells which are undergoing endomitotic duplication of their DNA will also become labelled and (as discussed later) this invalidates many of the conclusions previously drawn about the rates of division in pteridophyte apical meristems.

There are, however, some plants where the labelling index is informative and these are plants in which there is a region of the apex which does not label at all with ^{3}H-thymidine whereas other regions do. The central zone at the centre of the *Helianthus* apex did not become labelled after 24 h or 48 h exposure to the label whereas cells on the flanks of the apex did become labelled (Steeves *et al.*, 1969). This was interpreted as showing that the central zone cells did not synthesize

DNA during the course of the experiment and therefore, if they were dividing at all, were almost certainly dividing much more slowly than the peripheral cells. A convincing demonstration that the central zone cells at the summit of the tobacco (*Nicotiana*) apex do not divide during the growth of the apex has been provided by a series of experiments with excised, cultured apices which initiated leaves at a steady rate (Sussex and Rosenthal, 1973). The peripheral cells all became labelled over a 72 h period of application of ^{3}H-thymidine, but the central zone cells remained unlabelled. It was possible to do the converse experiment and show that when the central zone cells were labelled, the label remained in them. Apices were cut into four, longitudinally, and supplied with ^{3}H-thymidine which was incorporated into all the cells as a new apex was regenerated. Once this was achieved and the apex was placed on unlabelled medium, the label became dissipated from the peripheral cells by the growth and division of these cells, but the cells of the central zone retained their label, indicating that they were not dividing.

The first direct measurements of the rates of division in different regions of the shoot apex were made on *Trifolium* by the use of colchicine (Denne, 1966a). If the applied colchicine results only in the inhibition of exit from metaphase, and has no other effects, then the rate of accumulation of colchicine-metaphases equals the rate of division (Evans *et al.*, 1957). If the cells are all exposed to colchicine for the same time, then the percentages of colchicine-metaphases in different regions of the apex will be proportional to the rates of division in these regions. In *Trifolium* the rate of division of the summit cells measured by this method was about one-half of that of the cells in the flank regions of the apex where leaves are initiated. This same method was used to measure in detail the rates of division in the pea apex, with very similar results (Lyndon, 1970a). Both these plants, *Trifolium* and *Pisum*, had the advantage of having a distichous leaf arrangement (i.e. the leaves arranged alternately, in two ranks on opposite sides of the stem) so that longitudinal median sections cut through not only the existing leaves and primordia but also the sites at which future leaf primordia would arise. Rates of division have also been measured by the colchicine method in *Datura* (Corson, 1969), *Chrysanthemum* (Berg, quoted in Gifford and Corson, 1971), *Solanum* (Leshem and Clowes, 1972) *Sinapis* (Bodson, 1975) and *Coleus* (Saint-Côme, 1973). In all cases, the rate of division in the cells at the summit of the apex was about half or a third of that on the flanks (Table I).

The lengths of the cell cycle in the different regions of the shoot apex have also been measured by labelling techniques, though these are more difficult to apply to shoot than to root apices. Firstly, there is difficulty in getting the label taken up by the apex. Secondly, the characteristically low mitotic index in shoot apices means that methods which depend on the scoring of the percentage of labelled mitotic figures require many apices, but because of the necessity of dissecting out each apex before application of the label this is not a requirement that can be met easily, if at all. In these circumstances it is not surprising that there have been only a few measurements made by labelling techniques of the cell cycle length in the shoot apex. In fact the only one to employ the technique of scoring the percentage of labelled mitotic figures was that for *Rudbeckia* (Jacqmard, 1970). This showed that the cell cycle was longer for the central zone (at the summit of the apex) than for the flanks of the apex (Table I), but an exact determination could not be made because the cell cycle in the central zone was longer than the length of the experiment (40 h). In *Isoetes* labelling was also used, but again the length of the cell cycle in the summit cells was longer than that of the cells of the flanks and longer than the period of the experiment (Michaux, 1969). To measure the length of the cell cycle in the summit cells of *Polytrichum* and *Coleus*, continuous labelling experiments had to extend over more than 2 weeks (Hallet, 1969; Saint-Côme, 1969).

All the data are consistent in showing that the rate of division of the cells at the summit of the apex (the central zone) is about one-half (or less) of that of the cells on the flanks of the dome. They are also consistent in showing that the mitotic index in these vegetative apices,

TABLE I. Length of the cell cycle (hours) at the summit (central zone) and on the flanks (the region of leaf initiation) of vegetative shoot apical meristems

	Region of apex		Reference
	Summit	Flanks	
Pisum	70	28	Lyndon (1970a)
Rudbeckia	>40	30	Jacqmard (1970)
Datura	76	36	Corson (1969)
Isoetes	>53	36	Michaux (1969)
Trifolium	108	69	Denne (1966a)
Chrysanthemum	140	70	Berg (see Gifford and Corson, 1971)
Solanum	117	74	Leshem and Clowes (1972)
Polytrichum	360[a]	96	Hallet (1969)
Coleus	237	125	Saint-Côme (1973)
Sinapis	288	157	Bodson (1975)

[a] Apical cell.

presumably growing in a steady state, is proportional to the rate of division. This is because the time spent in mitosis remains constant, and changes in the length of the cell cycle as a cell is displaced down the apex are almost entirely due to changes in the duration of interphase (Denne, 1966a; Corson, 1969; Lyndon, 1973). The estimates of relative division rates based on the mitotic index are therefore, luckily, probably correct. Whether it is safe to assume that this is always so in the shoot apex is more doubtful. Until it is known in each case how the mitotic index is related to the cell cycle it is unsafe (as the data from the root meristem have shown) to assume that it is an indicator of the cell division rate.

It is also unsafe to use cytological characteristics as indicators of division rates. The central zone in the pea consists of cells at the summit of the apex which can be distinguished by their staining characteristics, by their slightly larger size compared to cells in the rest of the apical dome and by their larger nuclei (Lyndon, 1973), and it corresponds to the region which has been thought of as having a slow division rate (Nougarède, 1967). It may be argued that perhaps the central zone of the pea is not so well marked as that in some plants. Nevertheless the region of low division rate is much more extensive than the central zone (Lyndon, 1970a). In this plant at least, the cytological characteristics do not correspond to the region of low division rate. In the pea, the central zone and the cytological characteristics which typify it are not a function of division rate but indicative of some physiological state of which we so far know nothing. There is no reason to think it would be any more valid in other plants to infer relative division rates from cytological characteristics than it is in the pea.

The clearest indication of relative division rates seems to be given by the colchicine method. It has the further advantage that, since the number of metaphases observed is usually much greater than mitotic figures in an untreated plant, it is easier to get more detailed information about the distribution of division rates in the shoot apex. By superimposing data from several apices of the same developmental age it is possible to construct maps showing the distribution of division rates in sections or on the surface of the apex (Fig. 1a). Using this method on the pea apex, with the apices sorted into developmental stages within a single plastochron according to the length of the youngest primordium, maps showing the distribution of rates of division throughout the apex (using all the sections of the serially sectioned apices) were made (Lyndon, 1970a). These showed that as the apex grew between the initiation of one leaf and the next a plate of rapidly dividing cells became established between the axil of the newly formed primordium and the next oldest primordium. That this was in fact a plate of cells could be shown by building up from the data sections in other planes than the median. Just before the next primordium appeared as a bump (at the maximal area phase of the plastochron), the structure of the apex in terms of rates of division was that depicted in Fig. 1a. There is a bowl-shaped area with a low division rate at the summit of the apex, indeed extending to most of the apical dome. The incipient primordium and the axial tissue subtending it have an intermediate rate of division and separating these two zones is the plate of faster dividing cells. This plate is extended downwards at the sides of the apex and links up with the

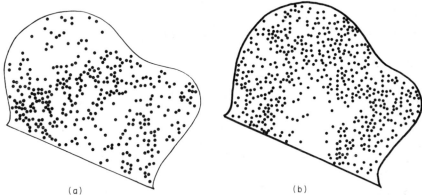

(a)   (b)

FIG. 1. Median longitudinal sections of *Pisum sativum* shoot apices. (a) Relative rates of division, proportional to the density of the points, each of which represents a colchicine-metaphase (Lyndon, 1970a). (b) Distribution of labelled cells after supplying ^{3}H-thymidine for 2 h (Lyndon, 1972a).

procambium which is only easily visible in sections a little lower down the stem. Although the position of the incipient procambium (which is hardly visible at this level) is readily indicated by regions of rapid divisions at the sides of the apex where the stipules will be formed, the procambium which can be seen differentiating in the youngest primordium does *not* seem to consist of cells having this high rate of division. It seems to have the same rate of division as the other cells around about it. The analysis was not extended into the developing nodes and internodes; it would be interesting to see whether the plate of rapidly dividing cells disappears as the node develops or whether it persists as a nodal plate. Is the development of the nodal structure in fact associated with a higher division rate than the cells around it? This seems quite possible, but as yet we have no data on this point.

All the data point to the existence in all or most shoot apices of a gradient in the rate of cell division and growth from a minimum at the summit of the apex to a maximum in the region of leaf initiation. This raises the questions of how the rate of division is controlled, whether the gradient in division rates is essential to the functioning of the apex, and what the significance of this gradient may be.

2. Cell Division and the Displacement of Cells from the Summit of the Apex

If there are groups of cells in the shoot apex which do not divide, or divide very slowly, one would expect them to be found somewhere near the summit, just as the quiescent centre in the root is located at the tip of the promeristem, otherwise they would soon be carried out of the meristem by the growth and division of the cells distal to them. Experiments on lupins have shown that when individual cells at the summit of the shoot apex were killed by being punctured with a fine needle they were invariably displaced down the apical dome (Soma and Ball, 1963). It may be argued that the puncturing stimulated in the summit cells divisions that would not otherwise have occurred and that this resulted in the displacement of the punctured cell. When a carbon particle was placed on the apex without apparently damaging the cells, in some plants the particle was still at the summit after 6 weeks, indicating that the surface cells had not been displaced and that division and growth were therefore very slow at this point in the apex.

The carbon particles which were displaced (and this was most of them) took a minimum of 28 days to move off the apical dome, a distance (judged from the illustrations in Soma and Ball, 1963) of less than 20 cell diameters. A movement of 20 cell diameters could be achieved in 4–5 cell generations if all the cells were dividing and giving rise to new cells only in the longitudinal direction. Since divisions give rise to cells laterally as well as longitudinally the rate of division to give this degree of longitudinal displacement would have to be about twice as great, or about 8–10 cell generations in the 4 weeks. This would be equivalent to a cell generation time of the order of 3 days, which is similar to the values obtained by direct measurements (Sunderland and Brown, 1956; Sunderland, 1961). In the apices in which a cell was punctured, the wounds moved off the apical dome in a minimum of about 3 weeks, again equivalent to a mean cell generation time of about 3 days. The occurrence of cell displacement in the apex of Lupinus shows that the absence of labelling by ^{3}H-thymidine in the cells at the summit of the Lupinus apex (Nougarède, 1967) is not a reliable indication of the histogenic function of these cells. It is worth noting (Fig. 2) that even if the rate of division of all the cells in the surface of the apical dome were exactly the same a cell would be displaced with a velocity which would accelerate exponentially with increasing distance

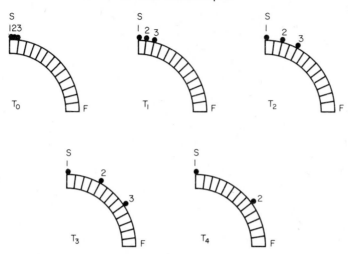

FIG. 2. The diagrams represent surface cells of an apical dome (as seen in longitudinal section) at successive equal time intervals (T_0-T_4). F marks the flank of the apex; S the summit, on which the marker spots 1, 2 and 3 have been placed at time T_0. It is assumed that the rate of division and growth of all the cells is the same. If the cells divide with equal frequency in all anticlinal planes, then each interval would represent a period somewhat longer than two cell cycles. Spot 1 has been placed right on the summit and has not been displaced during the experiment. Spots 2 and 3, placed just off centre, are displaced with a velocity which increases exponentially because of the division and growth of cells in the more distal parts of the apex. By T_4, spot 3 has been displaced off the apical dome and on to the developing stem or a young leaf.

from the summit. The observation that cells are displaced less rapidly at the summit of an apex than further down the flanks (Loiseau, 1962) does not in itself allow us to infer that the rate of cell division at the summit is also less.

In the lupin about nine new leaves were initiated during the time taken for a cell to be displaced down the apical dome and into a leaf primordium (Soma and Ball, 1963). In plants which have a long vegetative phase there would be a greater possibility that any cell at the summit of the apex could be displaced off the dome and into the leaf primordia. Plants which have an absolute photoperiodic requirement in order to flower can be kept vegetative for long periods. Such a plant is *Silene*, a long-day plant which has been maintained vegetative for more than 6 months during which time more than 40 pairs of leaves were initiated at a steady rate (Lyndon, unpublished data). A cell would have to be displaced about 14–20 cell diameters' distance from the summit to be incorporated into the leaf primordia and this would be achieved in about 8 cell generations. Even if the cells of the apical dome were dividing on average only about once every 3 weeks this would be sufficient to result in displacement of cells down from the summit of the dome and into leaf primordia. In fact the cells in the *Silene* apex divide on average once a day (Miller and Lyndon, 1975), so that even if the cells at the summit were dividing at only about one-twentieth of this rate they could still be displaced down the apex and incorporated into leaf primordia. The fact that the cells at the summit of the apex may be dividing slowly compared to cells in the rest of the apex is therefore not a barrier to their being functional as initials which can contribute cells to the leaves. Although extremely low rates of division in the cells at the summit of the apex may not be detected by histological examination of apices they can be demonstrated by the use of chimaeras, which show that these summit cells, even when they divide infrequently, may in fact play an important role in the growth of the vegetative plant (Stewart and Dermen, 1970).

The concept of the "méristème d'attente" is therefore not likely to be of general application. It may be descriptive of the situation in plants such as *Helianthus* and *Nicotiana* in which the cells of the central zone do not appear to divide in the apices which have been examined. It may also have a use in describing the situation in plants such as *Zea*, *Pisum* and *Lupinus* in which the apical dome grows during the vegetative phase. If the vegetative phase is restricted and the plant fairly quickly makes the transition to flowering then it could well be that the cell divisions in the apical dome could give rise only to growth of the dome itself, during vegetative growth. Such a situation has not yet been quantitatively and rigorously demonstrated, but there seems no reason why it should not occur in some plants.

The displacement of cells from the summit of the apical dome has also been observed directly (Newman, 1956; Ball, 1960; Ball and Soma, 1965). *Tropaeolum* apices were observed by reflected light, under a dissecting microscope, tracings were made (with a camera lucida) of the outlines and positions of the apical cells, and the displacement and division of cells at the summit were observed (Newman, 1956). More complete records were obtained of growth and division of cells in the *Vicia* apex (Ball, 1960). The displacement of cells from the apical summit and into the developing primordia was clearly recorded. In Newman's experiments the apices began to dry out between observations and in Ball's experiments the apices were excised and cultured on media which included coconut milk and gibberellin. It has been argued that these apices may therefore not have been behaving in the same way as apices in the intact plant (Nougarède, 1967). However, they were, as far as could be judged, quite normal and comparable to apices in intact plants.

Once a cell leaves the summit of the apex it becomes displaced into regions where the rate of division is faster and the cell cycle is shorter. However, it is possible, as has been illustrated for the pea (Lyndon, 1973), that the rate of displacement is relatively great compared to the length of the cell cycle so that a cell may always be in the process of speeding up its cell cycle and may not reach a steady state, so far as cell cycle length is concerned, until it is incorporated into a leaf primordium. In those cells which form the pith the cell cycle may no sooner get to minimum length than it begins to lengthen again as the cells mature and pass out of division. The only cells in a reasonably steady state of cell cycle length may be those in the primordium, those in the procambium (for a short time) and those at the extreme summit of the apex.

The observation that sometimes a carbon particle was not displaced from the summit of the apex over a period of several weeks (Soma and Ball, 1963) is not inconsistent with displacement of cells being the norm. For as Newman (1961) has pointed out, there must be at least three cells at the summit of the apex which are the descendants of previous apical cells and at least one of these must be the ancestor of subsequent apical cells. Newman has called these cells the "continuing meristematic residue". Presumably the immobile carbon particles have been placed fortuitously over these particular cells.

Ball's view (and Newman's) seems to be that the cell which will be found at the centre of the dome cannot be predicted very far in advance, and may depend on the nutritional, hormonal and cellular environment. The likelihood of all cells eventually finding themselves in this position may not be the same, because in the steady state growth of the shoot the future roles of cells in different parts of the apical dome will be relatively fixed if the system is not perturbed.

In some pteridophytes there seems to be a specialized situation in which the cells at the summit of the apex may have lost their histogenic potentialities. The shoot apex of *Equisetum* resembles the root apex of other pteridophytes in having an apical cell which may divide but is polyploid (D'Amato and Avanzi, 1968; Avanzi and D'Amato, 1967). The structure of the apex suggests that when the apical cell has become polyploid it gives rise to adjacent polyploid cells by division but that the growth of the shoot and the initiation of leaves occurs as a result of the growth and division of the diploid cells less distal in the apex. This is at first sight comparable to a "méristème d'attente" except that it is not known whether the polyploid pteridophyte apex has any histogenic contribution to make during subsequent development of the plant or whether this signals its end as an effective contributor to the growth of the shoot. The fern, *Dryopteris*, has been used for many experiments and observations on the functioning of the shoot apex. It would be of interest to know for this plant what contribution, if any, the apical cell and its immediate derivatives make to the regions of the apex where the leaves are initiated.

3. Cell Division and the Shape of the Apical Dome

The apical dome of many plants approximates to a hemisphere or a portion of a hemisphere. It can be demonstrated that this shape could be maintained at a growing point showing isotropic growth (equal in all directions) only if there is a gradient (a cosine gradient) of decreasing relative growth rate from a maximum at the summit of the dome to zero at the base (Green *et al.*, 1970). This is the opposite of what is characteristically found in shoot apices: a minimum growth rate at the summit increasing to a maximum growth rate on the flanks of the apical dome where the leaves are initiated. The growth of the apical dome therefore cannot be isotropic; there must be some anisotropy in its growth. One would expect that the longitudinal component of growth would increase with distance from the summit of the apex at the expense of the latitudinal component. These two components one would find to be equal if growth were isotropic.

Despite a claim to the contrary (Lyndon, 1970b), longitudinal growth does seem to predominate in the pea apex. Re-examination of my earlier data for orientation of mitotic spindles on the surface of the pea apical dome can be made from Fig. 3. It is difficult in view of the small number of divisions to know whether growth at the extreme summit is isotropic or not. Over most of the surface of the apical dome the longitudinal (radial) component of growth is about twice as great as the latitudinal component. The anisotropic growth on the surface of the pea shoot apical dome therefore appears to be consistent with that expected for a markedly hemispherical apex with a gradient of increasing rate of growth from the summit downwards. The predominance of the longitudinal component of growth was also seen in *Impatiens* apices which had had marks of carbon black placed on their summits. As the marks became displaced down the apical dome they often became elongated radially (Loiseau, 1962). In apices which are less markedly domed than the pea one might expect growth to be less anisotropic and in apices like the *Helianthus* apex, which are essentially planar, to be isotropic.

These considerations lead to the conclusion that the shape of the

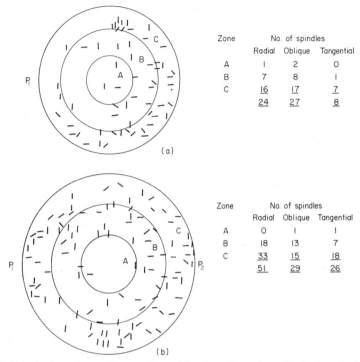

Zone	No. of spindles		
	Radial	Oblique	Tangential
A	1	2	0
B	7	8	1
C	<u>16</u>	<u>17</u>	<u>7</u>
	24	27	8

(a)

Zone	No. of spindles		
	Radial	Oblique	Tangential
A	0	1	1
B	18	13	7
C	<u>33</u>	<u>15</u>	<u>18</u>
	51	29	26

(b)

FIG. 3. Orientation of growth as indicated by the orientation of mitotic spindles in the surface cells of the shoot apex of *Pisum sativum*. (a) Minimal area stage (just after initiation of a leaf at P1). (b) Maximal area stage (just before initiation of a leaf at P2).

The diagrams represent the surface of the apical dome as seen from a point directly above the highest point of the dome. As seen from the tables which summarize the diagrams, the orientation of the mitotic spindles is predominantly longitudinal (radial) rather than latitudinal (tangential). (Data from Lyndon, 1970b.)

shoot apex is probably governed not by variations in the growth rates of the cells in the apical dome but by the relative amounts of longitudinally orientated and latitudinally orientated growth, i.e. by variations in the directions of growth. Experimental observations which support this view are those which were made on *Chrysanthemum* apices treated with tri-iodobenzoic acid (Schwabe, 1971). The rate of growth, measured as the cell doubling time, did not change in the treated plants (compared with the untreated controls), although their apices became narrower and more elongated. Unless there had been a complete reversal in the growth rate gradient in the apex (and this seems unlikely) this change in shape could only have been achieved by a change in the direction of growth with the longitudinal component being emphasized at the expense of the latitudinal component.

If the gradient of increasing rates of cell division and growth from the summit of the apex downwards is not involved in maintaining the shape of the apex, then its significance must be looked for in other aspects of apical structure and function, of which leaf initiation is the most likely.

C. Control of Division Rates

1. Control Points in the Cell Cycle

The information from the colchicine data in conjunction with measurements of mitotic index indicates that in shoot apical cells with different cycle lengths the absolute time spent in mitosis is constant, and it is interphase which varies (Lyndon, 1973). With more detailed information about the lengths of the various parts of interphase (G_1, S, G_2) it is possible to see what stage in the cell cycle is blocked in slowly dividing cells. For instance, in the quiescent centre of the root the cells are held in the G_1 phase so that the bar to progress through the cell cycle is a block at the stage of DNA synthesis (see Chapter 7, this volume). Despite the experimental difficulties involved in labelling shoot apices, Jacqmard (1970) succeeded in measuring the lengths of the component phases of the cell cycle in the cells of the flanks and the incipient pith of the *Rudbeckia* apex, although complete measurements could not be made for the summit cells (central zone). G_2 and S were longer in the summit cells than in the cells on the flanks of the apex, but neither the length of the whole cycle nor G_1 could be obtained for the slowly dividing summit cells. However, these results showed that the cells of the shoot were probably different from those of the root in having both G_2 and S extended, whereas in the root the major extension was in G_1.

The only estimates of the lengths of all the phases of the cell cycle in each region of the shoot apex are those for the pea (Lyndon, 1973). The method used was that of Mak (1965). This involves labelling the cells with a short pulse of ^{3}H-thymidine so that the cells in S become labelled. The apices are then sectioned and stained with Feulgen. Autoradiographs are made of the sections and the percentage of labelled nuclei (those in S) is scored. The DNA content of each unlabelled interphase nucleus is then measured by microdensitometry so that the proportions of 2C (G_1) and 4C (G_2) nuclei can be found. The proportion of cells in mitosis is given by the mitotic index. From these data the proportions of cells in each phase of the cell cycle (G_1, S, G_2

TABLE II. Lengths (hours) of the cell cycle and its component phases in the shoot apical meristem of *Pisum sativum* (data from Lyndon, 1973)

	Whole cycle	Phases of the cell cycle			
		G_1	S	G_2	M
Central zone	69	37	13	18	1
Flanks of the apical dome	29	15	8	5	1
Leaf primordium	29	15	9	4	1

and M) can be calculated, and when a correction is made for the age gradient in a population of exponentially dividing cells (Nachtwey and Cameron, 1968) the relative lengths of time spent in each phase can be calculated. If the length of the whole mitotic cycle in each region is also known (e.g. from colchicine-metaphase data) then the absolute lengths of the phases in each part of the apex can be calculated. A simpler method for finding the proportions of cells in G_1 and G_2 can be used which requires only the measurement of average values for DNA per cell (Lyndon, 1973). Both methods gave similar values for the lengths of the phases of the cell cycle in the pea apex and the means of both sets of measurements are given in Table II. The cell cycle in the central zone was more than twice as long as in the rest of the apex. This longer cycle was not because of the extension of any one part of interphase; G_1 and G_2 were extended by about 2·5 times and 3·6 times respectively, and S was also extended but only by about 1·5 times. Since the amount of DNA to be synthesized during S was the same in all parts of the apex (from 2C to 4C) the longer period in S means that the actual rate of DNA synthesis is slower in the more slowly dividing cells.

It remains to be seen whether the plants so far examined are typical. If they are, the results suggest that there is no one point in the cell cycle which is blocked in the cells at the summit of the shoot apex but that the whole cycle is slowed down, except for the actual process of division (M) itself. The lengthening of G_1 and G_2 to a greater extent than S suggests that the main control points are at the transition from G_1 to S, and G_2 to M. Gibberellic acid, when it affects the cell cycle (probably shortening it) in the apical meristem, increases the proportion of cells in M and S as indicated by the increase in the mitotic index and labelling index respectively (Bernier *et al.*, 1967a). These observations suggest that gibberellic acid may also act at the points of entry into S and M. These are the main points at which carbohydrate level also controls the rate of progress through the cell cycle in roots (Van't Hof and Rost, 1972). The rate of cell division in cultured *Vicia* shoot apices is a function of the glucose concentration (Ball and Soma, 1965) which could presumably be controlling entry into S and M as in the roots.

What the relative concentrations are of substances such as gibberellic acid and sugars in the different regions of the shoot apex is unknown. However, the concentrations of macromolecules such as protein and nucleic acid are less in the slowly dividing cells at the summit of the pea apex than on the flanks of the apex where cell division is faster (Lyndon, 1970c). This is also shown by the ubiquitous occurrence of zonation patterns, visible after staining, in which the lightest staining, indicating the lowest concentration of protein and nucleic acid, is in the cells at the summit of the apex. It could possibly be that the concentration of small molecules is also less at the summit in the slowly dividing cells, but this is complete speculation.

2. Replication of Sub-cellular Components

Measurements of the amounts of RNA and protein per cell show that in the pea apex all cells have the same gross composition and so the rate of increase in RNA and protein is a function of the rate of cell division (Lyndon, 1970c). Whether this means that the rates of *synthesis* of these substances are also a function of the rates of cell division cannot be demonstrated until adequate experiments are done which take into account the possible complications of differential rates of uptake of precursor molecules and the sizes of the pools of endogenous metabolites (Lyndon, 1972b).

But the cells in different regions of the apex do not have the same ultrastructural composition (Gifford and Stewart, 1967) and this implies that the rate of formation or replication of subcellular components may not be correlated with the rate of cell division but may be controlled independently. A quantitative examination of the ultrastructure of the pea apex has shown that as cells are displaced down the apical dome and become incorporated into the leaf primordia the number of organelles per cell tends to increase, implying that the replication of organelles is going on faster than cell division (Lyndon and Robertson, 1975). In other cells, which eventually form the pith of the stem, the number of mitochondria per cell increases but the number of plastids per cell remains the same, suggesting that the rate of replication of at least mitochondria and plastids may be controlled by different mechanisms (see Leech, Chapter 4). The control of the rate of cell division is therefore only one facet of the problem of the control of replication at all levels of organization from the molecule to the cell.

3. Diurnal Rhythms

A plant growing under natural conditions is exposed to a diurnal fluctuation of light intensity and often of temperature as well. The plant might be expected to alter in carbohydrate and metabolic status, and it would not be surprising if the rate of growth and cell division at the shoot apex also varied diurnally.

There are a number of instances in which diurnal peaks of mitotic index have been found. In many cases (Bünning, 1952; Denne, 1966b; Karsten, 1915; Lance, 1952) the experiments have been done under normal conditions, i.e. alternations of light and dark periods and uncontrolled temperatures. The mitotic peaks are usually found during the dark period, but Denne (1966b, c) found the maxima to be in the light, in the afternoon. Other workers have found no evidence of a mitotic rhythm (Savelkoul, 1957; Popham, 1958; Jacobs and Morrow, 1961). Rotta (1949) and Karsten (1915) were able to demonstrate that the mitotic peak recurred after 24 h even when the plants were held in constant darkness at constant temperature. Rotta's data showed the peak mitotic index to be about 90%. This can only mean that the cells were highly synchronized. This being so, the recurrence of the peak after 24 h in constant conditions implies that the cell cycle was 24 h in these plants. In *Tradescantia* (Denne, 1966c) the average cell cycle was 4 days, and this may be partly why the maximum values for mitotic index were only about 7%. Similarly, in *Trifolium* (Denne, 1966b) with a cell cycle·of 64 h, the maximum mitotic index was only about 5% but was again in the light.

The occurrence of periodicity has been linked to the occurrence of the light/dark cycle, but it is puzzling why it is sometimes not found and why, when it is, the peak is in some plants in the dark and in others in the light. It may be of significance that in the plants in which maxima have been found only in the dark, the light has been provided by artificial illumination, of the order of a few hundred foot-candles, but that in the other instances, when the maxima have been smaller or occurring in the light, or non-existent, the plants have been grown under the much greater intensity of natural illumination. It is relevant here to consider the situation in unicellular algae, in which cell division is usually restricted to the dark period (Jones, 1970). In *Euglena gracilis*, which can be autotrophic or heterotrophic, the nocturnal periodicity of division can be removed, so that the cells will divide during the light as well, by the provision of a rich food supply (Leedale, 1959). If the periodicity which has been observed in the feebly lit shoot apices is a similar phenomenon, then it would be expected that strong illumination (as with natural light) could allow sufficient photosynthesis and thus increased food supply to suppress the periodicity of division. If this is so, and the observations strongly suggest that it is, then a periodicity in cell division might occur if the plants are grown under artificial conditions and insufficiently illuminated, but under adequate conditions of illumination one might expect a periodicity to be absent.

The other usual variable in these experiments is temperature. Bünning (1952) found that periodicity of mitosis was absent in fluctuating temperature in *Vicia* roots. This does not necessarily imply that there was no effect on the rate of division, because Burholt and Van't Hof (1971) showed that the mitotic index remained constant, at 7%, in *Helianthus* roots over the temperature range 10–35°C even though the cell cycle varied from 46 h at 10°C to 6 h at 35°C. Lack of change in the mitotic index does not therefore imply a lack of change in division rates if the cells remain asynchronous. On the other hand, if the cells are synchronous, then the cell cycle can remain constant even though there are peaks of mitotic index (which is what one would expect). If a mitotic maximum occurs for the first time, the occurrence of a peak implies that either cells have been speeded up in interphase, so that a greater than usual number of cells reach mitosis together or else that mitosis itself has been retarded so that cells tend to pile up in mitosis. The occurrence of a single mitotic peak gives no information about how the cell cycle has temporarily changed, but it does indicate that some degree of synchrony has been introduced into the system (see Chapter 3). Even if there are successive or diurnally recurring peaks of mitotic index these are not necessarily at intervals of one cell cycle unless each peak is large enough to account for the division of all the cells.

Although the evidence suggests that there are plants, grown under some conditions, which show a diurnal periodicity in cell division it is doubtful whether this is a universal feature of shoot apices. When it does occur it may be accompanied by diurnal fluctuations in growth rate at the apex, but this possibility has yet to be investigated.

III. CELL DIVISION AND LEAF INITIATION

A. Is an Increase in Division Rate Necessary?

In considering the growth of the apex as a whole (p. 287) it was demonstrated that once the young primordium is initiated its relative growth rate need be no greater than that of the apical dome, and could in fact be less, and yet the form of the apex, with the young leaves overtopping it, would not be fundamentally changed, because the primordium grows on through several plastochrons whereas the dome is re-defined every plastochron so that its maximum size is fixed (Lyndon, 1972a). An increased rate of cell division may not, therefore, be necessary to *maintain* the growth of the primordium once it is initiated, but is it necessary for the initiation process? Since the primordium grows out in a direction different from that of the continued growth of the apical dome then this different direction of growth is somehow brought about during primordial initiation and first becomes obvious with the occurrence of periclinal divisions in an inner layer of the tunica, which otherwise shows only anticlinal division. But is this the primary event or is it a secondary event, as it would be if the change in the direction of growth were brought about by pressure of a growing mass of cells inside it in the corpus? To what extent is the inception of a primordium the result of a change in the polarity of growth and to what extent a change in the rate of growth?

The changes in the rate of cell division that are associated with leaf initiation can be followed by comparing the rates of division in those regions of the shoot apex which represent a developmental sequence leading to leaf formation. The position occupied by the cells which form the incipient primordium are, in the plastochron before the primordium becomes visible, designated I_1 according to the terminology of Snow and Snow (1931) and in the plastochrons before this, I_2, I_3, etc. Since these represent the positions of future leaves, which are predictable, the cells in these stages of the developmental sequence can be identified by their position alone. Plants so far examined in sufficient detail are *Trifolium* (Denne, 1966a), *Datura* (Corson, 1969) and *Pisum* (Lyndon, 1970a). In each case the rates of division in the I_2 and I_1 regions were similar, differing at the most by no more than about 30%, so that the increase of cell division rate leading to leaf initiation was not very marked. However, these represent measurements of regions of the apex of which the delimitation is necessarily arbitrary, and if there are variations in division rates within these regions then the overall values may be misleading.

The detailed maps of division rates in the pea apex (Lyndon, 1970a) showed that the I_2 and I_1 region as delimited each consist of distal regions of slow division in the apical dome and regions of much faster division in the more proximal region at the base of the apical dome (Fig. 1a). Hussey (1972) pointed out that in fact there was in the I_1 region of the pea a small part which had an apparently higher rate of division than anywhere else in the apex and he suggested that this was a growth centre responsible for a rapid growth which resulted in the formation of the primordial bulge. A similar growth centre consisting of cells with a higher rate of division was also observed in the apex of the tomato (Hussey, 1971). It was suggested that the initial increase in division rate which led to the formation of the primordium in the pea was the increase in division rate from I_3 to I_2. Hussey (1972) is probably right in concluding that this increase in division rate in I_2 is an essential part of the functioning of the shoot apex, but it seems less certain that it is an obligatory part of the processes leading to formation of the primordial bulge itself, especially since the initial increase in division rate occurs (at the transition from I_3 to I_2) two plastochrons, or approximately four cell cycles, before the primordial bulge is formed.

Would the existence of a greater rate of division at the base of the I_1 region necessarily lead to the formation of a bulge? This would seem to depend on the cellular structure of the apex. If the apex consists of a tunica of two or three layers of cells, overlying a corpus in which divisions are in all planes, then an increase in division rate in the corpus would result in a nodule of cells which could be visualized as exerting pressure on the tunica cells and so stimulating them into division (Fig. 4a). Alternatively, if the apical dome in the first half of the plastochron is all tunica, as claimed for the pea (Lyndon, 1970b), then an increase in division rate should still give anticlinal growth but with some outward buckling of the surface, which could be visualized as resulting in a tension on the inner layers and thus stimulating periclinal divisions

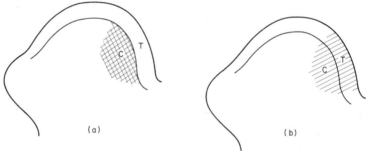

Fig. 4. Possible modes of growth of the apex during leaf initiation. (a) An increased rate of division in the shaded part of the corpus (C), giving growth of the shaded tissue, could result in pressure on the tunica in the region T so that it would be forced out as a bulge. On this model the divisions in C would be typical corpus ones, in all planes, throughout the plastochron. (b) An increased rate of anticlinal division in the shaded regions of the corpus, C, and tunica, T, could result in a tension on the cells at C which might result in outward growth to relieve the tension. The divisions in C would then contain a periclinal component, as a result of the outward growth, just before the leaf primordium began to be visible.

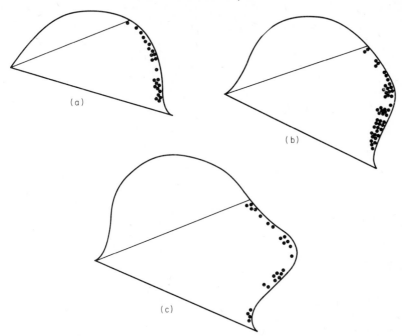

FIG. 5. Relative division rates, as shown by the distribution of colchicine-metaphases, in the epidermis of the median part of the pea apex at the point of leaf initiation. (a) Just before the leaf is formed. (b) At the initiation of the leaf primordium. (c) After one plastochron's growth of the leaf primordium.

The density of the dots is proportional to the rate of division. The greater rate of division in the basal part of the region (a and b) leads to faster growth of the abaxial surface of the leaf primordium in the early stages of its growth. The rates of division are less different on adaxial and abaxial surfaces once the primordium has become established (c).

in these inner layers (corpus). Note that the first possibility means that periclinal divisions are present in the corpus at all times, the second, that the formation of the bulge would be accompanied by tensions in the tissues at the onset of bulge formation (Fig. 4b). Neither of these explanations accords with the facts. The occurrence of a dome which is all tunica for part of the plastochron finds support in a number of observations (see below) and in several apices, including the pea, the tissues are not under tension (Hussey, 1973). The most plausible explanation is that the primary process in leaf initiation is the change in the plane of growth and that changes in the rate of growth are associated, but secondary, processes.

A characteristic feature of the early growth of the leaf primordium is that the growth of the abaxial surface is greater than that of the adaxial surface so that the young leaf tends to grow upwards (Gifford, 1951; Girolami, 1954; Hussey, 1971). The increased division rate in the basal part of the I_1 region may be important in this upward growth of the primordium when it has been initiated, and Hussey (1972) gives measurements for the pea apex to support this contention. He shows that the increase in cell number in the longitudinal plane is because of a higher rate of cell accumulation (1·8 cells per cell per plastochron) on the abaxial side of the young primordium than on the adaxial side (1·1 cells per cell per plastochron). Since these are data of increases in cell number in only one plane (in the epidermis) they would accurately reflect the actual rates of division only if the orientation of growth is equally longitudinal and transverse on both the adaxial and abaxial sides of the primordium. If the proportions of longitudinal and transverse growth differed on the two surfaces of the primordium then these values would not represent the true division rates. Values for the actual division rates in the epidermis are available from measurements made by Lyndon (1970a) and from supplementary data, and are shown in Fig. 5. These confirm that the rate of division is indeed greater on the abaxial surface of the incipient primordium than on the adaxial surface. This difference can be regarded as a natural extension of the difference in division rates which is established as early as I_2 between the upper and lower parts of this region. It should, however, be noted that this process is not necessarily involved in the formation of the primordial bulge, but in the establishment of the shape of the bulge while it is being initiated. This may well be a process that is concerned with the establishment of the dorsi-ventrality of the leaf.

B. Planes of Division and Growth

Whether or not there is an increase in the division rate of the cells at the point of initiation of the new primordium at the time it emerges as a bump on the apex, there are changes in the directions of growth both on the surface and within the tissues of the apex which are associated with leaf initiation.

A detailed examination of the pea apex at intervals throughout a single plastochron showed that essentially no mitotic spindles which would give periclinal divisions could be found in the apical dome in the first half of the plastochron, whereas one-third of all mitotic figures would give periclinal divisions during the second part of the plastochron (Lyndon, 1970b, 1972a). This implies that for about 30 h out of the 46 h of the plastochron, i.e. for almost two-thirds of the time, there were no periclinal divisions in the whole of the summit of the apex and that this was therefore all tunica, with only anticlinal divisions, so that the tunica at this time effectively extended inwards about six layers of cells from the surface. In the second part (one-third) of the plastochron, periclinal mitoses suddenly appeared and could almost always be found, so that the tunica became reduced to the typical single layer at the point of leaf initiation and to two layers at the extreme summit of the apex (see Fig. 3 in Lyndon, 1972a). In the pea the number of tunica layers at the summit (and not just where the leaf forms) therefore seems to vary according to the stage of the plastochron. Is this an observation of general validity? There is in fact quite a considerable amount of evidence that suggests that it might be. Gifford (1954) cites a number of examples (Gifford, 1950; Cross and Johnson, 1941; Schnabel, 1941; Reeve, 1942, 1948; Rouffa and Gunckel, 1951; Zimmerman, 1928) in which a more stratified appearance of the apex has been noted at certain times of the plastochron, often apparently at the maximal area phase, just before initiation of the new primordium. This would tend to correspond to the second half of the plastochron in the pea in which the periclinal divisions are found. However, it must be remembered that the most stratified appearance will be only after anticlinal divisions have gone on for some time, i.e. about half-way through the plastochron and at the time when periclinal divisions begin. Other workers have also noted changes in the number of tunica layers with the plastochron stage. In *Cosmos* periclinal divisions were found only in the corpus cells of the flanks (peripheral region) and only during the minimal area stage of the plastochron (Molder and Owens, 1972). The maximum degree of stratification was seen during the maximal area phase of the plastochron, i.e. when periclinal divisions were not found. In *Acorus* the stratification was also greatest during the maximal area phase of the plastochron, just before a new leaf emerged, and extended into the corpus so that as many as seven layers of cells were apparent at this stage (Kaplan, 1970). Soma (1958) also found in *Euphorbia* that maximal stratification occurred "in the stage prior to the maximal area phase" with minimal stratification "just after the initiation of leaf primordia". Observations on other plants can be interpreted as showing maximal stratification in the minimal area phase of the plastochron (Shushan and Johnson, 1955; Sterling, 1949). Although changes in stratification have not always been seen, there is quite a body of evidence which strongly suggests that the absence of periclinal divisions in the apical dome for about a third or a half of the plastochron could be a characteristic feature of shoot apices. If so the pea would be typical, and the change in plane of growth would be, as suggested (Lyndon, 1970b, 1972a), the primary event in leaf initiation.

The occurrence of periclinal divisions in the I_1 region during the 16 h before the leaf buttress forms in the pea has been regarded not as the imposition of a new direction of growth but rather the lifting of a constraint so that the cells can divide and grow in all directions. The occurrence and maintenance of a 1 : 1 : 1 ratio of divisions in the three planes into which the data were classified are most readily understandable as a random orientation (Lyndon, 1972a). The control of the planes of growth and division is therefore thought to be exerted in the apical dome, to allow only anticlinal divisions during the first part of the plastochron. Leaf initiation is visualized as a process which depends on the temporary lifting of a restraint so that random growth can occur.

C. Significance of the Gradient in Division Rates: Models of Apical Growth

One way in which the gradient in the rate of division down the apical dome may be of importance is in its providing a difference in growth rates between the upper and lower surfaces of the primordium, as discussed on p. 308. But this does not seem a sufficient explanation if the growth of the abaxial surface of the primordium results from the establishment of a locus of rapid division only in the I_1 region of the apex.

It would be difficult, if not impossible, to alter experimentally the rates of growth and division in different parts of the apex, or to reverse the gradient, to see what effects this might have on the growth of the apex and the initiation of leaves. However, it might prove possible to gain some insight into the functioning of the apex if it were possible to build a convincing model in which the effects of altering division rates in different parts of the apex could be tested. With this in view, a computer simulation of the growth of the pea apex has been attempted, using as its basis the data obtained from the growth of living apices (Lyndon, 1975). The rate of division was programmed to increase as a function of distance from the summit of the apex to the region (on the flanks of the apex) about ten cells from the summit and then to decrease to a constant value with further distance from the summit of the apex, thus simulating the situation in a growing apex (Lyndon, 1970a, 1973). The plane of division was programmed to be entirely anticlinal throughout the whole of the apical dome for 30 h and then random (and so including periclinal divisions) in all cells, other than the epidermis, as apparently happens in the growing apex during the plastochron immediately before and during the emergence of a new primordium as a bump (Lyndon, 1970b, 1972a). The result of this simulation was the formation of a bulge on the flanks of the apex which after two plastochrons of growth was comparable in size and position to the primordium which develops in the living apex.

The results obtained by varying the input parameters in this model suggest that, if the rate of division and growth did not increase with distance from the summit of the apex, primordia would not be formed on the apex at all. The apex would simply grow on like a filament showing tip growth. The increasing rate of growth and division of cells displaced down the apical dome ensures that, when the plane of growth is not restricted, the apex below the summit can grow out into lateral protuberances, the leaf primordia.

D. Rates of Division and Growth in Relation to the Length of the Plastochron

So far we have considered the factors involved in the initiation of a leaf and the processes leading to the growth of the bulge which is the young primordium. Other factors which require consideration are those which determine the site of leaf initiation and the length of the plastochron, which is a measure of the frequency of leaf initiation. Richards (1951) showed that the phyllotaxis is the resultant of three things: the rate of growth at the apex (i.e. the radial relative growth rate); the shape of the apical dome (which modifies the rate of radial displacement of cells and the rate of their tangential separation); and the size and shape of a primordium on initiation in relation to the size of the apical dome. The plastochron, the interval between the initiation of one leaf and the next, is usually considered to be the resultant of these factors.

The shape of the apex is probably determined by the directions of growth, as reflected in the planes of cell division. To what extent the rates or planes of cell division are involved in determining the size of a primordium on initiation is not known. The rate of cell division is a function of the relative growth rate, which is the third determinant of the rate of leaf initiation. One might expect that the faster the growth rate the faster would leaves be initiated. This is what was found for the maize, in which the relative growth rate of the apex increased during vegetative development (Abbé et al., 1951). The rate of leaf initiation also increased so that the plastochron decreased from 4·7 days to 0·5 day. The growth rate increased sufficiently also to allow an increase in the size of the apical dome as well as a corresponding increase in the size of each primordium on initiation (Abbé and Phinney, 1951; Abbé et al., 1941).

The rate of growth of the apex and the rate of leaf initiation are not always positively correlated. In both the lupin and the rye while the apex was still vegetative the relative growth rate of the apex decreased, but the plastochron remained unchanged. Without other compensatory changes the apical dome would have become smaller and smaller, a trend which could not have been kept up for very long. In fact the apical dome actually increased in size. This was possible because "the size of each primordium, and the number of cells in it, progressively decrease as the dome develops. The size of the dome therefore increases because at the differentiation of each primordium a progressively smaller proportion of the cells in the dome is given over to the new primordium and a correspondingly larger proportion remains to constitute the new dome" (Sunderland, 1961). It is curious that in each case these factors compensated for each other so precisely that the plastochron remained constant.

The spruce (*Picea*) is similar to the lupin and the rye in that the relative growth rate of the apex decreased during development and the apical dome increased in size, but in *Picea* the rate of leaf initiation did not remain constant but became much faster, so that the plastochron shortened (Gregory and Romberger, 1972). Presumably the size of the primordia on initiation became smaller; unfortunately no data about this were given.

Another case in which the rate of growth and the rate of leaf initiation were not correlated was in *Chrysanthemum* apices, some of which were treated with tri-iodobenzoic acid (Schwabe, 1971). In the treated plants the relative growth rate of the apex remained unchanged, but the plastochron became longer. This was because the shape of the apex changed so that primordia were initiated further down its flanks.

These examples show that the relative growth rate, and the rate of cell division, is only one of the components of the apical system that determine the rate of leaf initiation. The site of initiation of a new leaf is also a function of these components, none of which can properly be considered in the absence of the others. So far, experimental analyses of the growth of the apex taking all these factors into consideration simultaneously do not seem to have been attempted, although the theoretical basis for their understanding has been carefully worked out (Richards, 1948, 1951, 1956).

IV. CELL DIVISION AND FLOWERING

The formation of the flower or inflorescence is generally thought to be accompanied by an increase in the rate of cell division in the apex.

The evidence for this is of several types. First, the increase in mitotic index which is often seen at the onset of flower formation has been cited as indicating an increase in the rate of cell division and a similar increase in the labelling index has been put forward as supporting evidence (Nougarède, 1967). Neither of these observations necessarily shows that the rate of division has increased. They only show that S and M occupy a larger part of the cell cycle in reproductive than in vegetative apices. Although this is consistent with a shortened cycle it could equally well be consistent with a lengthened or unchanged cycle time. This evidence is therefore not conclusive.

The increase in mitotic index which occurs in a number of plants just after floral induction (Bernier, 1971) has been studied in detail in *Sinapis* (Bernier *et al.*, 1967b). The increase in mitotic index which occurs 30 h after the beginning of the inductive long day is associated with a release of cells from G_2 (Kinet *et al.*, 1967) and is followed by an accumulation of cells in G_1 (Jacqmard and Miksche, 1971), indicating that the cells have been synchronized by the floral stimulus. A further peak of mitotic index occurs 32 h later, when floral morphogenesis is beginning, and indicates that a further degree of synchrony has been imposed on the cells (Bernier *et al.*, 1967b). These changes are accompanied by increases in the rate of cell division as shown by the rate of accumulation of colchicine metaphases (Bodson, 1975). The first increase of mitotic index does not always need to be followed by induction. If *Sinapis* plants are given a 12 h photoperiod, which is not sufficient to induce them, the first rise in mitotic index still occurs, showing that it can be separated from the flowering process (Bernier *et al.*, 1970). Although data are not available, it seems probable that the rise in mitotic index in this case is indicative of a temporary increase in division rate.

The rates of division during and subsequent to induction were measured in *Datura* by the rate of accumulation of colchicine metaphases (Corson, 1969). The cell cycle decreased from 46 h to 26 h on induction, but, as in *Sinapis*, the differential between the lateral and summit regions of the apex was maintained so that the summit cells were still the slower dividing cells as they are in the vegetative apex. In *Coleus*, however, the rate of division of the summit cells increased so that in the flowering apex it was the same as in the cells on the flanks (Saint-Côme, 1971). The changes in the MCGT during and after transition to flowering in lupin and vernalized rye were measured by rates of increase of cell number (Sunderland, 1961). The MCGT of the shoot meristems and the developing leaf primordia fell during vegetative growth, increased again on transition to flowering and, at least in the lupin, fell again when this transition had been accomplished.

This evidence is consistent with the idea that the rate of division increases during induction and may increase even more when the flower is formed. All these plants considered so far produce inflorescences, and an integral part of the flowering process is the growth of the axis to support the many flowers. It could be argued that in these plants there is a growth of the shoot apex during flowering which is not concerned with the formation and morphogenesis of the flowers themselves but which is concerned with the formation of the structure which subsequently bears the flowers. An attempt was made to separate these two processes—growth of the axis (essentially a continuation of the vegetative mode of growth) and formation of the flower itself—by the use of a plant in which the first-formed flower was terminal so that the apex itself became transformed into a flower as a first and direct result of induction. This plant was *Silene coeli-rosa*, a long-day plant. The MCGT of the cells in the apical dome (the apex distal to the youngest primordia) was measured by counting the number of cells in it at daily intervals over the course of several plastochrons, a plastochron being about 3·7 days (Miller, 1975). The MCGT was calculated to be 20 h in vegetative (short-day) plants and in plants during the 7 days required for induction, and 9 h in plants in which flower morphogenesis was just beginning (2–4 days after the end of the inductive period). This result, that the rate of division remained unchanged during induction and increased only during the process of flower morphogenesis, was confirmed by measuring the length of the cell cycle by a double labelling technique, which gave values of 20 h for plants which were vegetative or were undergoing induction and 10 h for plants in which flower formation had begun (Miller and Lyndon, 1975). The conclusion to be drawn from these experiments is that although an increased rate of cell division appears to be a necessary part of flower formation and morphogenesis it does not necessarily accompany the inductive process.

The rates of cell division and the changes in the planes of growth which accompany flower development and the formation of the floral parts have not yet been measured or described quantitatively. Only when this has been done, and the rates of formation of the primordia have also been measured, will there be a basis for interpreting the growth and morphogenesis of the flower in a way which has now become possible for the vegetative shoot apex.

9. Cell Division in Leaves

J. E. DALE

Department of Botany, University of Edinburgh, Scotland

I. INTRODUCTION

The leaf is an organ of fundamental importance and great interest. Its importance lies in the fact that as the major carbon-assimilating unit of the green plant the whole basis of crop production depends on its activities. But apart from the interest intrinsic in it as an organ for the production of dry matter, the leaf presents basic problems relevant to our attempts to understand the generation and control of form and size in biological systems. These aspects are central to this chapter.

Three questions relating to leaf morphogenesis have attracted a great deal of investigation and discussion. These are: firstly, how it is that leaves are initiated in regular and highly determinate sequence at the stem apex; secondly, how it is that leaves are usually, but not always, dorsi-ventrally flattened, and of a shape which while varying enormously between species is nevertheless often sufficiently constant to be used as a taxonomic character, when they arise as lateral structures on a radially symmetrical stem; and thirdly, why leaves, in contrast to other lateral organs, are of determinate growth, showing only a limited period of meristematic activity, with no cambial development.

The first of these aspects, primordial development, has been discussed by Lyndon in Chapter 8; the remaining two are considered here. The generation of form involves cell division and cell expansion, and the processes are so intimately related that consideration of the one must involve some consideration of the other. This applies especially to the histological analyses which together with more experimentally based approaches to the problem of form are considered in the following section. A complex series of interactions between environmental and endogenous factors appears to affect the extent and duration of cell division in the leaf and hence ultimate size. These are discussed in Sections III and IV in which descriptions of the extent and rate of cell division in the expanding leaf are arbitrarily separated from consideration of studies in which effects of light and growth substances on cell division and cell number increases have been examined experimentally.

II. THE GENERATION OF LEAF FORM

A. Primordial Growth and the Pattern of Meristematic Activity

In its early development the leaf is characterized by a complex, changing pattern of meristematic activity and an appreciation of this is fundamental to any consideration of the generation of form. The generalized description that follows does not set out to duplicate the more extensive accounts of the histology of leaf initiation and development given by Esau (1965) and Cutter (1971) to which the interested reader is referred. Mention may also be made of a number of accounts for individual species including those for *Nicotiana tabacum* (Avery, 1933), *Zea mays* (Sharman, 1942), *Pyrus malus* (MacDaniels and Cowart, 1944), *Drimys winteri* (Gifford, 1951), *Linum usitatissimum* (Girolami, 1954),

Oryza sativa (Kaufman, 1959), *Trifolium wormskioldii* (Lersten, 1965) and *Acorus calamus* (Kaplan, 1970), while Roth (1957) has considered the histological basis for the origin of the bizarre foliar structures in such species as *Sarracenia* and *Anthurium*.

Following initiation, growth in the mound-like primordium, often termed the foliar buttress, becomes localized at a distal point which extends to form a peg- or awl-shaped structure, the leaf axis. At first, growth of the axis, which is principally by cell number increase, is localized at its apex, but very soon divisions in the intercalary region become general, leading to further increase in axis length; in the cereal grasses this process leads ultimately to the establishment of a basal meristem which gives rise distally to the cells of the blade and proximally to those of the leaf sheaf. Increase in axis length is accompanied by further development of the procambial strand which extends into the axis from the foliar buttress. By the time that the axis is a few hundred micrometres in length the pattern of growth changes again as the form of the leaf begins to be determined. This involves the establishment of sub-axes in the cases of species having lobed (e.g. *Cannabis, Tropaeolum, Helleborus*) or palmately compound leaves (e.g. *Lupinus, Trifolium, Phaseolus*), and the proliferation of cells at the lateral margins of the axis to commence lamina formation. In the case of entire, sessile leaves activity of the marginal meristem is general along the axis, but for petiolate and pinnate leaves this is not so and meristematic activity is confined to, or substantially greater in, particular regions along the axis, often associated with branching of the procambial strand. Once lamina proliferation is under way cell division in the marginal region is followed by general mitotic activity over the whole lamina region, which thus functions as a plate meristem. This stage is reached before, or by the time that, the leaf is 1 mm in length and it must be strongly emphasized that the morphogenetic processes which lead to the establishment of leaf form are begun very early in the life of the primordium.

Observation of sectioned material is the basis for most descriptions of early primordial growth; it is of limited usefulness since it tells us nothing of the rates of division and because in the absence of visible mitotic figures little can be reliably inferred of the location and planes of cell division without a great deal of laborious analysis which is seldom attempted. Consequently the cellular mechanisms involved in the two critical phases of the determination of leaf form, establishment of the leaf axis and blade initiation, are not understood. It is possible that the differential growth, either by local enhancement of rates of cell division or by changes in the plane of division (or by a combination of both), which leads to primordial initiation, continues to result in the establishment of the apical growth centre that marks the first stage of axis development. The superseding of apical growth by more general divisions along the axis does not of itself imply the development of new meristematic activity there; clearly, if a constant proportion of cells go into successive divisions and if division times remain constant then the size of the meristem as well as cell number in it will increase exponentially, as is often assumed. Elongation of the axis is accompanied by extension of the procambial strand and it is probable that the supply of metabolites essential for processes of cell growth and division are controlled by the pattern and extent of strand development. That is not to imply a causal connection between strand development and meristematic activity, merely an inevitable interrelationship becoming progressively more important as the primordium increases in size and hence in its nutritional requirements.

Very little is known about the factors which determine at what point along the leaf axis formation of the lamina occurs. The buttress at the base of the primordium is subject to lateral stretching as it is displaced further down the flank of the extending stem apex, and it is possible that this stress modifies the rates and planes of division in cells of the axis, as was found by Yeoman and Brown (1971) for parenchyma cells in artichoke tuber explants subjected to mechanical tension. However, while such tangential stresses could promote a tendency to growth in the lateral plane, the development of petiolate and pinnate leaves, in which the marginal meristems do not extend all the way to the base of the leaf axis, suggests that factors other than stress also control the points at which lateral, dorsiventrally orientated growth occurs. It may be mentioned here that experiments in which stresses at the stem apex have been altered by means of surgical incisions have not yielded consistent effects on leaf form. In the fern *Dryopteris dilatata* (Wardlaw, 1955b) and in *Solanum tuberosum* (Sussex, 1955) various patterns of incision at the stem apex were found to induce production of buds or radial or centric leaves at the positions normally occupied by dorsiventral leaves. This is not a general finding though, and for *Nuphar* and *Nymphaea* Cutter (1958) found isolation of primordia or incipient primordia by incisions at the apex to have no effect on leaf form.

If dorsi-ventrality cannot be entirely explained on the basis of tissue tensions then another mechanism must be invoked. The involvement of chemical factors is suggested by the work of Rijven (1968), who found that when embryos of *Trigonella foenum-graecum* were cultured on nutrient agar containing 2 mM arginine the leaves formed were exclusively trifoliate whereas in normally developing plants the first leaf was unifoliate. However, this apart, there is little direct evidence to implicate chemical factors in lamina initiation, although the work of Kuehnert (1969, 1972) and his collaborators with cultured primordia of *Osmunda cinnamomea* strongly indicates control of form by the shoot apical meristem and by older primordia; such control must be of a chemical nature. It could well be that growth-active nutrient factors supplied through the procambial strand may reach the presumptive marginal meristem, by diffusion, in concentrations such as to allow locally enhanced rates of division which, coupled with a predisposition for the planes of division to be anticlinal, lead to the origin of a wing-like projection from the leaf axis. The role of nutritional factors in the generation of form in the young primordium requires more detailed systematic examination, and the use of stem apex cultural methods as employed by Steeves (1957) for dicotyledonous leaves and by a number of workers for fern primordia (e.g. Kuehnert, 1972) could be a means for attacking the problem of lamina initiation experimentally. As yet the potentials of this approach remain largely unexplored for angiosperm material.

It is a feature of the expanding lamina that in cross section it is seen to be composed of regularly arranged layers of cells in the inter-veinal regions. In the older anatomical accounts (e.g. Foster, 1936) it was considered that the layered structure arose from a regular sequence of divisions in two rows of initials located at the edge of the lamina. The outer file of cells, the marginal initials, were considered to give rise to the two epidermal layers by regular anticlinal divisions, while a file of cells immediately beneath these, the sub-marginal initials, were considered to divide in a regular sequence, differing according to species, to give abaxial, adaxial and inner layers of cells. Consideration of the irregular patterns formed at the margin of variegated leaves of chimaeras of the white-on-green type suggests that this view is too facile. On the classical view regular patterns of divisions in marginal and sub-marginal initials should result in a regular arrangement of white and chlorophyllous tissue at the margin, but this is not so; as Tilney-Bassett (1963) wrote: "The highly variable modes of development of the two or three green layers, in the formation of the body layers of the leaf, are responsible for the different structural appearances."

Recent work has confirmed that the pattern of divisions in the margins of leaves of a number of species is not regular. A detailed analysis of the angle of the cell plate in mitotic divisions in the marginal meristem and adjacent plate meristem region for leaves of *Xanthium pennsylvanicum* was made by Maksymowych and Erickson (1960; see also Maksymowych, 1973). It was shown that in the presumptive epidermis the majority of divisions (92%) were anticlinal with the few periclinal divisions giving rise to epidermal hairs. Many more periclinal and oblique divisions were found in the sub-epidermal zones (Table I). The analysis showed that in both zones substantial numbers of divisions were anticlinally oriented, the value for the plate meristem being somewhat higher. Periclinal divisions accounted for 30% of those seen in the plate meristem but for the marginal region the figure was only 15%, although this was compensated by a very much greater number of oblique divisions in this zone. Even with the system used by Maksymowych and Erickson the classification of a division figure as oblique, especially in the curved marginal region of the leaf, reflects the impossibility of assigning it with any certainty to the anticlinal or periclinal category.

TABLE I. Planes of division in the marginal and proximal plate meristem regions in leaves of *Xanthium*. Data are those of Maksymowych and Erickson (1960) expressed as percentages.

| | Plane of division | | |
	Anticlinal	Periclinal	Oblique
Marginal meristem	42	15	42
Plate meristem	58	30	12

Yet the regularity of the layered structure of the leaf suggests that oblique divisions contribute either to an increase in surface or to an increase in thickness, i.e. to extension either anticlinally or periclinally. If divisions classified as oblique are partitioned equally between the other two classes then in both marginal and proximal regions 36% of the divisions contribute to growth in the periclinal plane and 64% to extension anticlinally. If this is a correct view of the data then the suggestion that differences exist between the patterns of division in the plate and marginal meristems is not supported; indeed the distinction between the two meristematic regions, which in any case cannot be distinctly defined, is seen to be arbitrary, if not misleading, the more so since there is no evidence from the *Xanthium* work to suggest that divisions are more abundant in one zone rather than the other.

Analysis of the frequency of mitotic figures in the young leaflet of *Lupinus* (550–600 μm long) has also shown divisions to be general across the blade with no evidence for a greater frequency in the marginal region, although in older leaflets (8500–9000 μm long) frequency of mitotic figures was found to decline from mid-vein to the leaf margin (Fuchs, 1966). Examination of the marginal region showed anticlinal divisions to be uniformly distributed in the sub-epidermal zones of both young and older leaflets. However, the distribution of periclinal divisions differed, being found in all parts of the sub-epidermal zone in the older leaf but absent from the upper zone in the younger leaflet (Fuchs, 1968). On the basis of this analysis Fuchs considered that there was no evidence to suggest that cell lineages could be derived from identifiable initials in the marginal meristem. It may be concluded that while periclinal divisions leading to increased numbers of cell layers are found in the marginal region they are not unique to that zone, nor is the pattern of such divisions rigidly and identifiably controlled. The characteristic and predominant plane of division in both marginal and proximal regions of the developing leaf is, however, anticlinal.

B. Cell Division and Expansion and Leaf Form

Although overall control of leaf shape is brought about by genetic mechanisms (Hammond, 1941; Stephens, 1944) it is clear that interaction with environmental factors can result in considerable plasticity in shape within a species (e.g. Njoku, 1956; Böcher and Lewis, 1962; Raper and Thomas, 1972), although this has seldom been analysed in cellular terms. It is also apparent that in spite of initial similarities in shape of the foliar primordium, shape of the mature leaf varies enormously between species. In general, shape and size of the stem apex are no guide to leaf shape (Wardlaw, 1956), although this is not necessarily unconnected with apical morphology; Abbé and Phinney (1951) showed a correlation between apical dome size and blade width for successive leaves of *Zea mays* and other correlations between apical size and the heteroblastic leaf sequence have been postulated and, in the case of the fern *Marsilea*, demonstrated (White, 1968).

In a restricted sense the more obvious features of growth can be measured in terms of cell number and cell size and these parameters have been thought of as criteria determining leaf form. Thus Ashby (1948) considered leaf shape to be determined by primordium shape and the number, distribution and orientation of cell divisions and the extent and direction of cell enlargement in the developing leaf. This view is implicit in recent work of Fuchs (1972a, b, c) on growth and the establishment of form of leaves of *Tropaeolum peregrinum*. Observations on sectioned material suggested that systematic differences in rates of cell division occur in various parts of the lobes of developing leaves and that local differences in the plane of division also occur. These differences in mitotic rates and polarities were found to change with time as the leaf enlarged and as cell expansion, itself varying in rate in different parts of the leaf, began to contribute increasingly to blade expansion. Unfortunately, Fuchs has so far presented few numerical data to support his careful histological analysis and there is urgent need for other more quantitative and experimental studies in this difficult area.

An alternative and perhaps more fundamental approach to the problem of leaf shape comes from the work of Haber (1962), who considered that the classical interpretation of leaf form could be summarized in the following hypothesis:

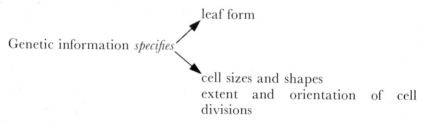

Genetic information *specifies*

cell sizes and shapes

and

extent and orientation of divisions

which uniquely determine form

Haber pointed out that an alternative hypothesis is possible whereby cellular characters and the establishment of form are separated in parallel rather than in series, thus:

Genetic information *specifies*

leaf form

cell sizes and shapes
extent and orientation of cell divisions

Evidence favouring the alternative hypothesis was presented by Haber (1962) and Haber and Foard (1963, 1964a, b) on the basis of work with wheat and tobacco leaves. Haber and Foard made use of the well known allometric relationship between leaf length and width which is expressed as

$$y = a \cdot x^k \tag{1}$$

where y and x are leaf lengths and widths and a and k are constants. It follows that the plot of log leaf length against log leaf width is linear since

$$\ln y = \ln a + k \cdot \ln x \tag{2}$$

Where leaf dimensions are increasing over the time interval t_1 to t_2 the allometric constant, k, is given by

$$k = \frac{\ln y_2 - \ln y_1}{\ln x_2 - \ln x_1} \tag{3}$$

Where the value of $k>1$ leaf shape becomes thinner and where $k<1$ leaf shape becomes broader with time.

It was shown by Haber (1962) that irradiation of the dry grain of

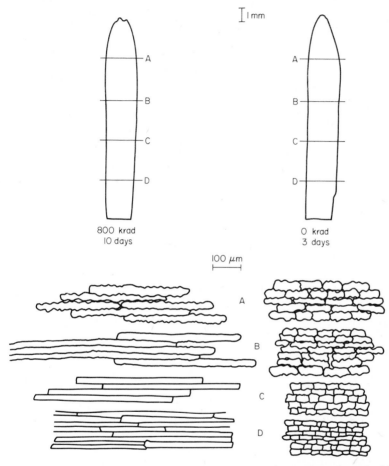

Fig. 1. Differences in mesophyll cell size and shape at various points along the blade of the first leaf of 10-day-old plants of wheat (left) irradiated with 800 krad γ-radiation, and of 3-day-old untreated plants (right). Redrawn from Haber (1962).

wheat with γ-rays did not prevent slow germination and development of the seedling; however, cell division in the basal meristem was prevented in the first leaf of these γ-plantlets. Yet in spite of irradiation, the first leaf, present as one of the three primordia in the dry grain, expanded as already-formed cells enlarged. Moreover, at similar lengths, reached much earlier in non-irradiated plants, values of k were similar; since $k > 1$ leaf shape changed with time, but for leaves of the same length, and hence of similar shape, cell number and size were very different (Fig. 1). Because cell characteristics were so different whereas form was similar it was argued that the alternative hypothesis rather than the first was supported, and that leaf form did not, in this case, depend upon the occurrence of cell divisions. An implication of this is that the mechanism controlling form is not sensitive to γ-irradiation in this material. However, values of k were similarly affected for both control and γ-plantlets when these were treated with either 0·3% colchicine, which decreased the value of k, or with 2×10^{-4} M gibberellic acid, which increased it. These observations indicate that leaf form may be altered without affecting or involving cell division.

It could be argued that the wheat leaf represents a rather simple system in that it develops from a basal meristem from which cells expand in files, principally in one direction, and that a closer relationship between leaf form and cell characteristics might exist in those leaves where a more complex meristematic organization exists. Haber and Foard (1963) considered this point by analysing cell numbers in epidermis and palisade of successive leaves of tobacco plants, together with measurements of leaf dimensions. The leaves showed a similar and constant value of k of less than 1, indicating that the lamina became broader during development. This change in form occurred over a period in which (a) cell division was initially widespread but finally ceased, (b) cell size increased although cell shape, seen in paradermal section, remained unchanged, and (c) separation of cells and formation of irregular air spaces also occurred. Again it was argued that cell division and the establishment of form were not closely and causally related. It was also concluded that polarization of the leaf, measured by k, was not closely and causally related to cell size and shape.

In their work Haber and Foard accept that form of an organ can be *described* in terms of cell number and size, but they argue that form is not *determined* by these parameters and that cell division plays a secondary role in influencing organ form. However, the wheat and tobacco leaf material used in their studies was already out of the early primordial phase, and extension of their interpretation back to cover the situation in the young primordium can only be made with great caution since lamina initiation is clearly dependent upon cell division and proliferation in particular parts of the axis margin. It seems certain that the major features of leaf form are determined at an early stage and consequently the changes in form of the developing leaf may be regarded as inevitable and reflecting an ontogenetic programme initiated in the young primordium. The changes in leaf form in wheat treated with colchicine or gibberellic acid, or those recently reported for leaves of tobacco plantlets in tissue culture treated with varying levels of a cytokinin and gibberellic acid '(Engelke *et al.*, 1973), may be regarded as indicating comparatively minor perturbations of an established mechanism.

III. THE EXTENT AND RATE OF CELL DIVISION

Just as leaf form can be described in terms of cell number so too can leaf size, recognizing that cell size is variable and that air spaces add a substantial if variable additional contribution to volume of the mature leaf. As will be discussed below, cell number increases throughout the period of leaf unfolding in many species. In these circumstances analysis of cell number changes with respect to both time and position in the leaf is clearly important, the more so since elucidation of the reasons for cessation of meristematic activity is of general interest to the cell physiologist.

A. Changes in Cell Number

Notwithstanding the considerable meristematic complexity in the primordium there is evidence from a number of species that cell number increase in the young foliar structure is exponential, or approximately so, over at least part of the period of early development (Maksymowych, 1959; Sunderland, 1960; Milthorpe and Newton, 1963; Hannam, 1968; Lyndon, 1968). Where cell number increase in the primordium is exponential we have that

$$N = N_0 e^{Rt} \tag{4}$$
$$\ln N = \ln N_0 + Rt \tag{5}$$

and

where N is cell number at time t, N_0 is the initial cell number at time t_0 and R is the relative, or specific, rate of cell number increase. We also have that

$$R = \frac{1}{N} \cdot \frac{dN}{dt} \tag{6}$$

and over the time interval t_1 to t_2 the mean value of relative growth rate is given by

$$\bar{R} = \frac{\ln N_2 - \ln N_1}{t_2 - t_1} \tag{7}$$

this result being applicable irrespective of the relationship between N and t. Of course, where growth is exponential values of $\bar{R}$ are constant and for the case where $N_2 = 2N_1$ equation (5) gives

$$t = \frac{\ln 2}{R} \tag{8}$$

where $t = T$, the time taken for cell number to double, i.e. the mean cell generation time; this will be constant for constant values of R. Hence where cell number increase is exponential the mean cell generation time should also be constant. But an exponential relationship can be brought about either if a constant proportion of cells, not necessarily all, continue into division, or if a variable proportion go into successive divisions. For the first contingency the ratio of cell generation time for those cells dividing to mean cell generation time will be constant, but this will not be the case for the second situation. Further analysis of the exponential increase thus requires information on the proportion of cells dividing and the actual division rates rather than mean values.

Exponential increase in cell number continues into the phase of leaf unfolding and lamina expansion, but thereafter rate of increase gradually declines to zero. Milthorpe and Newton (1963) analysed cell number changes with time for unfolding leaves of *Cucumis* and suggested that their data were consistent with the hypothesis, which they were unable to test, that the cell generation time was constant and that the proportion of cells going into successive divisions was a negative logistic function of time such that

$$p = 1/(1 + e^{Bt})$$

where p is the proportion of cells going into division after time t, and B is a constant. However, since $e^{Bt} \simeq 1 + Bt$, values of p must always be less than 0·5, i.e. less than 50% of the cells present initially go on to divide. An alternative model, that

$$p = 1/e^{Bt} \tag{10}$$

does not suffer from this limitation since p can assume a maximum value close to unity.

Both of these models were examined by Dale (1970), using data for cell number changes in primary leaves of *Phaseolus vulgaris* (Fig. 2). These leaves are well developed in the dry seed and contain about 10^6 cells. This value rises to about 50×10^6 during development which both precedes and follows emergence of the seedling above ground (Dale,

1964); the rate of increase in cell number is only approximately exponential over the early phase of development (Dale, 1964; Wilson and Ludlow, 1968). Examination of sectioned material showed that early in germination cell division figures could be found in all parts of the lamina and midrib tissues and it was considered that initially values of p may be close to unity. Using a value of B of 0·148 the model of equation (9) was fitted to the observed data. However, analysis showed that over the final 24 h during which divisions occurred 11 cycles of division would be required. This would imply a mitotic cycle

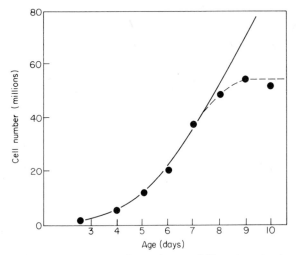

FIG. 2. Changes in cell number for primary leaves of *Phaseolus vulgaris* cv. 'Canadian Wonder' (points and broken line) and the curve for cell number generated by equation (10). Data of Dale (1964, 1970) replotted.

time of only about 125 min and was thought to be unlikely, as well as being incompatible with the very low frequencies of division figures found in sectioned material.

A more satisfactory analysis resulted from use of the second model (equation (10)) with similar value of B. Here calculations of cell number were found to be in close agreement with observed values until the very end of the period of division. The cycle time for dividing cells was calculated at 12 h, which in the light of other values (e.g. Loewenberg, 1955) is not unreasonable (see below). An interesting feature of the results was that the model predicts that about 20% of the cells in the leaf should retain the capacity to divide after the period when divisions no longer occur; the possible significance of this is discussed in the next section. Over the period where the model fits the data the proportion of cells going into successive divisions was found to decline showing a negative exponential relationship with time, as an increasing proportion of cells underwent transition from the meristematic to the non-meristematic condition. The constant B may be thought of as an index of the morphogenetic constraints bringing about this transition.

B. Distribution of Divisions throughout the Leaf Blade

Mention has already been made of work on *Lupinus* and *Xanthium* which showed cell divisions to occur generally over the developing leaf blade. However, as the leaf matures the pattern of cell divisions varies along the blade and also between the tissues of the lamina.

In the grass leaf the cells originate from a basal meristem so that heterogeneity in the distribution of mitotic activity along the leaf is an inevitable feature of development. However, divisions do continue in the epidermis to give rise to stomata (Haber and Foard, 1964b); these divisions also cease first in the distal, more mature region of the leaf at the tip. For dicotyledonous species there is also evidence that cell divisions cease first in the tip of the leaf and continue longest in the basal regions of the blade (Avery, 1933; Maksymowych, 1963; Saurer and Possingham, 1970). This basipetal trend is correlated with the earlier maturation and differentiation of cells, including vascular elements, in the leaf tip (Avery, 1933; MacDaniels and Cowart, 1944; Heslop-Harrison, 1962; Denne, 1966; Steer, 1971).

There are also differences in the period over which cell division occurs in the various layers of the lamina. Direct estimates by cell counts and analyses of cell division figures in sectioned material have shown that divisions cease first in the epidermis and continue longest in the palisade mesophyll (Avery, 1933; Heslop-Harrison, 1962; Haber and Foard, 1963; Dale, 1964; Denne, 1966; Maksymowych and Blum, 1966; Steer, 1971). An exception to this general statement must be made for stomatal initials for there is evidence that divisions in the stomatal complex continue after other divisions in the epidermis have ceased (Stebbins and Shah, 1960; Shanks, 1965; Denne, 1966). Because general divisions cease in the epidermis as a whole before they do in the palisade the use of epidermal cell number as an index of leaf cell number must be regarded as unsatisfactory.

The prolonged division in the palisade mesophyll is of especial interest since there have been a number of reports of successful culture of isolated palisade cells (Kohlenbach, 1965; Joshi and Ball, 1968) and more recently mesophyll cells of a number of species have been used for protoplast culture (Durand *et al.*, 1973; Harada, 1973; Koehler,

1973). Clearly these cells can be brought back to the meristematic condition relatively easily and one is reminded of the model analysis of the previous section which suggested that some cells in the leaf retained the capacity for further division even though divisions had actually ceased. If the model is correct it is possible that these are in fact palisade cells which account for 20–25% of the cells in mature primary leaves of *Phaseolus*.

The older view that cell division occurs only in the primordial phase of growth while expansion of the lamina results solely from cell expansion is now known to be incorrect; both cell division and expansion occur concurrently during a substantial part of the period of expansion of the leaf, although it is correct that the final stages of area increase are not associated with general cell division (Sunderland, 1960; Milthorpe and Newton, 1963; Dale, 1964; Saurer and Possingham, 1970; Steer, 1971). Furthermore, even though the rate of cell number increase declines following unfolding, the majority of cells in the mature leaf are derived from divisions occurring during this period. Thus, more than 90% of the cells in fully expanded leaves of *Lupinus* and *Helianthus* were found to originate from divisions as the lamina unfolds (Sunderland, 1960), while for *Cucumis* comparable values were 70–98% according to the radiation levels under which the plants were grown (Milthorpe and Newton, 1963). Because divisions continue longest in the palisade the proportion of these cells originating after unfolding is especially large; for tobacco it has been estimated that 99% of palisade cells originate during lamina unfolding (Hannam, 1968).

Thus the spatial and temporal variation in meristematic activity that is characteristic of the primordium is found also in the developing blade where the course and extent of division vary from tissue to tissue and from point to point along the length. The meagre evidence available indicates that variation in rate of cell division also occurs in the tissues of the developing leaf; this is now considered.

C. Rates of Cell Division

If values for R can be derived for changes in cell number, equation (8) may be used to calculate values for mean cell generation time. However, T only has general significance if all the cells in the leaf continue to divide; if and when this is not the case then the actual generation time for those cells which are dividing will be overestimated.

Analysis of data for cell number of seven successive primordia in *Lupinus* (Sunderland *et al.*, 1956; Sunderland and Brown, 1956) gives a value for $\bar{R}$ of 0·24 day^{-1} and from this a value of 69 h can be obtained for mean cell generation time. This indicates a comparatively slow rate of doubling of cell number, but a similar value was found by Williams (1960) for young foliar primordia of wheat. In *Lupinus*, the mean cell generation time for primordial cells is shorter than that for the adjacent cells in the apical dome and internodes; similar differences between tissues were found by Lyndon (1968) for the apex of *Pisum*.

Substantially lower values of mean cell generation time have been found for *Cucumis* and *Nicotiana* plants grown under higher temperatures and light intensities than the lupin plants. Wilson (1966) found a value of T of 17 h for expanding leaves of *Cucumis* grown in light in-

TABLE II. Changes with time in total cell number and number of vascular and lamina cells (millions) for the eighth leaf of tobacco (Hannam, 1968). Calculated values for mean relative growth rate, R (day^{-1}) and cell doubling time, T (h), are also shown.

Time (days)	Total cells			Vascular cells			Lamina cells		
	Cell no.	R	T	Cell no.	R	T	Cell no.	R	T
0	0·43			0·43			—		
		0·87	19·0		0·69	24·0			
2	2·46			1·72			0·74	—	—
		1·11	14·9		1·07	15·8		1·20	13·8
4	22·78			14·60			8·18		
		0·95	17·4		0·87	19·0		1·06	15·6
6	151·8			83·12			68·72		
		1·14	14·5		0·96	17·3		1·30	12·7
8	1496			570·3			925·9		
Mean values			16·5			19·0			14·0

tensities of about 97 W m^{-2} compared with a value of 25 h for leaves in a lower light intensity (12 W m^{-2}). These values straddle estimates of 22 h for generation time of dividing cells, based on autoradiographic studies with labelled thymidine in *Xanthium* leaves (see Makswmowych and Blum, 1966; Maksymowych, 1973), although consideration of cell number data indicates an apparent doubling time of over 50 h for this tissue. Mention has already been made of an estimate for T of 12 h using *Phaseolus* primary leaf data processed by the model of equation (10).

Much more rapid rates of cell number doubling were found for the primordial stage of the eighth leaf of tobacco. Table II shows calculations of R and T based on values of cell number presented by Hannam (1968) and categorized for either vascular cells, which represent the main primordial axis and veins, or cells of the lamina. The considerable heterogeneities within the data for a particular category probably represent sampling variations and if these are disregarded it appears that T is about 5 h shorter for cells in the lamina than for those in vascular tissues. Again the calculations are based on the assumption of a general and continuing ability of all cells in the primordium to divide. This is almost certainly unrealistic for vascular tissues since it is known for tobacco (Esau, 1938) and for other species (Lersten, 1965) that phloem differentiation commences very early in primordial growth so that the transition of a proportion of vascular cells to the non-meristematic condition seems inevitable. The calculated differences in T could

TABLE III. Variations in rates of cell division in various tissues of 10–15-day-old leaflets of *Trifolium repens* treated with colchicine; data of Denne (1966).

	Metaphases accumulated in 8 h (%)	Duration of mitosis (h)	Mitotic cycle (h)
Marginal third of lamina	11·86±0·65	2·4	48·1
Proximal remainder of lamina	12·88±0·65	2·3	42·1
Upper epidermis	10·69±0·97	2·9	53·0
Lower epidermis	11·38±0·68	1·6	50·6
Mesophyll	14·83±0·84	2·4	38·9

be explained if 26% of the progeny of successive divisions in the vascular tissues differentiated into the non-dividing state while the remainder continued to divide with a generation time similar to that of the lamina cells.

Although the tobacco data do not demonstrate unequivocally differences in division times in different parts of the primordium, other results are more clear-cut. Denne (1966) studied development of leaflets of *Trifolium repens* which were from 2·5 to 3·8 plastochrons old and increasing almost exponentially in cell number. She used colchicine to bring about accumulation of nuclei in metaphase and went on to calculate duration of the mitotic cycle in different parts of the leaflet; again it was assumed that all cells divided or had the potential to divide. Her data (Table III) show mitotic cycle time for the marginal tissues of the lamina to be about 6 h longer than that for the inner tissues. This could well represent genuine differences in the cell cycle time since it is unlikely that a smaller proportion of the cells in the margin were meristematic than in the inner region. The difference is not explicable in terms of time spent in mitosis for the data show this to vary by only 0·1 h. The longer interphase and prophase shown by the cells at the margin could be due to a slower, limiting supply of metabolites for nucleic acid synthesis, these cells being further from the vascular system supplying the lamina as a whole (cf. Maksymowych *et al.*, 1966). The differences in mitotic cycle time between the epidermis and the mesophyll may also reflect proximity to vascular tissue and the source of nutrient supply. Cell expansion data presented by Denne give no reason to suppose that at this stage the epidermal cells are passing out of the meristematic state and it must be concluded that the differences in the mitotic cycle time are probably real.

IV. FACTORS CONTROLLING CELL DIVISION IN THE LEAF

Since differences in extent and rate of cell division are to be found throughout the leaf, examination of factors affecting division should take account of these. In practice, because of the technical difficulties inherent in determining division rates and cell numbers in specific parts of the leaf most experimental studies have been restricted to analysis of total cell number of leaves subjected to various experimental treatments either from primordial initiation onwards or throughout the period of lamina expansion. The most extensive of these investigations are those in which effects of light and of growth substances have been examined.

A. Light and Photomorphogenic Effects

The effects of lack of light on leaf growth are well known and the response of a number of species to etiolating conditions was considered by Macdougall (1903) in a fascinating monograph. More recent work (e.g. Butler, 1963; Murray, 1968) has shown light to affect leaf cell number through at least two mechanisms, one of which involves low light intensities, the other involving much higher light energies.

1. Low-energy-requiring Mechanisms

The low-energy mechanism has been investigated most extensively using legume species. These show a well marked etiolation response in which leaf development in darkness is slight and, because of their large seed size, they can be grown for long periods in darkness. The following description is taken from work on the primary leaf of *Phaseolus vulgaris* by Murray (1968) and Dale and Murray (1968, 1969).

In complete darkness, cell number of the primary leaves reaches a maximum value of $12–18 \times 10^6$ in contrast to values of $45–50 \times 10^6$ for plants grown in 12 h days. For both dark- and light-grown leaves maximal rate of cell number increase occurs between days 4 and 7 from planting and divisions cease after day 8; at this stage mean cell volume in light-grown leaves is 4–5 times greater than that of leaves kept in darkness which show no signs of unfolding. Thus in darkness both cell number and cell volume increase are restricted.

However, treatment of dark-grown plants with low-intensity white light for periods from 1 min to 2 h results in an increase of cell number of 55% if the treatment is given on day 6 from planting, or of about 15% if it is delayed until day 10. The latter response is particularly interesting since divisions cease on day 8 and the response indicates divisions to be initiated in cells which would otherwise not have divided. It may be emphasized that even after treatment cell number in the leaves remains well below the level found for plants grown throughout in high light conditions. Effects of short irradiations on cell volume are slight so that treated leaves retain their etiolated and unexpanded appearance. The response to light is independent of intensity over a wide range and has been shown to be mediated by the phytochrome system, known to be present in the primary leaves of *Phaseolus* (Downs, 1955; Briggs and Siegelman, 1965), since the effects can be brought about by red light but not by blue, and are reversible by far-red light.

Available evidence favours the view that light does not affect the mechanism of cell division directly. For instance, it has been found that irradiating leaves *or* cotyledons brings about the response of cell number increase, implying that perception of the stimulus at the cotyledons can lead to action at the leaves; this is incompatible with a direct effect of light on the cell division mechanism. The importance of the cotyledons is also shown by the fact that their removal within 12 h of irradiation prevents cell number increase in the leaves. The simplest interpretation of these findings is that metabolites from the cotyledons are necessary for divisions to go to completion in the leaves and that light may affect the supply of these factors either by promoting breakdown of stored material in the cotyledons, as has been shown for *Sinapis* (Mohr, 1966, 1972), or by accelerating transport of material out of the cotyledons to the leaves. Within the leaf the phytochrome mechanism may operate by affecting accumulation of metabolites, perhaps by changing cell permeability characteristics or by affecting their utilization in the syntheses required in preparation for division. It appears that light does not operate through a mechanism involving DNA synthesis since Murray (1968) showed DNA content per cell to go down when dark-grown leaves were induced to divide by light. This implies that the cells were already in the 4C condition and that failure to divide in darkness is not caused by failure to synthesize DNA.

It has also been found that short irradiations with white or red light do not cause cell number increase in disks taken from dark-grown leaves even though the disks are cultured on a medium known to support both cell division and expansion for similar disks cut from light-grown leaves (Dale, 1966, 1967). Again this implies that the effect of light on cell division is indirect.

Since in normal development plants will be exposed to prolonged and higher levels of light, the significance of the low-energy response may be questioned. Under most conditions of growth, seedlings are subject to low intensities of light some time before the plumule emerges above the soil, since the surface of the culture medium tends to crack as the young plant grows. There is thus the possibility that phytochrome-mediated responses may affect cell number in the early-formed leaves. It is also possible that phytochrome is involved in controlling cell division in later-formed leaves prior to unfolding. Within the apical bud light quality will tend to be rich in far-red with a low ratio of red : far-red resulting from absorption of red light by chlorophyll in the investing leaves at the outside of the bud. The light regime at the young leaf will thus favour the presence of low levels of P_{fr} over long periods and in these circumstances the phytochrome-mediated responses brought about by prolonged exposure to far-red light (Mohr, 1972) may be expected to occur. These responses include those associated with the development of the photosynthetic apparatus (e.g. Margulies, 1962, 1965; Filner and Klein, 1968) and may well also affect the supply of metabolites essential for cell division or their utilization in the process.

2. High-energy Processes

The effects of prolonged high light intensity which are a pre-requisite for the attainment of maximum cell number have been examined in most detail for leaves of *Phaseolus* and *Cucumis*. A single exposure to high intensity light for as long as 20 h was shown by Murray (1968) to be insufficient to give maximum cell number in *Phaseolus* primary leaves, at least two cycles of high intensity light being necessary for this purpose. This is in contrast to the low-energy response which was shown to be induced in full by a single brief exposure.

Another difference between low- and high-energy treatments is that cell volume is substantially increased by the latter. However, the effects on cell number and cell volume are not closely correlated (Murray, 1968; Dale and Murray, 1968), suggesting that they are affected by light either through different mechanisms, or alternatively that the processes are affected differentially through a common mechanism.

Estimations of cell number in leaves of *Phaseolus* have shown this parameter to be remarkably constant for the primary leaves of plants grown in two light intensities and a variety of photoperiods (Fig. 3). This suggests that it is unlikely to be the products of carbon assimilation that restrict cell number since if this were the case then the number would be expected to be higher in high light intensities and in long days. What then is the role of light in the high-energy response? Results from studies using leaf disks suggest that here the importance of light may be in the products of the light reaction rather than through carbon

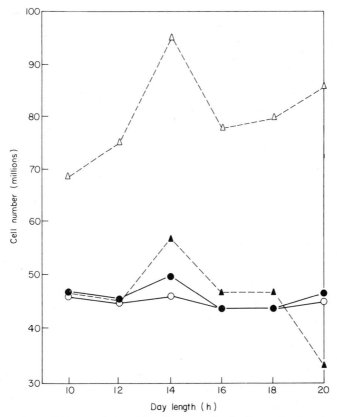

Fig. 3. The effects of photoperiod and irradiance on final cell number in primary and trifoliate leaves of plants of *Phaseolus vulgaris* cv. 'Canadian Wonder' grown at 22·5°C. Data of Dale (1965) replotted.

   Circles—primary leaf; triangles—trifoliate leaf. Open symbols—58·0 W m^{-2}; closed symbols—11·6 W m^{-2}.

assimilates. When disks were cut from dark-grown leaves and cultured on a defined medium in continuous light, cell number increased. This increase was prevented if the inhibitor of photophosphorylation 3(3,4-dichlorophenyl)-1,1-dimethyl urea, DCMU, was included in the incubation medium. It was also found that increase in cell number was not prevented by CO_2-free conditions provided that sucrose was included in the medium, although if this was absent divisions quickly ceased (Dale and Murray, 1968). These results are compatible with the hypothesis that products of photophosphorylation are involved in processes which culminate in or are involved in cell division. It may be significant that activity in the Hill reaction develops quickly in leaves transferred from dark to light conditions and is found in young light-grown leaves before they are half expanded (Dodge and Whittingham, 1966; Gyldenholm and Whatley, 1968; Dickman, 1971). Also, Dale and Heyes (1970) found cell number to be reduced in a *virescens* mutant of *Phaseolus* showing poor chloroplast development and low photophosphorylating capacity.

In his pioneer work on *Cucumis*, Gregory (1928, 1956) considered that

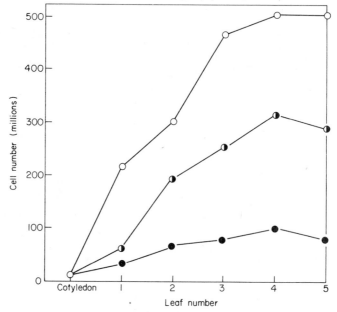

FIG. 4. The effect of irradiance on final cell number of cotyledons and successive leaves of *Cucumis sativus* cv. 'Butcher's Disease Resister'. Data of Milthorpe and Newton (1963).
Open   circles—81·2 W m⁻²;   pied   circles—34·8 W m⁻²;   closed   circles—11·6 W m⁻².

the significance of light lay in the synthesis of factors which affected leaf growth but were not the products of carbon assimilation. This view is compatible with the present interpretation. However, recent work with *Cucumis* has been interpreted to indicate that the greater cell number of leaves kept in high irradiances (Fig. 4) is the result of enhanced photosynthesis and carbohydrate production. Milthorpe and Newton (1963) and Wilson (1966) showed that the duration of the period over which divisions occurred in leaves of *Cucumis* was similar at all levels of radiation and from this it follows that the smaller cell number in low irradiances can only result from a smaller proportion of the cells continuing into successive divisions. This conclusion is not easy to interpret. If it is argued that high light intensities promote the formation of factors necessary to maintain leaf cells in the actively dividing condition then it might be expected that in low intensities the level of these factors will be lower and cell divisions will cease earlier, but this does not happen. This suggests either a non-uniform distribution of the factors between the cells or a differential sensitivity of the cells to a uniformly lower level. A further paradox is also apparent. If photosynthetic production of carbohydrate controls the extent of cell division in leaves, why do divisions cease at a time when rates of photosynthesis are maximal and export of sucrose from the leaf is rapidly increasing (Hopkinson, 1964)? An attempt to resolve this paradox is given in the following section.

B. Carbon Nutrition and Cell Division

During development the leaf undergoes a period when it is dependent upon a supply of metabolites from other leaves or from specialized storage organs such as the cotyledons. Ultimately, as it becomes increasingly autonomous for assimilated carbon, the period of import comes to an end and the expanding leaf itself begins exporting to other centres of growth. There is evidence for basipetal development of photosynthetic capacity in the leaf (Larson *et al.*, 1972) to parallel the basipetal changes in cell division and cell maturation already discussed. The early maturation of the leaf tip is associated with it entering the phase of export while the basal regions are still importing carbon and some data indicate that assimilates from the tip can be utilized in the immature basal regions (Jones and Eagles, 1962; Turgeon and Webb, 1973). The transition of the leaf from an importing organ to one exporting assimilated carbon occurs at about the same time that cell divisions cease, and the nature of this correlation is now examined.

The importance of cotyledonary reserves for cell division in primary leaves of *Phaseolus* has already been mentioned in connection with the low-energy light response. It has also been found that removal of the cotyledons from light-grown seedlings lowers the final cell number in the leaves, which is also proportional to seed weight (Table IV). Irrespective of seed size the cotyledon reserves are exhausted by day 8–9 (Dale and Felippe, 1968) and it is at this time that the primary leaves begin to export carbon assimilates (Köcher and Leonard, 1971; J. E. Dale, unpublished). Coincidentally with this, cell division ceases in the leaves.

So far it has not been possible to extend the period of cell division experimentally, and it is of interest that Köcher and Leonard showed

TABLE IV. The relationship between seed weight and final cell number of primary leaves of *Phaseolus vulgaris* cv. 'Canadian Wonder'. Plants were grown at 22·5°C in 14 h daylength with irradiance of 58·0 W m^{-2}.

Seed weight range	0·41–0·45 g	0·56–0·61 g	0·71–0·80 g
Cell number per leaf	$40·1 \times 10^6$	$45·9 \times 10^6$	$51·7 \times 10^6$

($LSD_{(p=0·05)}$ for cell number $\pm 4·8 \times 10^6$)

the transition to the exporting condition to occur even in detached leaves since in these there was an accumulation of photosynthetically produced carbon in the petiole base. Mention has already been made of the fact that cell divisions will occur in disks cut from young leaves and cultured in the light on a sucrose-containing medium. Under these conditions the rate of cell number increase in disks is very close to that in intact leaves, and division ceases on day 8, at the same time as in intact leaves (Dale, 1966). It is tempting to suggest that parallel changes occur in disks and leaves which lead to a cessation of division and that these are related to changes in the autonomy of the tissue with respect to carbon and to the initiation of export.

If cell number in a leaf is governed by the amount of essential metabolites that can be imported by that leaf over a limited period, then reducing the amount should also lower cell number. Thus removal of the cotyledons lowers primary leaf cell number in *Phaseolus*. But in this species the trifoliate leaf is supplied with assimilated carbon from the primary leaf and if the photosynthetic activity of the latter is reduced then cell number in the former should also be affected. This has in fact been found to be the case. Photomorphogenic effects on area of the primary leaves, not connected with cell number, lead to maximum dry matter production by these leaves in 14 h days (Dale, 1965). Although variations in photoperiod and irradiance do not affect cell number in the primary leaves this is not so for the first trifoliate leaf (Fig. 3) in which the cell number is least in low irradiances and maximal in 14 h photoperiods where the primary leaves are most effective photosynthetically. This is compatible with the interpretation that cell number in the younger leaf is controlled, at least in part, by photosynthetic activity in the older leaves which supply it. This view has been tested experimentally by holding primary leaves in low light intensities (11·6 W m^{-2}) while at the same time exposing the developing trifoliate leaves to higher levels of irradiance (58·0 W m^{-2}). Under these conditions cell number in the trifoliate leaves was reduced by 35% when both

TABLE V. The effect of shading the primary leaf of *Phaseolus vulgaris* cv. 'Canadian Wonder' on final cell number in the unshaded first trifoliate leaf. Plants were grown at 22·5°C in 14 h daylength with irradiances of 58·0 or 11·6 W m^{-2}.

Treatment of primary leaf	Unshaded	One shaded	Both shaded
Cell number per leaf	95×10^6	90×10^6	62×10^6

(LSD$_{(p=0·05)}$ for cell number $\pm 5·3 \times 10^6$)

primary leaves were shaded and by an insignificant amount when only one of the leaves was treated (Table V). These findings are entirely consistent with the idea that material produced in high light by the primary leaves is required for cell number increase in the trifoliate leaf. Clearly, this material can be provided in sufficient quantity if only one of the primary leaves is held in high light. Attempts have been made to feed shaded primary leaves with sucrose and thus to increase trifoliate leaf cell number, but these have been unsuccessful. Whether this means that the factor required for cell division in the trifoliate leaf is a specific one which cannot be metabolized from sucrose is not clear.

Other evidence of an association between cessation of cell division in leaves and commencement of active export comes from studies on *Cucumis*. Here the data of several investigators (Table VI) indicate that entry to a phase of export of sucrose coincides with or precedes cessation of cell division in each of the first three leaves of this species.

We can now consider the paradox, outlined in the previous section, that cell divisions cease at the time that photosynthetic production in the leaf is maximal. Resolution of the paradox rests on the fact that it is the supply of metabolites from older leaves that governs the extent of division in the younger, and that when this supply ceases so too will

TABLE VI. The relationship between time of commencement of export of carbohydrate and time of cessation of cell division in the first three leaves of *Cucumis sativus* cv. 'Butcher's Disease Resister'.

	Commencement of export	Cessation of cell division
First leaf	Day 11[a]	Day 12[b]
Second leaf	Day 14[a]	Day 17[b]
Third leaf	Day 19[a]	Day 19[c]

[a] Hopkinson, 1964; [b] Milthorpe and Newton, 1963; [c] Wilson, 1966.

cell division. Thus the larger number of cells in a leaf in high light intensity is the result of more material being supplied to that leaf from the older leaves below it and not of a higher photosynthetic activity locally. The higher level of photosynthesis in the leaf which is continuing to show divisions is irrelevant, being important only to leaves developing subsequently. In this connection it is notable that the effects of light intensity on cell number in *Cucumis* become progressively larger for successive leaves (Fig. 4). It is also of particular interest that cell number in the cotyledons, which expand to become the first photosynthetic organs, is not affected by light intensity, in contrast to that of the later-formed leaves. In this the cotyledons resemble in their behaviour the primary leaves of *Phaseolus* and such a result might be expected by analogy. Again it confirms the idea that local carbon assimilation does not affect cell number in a leaf.

On the basis of the foregoing discussion a working hypothesis is proposed to the effect that cell division in the leaf depends upon the continued import of metabolites from elsewhere in the plant, usually from other leaves and produced through photosynthesis, and that when this supply ceases, coincident with the development of the leaf as an exporting organ, divisions cease also. A number of points of interest arise from this hypothesis.

The first point is that material produced in one leaf can, it is postulated, be imported and utilized in cell division in another, yet the same material produced in the second leaf is apparently not utilized for divisions by that leaf. There is no ready explanation for this. It could be that photosynthetic products destined for export are kept in compartments separate from the centres where they are used in the syntheses associated with and necessary for division. Alternatively, but not necessarily exclusively, the metabolic qualities of the cells could change during growth coincidentally with the availability of locally produced material so that this can no longer be used by the leaf producing it.

The second point arising from the hypothesis is that any factor which affects the level of photosynthesis, and hence the supply of metabolites to a developing leaf, might be expected to affect division. This could explain how factors such as level of supply of mineral nutrients can influence leaf cell number (Milthorpe and Newton, 1963; Humphries and French, 1963); it is not necessary to invoke a direct effect on the division process, merely an indirect action through photosynthetic activity in the older leaves supplying those which are meristematic.

A final point is that the hypothesis involves consideration of import and export by the leaf as a unit. But it has already been mentioned that divisions cease earlier in the epidermal cells than in those of the mesophyll. Interpretation of this difference solely in terms of the import of essential metabolites implies that the mesophyll is favoured over the epidermis, perhaps because of closer proximity to the vascular tissues through which import occurs (see also Parkhurst, 1972). Yet an interaction between products of the light reaction and imported metabolites could be envisaged which would allow continuing division in the palisade but not in the non-photosynthetic epidermis. It would be of interest to know the duration of the phase of division in the leaf tissues of those species which have a green, photosynthetically active epidermis.

The diverse effects of light on cell division in leaves are brought together in Fig. 5. This diagram summarizes the interrelationships,

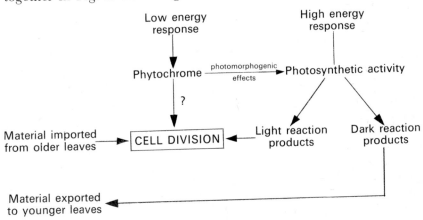

Fig. 5. Summary diagram showing interaction of factors affecting cell division in leaves.

outlined in this and the preceding section, between the low-energy responses affecting development of the photosynthetic mechanism and with indirect effects on division, and the high-energy effects which through the light reaction of photosynthesis control local levels of division and through the production of carbon assimilates control division in younger leaves.

C. The Role of Growth Substances

From time to time it has been proposed that specific leaf-forming substances are involved in the initiation of primordia at the stem apex. Such substances have not been extracted and there is little evidence for their existence. Indeed, careful consideration of the concept of specific organ-forming substances, with special reference to the flower-inducing hormone, florigen, has led Evans (1969) to conclude that their existence is unproven and, on the evidence available, unlikely. It has also been suggested that specific leaf-forming substances are involved in leaf expansion. The phyllocaline concept proposed by Went (1938), largely through an interpretation of his data in terms of Gregory's (1928) work on *Cucumis*, has been widely quoted, and Humphries and Wheeler (1963) went so far as to suggest that "gibberellins partly fulfil the concept of the leaf-growth hormone postulated by Went". Nevertheless it has been found that expansion of leaf tissue, not necessarily involving cell division, can be brought about by a large number of chemicals from oxidizing and reducing agents to sucrose and a variety of growth substances. In view of this the concept of a specific phyllocaline is best abandoned.

   In considering whether growth substances affect cell number and cell division in leaves two approaches have been followed. These are to correlate changes in endogenous growth substances with changes in cell number, and to examine the effects of exogenous application of particular substances on cell number and leaf size. Using the first approach, Humphries and Wheeler (1964) followed levels of free indole acetic acid and gibberellin in leaves of *Phaseolus*. They found that young primary and trifoliate leaves showed high levels of IAA, but in leaves in which cell division had ceased zero or very low concentrations of IAA were found. The correlation between IAA level extracted and the rate of increase of cell number does not appear to be particularly close from their data, but Humphries and Wheeler suggested that free IAA could be necessary for cells to divide in leaves. They also found levels of gibberellin to be highest in young leaves, but in contrast to IAA the levels dropped more slowly after the cessation of the phase of cell division. Despite the basic similarities in trends for both IAA and gibberellins and notwithstanding the demonstration by Wheeler (1960) that gibberellin level in primary leaves was highest when cell divisions were most abundant, Humphries and Wheeler considered that gibberellins were not directly involved in cell division but were involved in processes of cell expansion. More recently, Wheeler (1973) has found the basal, meristematic, portion of expanding wheat leaves to contain significant amounts of gibberellin although "young, elongating leaves did not contain any significant amount of auxin".

Where growth substances have been applied to leaves variable results have been obtained. There appear to be no recorded cases of IAA application increasing cell number in leaves, but there are a number of reports showing meristematic activity in developing leaf primordia to be adversely affected by the growth hormone 2,4-dichlorophenoxy acetic acid (2,4-D). Leaves of *Gossypium* and *Phaseolus* are particularly sensitive to treatment with this compound which inhibits formation and development of cells of the inter-veinal areas without affecting vein extension to the same extent; the result is that the leaves become deformed and strap-shaped (Watson, 1948; Eames, 1951; Gifford, 1953). In these cases, treatment clearly affects the development and function of the plate meristem and is only seen when leaves are treated in the primordial state, and not if treatment is given to unfolding leaves. It is of interest to note that substances as diverse as phenylboric acid (Haccius and Massfeller, 1961) and 2-thiouracil (Heslop-Harrison, 1962) also inhibit lamina initiation and development in the young primordium. These findings are entirely consistent with the idea that leaf form begins to be determined very early in primordial development.

Where gibberellic acid is concerned, applications to *Phaseolus* plants have been found to have no effect on cell number in either primary or trifoliate leaves, although for the latter the length : breadth ratio was affected (Felippe, 1967; Felippe and Dale, 1968a). On the other hand disks of primary leaf tissue cultured in presence of GA_3 did show a significant increase in cell number which was not found with IAA (Dale, 1966).

Results of experiments with the growth retardant chloro-choline chloride, CCC, are of considerable interest. This compound is known to inhibit the biosynthesis of gibberellin in the fungus *Fusarium moniliforme* and to affect the metabolism of gibberellins in a range of higher plants (Reid and Carr, 1967; Felippe and Dale, 1968b). Application of CCC to light-grown plants of *Phaseolus* caused a slight reduction of primary leaf cell number and a larger reduction in cell number of the trifoliate leaves; these effects were reversible on application of GA_3

(Felippe, 1967; Felippe and Dale, 1968a). However, application of CCC to dark-grown plants at any time up to day 4 from planting caused an increase in primary leaf cell number to values comparable with those found in light-grown plants. This effect could be prevented by application of GA_3 and did not involve cell expansion as the leaves remained folded about each other and unexpanded.

These remarkable results are of interest on two counts. Firstly they show CCC at the same concentration and on the same species to have two opposing effects on leaf cell number; at the same time they show that gibberellic acid treatment in reversing the CCC effect can also lead either to a reduction or to an increase in cell number in the leaf. It is difficult to see GA_3 as acting at all directly on the process of cell division in these cases.

The results are of interest for another reason. They show the light requirement for cell division to be completely circumvented by CCC treatment. This has been interpreted as resulting from a suppression of gibberellin synthesis at the stem apex by CCC, leading to a reduced capacity of the stem tissues for growth. In consequence cotyledonary materials normally destined for the stem are diverted to the primary leaves, thus facilitating divisions which would normally require the intervention of light. Changes in dry weight of the plant were in keeping with this interpretation.

In his studies on etiolated pea seedlings Went (1938) showed that removal of the root system reduced leaf expansion. This can readily be shown also for the primary leaves of *Phaseolus*, and Murray (1968) found that removal of the root system for dark-grown plants prevented the usual cell number increase in response to short irradiations with white light. It was also found by De Ropp (1945, 1946a, b) that when rootless stem portions of the embryo of rye were cultured on agar the first leaf expanded but without cell divisions in the basal meristem, whereas if the roots regenerated, mitotic activity was found in the meristem. It was concluded that some metabolite was passed from root to leaf and that this was necessary to maintain meristematic activity in the latter. Since it is known that exudates of xylem sap often show high levels of cytokinin activity (Sitton *et al.*, 1967) it is tempting to suggest, as others have done, that roots normally supply such substances to the shoots where they may be involved in growth of the leaf. However, in spite of many reports establishing the involvement of cytokinins in nucleic acid and protein metabolism in leaves there is no clear evidence that cytokinins are directly involved in the division process there.

Indeed there is little unequivocal evidence for a direct effect of any known natural growth substance on the process of cell division in leaves.

It seems likely that the importance of these compounds lies in their involvement in the general control and organization of metabolic activities of the tissue which is a necessary pre-requisite for cell division. Just as affecting gibberellin metabolism in the dark-grown bean alters the partition of metabolites from the cotyledons, and hence their availability, so too the high levels of IAA and gibberellin found in young leaves may reflect the fact that these organs are centres of import. The control of import almost certainly involves complex interactions between growth substances, and without this, meristematic activity does not occur.

V. CONCLUDING REMARKS

In spite of a considerable amount of work, often resulting in elegant description, our knowledge of the mechanisms controlling leaf form remains slight. These mechanisms become operative in the very young primordium and hitherto this has proved an unattractive experimental system for most workers on leaf growth. However, methods for the study of leaf primordia do exist and it is to be hoped that cell physiologists as well as experimental morphologists will extend investigations in this area. It may be reiterated that even accepting that the genetic specification of form does not imply a rigidly controlled programme of cell division, it is still the case that explanation of leaf dorsiventrality and the initiation of leaf shape must be sought in analysis of local changes in rates and planes of division and the factors, physical and chemical, involved in these.

For the expanding leaf blade the basipetal and inter-tissue differences in rates and duration of cell division are somewhat easier to examine and investigate. Yet even here a general interpretation of the available data does not exist and it is against this fact that the scheme outlined in Fig. 5 should be seen. It is to be hoped that the ideas and assumptions implicit in this scheme will be examined further and that ultimately a fuller and more detailed explanation can be advanced.

10. The Cambium

I. D. J. PHILLIPS

Department of Biological Sciences, University of Exeter, England

I. INTRODUCTION

A description of the form of any species of higher plant has of necessity to be couched in less precise terms than can be applied to the vast majority of animal species. The reason for this is that, unlike a higher animal, each higher plant can be constituted of indefinite numbers of separate organs (leaves, branches, roots, etc.). This does not, however, mean that the development of plant organs and tissues is unregulated. Characteristic plant form exists and is readily recognized by many criteria, such as the shape and positions of leaves on the stem axis, branching habit of the shoot, floral structure, and many internal anatomical details. Indeed, of course, without genetically determined plant form it would be impossible to classify plants according to any taxonomic system.

This difference in developmental behaviour between higher plants and animals does not reflect any relative "primitiveness" in control systems of plants compared with animals, but adaptation to different life styles. Higher plants, unlike higher animals, are unavoidably sedentary, tapping the soil for water and nutrients. Their construction from individual cells each bounded by a rigid cell wall (which solves the problems of osmoregulation and skeletal support very neatly) means that once growth ends, then the ability to explore new regions for sources of nutrients and energy also ceases. In other words, a non-growing plant is completely dependent on the presence or arrival at its surface of water, inorganic elements and sufficient light. It is no longer responsive enough to its environment to explore pastures new when it has fully exploited the immediately available resources or is shaded by other vegetation. Consequently, whereas animals characteristically exhibit a determinate pattern of development (the basic plan of the animal body is determined during embryogenesis, and further development consists of the growth and elaboration of the embryonic structures already laid down), in plants development is *indeterminate*. That is, plants retain embryonic activity for the whole of their lives. We have seen in other chapters how the root (see Chapter 7) and shoot (see Chapter 8) apical meristems repetitively give rise to new tissues of the longitudinal axis, and also to lateral organs such as leaves and lateral shoots. In plants which are relatively long lived, particularly those that attain considerable heights, it is of clear mechanical advantage that the capacity for increasing the diameter of the longitudinal axis also be retained throughout life.

Increased stem and root girth occurs through processes involving cell division, expansion and differentiation. In dicotyledonous plants, and in most gymnosperms, radial growth of stem and root is an expression of the activities of the *lateral meristems*: so called due to their position in the plant body, lying as cylindrical layers of cells around the axis. Two types of lateral meristem are generally recognized, the *vascular cambium* and the *cork cambium* or *phellogen*. In those monocotyledonous species that undergo radial growth to any extent, cambial cells still play a part, but their activities are more restricted (Philipson *et al.*, 1971).

II. CAMBIAL MORPHOGENESIS IN SHOOT AND ROOT

Vascular and cork cambia are similar to the apical meristems of root and shoot in that they retain indefinitely the capacity for cell division. Cambial cells do nevertheless differ significantly from apical meristems. Not only are they differently situated within the plant compared with apical meristematic cells, but they also differ in general organization and cytological features, and give rise to only a few specific tissues, whereas the apical initials are responsible for the origin of many tissues and whole organs. In marked contrast to the apical meristems, the structure of cambial initial cells (*fusiform* and *ray initials*) is very similar to the tissues derived from them: secondary xylem and phloem. A further point of distinction between apical and lateral meristems concerns their morphological origins. Thus, the apical initials are already present at a very early stage in the development of the embryo, and never lose their embryonic status as actively dividing cells. In contrast, the cambium always arises after embryogenesis and from cells which have already undergone at least a certain measure of differentiation. Cambium initiation therefore invariably involves a process of de-differentiation of cells and their return to an embryonic condition.

A. Procambium Initiation and Differentiation

The first indication of cambium to appear during plant development is the formation of the *procambium*, and this tissue develops later into vascular tissue and vascular cambium. The procambium always arises very close to the stem or root apical meristem, but the details of pro-cambium morphogenesis have been a matter of some conjecture for many years (see Cutter, 1971; Wetmore and Steeves, 1971; Steeves and

Sussex, 1972). In shoots, it is generally recognized that the procambium first differentiates as either a solid core or a ring of cells with a smaller size and greater cytoplasmic density than the cells of the developing cortex lying immediately to its outer side. At this stage of development the procambium has been variously called the meristem-ring, residual meristem, incipient vascular tissue or prestelar tissue (Esau, 1954, 1965; Wetmore and Steeves, 1971). The procambium when fully differentiated is recognized as consisting of cells which are longitudinally elongate compared with surrounding cells at the same distance behind the apical meristem.

Whilst the initiation of the procambium appears to be closely associated with the activities of the apical meristem of either root or shoot, its subsequent development in the shoot is markedly influenced by regulatory influences from newly formed leaf primordia. Thus, in microphyllous vascular cryptogams such as *Lycopodium*, a central core of procambium exists at levels above that of the youngest leaf primordia (Freeberg and Wetmore, 1967). In the megaphyllous Filicales it has similarly been found that the procambium is first formed either as a solid central column or as a ring, prior to the initiation of leaf primordia on the flanks of the apical meristem (Wardlaw, 1944; Steeves, 1963). In *Lycopodium* the solid core of procambium matures first at its periphery, differentiating acropetally into protophloem and protoxylem, and later-formed metaphloem and metaxylem differentiate from the inner parts of the procambial core, giving rise to the characteristic protostele. In stems of megaphyllous ferns and in seed plants, the core or cylinder of procambium becomes broken up into separate strands, each associated with a developing leaf, with interfascicular parenchymatous leaf gaps between. In consequence, the vascular tissues differentiating from the procambium in shoots of megaphyllous plants acquire the form of an anastomosing network of leaf traces, the particular pattern of which is determined by the phyllotaxis of the shoot, and is first laid down in the procambium.

In roots of both microphyllous and megaphyllous plants, the pattern of procambium initiation and differentiation is very similar to that seen in the shoot of a microphyllous plant such as *Lycopodium*. In other words, the absence of lateral appendages such as leaves is associated with the transformation of the procambial core into the typical protostelar vascular system of roots.

Protophloem differentiates acropetally from the outer cells of each procambial strand, in both shoot and root. At first only the outermost file of cells differentiates into phloem elements, but successively deeper layers of procambial cells subsequently undergo transformation into protophloem (Steeves and Sussex, 1972). Phloem development appears, therefore, to follow a pattern which results in continuity between mature phloem elements and newly differentiating protophloem (Esau, 1953; 1960), although there is a certain amount of evidence for occasional discontinuities in differentiating protophloem of leaf traces (Jacobs and Morrow, 1967).

The development of protoxylem in shoots from procambium appears to follow a rather more variable pattern than that of protophloem. In angiosperms, it has generally been found that protoxylem shows a discontinuous progress of initiation. In a number of species it has been observed that protoxylem first appears at the base of each developing leaf at the approximate time that protophloem, differentiating acropetally, first reaches there. Further differentiation of the protoxylem from procambium then proceeds both acropetally into the developing leaf and basipetally to unite eventually with the mature xylem of the stem (Esau, 1938). There are some reports that xylem differentiation from procambium may be initiated at more than one locus. For example, Jacobs and Morrow (1957) observed that in *Coleus* protoxylem was initiated not only at the leaf base but also more basally in the stem; further differentiation took place both acropetally and basipetally from each of the two sites of xylem initiation, resulting in the union of the two strands, and their connection with the mature xylem of the stem.

In roots, protoxylem initiation from the procambial core occurs later than protophloem initiation, with the result that the first observable protoxylem elements lie behind the region of root extension growth. The faster the rate of root elongation, the further behind the root tip does vascular tissue differentiation occur (Esau, 1965). Both phloem and xylem initiation in roots proceed acropetally without discontinuities.

The general pattern of vascular tissue development from the root procambium described in the previous paragraph holds for those vascular cryptogams which have received study, as well as for seed plants. In the shoot of vascular cryptogams, however, the limited available evidence suggests that several patterns of phloem and xylem differentiation occur, some of which differ from that seen in angiosperms. Thus, in *Lycopodium* it was observed by Freeberg and Wetmore (1967) that xylem is initiated earlier than phloem, and that both differentiate acropetally with continuous connection with the mature stem vascular tissues. On the other hand, in *Equisetum* each node represents a site of discontinuous initiation of both phloem and xylem, and further differentiation of the two proceeds basipetally to unite with the vascular tissues at the next older node. In the ferns, various patterns of vascular tissue initiation from procambium have been reported. In some species the xylem shows discontinuous initiation followed by both acropetal and basipetal differentiation, whilst the phloem is initiated continuously and differentiates acropetally as in seed plants, whereas in other ferns both xylem and phloem initiation and differentiation are continuous and acropetal, phloem forming earlier than xylem (Wetmore and Steeves, 1971).

B. Vascular Cambium Initiation

Development of procambial cells into vascular tissues leads to the appearance of typical collateral vascular strands as seen in the angiosperm shoot, or to the protostelic condition of most roots and stems of more primitive cryptogams. In this way the procambium gives rise to the *primary vascular tissue* of the *primary plant body*. This is not, however, necessarily the end of meristematic activity associated with vascularization, for in all woody and many herbaceous plants the primary vascular bundles consist of not only phloem and xylem, but also of an intervening layer of meristematic cells—the *fascicular cambium*. The fascicular cambium represents of course the first true vascular cambium. A fundamental distinction between procambium and cambium proper exists. Thus, whereas procambial cells themselves may differentiate into vascular elements, true vascular cambial cells to not differentiate but at each division yield two cells of differing fate. One daughter cell remains meristematic and the other undergoes differentiation into a cell of the vascular system. In stems with bicollateral bundles, cambium forms only between the xylem and outer phloem, and not adjacent to the inner phloem.

It is very difficult to describe accurately the actual process of transition between the procambial condition and the state where it may be said that a true fascicular cambium exists. The reason for this is that very few studies of the transition have been conducted, and that the limited available evidence is often contradictory. For example, it was found by Esau (1965) that in *Nicotiana* stem a fairly clear anatomical distinction could be drawn between procambium and cambium, the latter showing an organized condition of fusiform and ray initials whereas the procambium was a more homogeneous tissue. The transition in sycamore, *Acer pseudoplatanus*, was found to be even more marked (Catesson, 1964), in that undifferentiated homogeneous procambium did not change form until internode extension ceased, but then each procambial cell elongated by intrusive growth and acquired pointed ends. Each elongated procambial cell became a fusiform initial of the true vascular cambium. Some of the fusiform initials so produced subsequently (at the time secondary growth commenced) divided transversely several times to form a longitudinal file of ray initials of the vascular cambium. On the other hand, studies of some other species have indicated a much greater similarity between the anatomy of the procambium and cambium. Cumbie (1967), for example, found that the procambium in *Canavalia* was already organized into longitudinal files of long and short cells analogous to the fusiform and ray initials of the cambium. Cumbie also pointed out several other features of such similarity between procambium and cambium in *Canavalia* that it was impossible to define a time of origin of the vascular cambium. It is clear that more anatomical studies of the development of cambium are required before one can accurately describe its origin, but already it

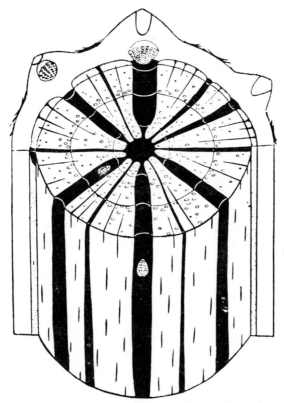

Fig. 1. A three-dimensional diagram illustrating that, in *Casuarina* stem, distinctive secondary tissues are formed by the fascicular and interfascicular sectors of the vascular cambium. (From E. C. Jeffrey, *The Anatomy of Woody Plants*. Chicago University Press, Chicago, 1917.)

would appear that the pattern of vascular cambium initiation differs from species to species.

In those species where secondary growth is very restricted, vascular cambial activity is usually confined to the primary vascular strands, but in many other species further secondary development of the vascular system occurs. If a complete cylinder of cambium does not already exist, then secondary development commences by the de-differentiation of parenchymatous cells lying between the vascular strands, which divide periclinally, leading to the appearance of a new interfascicular cambium linked to the fascicular cambium of the vascular strands. Thus, a complete cylinder of vascular cambium is formed (see Fig. 6), consisting of fusiform and ray initials, which give rise respectively to vascular elements and parenchymatous ray cells. The stimulus for the initiation of interfascicular cambium appears to arise in the fascicular cambium, for the first interfascicular cambial cells form immediately adjacent to the fascicular cambium of each vascular strand. Most commonly the cylinder of cambium gives rise by meristematic activity to similar tissues over its whole circumference, although in some regions of the plant, and/or in certain species, xylem produced by fascicular cambium may be different structurally from that formed from interfascicular cambium (Fig. 1).

C. Vascular Cambium Structure and Activity

The vascular cambium is made up of two types of meristematic cells, fusiform and ray initials, which are sandwiched as a thin sheet between the xylem and phloem tissues. The fusiform initials are often axially elongated to a great degree, being up to several hundred times longer than their radial diameter, and have tapering pointed ends to give an overall prismatic form. Ray initials, in contrast, are more nearly isodiametric. Although the detailed architecture of cambial initials varies, it is a general feature that fusiform initials give rise to the vascular elements of xylem and phloem, and ray initials to the cells of medullary rays.

Almost all studies of cambial structure and organization have been conducted on tree stem material. Much of our knowledge of cambial cell architecture was obtained through the work of I. W. Bailey on *Pinus strobus* during the 1920s and 30s, which is summarized by Bailey (1954).

In *P. strobus*, as in other conifers, such as *Pinus sylvestris* (Dodd, 1948), fusiform initials are normally very elongated, in some cases to almost 9 mm per cell, though to considerably varying degrees, whereas in woody dicotyledons they are usually much shorter than in conifers. In the course of evolution, there appears to have been a reduction in fusiform initial length with the development of xylem vessels, which are made up of individual elements shorter than the separate tracheids of more primitive plants (Bailey, 1954). In some highly evolved woody dicotyledons, the fusiform initials are very short and arranged in horizontal tiers, in which case the plant is described as having *storied* or *stratified cambium* (Fig. 2). The cell walls of fusiform initials are relatively

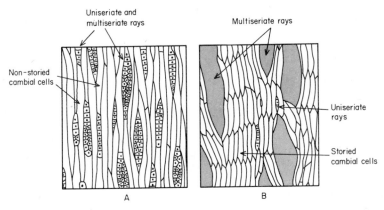

FIG. 2. Semi-diagrammatic tangential longitudinal views of the non-storied cambium of *Juglans cinerea* (A) and the storied cambium of *Robinia pseudacacia* (B). Note that both uniseriate and multiseriate rays occur in each species.

thick and contain numerous primary pit fields, and the cytoplasm is highly vacuolated. Even in the very large fusiform initials of some conifers, only one nucleus is present per cell.

The almost isodiametric ray initials are distributed among the fusiform initials in species-specific patterns, and give rise to the radially arranged rays which can be seen traversing the wood of the stem in transverse section. The ray initials themselves are, although much shorter, like the fusiform initials in being highly vacuolate.

The ultrastructure of cambial initials has received increasing attention over the past few years. Srivastava (1966) and Srivastava and O'Brien (1966) found essentially similar fine-structural features in the cambia of the dicotyledonous angiosperm *Fraxinus americana* and of the gymnosperm *Pinus strobus*. They noted that during summer the actively dividing fusiform and ray initials were very alike, each containing a high density of rough cisternal endoplasmic reticulum, ribosomes, golgi bodies, coated vesicles, a large nucleus with a single nucleolus, a nuclear membrane containing marked pores, and microtubules in the peripheral part of the thin layer of cytoplasm in these highly vacuolated cells. During winter, the dormant cambium of *F. americana* differed from the active summer form in containing more than one vacuole, lower densities of coated vesicles and endoplasmic reticulum in the rough cisternal form. Similar studies on *Salix fragilis* by Robards and Kidwai (1969) revealed very much the same picture, except that the dormant cambium contained considerable numbers of protein bodies and lipid droplets, together with vesiculate smooth endoplasmic reticulum. It was suggested that these represent storage materials for utilization during early spring growth.

D. Patterns of Cell Division Activity in the Cambium

Cell division in the cambium occurs in several planes, and at varying rates depending on factors such as season, water and nutrient supply. Derivatives of the cambium destined to differentiate into new xylem, phloem, or ray cells, all arise from periclinal divisions (i.e. in the tangential plane) of cambial cells. Xylem cells are normally laid down towards the centre of the stem or root, and phloem to the outer side. In either case a longitudinal cell plate is laid down in the tangential plane along the full length of the fusiform initial. One of the two daughter cells proceeds to differentiate into a xylem or phloem element while the other remains as a fusiform initial.

The production of new cells to the inner surface by the cambium, and the enlargement of these during their development into xylem tissues, inevitably result in an increase in the girth, or circumference, of the cambial region. In consequence, it is essential that new cambial cells be formed to accommodate the increased diameter of the xylem. The means by which such an increase in cambium circumference is obtained varies between species (see Bailey, 1954), but in all cases an essential component of the process is the occurrence of anticlinal divisions (i.e. in the radial plane) in fusiform initials. In those highly evolved woody dicotyledons with a storied cambium, the anticlinal divisions of fusiform initials are truly vertical and each daughter cell undergoes tangential expansion (Fig. 3A). Repeated divisions of this type lead to the appearance of typical horizontal rows, or storied, fusiform initials. The length of fusiform initials in such storied species remains constant. By contrast, in evolutionarily more primitive species with non-storied cambium, it has been observed that both the length and tangential width of fusiform initials gradually increase during the early years of cambial activity (Bannan, 1960), and that this period

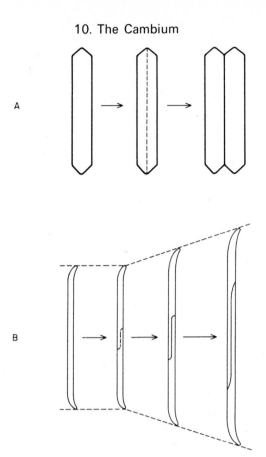

A

B

Fɪɢ. 3. Diagrams illustrating modes of anticlinal cell division in fusiform initials. A: True anticlinal (radial longitudinal) cell division, typical of storied fusiform initials. B: Pseudo-transverse division, typical of non-storied fusiform initials. In both A and B the result is to increase the circumferential dimension of the vascular cambium by an increase in number of fusiform initials.

may last as long as 60 years or more in conifers (Bailey, 1954). Now, it is clear that since longitudinal growth of the stem does not occur during the period over which fusiform initials increase in length, then the ever longer and wider initial cells can only be accommodated by their sliding past one another during intrusive longitudinal growth. In this way the cambium in non-storied species continues to increase in circumference and completely enclose the continuously enlarging surface of the xylem tissues for many years. However, these processes of cambial initial cell enlargement do not suffice to explain the continuing increase in cambium diameter in non-storied cambium species, for it has been shown that even after fusiform initial enlargement ceases the diameter of the cambium continues to increase, as does also the number of fusiform cells seen in transverse section (Bailey, 1954). Because there is a persistent increase in the number of initials seen in transverse section of the cambium, it is clear that anticlinal divisions must occur in the fusiform initials. But in conifers and more primitive woody angiosperms, no sign is ever seen in tangential sections of the horizontal rows of initial cells which characterize the highly evolved species with storied cambium. The question as to why anticlinal divisions in the cambium of conifers and more primitive woody angiosperms do not lead to a storied arrangement of the fusiform initials was answered by observations that in these species anticlinal divisions in the fusiform initials are almost transverse in orientation and have hence been termed *pseudo-transverse divisions* (Bailey, 1923) (Fig. 3B). The two approximately equal-sized daughter fusiform initial cells produced by such a division, an upper and a lower, proceed to elongate by tip-growth alongside one another and the number of cambial cells in transverse section is thereby increased.

Thus, in both storied and non-storied forms of vascular cambium, anticlinal divisions (true vertical or pseudo-transverse) are essential to the maintenance of the integrity of the meristem. It must be noted, however, that the frequency of anticlinal divisions is probably very much less than of the periclinal divisions which give rise to new vascular tissues. There is some evidence that the rate of anticlinal cell division activity in the cambium is at its highest when the rate of periclinal division is subsiding at the end of the growing season. Several workers have made estimations of the rates of new fusiform initial formation, and the overall picture which has emerged is that many more new initials are formed than are actually required for the actual increase in cambium circumference. Excess fusiform initials are either completely lost from the cambium (by either gradual obliteration, or abrupt differentiation into xylem or phloem elements) or are transformed into the smaller ray initials by true transverse divisions (see Philipson *et al.*, 1971). For example, in *Chamaecyparis* cambium it was found that only 162 functional fusiform initials persisted after a total of more than 1100 anticlinal divisions (Bannan, 1950).

This leads us to consider how, during the radial expansion of the stem or root, the observable number of rays increases. Since the number of rays does increase, then the question to be answered concerns the mechanism whereby the number of groups of ray initials in the cambium increases, a problem addressed particularly by Barghoorn (1940a, b, 1941). One route is through the conversion of fusiform initials as mentioned in the previous paragraph. This occurs in at least three

ways: (a) by the cutting off of a single new ray initial by a true transverse division near the tip of a fusiform initial; (b) by a series of transverse divisions along the length of the fusiform initial yielding several
ray initials which then proceed to give rise to a uniseriate ray; and
(c) by a so-called lateral division (Cheadle and Esau, 1964) of a fusiform
initial, which occurs by the laying down of a cell plate longitudinally
and to one side near the centre of the fusiform cell, thus cutting off
a new cell to the side (Fig. 4); such lateral divisions of fusiform initials

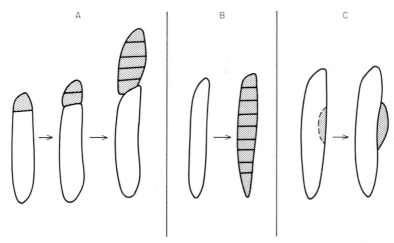

Fig. 4. Some of the ways in which new ray initials can be formed from fusiform initials.
In each case the shaded cells are new ray initials. A: By the cutting off of one new
ray initial cell to one end of a fusiform initial; the new ray initial may then divide
further as shown. B: The conversion of the whole of a fusiform initial by successive
divisions. C: By a lateral division cutting off a new ray initial to one side of a fusiform
initial.

tend to be more frequent in dicotyledons than in conifers (see Philipson
et al., 1971).

The number and/or width of rays can be increased in other ways
than by conversion of fusiform to ray initials. A group of existing ray
initials may, for example, be separated into two by intrusion of the elongating tips of fusiform initials. In dicotyledons, uniseriate ray initials
may divide anticlinally into two longitudinal files and consequently
proceed to form a multiseriate ray, or two groups of ray initials may
fuse and subsequently form a wider ray than previously. Various other
patterns of ray modification are known to take place during ontogeny,
particularly in dicotyledons, all of which emphasize the highly dynamic
nature of the cambial region.

E. Regional Variations in the Vascular Cambium and Anomalous Vascular Cambia

The general form and function of vascular cambium so far described relates primarily to the stems of the majority of gymnospermous or dicotyledonous species. To a large extent the description is also true for the root systems in plants of these groups. However, the vascular cambia in different regions of any plant can and do vary in structures and functions.

As we have considered earlier in this chapter, the cambium forms a complete ring in species showing secondary development, either through a process involving differentiation from a complete cylinder of procambium or by development of interfascicular cambium linking originally separated groups of fascicular cambium. In either instance it is most usually the case in trees that the vascular tissues produced from the ring of cambium are identical over the whole of its circumference. On the other hand, in a large number of species of herbaceous plants, and also a few woody ones (e.g. beech, *Fagus sylvatica*), the interfascicular sectors of the cambium produce inner derivatives which do not differentiate into normal metaxylem elements, but remain largely or completely parenchymatous in appearance, though quite frequently possessing heavily lignified walls. Such interfascicular sectors thus have the superficial appearance of medullary rays, but differ from typical rays in that they always extend from pith to cambium over the whole length of an internode, are wider than normal rays, and are not formed from groups of ray initials but from a whole sector of the cambium (Philipson *et al.*, 1971).

Other modifications to typical cambial structure and behaviour include changes associated with the development of lateral organs such as leaves and branches. In the region of separation of a leaf-trace from the central stele, a sector of cambium often produces derivatives which differentiate not into vascular tissues but into parenchyma, leading to the appearance of a parenchymatous leaf-gap. Depending on species, a leaf-gap may persist for some time, or be soon obliterated as cambium near the leaf-trace commences to form normal secondary tissues. At the point of union between two branches in trees, the vascular cambium is frequently distorted into folds of various sorts; this results in the formation of markedly abnormal wood structure (Jost, 1901).

The rates of cell division activity in the cambium, and pattern of differentiation of its derivatives, are also very much influenced by gravity. This matter is discussed later in this chapter, but briefly it is known that when gravity acts across the cambial ring (as when a branch is inclined from the vertical), periclinal and anticlinal division rates vary in different sectors of the cambium, and also the form of fully differentiated xylem derivatives is modified depending on the sector of cambium from which they were derived. This results in the appearance of *reaction wood* of various forms (Scurfield, 1973).

The vascular cambium of roots shows generally similar features and variations to those in shoots, except that the cambial initials in roots often differ in size and activity from those in the stems of the same plant (Patel, 1965; Fayle, 1968).

We will now very briefly consider arrangements of the vascular cambium in dicotyledonous plants which are often referred to as *anomalous cambia*, although it must be remembered that such arrangements are in fact perfectly normal in those species which show them. Anatomical aspects of anomalous cambia have been reviewed in some detail elsewhere, most recently by Philipson and Ward (1965) and Philipson *et al.* (1971), and the interested reader is referred to these publications for details. These authors have conveniently categorized anomalous cambia into two groups, each of which is sub-divided. They distinguish between cambia which are in the normal position but show unusual features in activity, and cambia which occur in unusual locations.

F. The Problem of the Permanency of Vascular Cambial Initials

In the foregoing account, an implicit assumption has been that the cambium consists of a single layer of initial cells, which divide periclinally to yield xylem mother cells to the inside and phloem mother cells to the outside. In other words, that at each periclinal division a fusiform initial, for example, separates into two daughter cells of different fates: one differentiating into a vascular element and the other retaining the properties of an initial cell. This view of the cambium therefore supposes that cambial initials have a permanency throughout the active life of the plant. However, histological studies of the cambium raise difficulties for this simplistic view, for only rarely is it found that the vascular cambium has the appearance in transverse section of a single layer of cells. Because of this, as early as 1853 it was suggested by Hartig that the cambium consists of two layers of cells, the outer of which give rise to phloem and the inner to xylem. Hartig's concept of a double-layered cambium has been abandoned in the face of various objections raised subsequently (Sanio, 1873; Bannan, 1955).

To explain the apparent multi-layered nature of the cambium, another proposal has been that the cambium is multiseriate, with each of the several layers of cells having the same properties as meristematic initials

(Raatz, 1892; Kleinmann, 1923; Catesson, 1964). This proposition is largely based upon studies of dicotyledons, in which the width of the layer of cambial initials is very variable. In coniferous tissues, on the other hand, much evidence points to the cambium consisting of a single layer of true initials. Work by Bannan (1955, 1968) on *Thuja occidentalis* and by Newman (1956) and Mahmood (1968) on *Pinus radiata* showed that it is possible, even where the cambial zone is very wide, to distinguish a single layer of functional initials. Identification of initial activity in these studies was based upon anatomical, histological, cytological and ultrastructural considerations.

Thus, so far as conifers are concerned, it does appear that the true vascular cambium consists of a single layer of fusiform and ray initials, but that in dicotyledons there is much variability in the number of layers of active cambial cells.

G. Phellogen

The meristematic activities of the vascular cambium, and differentiation of its derivative cells, lead to a progressive increase in diameter of the stem or root. Clearly, if the tissues which lie outside the phloem were not added to, rupture of the epidermis and cortex would occur, with the appearance of longitudinal fissures exposing previously internal tissues to the external environment. Water loss and susceptibility to invasion by pathogens would consequently both be greatly increased. These dangers are commonly avoided in gymnospermous and dicotyledonous trees by the production of layers of suberized protective cells to the surface, often referred to as "bark".

The *phellogen*, or *cork cambium*, appears early in secondary growth of both stems and roots. It is a meristematic layer, situated near the surface, which divides periclinally, initiating new cells to both outside and inside. The initials formed to the outside differentiate into dead cork cells (the *phellem*), and the inner initial cells into parenchyma-like cells (the *phelloderm*). The three layers, phellem, phellogen and phelloderm, together constitute a *periderm*—a protective region replacing the original epidermis of the primary plant body.

Like the vascular cambium, the phellogen is initiated by a process involving dedifferentiation of cells which had earlier ceased meristematic activity. Most commonly the cells of the sub-epidermis become transformed into the phellogen, but sometimes the epidermis itself, or alternatively more deep-seated cells (even, in some cases, those of the phloem region), resume meristematic activity and become a phellogen. The initiation of phellogen commences either discontinuously in separate regions around the circumference of the organ, or around the entire circumference at approximately the same time. In both cases a complete ring of phellogen is formed, which forms radial files of suberized, dead, cork cells to the outside. The phellem is often interrupted by lenticels, which allow gas exchange between the environment and living tissues of shoot and root.

Cell division rates in the phellogen are normally very much lower than those of the vascular cambium (Waisel *et al.*, 1967). Seasonal variations occur in phellogen activity, but these do not necessarily coincide with the seasonal changes in vascular cambium activity (Waisel *et al.*, 1967; Kozlowski, 1971). Anatomically distinct early phellem and late phellem, analogous to early wood and late wood, may be formed in many tree species (Srivastava, 1966; Waisel *et al.*, 1967; Kozlowski, 1971).

More than one periderm is formed during the life of many, but not all, trees. The first periderm is usually replaced within a few years, or even after only one year, by a periderm formed from a deeper layer of cells. The old periderm then dies completely. Further periderms form from successively deeper layers of cells, and may survive for one or many years (Bloch, 1965).

A periderm is very commonly formed in response to wounding of plant tissues. After development of a callus from the wounded tissues a phellogen, and also perhaps a vascular cambium, differentiates at particular levels from the callus surface (see Chapter 12).

The physiology of phellogen initiation and activity has received very little study. However, it seems likely that chemical and physical factors determine at what level a phellogen forms, in a similar manner to that in which they appear to be concerned in vascular cambium initiation. As long ago as 1921, Haberlandt reported that extracts of wounded plant cells induced formation of a periderm in potato tubers and other organs, and was led to conclude that a "wound hormone" from the injured tissues and a "leptohormone" from the phloem interacted to induce phellogen initiation. Subsequent discoveries of plant growth hormones and their physiological roles, particularly the auxins and cytokinins, suggest that these substances may be the active constituents of Haberlandt's hypothetical phellogen-inducing hormones (Haberlandt, 1921). Leroux (1954) certainly found that an auxin application to *Salix* stems stimulated periderm formation in the cortex.

III. PHYSIOLOGICAL ASPECTS OF CAMBIUM INITIATION AND THE DIFFERENTIATION OF ITS DERIVATIVES

The origins of the vascular cambium can be traced in morphological terms as discussed above. The primary vascular system arises under the principal controlling influence of the apical meristem of shoot or root, but with modifications imposed by the presence and development of lateral appendages such as leaves in the shoot. The earliest sign of procambial development is the appearance of differential staining reaction of cells near the apical meristem. This is soon followed by visible morphological differentiation of the longitudinally elongated procambial cells, which is a consequence of either fewer transverse divisions in these cells compared with neighbouring cells, or of a greater frequency of longitudinal divisions. Surgical experiments of various sorts have amply demonstrated that this pattern of procambial initiation is determined by influences arising from the apical meristem and developing leaf primordia (Wardlaw, 1965). Although this much is clear, we still have only a very rudimentary understanding of the physiological basis for the regular normal pattern of procambium initiation at the apex. Our knowledge of the physiology of initiation of the true vascular cambium is equally vague.

A. The Inception of the Stele

The apical region of shoot or root is an organizationally complex region which forms new cells, some which differentiate into procambial tissue according to a characteristic pattern which determines the general architecture of the future stele. The development of only some of the apically derived cells into procambium is merely one example of the fact that the developmental fate of any cell appears to be principally determined by its environment within the plant body, a concept first formulated as early as 1877 by Vöchting in the words, "the fate of a cell is a function of its position". In physiological investigations the aim is to characterize the nature of, and interactions between, the forces which act upon newly formed cells, inducing them to differentiate into procambium.

By the use of various surgical techniques, the stimulus for procambium initiation has been shown to originate in the apical meristem. For example, in a series of experiments on fern shoot apices by Wardlaw and his associates (see Wardlaw, 1965), the apical meristem was isolated by vertical incisions not only from leaf primordia but also from existing vascular and procambial tissues, and left supported upon only

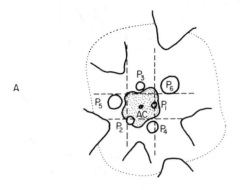

Stippled area = apical meristem
AC = large apical cell
$P_1 - P_6$ = developing leaf primordia

A

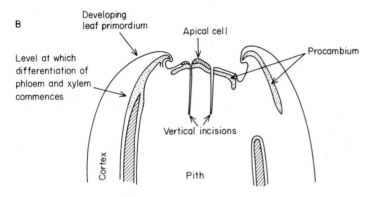

Fig. 5. Experiment involving surgical isolation of the shoot apical meristem of *Dryopteris aristata* from existing vascular tissues.

A: View from above showing the positions of the four vertical incisions (dashed lines) that isolated the apical meristem (stippled area) from surrounding developing leaf primordia. AC=large apical cell of the fern apex.

B: A longitudinal section showing two of the vertical incisions which isolated the apical meristem on a plug of parenchymatous pith cells. Consequently, the only connection left between the apical meristem and existing vascular tissues was *via* parenchyma, yet the apex continued to grow, produced new leaf primordia, and developed normal vascular tissues. (Re-drawn from C. W. Wardlaw, 1947.)

a plug of parenchymatous pith tissue (Fig. 5). The result was that although meristem growth rate was initially reduced, it soon recovered and initiated new normal leaf primordia and stem tissues and, furthermore, vascular tissue differentiation occurred basipetally through the originally parenchymatous plug of the pith. The vascular tissues differentiating basipetally from the isolated apex usually established connection with the established vascular system of the shoot. Thus, when the apical meristem was at least partially isolated from other parts of the shoot, procambium development appeared to be initiated from above rather than from below.

Even more dramatically, it has been found that completely excised shoot apical meristems, lacking both leaf primordia and vascular tissue, can be grown in sterile culture, giving rise in many cases to complete new plants with normal vascular systems. This was first demonstrated for isolated shoot apical meristems of various vascular cryptogams maintained on an agar medium containing mineral salts and sugar (Wetmore, 1954). More recently similar excised angiosperm shoot apical meristems have also been successfully grown in sterile culture (Ball, 1960; Smith and Murashige, 1970), although not without the provision of a more complex medium containing not only mineral salts and sugar but also thiamin hydrochloride, myo-inositol and a growth hormone such as indole-3-acetic acid or gibberellic acid. Nevertheless, it is clear from such experiments that the shoot apical meristem has a marked degree of autonomy, in that it retains the ability to initiate normal patterns of development of vascular and other tissues after separation from the plant. Similarly, successful sterile culture of excised root apical meristems (Torrey, 1965b) also indicates that the basis for the initiation and organization of procambium lies within the apex itself.

The actual nature of the physiological stimulus to procambium initiation emanating from the apical meristem is not, however, by any means fully understood at present. A tempting proposition is that it is of a hormonal nature, as it is known that regeneration of vascular tissue in many circumstances (see below) can be initiated by auxins. There is little or no evidence that the apical meristem itself secretes auxin (Sheldrake, 1973), and excised cultured angiosperm shoot apical meristems even require an exogenous supply of auxin for their development (Smith and Murashige, 1970). Thus, it is difficult to see how apically derived auxin can be responsible for procambium initiation. On the other hand, it is well known that developing leaf primordia synthesize auxin that is transported into and through the stem tissues (Wetmore and Jacobs, 1952), which perhaps induces cells to develop into procambium. Supporting this is the observation (McArthur, 1967) that, in isolated or completely foliated shoots of Geum chiloense, procambium developed only when auxin was applied to the apical dome, in conjunction with an adequate supply of sugar to the lower cut end of the stem. It is possible then that auxin derived from leaf primorda regulates the initiation of procambium, but against this suggestion there is the evidence that auxin transport is basipetally polar (Goldsmith, 1969), and that procambium is initiated within the apex at a higher level than the youngest leaf primordia (Wardlaw, 1944; Steeves, 1963; Freeberg and Wetmore, 1967). Also that procambium initiation in roots is under the control of the root apical meristem, where there are no lateral appendages such as leaves present to supply auxin. Nevertheless, Torrey (1957) demonstrated that an exogenous auxin supply to excised pea root apices in sterile culture regulated the pattern of initiation of the procambial cylinder, and concluded that procambium formation in roots is normally regulated by endogenous auxin. The source of endogenous auxin in or near the root apical meristem is highly debatable, particularly in view of reports that auxin transport in roots is preferentially acropetal (see Davies and Mitchell, 1972).

B. The Transformation of Procambium to Vascular Cambium

The physiology of conversion of procambium to fascicular cambium has not received study. Somewhat more attention has, however, been paid to the secondary development of interfascicular cambium and the continuing activity of the entire cambial cylinder.

The vascular cambium, both fascicular and interfascicular, always differentiates at a fairly definite distance in from the outer surface of the organ. This is true in both normal development and regenerative development after wounding. The most spectacular demonstrations of the constancy of distance from the outer surface at which cambium differentiates have come from wounding and grafting experiments. For example, Snow (1942) split sunflower hypocotyls into two longitudinal halves. Each half first regenerated a parenchymatous callus from the cut surface, and a new vascular cambium formed within this at a similar distance from the new callus surface to that between the original hypocotyl epidermis and the previously existing cambium (Fig. 6). A number of experiments have indicated that an important factor in determining the initiation of cambium is mechanical pressure. Thus, although cambial cells can be kept dividing relatively easily in sterile culture, it is only when they are subject to a degree of physical pressure (which they must be *in vivo*, as they lie between the expanding central stem core and the restraining influence of the outer tissues) that the planes of cell division are organized and regular, as normally occurs in the cambial zone (Brown and Sax, 1962; Brown, 1964; Steeves and Sussex, 1972).

Cambium regeneration experiments such as those illustrated in Fig. 6 also indicate that the stimulus for the formation of interfascicular cambium from cortical parenchymatous cells originates in the fascicular cambium: an example of homeogenetic induction in differentiation (see Lang, 1965). But, homeogenetic induction of interfascicular cambium by the fascicular cambium does not occur in the absence of an auxin supply from the more apical regions of the shoot, particularly the young growing leaves. This requirement for auxin was first demonstrated for *Helianthus annuus* by Snow (1935), who showed that secondary vascular development did not occur in a decapitated seedling stem

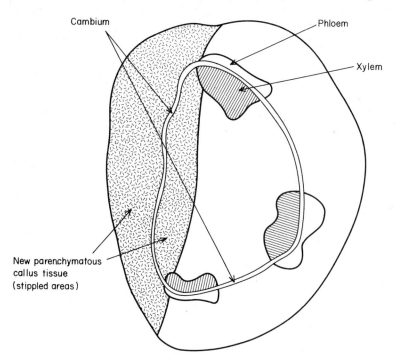

Fɪɢ. 6. Regeneration of cambium through the wound callus (stippled area) formed from the cut surface of a longitudinally split sunflower hypocotyl. Note that the new cambium differentiates at approximately the same distance from the callus outer surface as lies the cambium of the original half hypocotyl. (Re-drawn from R. Snow, 1942.)

unless the auxin indole-3-acetic acid was supplied to the cut stem surface. Many subsequent, similar studies on a number of both herbaceous and woody species have amply confirmed Snow's results and conclusion. In addition, experiments on cambium and vascular tissue initiation in a variety of regenerative systems, such as in isolated portions of stem (Clutter, 1960; Dalessandro, 1973) or root (Torrey, 1957), in and around wounds (Jacobs, 1952), or in callus cultures (Wetmore and Sorokin, 1955; Wetmore and Rier, 1963), have all indicated the

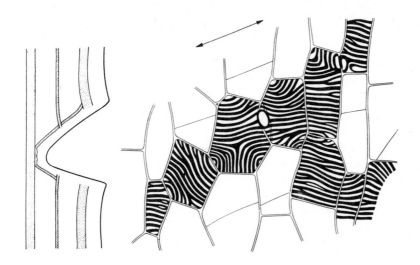

Fig. 7. Development of a new vascular connection in *Coleus* stem following interruption by wounding of the original vascular system. Left: Diagram showing the regeneration of connection between several vascular strands. Right: A detailed view of the upper point of junction illustrating that individual parenchyma cells have differentiated into reticulate xylem elements. The arrow shows the direction of development of the new vascular strand. (From Sinnott and Block, 1945.)

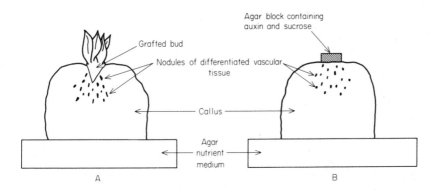

Fig. 8. Differentiation of vascular tissues in sterile cultures of callus tissues. A: Vascular tissues induced to differentiate in a callus by the presence of a grafted bud of the same species. B: Induction of vascular tissue in a similar callus by an agar block containing auxin and sucrose. (A adapted from Wetmore and Sorokin, 1955; B adapted from Wetmore and Rier, 1963.)

necessity of an auxin supply (Figs 7 and 8), but other stimuli such as cytokinin (Dalessandro, 1973) and/or sugar (Wetmore and Rier, 1963) may also be required in addition to auxin.

C. Differentiation of Cambial Derivatives

The numerous light and electron-microscopic, and biochemical studies which have been performed on the nucleus, cytoplasm, organelles and cell walls of differentiating vascular tissues have yielded valuable information, but go beyond the scope of this account of the cambium itself. A voluminous literature has been built up to which it is impossible to do justice here. In essence, however, it has been found that differentiation of xylem and phloem elements, and of fibres and xylem- or phloem-parenchyma, involves complex changes in cell architecture and biochemistry. Particular attention has been devoted to features of cell wall synthesis, and it has been repeatedly suggested that the deposition of cell wall material in an ordered specific manner involves the participation of the golgi apparatus, the endoplasmic reticulum, and microtubules (see Kozlowski, 1971; Northcote, 1972).

IV. SEASONAL AND OTHER FACTORS IN THE REGULATION OF CAMBIAL FORM AND ACTIVITY

In any woody perennial species, the structure and activity of the cambium are neither spatially nor temporally uniform. Very great differences occur in both the structure and the rate of cell division of the cambium. The causes of these variations are not by any means always understood, but we need to relate them to environmental or genetic parameters as far as we are able, in order to begin to understand their physiological bases.

Some of the factors which may regulate cambial activity can be recognized as being external to the plant in origin (e.g. light, water and nutrient supply, gravity, temperature), whereas others appear to be internal factors (e.g. growth correlations within the plant, and an inherited tendency for periodicity in cambial activity). In this section we will consider patterns of cambial activity and as far as possible relate them to external and internal regulating factors, to provide a basis for an evaluation in the final section of what is known concerning the physiological mechanisms governing the meristematic activity of the cambium, and the differentiation of its derivatives.

A. Periodicity of Cambial Activity

It is well known that temperate zone trees do not form wood (secondary xylem) continuously throughout the year, and that because of this there is usually a clear indication of the line of demarcation between each season's increment in wood production—the so-called annual, or growth, rings. Each annual ring represents one season's production of secondary xylem from the cambium. The actual seasonal pattern of secondary xylem formation differs from species to species. Xylem elements formed early in the growing season usually have thinner walls than those formed later on. In gymnosperms and diffuse-porous angiosperm species (e.g. *Acer, Betula, Populus*), the final diameter achieved by xylem elements remains more or less constant throughout the growing season, but in the ring-porous angiosperm trees (e.g. *Quercus, Fraxinus, Ulmus*) the xylem elements formed early in the season not only have thinner walls than those produced later on, but they are also of much greater overall diameter (Fig. 9). Not uncommonly in temperate zone trees there is more than one growth period per year, and this leads to the appearance of "false rings". However, formation of false rings is usually a response to unfavourable environmental conditions during the normal growing season, such as temporary drought or low temperatures (Kozlowski, 1971), whereas the true annual rings reflect an endogenous periodicity in the meristematic activity of the cambium. Thus, in temperate zone trees showing winter dormancy, the cambium itself is in a dormant condition during winter, and will not resume activity until it has emerged from this state by the following spring. At first sight it may be felt that the coincidence of cambial inactivity and low winter temperatures indicates a direct environmental suppression of meristematic activity. But, as has already been indicated, the regulation of the annual periodicity in cambial activity is primarily endogenous. Thus, with most temperate woody species, even if the plants are transferred to optimum growing conditions in late autumn, the cambium remains dormant. Another indication that periodicity in cambial activity is largely regulated by internal factors has come from studies of various tropical trees. In the tropics, even where conditions are suitable for growth all through the year, it is found that many, but not all, indigenous trees exhibit discontinuous cambial activity with a consequent formation of growth rings (these are not necessarily annual, and are usually relatively indistinct compared with the annual rings of temperate zone trees), thus apparently indicating the presence of an inherited rhythm of cambial activity (Koriba, 1958; Chowdhury, 1961; Langdon, 1963; Alvim, 1964). Certain tropical trees, for example

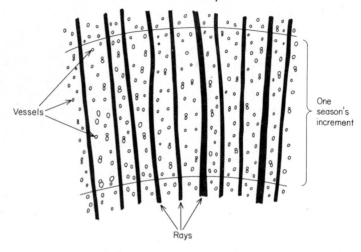

Vessels

One
season's
increment

Rays

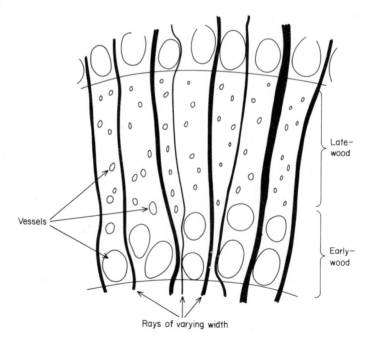

Late-
wood

Vessels

Early-
wood

Rays of varying width

FIG. 9. Diagrammatic representation of transverse sections of stems of a diffuse-porous woody species (above) and ring-porous species (below).

mango, also demonstrate that endogenous periodicity of growth, including cambial activity, in different branches may be asynchronous; in other words the cambium of some branches can be active at the same time that the cambium in other branches is dormant (Alvim, 1964).

However, there is also much evidence that annual periodicity in cambial activity is at least partially regulated by influences of the external environment. For example, in many species of temperate zone trees it has been found that daylength conditions have a marked influence on the duration of cambial activity. In general, long days (approximately 15 h or more continuous illumination each 24 h) favour the continuation of meristematic activity of the cambium, whereas short days (less than approximately 10 h continuous light per 24 h) favour the cessation of cambial activity (Wareing, 1951; Wareing and Roberts, 1956). These daylength effects appear to be true photoperiodic responses, in that only low supplementary illumination is required to extend a short day of high intensity light to elicit the typical long-day response. Thus, there seems little doubt that the progressive decrease in natural daylength which occurs between mid-summer and autumn in temperate zones, plays a significant part in the onset of cambial dormancy. Even in the tropics, Alvim (1964) noted that growth periodicity in trees was more marked in regions where seasonal fluctuations occurred in temperature, rainfall and daylength, and concluded that environmental factors probably play an important part in determining periodicity in cambial activity in tropical, as well as temperate zone, trees.

B. Cambium Structure in Relation to Season

Earlier in this account it was mentioned that the anatomy of the cambial zone varies during the year. The dormant cambium consists of a relatively narrow cylinder of radially compressed cells, compared with the wider more diffuse zone of meristematic cells visible in actively dividing cambia, particularly in angiosperm trees (Ladefoged, 1952; Bannan, 1962). In the spring, the first morphologically visible event is usually, but not always, the swelling of cambial cells, which expand radially and acquire a lighter appearance. This swelling of the cambium in spring commonly occurs rapidly and simultaneously in all parts of the tree, and is probably initiated by the increase in temperature at this time of the year (Priestley, 1930; Ladefoged, 1952; Wareing, 1958a). The individual cambium cell walls also become more plastic, and this is the cause of the well known "peeling" or "slip" which can be induced between bark and wood in woody stems with an active cambium, i.e. the "bark" easily separates from the stem at the junction between cambium and xylem.

The fine structure of cambium in relation to the dormant and active states has received some attention. Dormant cambium cells contain vesiculate smooth endoplasmic reticulum, free ribosomes, oil droplets and protein bodies which presumably serve as storage materials, and only few and small vacuoles. Active cambial cells, on the other hand, are highly vacuolate, and contain active vesicle-producing golgi apparatus, rough endoplasmic reticulum and polyribosomes (Srivastava, 1966; Srivastava and O'Brien, 1966; Robards and Kidwai, 1969).

The onset of cell division activity in the radial files of cells in the cambial zone varies. The first divisions may occur in the cells adjacent to the xylem (Bannan, 1955), or they may be evenly distributed across the zone (Evert, 1963; Derr and Evert, 1967). Within the whole tree, the cambium normally commences dividing in twigs, immediately below buds which are starting to swell in early spring, and the onset of cambial cell division activity spreads basipetally from these regions down the branches and trunk (Priestley and Scott, 1936; Wareing, 1958a; Wilcox, 1962), and onwards in an acropetal manner along the root system (Fayle, 1968). The velocity of basipetal onset of cambial activity from behind the expanding buds varies greatly from species to species; it has, for example, been reported to be much more rapid in ring-porous than in diffuse-porous species (Priestley and Scott, 1936; Wareing, 1958a).

C. Duration of Cambial Activity and Correlation with Growth in Other Regions

Once started, periclinal cell divisions normally continue to take place in the vascular cambium for a period of months. In those species showing a distinct periodicity in cambial activity, it is generally found that the period of active cell division in the cambium is shortest in higher latitudes, and lengthens towards the equator. The range of duration of cambial activity is from approximately 4 weeks to 6 months. It is usually the case that conifers show a more extended period of cambial activity than do dicotyledonous trees (Studhalter *et al.*, 1963; Philipson *et al.*, 1971). A common feature is that a rapid rise in cell division activity occurs early in the season, which is followed by a levelling off and eventually a gradual decline (Fig. 10).

Over the whole growing season, it is often the case that rather more xylem than phloem is formed from the vascular cambium. This is partially due to differing rates of periclinal divisions cutting off xylem and phloem initials to the inner and outer faces of the cambium, but perhaps more importantly to the fact that xylem mother cells undergo several successive divisions whereas each phloem mother cell normally redivides only once before differentiating (Philipson *et al.*, 1971). Nevertheless, the net result is that more xylem than phloem is laid down during

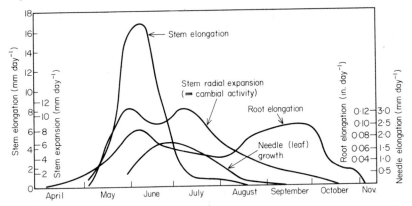

FIG. 10. Seasonal growth activities in different regions of pine trees situated outdoors in the U.S.A. (Adapted from Keinholz, 1934.)

the year. The ratio of xylem to phloem production is not, however, constant; for example, Bannan (1955) found that in *Thuja occidentalis* the period of maximum xylem formation occurred earlier than the peak of phloem production. A similar pattern was found in other species by various workers (see Philipson *et al.*, 1971), but in some species different relationships have been observed, sometimes with maximum phloem production preceding that of xylem (Evert, 1963; Catesson, 1964; Derr and Evert, 1967; Tucker and Evert, 1969). There are some indications that there may be correlations between seasonal patterns of xylem and phloem formation, and wood porosity (Evert, 1963; Derr and Evert, 1967; Tucker and Evert, 1969).

Periodicity in growth activity of the cambium is not only correlated with seasonal environmental factors, but also with other growth activities in the plant, such as those concerned with shoot and root elongation, and leaf initiation and development. This is not surprising, since growth correlations are basic features of integrated plant development (Audus, 1959; Sinnott, 1960; Steward, 1968; Phillips, 1969). As we have already considered, there is a positive correlation between the emergence of shoot buds from dormancy in the spring and onset of cambial cell division activity. The relationship between rate of radial growth, which is the consequence of cambial activity and differentiation of its derivatives, and rate of shoot extension growth appears to vary from species to species and season to season. However, there is usually, but not always (see Studhalter *et al.*, 1963), a positive correlation between total extension growth and radial expansion in each season, and that this correspondence is attributable to environmental influences upon availability of photosynthate, water and nutrients (Philipson *et al.*, 1971). In many trees, the cambium continues cell division activity for longer into the summer or autumn than does shoot extension growth (Fig. 10) (Kozlowski, 1971; Philipson *et al.*, 1971). The earlier cessation of stem elongation may often be attributable to the episodic nature of shoot extension, where the terminal winter resting bud contains all, or most, of the unexpanded internodes and leaf primordia which are capable of growth in the following spring and summer. Thus, once these have developed, further shoot extension can only occur following the initiation of new internodes and leaf primordia (Romberger, 1963).

The relationships between timing of maximum cambial activity and maximum root elongation are even more obscure. Any attempt to relate shoot or root cambial activity with root extension is bedevilled by the conflicting and confusing nature of evidence concerning annual cycles of root growth. Even in temperate zone trees which exhibit a regular cycle of dormancy and growth in their shoots, there is no conclusive evidence as to whether their roots undergo a period of true winter dormancy similar to shoot buds, or whether roots only cease growth when soil conditions are unfavourable (e.g. low temperature, or drought) but retain the capacity for growth at all times of the year (see Romberger, 1963; Kozlowski, 1971). Fairly recent experiments indicate that *individual* roots of the whole root system of a tree may undergo alternating cycles of dormancy and growth activity, but that at any one time all through the year, some roots are actively extending (Merritt, 1968; Wilcox, 1968).

D. Environmental Effects on Cambial Activity

Meristematic activity and cellular differentiation in the cambial region is but one expression of the total growth and differentiation activities of the whole plant. All these are dependent upon adequate supplies of warmth, photosynthate, water and inorganic nutrients. It is no surprise, therefore, to find that the rate of cambial activity is susceptible to environmental factors that influence the availability of nutritional materials to the cambium. It should, however, be noted that the role of climate in the synthesis and storage of food materials in the growing season can be carried over into the following season and be reflected in cambial and other growth activities. The external environment does, however, influence cambial activity by affecting internal processes other than, and additional to, purely nutritional ones. Further, all environmental factors interact in their effects on cambial activity, as on most physiological events in plants.

1. Light and Carbon Dioxide

The influences which light has upon cambial activity are complex. Light intensity can markedly affect the rate of photosynthesis, and thereby determine availability of organic nutrients to the cambium with consequent growth responses occurring there. Light quality can be of importance by influencing phytochrome-mediated responses such as leaf and stem expansion, dormancy and flowering. A further aspect of the effect of light on cambial activity is seen in photoperiodic responses of the cambium. The vascular cambium of many tree species, particularly those of the higher temperate latitudes, tends to go on dividing for much longer under long days than under short daylength conditions. The positive response of the cambium to long days occurs both in species in which cambial activity is closely related to shoot extension, and in those in which cambial activity normally persists after shoot extension has ended (Wareing, 1951, 1958a; Wareing and Roberts, 1956; Romberger, 1963; Kozlowski, 1971; Philipson et al., 1971). Photoperiod often influences the density of wood formed, in addition to its quantity (Larson, 1962a; Waisel and Fahn, 1965).

The availability of carbon dioxide to the leaves also can influence cambial activity, primarily due to the requirement for CO_2 in photosynthesis. Thus, any factor, such as light, wind, or water availability, which affects stomatal opening and closing can in turn regulate cambial activity by influencing the rate of photosynthesis. For similar reasons, defoliation by wind, insects, disease or chemicals suppresses both availability of photosynthate and cambial activity.

2. Water and Soil Nutrition

As essential factors for the maintenance of all metabolic processes, it is clear that water and inorganic nutrients are required for continued cell division in the cambium and development of its derivative cells.

However, much research has amply demonstrated the very great sensitivity of cambial activity to available water. As already documented by Kozlowski (1971), soil water deficits in forests are frequently prolonged, particularly at those times of the year when other environmental factors, especially light and temperature, are at optimal levels.

As a consequence close correlations between cambial activity and internal water status of the tree have been recorded (Shepherd, 1964; Alvim, 1964). It has been suggested that internal water potential plays a greater regulatory role in cambial activity than do endogenous hormones (Whitmore and Zahner, 1967; Doley and Leyton, 1968; Zahner, 1968). All cell expansion and synthetic processes require water, and it would be surprising if such a delicate and active meristematic tissue as the cambium did not exhibit a rapid response to water deficits. Quite clearly, all the hormonal and nutritional stimuli in the world are not in themselves sufficient to allow cambial activity should water be lacking.

There is, nevertheless, evidence that water stress may rapidly influence cambial activity by affecting the hormonal status of the plant. Cell division activity in sugar beet leaves (Terry *et al.*, 1971) and in radish (Kirkham *et al.*, 1972) can be inhibited by an extremely slight water deficit—much smaller than would significantly alter the photosynthetic rate, for example. Marked changes in hormonal levels in response to water stress have been recorded in a range of plant species, and include decreases in endogenous cytokinins (Itai and Vaadia, 1965, 1971; Itai *et al.*, 1968) and increases in abscisic acid (Wright, 1969; Wright and Hiron, 1969; Zeevaart, 1971; Most, 1971; Simpson and Saunders, 1972; Hoad, 1973). As discussed later (see Section VA below), cytokinins are known to have stimulatory effects on cambial activity whereas abscisic acid can be inhibitory. Thus, it is certainly possible that the relatively rapid reduction in cambial activity which occurs in response to water deficits may be a consequence of depressed cytokinin, and elevated abscisic acid, levels.

Mineral nutrient availability to the roots is naturally of significance in cambial activity. The cambium does not respond so rapidly to changes in soil nutrient levels as it does to soil water status, and this is probably attributable to the relatively greater stability of internal nutrient levels compared with internal water potential, and also the possibility of mobilization of stored reserves from other parts of the plant.

3. Temperature

Like all other metabolically dependent processes, developmental events in the cambial zone are affected by temperature. The effect of a change in temperature can be directly upon the cambial cells, or indirectly through influences upon photosynthesis, transpiration, leaf development, root growth, etc.

In temperate regions of the world, air temperature is the single most important environmental parameter determining the time of renewed cambial activity in the spring (Priestley, 1930; Fraser, 1956; Wareing, 1958a; Kozlowski *et al.*, 1962). Thus, not until the temperature has risen sufficiently does cambial activity commence, and this is attributable to effects of temperature on bud and root growth, and water uptake. Similarly, during the growing season cambial activity and temperature can show a positive correlation, provided other factors such as water are not limiting (Ladefoged, 1952; Glock, 1955; Fritts, 1958; Wareing, 1958a). Should late frosts occur in the spring after radial growth has commenced, then "frost rings" of various types may be formed by death or injury of newly formed xylem initials or of some of the cambial cells themselves (Kozlowski, 1971).

Temperature is also of very great importance in many temperate zone trees, in being the environmental factor most concerned in the emergence of both buds and cambium from winter dormancy. Thus, many tree species require a period of chilling, ranging from a few weeks to several months, at temperatures between 2°C and 5°C for the breaking of both bud and cambial dormancy (Larson, 1962; Romberger, 1966; Wareing, 1969).

4. Gravity

Whereas in upright stems of woody species the rate of cambial activity is approximately equal at all radii, in branches or trunks inclined from the vertical cambial activity is consistently greater in either the upper or lower sides. Differential cambial activity in a non-vertical stem leads to the appearance of irregular growth rings, each of which is wider in one half of the stem than the other. Further, the structure of the xylem elements differs between upper and lower sides. Together, the different rates of cell division activity in the cambium, and the patterns of differentiation of its derivatives, lead to the formation of "reaction wood". In angiospermous trees reaction wood is formed along the upper side of an inclined branch, but in conifers it is laid down along the lower side. For this reason, reaction wood in angiosperms is referred to as "tension wood" and in conifers as "compression wood".

In both conifers and angiosperms, therefore, there is an eccentricity in cambial activity which leads to reaction wood formation. The rate of cell division in the cambial layer is accelerated on one side, and correspondingly slowed on the opposite side. In both angiosperms and conifers, reaction wood formation involves an increase in the number of anticlinal divisions in the cambium initial cells. There is, in angiosperms, an increase also in periclinal divisions of cambial derivatives (Wardrop and Dadswell, 1952; Scurfield, 1973).

Anatomically distinctive reaction wood is usually formed on the side where growth is increased (Robards, 1965). There are, however, numerous exceptions to this; in some species typical reaction wood is formed with no asymmetry of the growth rings, and in many others it is found that in certain regions, such as the basal end of a branch, typical reaction wood is formed on both upper and lower sides, or even only in the side of least cambial activity (Wardrop, 1964; White, 1962; Robards, 1965; Philipson et al., 1971).

Numerous detailed accounts of reaction wood anatomy have been published (Berlyn, 1961; Wardrop, 1964, 1965; White, 1965; Westing, 1965a; Côté and Day, 1965; Kozlowski, 1971; Scurfield, 1973). In dicotyledonous trees, tension wood contains wood fibres which are much more numerous and shorter than those in normal wood. The walls of tension-wood fibres are modified by the presence of an unlignified, or only slightly lignified, gelatinous layer in which the cellulose microfibrils are orientated longitudinally in a parallel manner. Xylem vessels and ray parenchyma in tension wood are fewer in number than in normal wood, and are also frequently very compressed by the presence of the numerous gelatinous fibres. Reaction phloem tissues also occur in the side of greater cambial activity, but these are generally less markedly modified anatomically than reaction wood elements (Scurfield and Wardrop, 1962; Kozlowski, 1971).

Compression-wood anatomy in conifers also differs from that of normal conifer wood. There are fewer resin canals than normal, and tracheid structure is very modified. Thus, compression-wood tracheids tend to be shorter than usual and have distorted tips, to be more nearly circular in transverse section and possess abnormally thick walls. Biochemically, the walls of conifer reaction wood tracheids contain more lignin and β-D-(1,4)-linked galactan, and less cellulose than in normal tracheids. In addition, the cellulose microfibrils are differently orientated and may be less crystalline in compression-wood tracheids (see Scurfield, 1973).

One morphological effect of reaction wood formation is a tendency for restoration of the branch or trunk to the vertical plane. In angiospermous trees tension-wood formation appears to be associated with stem contraction on the upper side, and in conifers compression-wood formation is related to increased longitudinal expansion on the lower side. The mechanical and physiological bases of these growth movements are not fully understood, but have been carefully considered most recently by Scurfield (1973), who suggests that the force required to bring about recovery of a branch to the vertical, or more nearly vertical, position is generated during the swelling of cell walls by deposition of lignin in the region of differentiating reaction wood cells.

V. THE PHYSIOLOGY OF CAMBIAL ACTIVITY

As we have considered, cambial activity is affected by a wide range of factors, some arising outside and some inside the plant. Together, they comprise a complex of interacting influences, which makes for great difficulty in evaluating the effects of any one factor.

Within the limits determined by genetic constitution, cambial activity is regulated by availability of water, mineral elements, photosynthate, vitamins and hormones. Any external or internal factor which influences the availability of one or more of these essential supplies will affect cambial activity.

In this final section a brief consideration is made of the physiology of a number of aspects of the control of cambial activity, particularly those which appear to involve internal regulatory mechanisms different from, or supplementing, those concerned with availability of gross nutritional factors and water.

A. Hormonal Regulation of Cambial Activity

The origin of the current concept of hormonal regulation of the meristematic activity of the cambium, and of differentiation of its derivative cells, can be traced back to observations by Priestley (1930) that the onset of cambial activity in temperate-zone dicotyledonous trees during spring-time commences at the base of expanding buds, and spreads downwards through the twigs, branches, trunk and root system (see also Fayle, 1968). The discovery of the plant growth hormone auxin in the late 1920s, its subsequent identification as indole-3-acetic acid (IAA), and observation of its polar basipetal transport (Wareing and Phillips, 1970; Phillips, 1971), all lead to conjecture that renewed cambial activity in the spring may represent a response to auxin transported from the swelling buds. The first demonstration of the regulatory properties of auxin on cambial activity came in work by Snow (1935) on sunflower, where it was found that the cambium of the stem was dependent upon a supply of auxin from young growing leaves.

Since Snow's work on sunflower, well documented evidence has accumulated to indicate that cambial activity and vascular tissue differentiation in both herbaceous and woody species are subject to marked influence by auxin and other more recently discovered growth

Mean xylem width e.p.u.

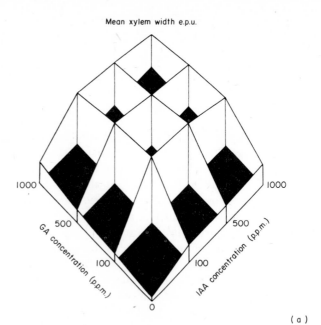

(a)

Mean phloem width e.p.u.

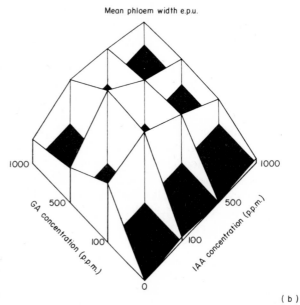

(b)

FIG. 11. Effects of various concentrations of gibberellic acid (GA) and indole-3-acetic acid (IAA) on (a) xylem and (b) phloem development in excised vertical disbudded twigs of *Populus robusta*. The concentrations indicated refer to hormone concentration in the applied lanolin preparations. (From Wareing *et al.*, 1964.)

hormones, such as cytokinin and gibberellin (Wareing, 1958b; Wareing et al., 1964) (Fig. 11).

Growing buds in spring have been found to transmit relatively large quantities of auxin to the stem tissues (Avery et al., 1937; Söding, 1937). Auxin, identified as IAA, has been detected in the cambial zone of Pinus radiata, the concentration there of the hormone showing a positive correlation with the rate of cambial cell division activity (Shepherd and Rowan, 1967). Similarly, auxin levels in both ring-porous and diffuse-porous dicotyledonous trees are positively correlated with seasonal changes in cambial activity (Larson, 1962b; Digby and Wareing, 1966b). The principal sources of auxin required for maintenance of cambial activity are young growing leaves, although in ring-porous species mature leaves also supply auxin to the cambium (Wareing and Roberts, 1956; Digby and Wareing, 1966b). There is also evidence that auxin may be released to the cambium from a stored form in conifer stem tissues in spring (Wort, 1962; Philipson et al., 1971).

Other known plant growth hormones, as well as auxin, also appear to play parts in the mechanism of regulation of cambial activity. Application of gibberellic acid to woody shoots can enhance the rate of initial cell production by the cambium (Bradley and Crane, 1957; Wareing, 1958b; Wareing, Hanney and Digby, 1964; De Maggio, 1966; Waisel et al., 1966; Digby and Wareing, 1966a) (Fig. 11). Similarly, cytokinin applications can result in increased cambial activity in shoot or root (Loomis and Torrey, 1964; Sorokin et al., 1962; Pieniazek, 1964; Street, 1966a; Radin and Loomis, 1971; Webster and Radin, 1972; Peterson, 1973). Endogenous levels of gibberellins and cytokinins, like those of auxins, are positively correlated with cambial activity (Digby and Wareing, 1966b; Radin and Loomis, 1971).

There is, therefore, ample evidence that the regulation of cambial activity involves the participation of endogenous growth-stimulating hormones. A certain amount of information is available which suggests that abscisic acid (ABA), a growth-inhibitory hormone, may also play a part in control of cambial activity. Thus, applications of ABA to Phaseolus vulgaris plants inhibited cambial activity, and also counteracted the cambial stimulating effects of both IAA and GA_3 (Hess and Sachs, 1972). Growth inhibitors, one of which is likely to be ABA, have been detected in cambial cells (Dörffling, 1963), and levels of growth inhibitors in the cambial region have been shown to be lowest when cambial activity is high (Wodzicki, 1964, 1965; Waisel and Fahn, 1965). On the other hand, there is clearly some confusion and doubt concerning possible involvement of ABA in cambial activity regulation, for whereas Hess and Sachs (1972) reported that ABA inhibited cambial activity and antagonized the stimulatory effects of auxin and gibberellin, other workers have noted that simultaneous ABA and IAA treatment can result in even greater cambial activity than when IAA alone is supplied in *Helianthus annuus* and *Acer pseudoplatanus* (Dörffling, 1964), and in *Larix decidua* (Wodzicki, 1965).

Overall, it is reasonable to regard hormonal control of cambial cell division activity as being similar to hormonal regulation of other developmental processes in plants, in that all classes of growth hormone interact in their effects, which are also subject to other factors such as water and nutrient availability to the meristematic cells.

The developmental fates of cambial derivative cells are also susceptible to regulatory influences of growth hormones originating in other regions of the plant. Wareing (1958b) and Digby and Wareing (1966a) demonstrated an interaction between IAA and GA_3 in regulating the relative proportions of xylem and phloem initials formed by the cambium in *Acer pseudoplatanus, Populus nigra, P. robusta, Fraxinus excelsior,* and *Vitis vinifera.* Gibberellic acid favoured phloem development and IAA xylem development, though maximum production of both xylem and phloem occurred when both IAA and GA_3 were supplied to disbudded stems. An interaction between IAA and GA_3 on differentiation of wound vessel members was found by Roberts and Fosket (1966), and Wodzicki (1965) obtained evidence that auxin-like growth promoters and endogenous growth inhibitors interacted to regulate xylem differentiation in *Larix decidua.* Hess and Sachs (1972) found that IAA, GA_3 and ABA interacted in their effects on xylem differentiation in stems of *Phaseolus vulgaris.* In a series of experiments on *Salix fragilis* (Robards *et al.*, 1969), complex interactions, in terms of patterns of xylem differentiation, were found between applied auxins, gibberellin and cytokinin, and also between these growth hormones, myo-inositol and sucrose.

All these observations suggest that normal patterns of cambial cell division activity, and of patterns of vascular tissue differentiation in stems, arise from the controlling influences of interacting growth hormones and nutritional factors. A similar situation probably pertains in the root system, for there have been numerous reports of changes in cambial activity and vascular tissue differentiation in roots induced by exogenous growth hormones, particularly auxins and cytokinins (Street, 1966; Loomis and Torrey, 1964; Peterson, 1973), and also of changes in endogenous auxin and cytokinin levels in roots correlated with degree of cambial activity (Street, 1966a; Radin and Loomis, 1971).

It is not difficult to visualize how, in general terms, correlative control of cambial activity occurs through the mediation of endogenous growth hormones. Young expanding leaves are known to synthesize both auxins (Wetmore and Jacobs, 1953) and gibberellins (Jones and Phillips, 1966), which are translocated down the stem. Older leaves synthesize ABA which can be transported into the stem tissues (Wareing and Saunders, 1971; Hoad, 1973). The relationship between bud growth and onset of cambial activity can therefore be understood in terms of regulation by auxins, gibberellins and ABA derived from young and mature leaves. The transition from early to late wood formation may therefore be a consequence of a change in internal hormonal levels and balance (Digby and Wareing, 1966b; Wodzicki, 1965; Balatinecz and Kennedy, 1968) determined by the relative proportions of young and mature leaves present and/or environmental conditions. There is considerable evidence that root tips are major sites of synthesis of cytokinins, from where they are translocated to the shoot system (Kende, 1965; Carr and Burrows, 1966) by an apparently polar transport mechanism (Radin and Loomis, 1971). Thus any factor which affects cytokinin synthesis or transport in the root system would be expected to influence cambial activity in both root and shoot.

B. Responses to Gravity

Reaction wood formation, and resultant bending of branches, is usually a response to a gravitational stimulus. Even very small deflections of stems from the vertical can induce reaction wood formation, though a larger response occurs at greater inclinations from upright; in *Salix fragilis* an inclination of 120° from vertical gave maximum reaction wood production (Robards, 1966). Centrifugal force can replace gravity in the induction of reaction wood (Jaccard, 1939).

By analogy with what is known of the perception mechanism in geotropic phenomena in plants (Audus, 1969), it was suggested by Westing (1965b) that a statolith system was involved in perception of the gravitational stimulus which leads to reaction wood formation. Iodine-stained transverse sections of horizontal stems of *Populus robusta* revealed that starch grains were clustered on the lowermost wall of certain cortical cells (Leach and Wareing, 1967). Thus, the classical starch-grain statolithic hypothesis of gravity perception in plants (see Audus, 1969) may also apply to the reaction wood responses of woody stems.

In geotropism, it has been well established that a greater concentration of auxin occurs in the lower than the upper tissues of a horizontally positioned coleoptile, stem or root (Audus, 1969). Very few studies appear to have been made of endogenous hormone levels in relation to reaction wood production, but Leach and Wareing (1967) found that endogenous auxin was present in greater concentration on the lower than on the upper side of horizontal *Populus robusta* stems. The same workers studied the distribution of ^{14}C along and across horizontal popular stem segments, after having applied [2-^{14}C]-IAA uniformly to the apical end, and found an increasing differential in ^{14}C level between lower and upper side with increasing distance from the apical end. At a distance of 4 cm from the source of [2-^{14}C]-IAA, the lower : upper ratio of ^{14}C was 60 : 40, and by 6 cm had risen to 68 : 32. Thus, horizontal orientation would appear to result in the deflection of basipetally travelling auxin towards the lower side. In view of the known cambial activity-enhancing effects of auxins, it is difficult to reconcile the apparent contradiction between greater auxin concentration in the lower tissues with reaction wood formation at the same time in the upper half of a horizontal dicotyledonous tree stem such as that of poplar. Earlier experiments by Wareing and his associates (Wareing *et al.*, 1964) demonstrated that apical application of IAA, or a mixture of IAA and GA$_3$, to horizontal *Populus robusta* stem segments resulted in marked enhancement of cambial activity along the upper side (Fig.

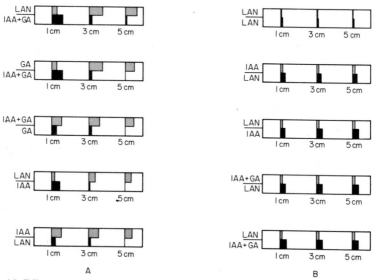

FIG. 12. Effects of applications of plain lanolin (LAN), indole-3-acetic acid (IAA) and/ or gibberellic acid (GA) to the apical ends of horizontal shoot segments. The stippled and black areas represent the mean width of new xylem tissues on the upper and lower sides, respectively, at various distances from the point of application. A: Horizontal shoots of *Populus robusta*. B: Horizontal shoots of *Pseudotsuga taxifolia*. (From Wareing *et al.*, 1964.)

12A), and that similar treatment of *Pseudotsuga taxifolia* induced considerable cambial activity in the lower half of segments (Fig. 12B). To date, therefore, there is no really convincing explanation of reaction wood formation in terms of hormone re-distribution in response to gravity. However, Leach and Wareing (1967) in their experiments on *Populus robusta* observed higher concentrations of endogenous growth inhibitors in the lower than in the upper tissues of horizontal stems, and suggested that the imbalance in cambial activity could be explained by suppression of cell division activity by the inhibitors on the lower side. Alternatively, Westing (1965) has proposed that cambium in the upper and lower stem halves could be differentially sensitive to auxin. Clearly, more work needs to be done on endogenous auxin and other endogenous growth hormones in relation to reaction wood induction by gravity.

It is clear that, as discussed earlier, reaction wood formation leads to a tendency for the branch or trunk to move towards a vertical condition (Scurfield, 1973). Only rarely, however, does a lateral branch move to such an extent as to become completely upright. On the contrary, it is a commonplace observation that branches of woody species are held at characteristic angles from the vertical; that is, they display plagiogeotropic behaviour. Many studies have shown that plagiogeotropism in lateral organs is an expression of correlative influences within the plant body (Phillips, 1969). Lateral branch orientation in trees is largely determined by rates of reaction wood production, and in turn this is subject to a controlling influence arising in regions of the shoot higher than the point of insertion of the branch to the main stem. Thus, excision of the leading shoot usually results in an upward movement of branches, until one of these, commonly the physically highest, assumes the role of leader by growing completely upright; thereafter angles of orientation of the other branches are regulated by the new leading shoot. Control by the main shoot of lateral branch orientation in woody plants is one expression of apical dominance (Phillips, 1969). As in other manifestations of apical dominance, synthesis of auxin in young leaves on the main shoot, and basipetal auxin transport down the stem, play major roles. Thus, the upward movement of branches in decapitated trees can be prevented or reduced by an application of auxin to the cut surface of stem above the branches, and downward movement of branches induced in intact trees by treatment with auxins (Verner, 1938, 1955; Preston and Barlow, 1950; Jankiewicz, 1956). The mechanism by which apically synthesized auxin exerts a regulatory influence over reaction wood formation in lateral branches is not understood, though when an explanation is forthcoming for correlative inhibition of axillary bud growth (Phillips, 1969, 1975), it is likely to be applicable with only minor modifications to the phenomenon of lateral branch plagiogeotropism.

C. Responses to Photoperiod

Photoperiod can affect both the rate of meristematic activity in the vascular cambium, and the course of differentiation of its derivative cells. Perception of the photoperiodic stimulus takes place in either the mature leaves or in buds, or in both (Wareing, 1951, 1954, 1958a; Wodzicki, 1960, 1961a, b). There takes place, therefore, transmission of a photoperiodically sensitive regulatory influence from leaves and buds to all regions of the cambium.

The importance of photoperiod in the natural regulation of cambial activity has been indicated experimentally in several ways. By growing young plants of *Pinus sylvestris* under controlled conditions, Wareing (1951) found that exposure to short days resulted in early cessation of cambial activity, whereas cambium in plants grown in long days continued in an active state. Wareing (1951) therefore suggested that cessation of cambial activity in *P. sylvestris* is controlled by the natural shortening of daylength in the autumn. Studies of cambial activity in nine species of trees growing outdoors in the same local area in Louisiana, U.S.A. (Eggler, 1955), revealed that radial growth in all individuals of each species stopped at the same time in the autumn, but that other species ceased radial growth at different times. The time of cessation of cambial activity was therefore determined genetically. In addition, cambial activity ceased in some of the species at a time when temperature, light intensity and water availability were clearly adequate for growth, but photoperiod was decreasing. These observations thus indicate the existence of genetically determined sensitivity to photoperiod in the regulation of cambial activity in woody species. It has been pointed out by Pauley (1958), that survival of perennial woody species in areas of the world with alternating frost-free and frost seasons depends upon the plants possessing a reliable timing mechanism, in order that growth be stopped to prevent frost injury. The regularity of seasonal changes in daylength makes this environmental parameter an excellent gauge against which plants can regulate physiological activities. Thus, many photoperiodic tree ecotypes exist, and an excellent demonstration of their adaptation to a particular annual photoperiodic regime was provided by Moshkov (1933, 1934, 1935), who found that trees transferred from southern latitudes to grow in Leningrad, Russia, where frosts occur in early autumn when daylengths are still relatively long, continued growth until damaged or killed by frost.

There is much evidence that the effects of photoperiod on the duration of cambial activity are mediated through influences on internal growth hormone synthesis, translocation or inactivation. In many species, particularly diffuse-porous dicotyledonous trees, the effect of photoperiod on shoot extension growth and cambial activity appear to be closely coupled, in that both cease at about the same time (Priestley, 1930). In many other tree species, notably conifers and ring-porous angiospermous trees, cambial activity persists for some time after shoot extension has ended. A study of endogenous auxins and gibberellins in ring-porous and diffuse-porous species by Digby and Wareing (1966b) revealed a number of interesting features in relation to the problem of photoperiodic regulation of the cessation of cambial activity. In diffuse-porous *Betula pubescens*, auxin synthesis by leaves fell very markedly after their maturation, resulting in a reduction in auxin supply to the cambium after the seasonal flush of stem extension and leaf development had ended. By contrast, in a ring-porous species, *Robinia pseudacacia*, even mature leaves synthesized considerable quantities of auxins, so that auxin continued to be supplied to the cambium after new leaf initiation and growth had ceased. Nevertheless, a fall in auxin level occurred after shoot extension ended even in *Robinia*, and Digby and Wareing (1966b) suggested that this change in level of auxin supply to the cambium caused the transition from early- to late-wood production. The synthesis of auxin by mature leaves therefore probably explains continued activity of cambium in the absence of shoot extension or leaf production in conifers and ring-porous dicotyledonous species. Other suggestions have nevertheless been made, such as the proposal that stem tissues may contain auxin in a "bound" form which is released to the cambium, or that autolysis of differentiating xylem elements results in auxin synthesis (Sheldrake and Northcote, 1968; Sheldrake, 1973), and in cytokinin synthesis by hydrolysis of tRNA (Radin and Loomis, 1971), and also perhaps gibberellin synthesis (Sheldrake, 1973).

Measurement of hormone levels after experimentally regulated photoperiodic treatments have been conducted on many species of plant in connection with a number of physiological processes in addition to cambial activity (see Black and Vlitos, 1972). There is no doubt that photoperiod does markedly influence levels of all classes of known plant growth hormone, and that these effects are usually phytochrome mediated. Thus, photoperiodic effects on not only total quantities, but also relative proportions, of endogenous growth hormones would be expected to result in the observed changes in cambial activity and patterns of vascular tissue differentiation which occur in response to changes in daylength (Digby and Wareing, 1966b; Waisel and Fahn, 1965).

D. Dormancy and Cambial Activity

Whilst there are clear indications that cessation of cambial activity in the autumn is a response to shortening daylengths, once the cambium and buds have entered true winter dormancy only relatively few species will respond to long days by renewed growth. The majority of temperate zone trees showing winter dormancy have a requirement for a period of chilling during the winter before bud growth and cambial activity can resume in the spring of the following year (Wareing, 1969). In late autumn and early winter, before the chilling requirement has been met, the cambium itself is dormant and unresponsive to treatment with exogenous auxins (Gouwentak, 1941; Reinders-Gouwentak, 1965). As with dormant buds of many species, cambial dormancy can be broken by artificial means, such as by exposure to ethylene chlorohydrin (Reinders-Gouwentak, 1949; 1965).

It therefore seems likely that the physiological and biochemical bases of bud and cambial dormancy are very similar and interrelated. The physiology of bud dormancy has been reviewed several times in recent years (see Wareing, 1969; Wareing and Saunders, 1971), making it unnecessary to consider the topic exhaustively here. The current weight of evidence favours the view that dormancy of buds and seeds is regulated through a mechanism involving the interacting influences of gibberellins, cytokinins, inhibitors such as abscisic acid and perhaps ethylene (Wareing and Saunders, 1971; Saunders et al., 1973). Cambial dormancy is likely to be similarly controlled, though very little work has been done on the physiology of dormancy in the cambium.

11. The Role of Cell Division in Angiosperm Embryology

W. A. JENSEN

*Department of Botany, University of California, Berkeley, California,
U.S.A.*

I. INTRODUCTION

Embryo development in angiosperms begins with the remarkable phenomenon of double fertilization. During the course of double fertilization one of the two sperm discharged into the embryo sac fuses with the egg. The eventual union of the egg and sperm nuclei gives rise to the zygote nucleus, the first nucleus of the new diploid generation. The second sperm unites with the central cell and its nucleus fuses with the two polar nuclei, giving rise to the triploid primary endosperm nucleus. There are many variations in the case of the endosperm formation and the number of the primary endosperm nuclei may range from two to many, but the most common number is three.

The first division of the zygote is unequal, resulting in a small, dense terminal cell and a large, vacuolate basal cell. The terminal cell through repeated divisions gives rise to the embryo proper, while the basal cell through a usually fixed number of divisions gives rise to the suspensor. The suspensor is a short-lived organ that functions, probably in a number of ways, in the development of the embryo.

The divisions of the terminal cell first result in the formation of a roughly spherical mass of cells known as the globular stage. Next a series of localized divisions results in a change in shape so that the embryo now resembles a small heart. Further divisions and some cell enlargement cause the embryo to elongate into a torpedo shape. Usually at this stage, or slightly after, cell divisions begin to decrease in number and eventually cease altogether, further development being the result of cell enlargement.

In contrast to the development of the embryo, the endosperm usually goes through a series of free nuclear divisions that results in the formation of a multi-nucleate mass. Only later in development are cell walls synthesized, but by this time no more divisions may take place. The endosperm in different species shows considerable variation in development. For example, in some cases, walls are formed after the first nuclear division and the endosperm is cellular throughout its existence, while in others, the endosperm remains free nuclear with walls never forming. In many seeds, the endosperm is no longer present as an organized tissue at maturity. However, in others, notably the monocotyledons, there are considerable amounts of endosperm in evidence in the mature seed and it is important in the germination of the seed and early development of the seedling.

This, then, is a brief overview of embryo and endosperm development in the angiosperms. Clearly, cell divisions are of great significance in understanding the course of this development. The place to begin a more detailed analysis of the role of cell division in embryogenesis is at the zygote and the factors that result in the first, unequal division so critical to later developments.

II. FORMATION OF THE ZYGOTE

Unraveling the details of fertilization in angiosperms has been greatly aided by the electron microscope (Jensen, 1973). Until a few years ago many significant questions concerning fertilization were unanswered (Maheshwari, 1950). We still lack answers to many of these, but our knowledge is steadily increasing. From the presently available data we can draw the following picture (Jensen, 1974).

The pollen tube enters the embryo sac from the micropylar end. It grows into one of the two synergids; these are cells closely associated with the egg that appear to have an important role in fertilization. The pollen tube discharges into the synergid and, by some as yet unknown method, one sperm nucleus enters the egg while the second enters the central cell where it will fuse with the two polar nuclei. This is the process of double fertilization found only in angiosperms.

Egg morphology varies greatly from one species to another. The egg may be highly vacuolate as in cotton (Jensen, 1965a) or densely cytoplasmic with few vacuoles as in corn (Diboll and Larsen, 1966), or have several vacuoles of varying sizes in relatively fixed positions as in *Capsella*

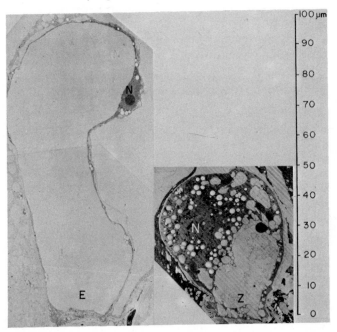

Fig. 1. Comparison of the egg (left) and zygote (right) in cotton.

(Schulz and Jensen, 1968a) or *Epidendrum* (Cocucci and Jensen, 1969). In all cases, the egg shows some degree of polarity and this polarity is reinforced by changes after fertilization and before the first division of the zygote. The point can be better made by examining two cases in detail: cotton and *Capsella*.

The egg in cotton is a large cell (Fig. 1) some 100 μm in length (Jensen, 1965a). At its micropylar end the egg is surrounded by a wall, but toward the chalazal end the wall thins and about half way along the egg it disappears completely. This means that half of the egg is surrounded only by a plasma membrane. The center of the egg is occupied by a large vacuole and there is only a thin peripheral layer of cytoplasm present in most of the cell. The nucleus is always found in the chalazal third of the egg and is somewhat flattened by the vacuole.

Fusion of the egg and sperm nuclei occurs when the two nuclei come into contact and the outer nuclear membranes fuse through short segments of ER (Jensen, 1964). The inner membranes next make contact and join so that a series of small bridges or channels is formed uniting the two nuclei. The ones toward the center enlarge and coalesce with the more peripheral, eventually resulting in the complete fusion of the two nuclei. In the case of the egg and sperm nuclei of cotton the fusion takes several hours. Following nuclear fusion a number of dramatic changes occur in the newly formed zygote (Jensen, 1968a).

The most striking of these is the reduction in volume of the cell (Fig.

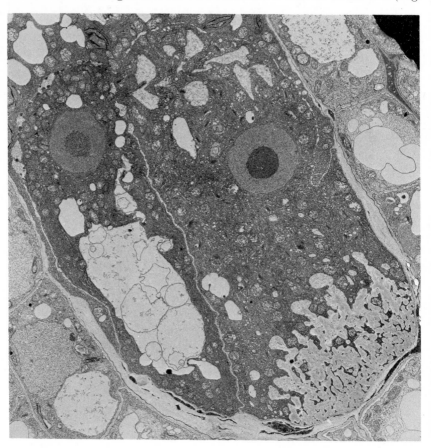

Fig. 2. Egg (left) and synergid (right) of *Capsella*. The filiform apparatus occupies the lower portion of the synergid. (Courtesy Dr P. Schulz.)

1). The zygote shrinks until it is half the size of the egg. This reduction occurs at the chalazal end with the final size of the zygote roughly corresponding to the extent of the wall present around the egg. As the vacuole decreases in size the nucleus becomes spherical but retains its relative position in the chalazal portion of the cell. The cytoplasm, which was thinly distributed over the chalazal end of the cell, accumulates around the nucleus. By the time the reduction phase is complete the zygote is a cell some 50 μm in length, with a vacuole occupying the micropylar half and the nucleus surrounded by a mass of cytoplasm occupying the chalazal end. A shell of mitochondria and plastids surrounds the nucleus. Huge polysomes formed from egg ribosomes and initiated in the zygote following nuclear fusion are associated with the plastids (Jensen, 1968c). A tube containing ER, also initiated following nuclear fusion, is found throughout the cytoplasm (Jensen, 1968b), but in greater concentration in the micropylar end of the cell. The division of the zygote takes place some $2\frac{1}{2}$–3 days after fertilization and by this time the zygote is a highly polarized cell (Jensen, 1963).

In *Capsella* the egg (Fig. 2) is completely surrounded by a wall, but at the chalazal end of the cell the wall is honeycombed (Schulz and Jensen, 1968a). Thus, there are extensive areas in which the plasma membrane of the egg is in direct contact with the plasma membrane of the synergid. The micropylar half of the egg is occupied by a good sized vacuole. The nucleus surrounded by cytoplasm is located at the chalazal end. There may be several, smaller vacuoles in the cytoplasm surrounding the nucleus. Thus, in comparison to cotton, the egg of *Capsella* is less vacuolate, more cytoplasmic, and surrounded by a wall.

Following fertilization, the zygote, in contrast to cotton, actually elongates (Schulz and Jensen, 1968a). The elongation of the zygote emphasizes the polarity of the egg (Fig. 3).

Thus, in both cotton and *Capsella* the polarity that already existed in the egg is reinforced in the zygote, but by completely different mechanisms. In one case it is by the cell reducing in size, while in the second the cell elongates.

III. DIVISION OF THE ZYGOTE

The division of the zygote is one of the most important in the life of the plant as it sets in motion the process leading to the mature sporophyte generation. This division is generally an unequal division in which the cell plate is perpendicular to the long axis of the zygote. The actual angle of the new wall varies from individual to individual and from species to species. In some cases the first division is parallel to

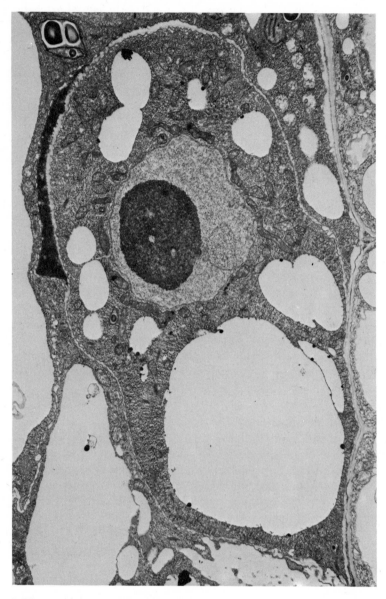

FIG. 3. The newly formed zygote in *Capsella* before elongation has occurred. (Courtesy Dr P. Schulz.)

the long axis of the cell, but there is little development until a second division occurs that is unequal and perpendicular.

The terminal cell usually is much smaller than the basal cell. It is densely cytoplasmic with few, if any, vacuoles, while the basal cell is vacuolate. The terminal cell appears to contain more plastids, mito-chondria and other cell organelles than the basal cell, but whether this is in fact the case remains to be proven. In the case of cotton, most of the large polysomes present in the zygote appear in the terminal cell, while the basal cell contains most of the tube containing ER.

In general, the terminal cell stains more intensely for RNA and pro-tein. In the case of cotton, the plastids accumulate starch during zygote

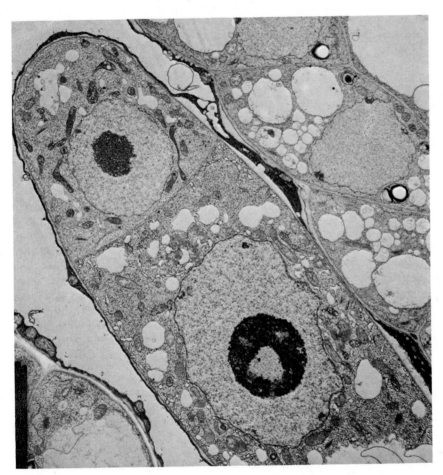

Fig. 4. The 2-celled embryo in *Capsella*. (Courtesy Dr P. Schulz.)

maturation and the nucleus becomes surrounded by starch grains (Jensen, 1963). Almost all of these plastids are present in the terminal cell following division of the zygote. Thus, both structurally and chemically, the basal and terminal cells are different and these differences will be expressed in the future development of the embryo.

In *Capsella* the picture is slightly more complex than in cotton (Schulz and Jensen, 1968a). The first division is unequal, but for a short time the basal cell cytoplasm actually stains more intensely for RNA and protein than the terminal cell (Fig. 4). The basal cell undergoes a division perpendicular to the long axis of the embryo, resulting in a 3-celled file of cells. The basal cell in the middle will give rise to a single file of relatively uniform cells that forms the bulk of the suspensor (Schulz and Jensen, 1969). The original basal cell, which retains that name, enlarges to form a huge vacuolate cell. The terminal cell stains deeply for RNA and protein, is rich in polysomes and contains numerous plastids and mitochondria.

The size of the terminal cell and the absence of large vacuoles appear significant in the future development of the embryo. In *Hibiscus*, which has an egg similar to cotton, it is possible to make crosses that fail to set seed even though a zygote is formed and the first divisions of the embryo take place. Ashley (1972) has shown that in the inviable crosses the zygotes do not shrink in size and, as a result, the first division results in two large vacuolate cells. Subsequent divisions are highly irregular, the embryo soon becomes a chaotic mass of cells and the seed aborts. In this case, the rearrangement of the zygote is a necessary condition for the subsequent normal development of the embryo.

IV. EARLY DIVISION OF THE EMBRYO

When the zygote divides the resulting cells do not enlarge, so that the two-cell embryo in cotton is no larger than the zygote. A similar trend continues through the early stages of embryo development. This reduction in cell volume during the early growth of the embryo in cotton is shown in Fig. 5 (Pollock and Jensen, 1964). Similar trends are true for *Capsella*, if the suspensor is not included in the analyses. The general conclusion from these data is that only when the cells have reached a minimal size is differentiation possible.

Patterns of cell divisions vary greatly from species to species during early embryo development. In cotton, the divisions seem almost at random until there are over 50 cells and a protoderm forms. In *Capsella*, in contrast, the divisions are extremely precise and appear carefully regulated (Fig. 6). *Capsella* is, in many ways, the classical plant embryo

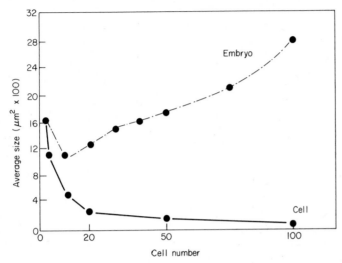

Fig. 5. Embryo and cell size in cotton.

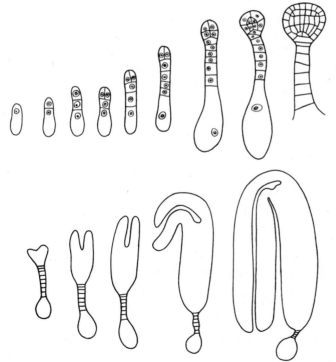

Fig. 6. Diagrammatic summary of stages of *Capsella* embryogenesis.

and its development has probably been more carefully examined than any other (Hanstein, 1870; Schaffner, 1906; Soueges, 1919; Pollock and Jensen, 1964; Schulz and Jensen, 1968a, 1968b).

After the first division in *Capsella*, the large basal cell divides again, producing two cells of unequal size. The smaller of these is located between the large basal cell and the original terminal cell. This cell will divide a limited number of times always in a transverse manner so that a single file of cells is produced. The large basal cell may divide once, or at most twice, adding to this file. The basal cell enlarges to a size of 150 μm × 70 μm and becomes highly vacuolate. An extensive network of wall projections develops on the micropylar end wall and adjacent lateral wall (Schulz and Jensen, 1969). The nucleus becomes highly lobed and suspended in a strand of cytoplasm traversing the large vacuole. In similar basal cells of other species (Clutter and Sussex, 1968; Nagl, 1969) it has been shown that they are highly polyploid and the chromosomes actually seem to be in a polytene condition.

In contrast to the huge swollen basal cell, the cells of the suspensor remain barrel shaped and arranged in a single file. The end walls of the suspensor cells contain numerous plasmodesmata, but there are no plasmodesmata in the walls separating the suspensor from the embryo sac (Schulz and Jensen, 1969). The lower suspensor cells fuse with the embryo sac wall. As the suspensor matures, the cells lose most of their ribosomes, ER, and other cell organelles, except the mitochondria, which remain numerous, and the nuclei.

The development and ultrastructure of the suspensor and basal cell of *Capsella* suggests that they function as an embryonic root in the absorption and translocation of nutrients from the surrounding tissue to the developing embryo.

The terminal cell divides when the embryonic filament consists of either three or four cells (Pollock and Jensen, 1964). The plane of the first division is longitudinal. This is followed by a second division at right angles to the previous division, to form a quadrant of four cells. The cells of the quadrant divide transversely to form the 8-celled embryo (Fig. 7). During this series of divisions there is a progressive decrease in cell size and a corresponding decrease in number of organelles, but if all the cells are taken together the number of organelles in them appears greater than the number in the original terminal cell. All the cells have similar ultrastructure, which is extremely simple (Schulz and Jensen, 1968b). Starch and oil reserves in the terminal cell are utilized during these divisions and disappear from the cell.

Each cell now undergoes a periclinal division and cuts off a proto-

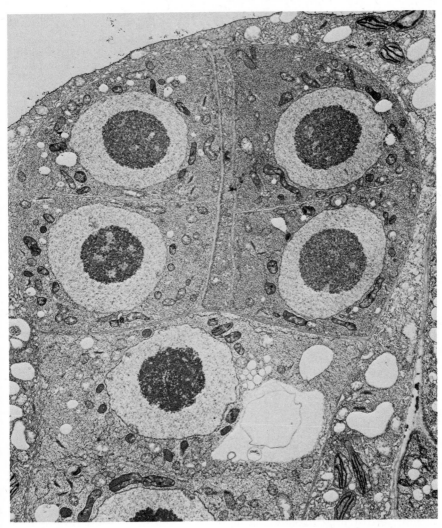

Fig. 7. The 8-celled embryo in *Capsella*. (Courtesy Dr P. Schulz.)

dermal cell to the outside. This division is important because it differentiates the protoderm, which will form the epidermis of the adult plant from the eight interior cells which will give rise to the procambium and ground meristem. Despite this further differentiation of tissue systems there are still no apparent ultrastructural differences in the cells of the embryo at this stage (Schulz and Jensen, 1968b). The only differences are those of cell shape and position in the embryo. The cells of the protoderm divide by anticlinal walls, and the eight inner cells by longitudinal walls, to form the 32-celled globular embryo composed of an upper and lower tier of eight cells each enclosed within the protoderm.

The general shape of the embryo is globular at this stage and it continues to enlarge by divisions within the sphere (Pollock and Jensen, 1964). The protoderm keeps pace with the expanding sphere by anticlinal divisions that increase in frequency in the terminal cells in the position of the future cotyledons. With the initial development of the cotyledons, the embryo assumes the shape of a heart. Only at the heart stage do distinct structural and chemical differences appear in the various parts of the embryo (Schulz and Jensen, 1968b). The cells of the procambium and ground meristem are more vacuolate than those of the protoderm. Cells at the tip of the cotyledons are extremely dense and have high concentrations of nucleic acids and proteins. The plastids in the cells of the ground meristem and protoderm develop chlorophyll and more extensive lamellar systems in contrast to those in the procambium.

Continued divisions of the cells of the cotyledons result in the growth of these organs. The same is true for the hypocotyl. In both the cotyledons and the hypocotyl some cell enlargement also occurs. The hypocotyl remains relatively straight, but the cotyledons curve back on the axis by differential enlargement of the cells.

The extreme regularity of cell division in *Capsella* is not apparent in cotton, although the same general pattern of development is followed (Pollock and Jensen, 1964). The first division is transverse, a small terminal and large basal cell being formed. There is no elaborate development of the suspensor in cotton, the mature suspensor consisting of a small pyramid of four to six cells. In the embryo proper the divisions are irregular and no precise cell lines can be established. The protoderm in cotton first becomes apparent when some 50 cells have been formed and it is well delineated when the globular embryo has some 100 cells. When the distribution of mitotic figures was followed during these early stages of embryo development (Pollock and Jensen, 1964), they were found to be relatively uniformly distributed in the first stages (Fig. 8). Next divisions become more numerous in the periphery as the protoderm forms and the globular embryo increases in volume. The beginning of the heart stage is heralded by a change in distribution of cell divisions. They increase in the regions where the cotyledons will form and decrease in frequency in the region in between. This pattern becomes more pronounced as the heart shape develops. Because of these divisions and cell enlargement, the embryo begins to elongate in the

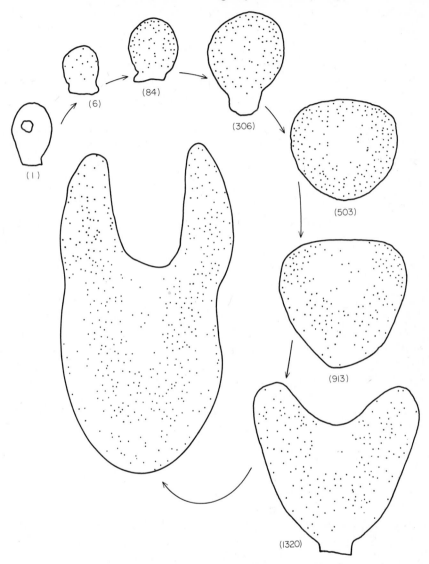

Fig. 8. Distribution of cell divisions at various stages of embryo development in cotton. The number of cells in the stage analysed is given on the embryo.

hypocotyl region. In a medium torpedo stage, the distribution of mitotic figures shifts to a Y-shaped pattern along the median long axis of the embryo. There is also renewed mitotic activity along the extreme periphery of the embryo, a relatively even distribution of mitosis in the cotyledons. Not long after this cell divisions cease and the remainder of embryo development is the result of cell enlargement and maturation.

In the monocotyledons the early stages of embryo development are much like those of the dicotyledons described above. One of the most thoroughly studied species of monocots is barley, *Hordeum vulgare* (Norstog, 1972). There is little change in shape in the egg-to-zygote transformation. The zygote contains many small vacuoles that are concentrated in the micropylar region of the cell. The shape of the egg cell is pyriform and this appears as one factor involved in the polarity of the zygote. The first division of the zygote is transverse, producing a basal cell and a terminal cell of approximately equal size. The second division also is transverse, but the third is longitudinal and divides the terminal cell. A subsequent division results in the formation of a terminal tier of four cells. There is also a longitudinal division that results in the formation of two suspensor cells.

After a series of four divisions the embryo consists of roughly twelve cells: two tiers of four cells each, two median cells (that will form a median tier of four cells) and several irregularly arranged suspensor cells. Cell divisions are not synchronous in the embryo as a whole, but are apparently synchronized in cells in the same tier. Several divisions later the embryo is no larger in volume, the cells having become more numerous but smaller. The data collected by Norstog (1972) indicate that the number of mitochondria in the total embryo have increased over the zygote, but that the mitochondria are smaller. They are also uniformly distributed throughout the cells, while in earlier stages they were arranged around the nuclei. Moreover, the vacuoles, which in the earlier stages tend to be peripheral, in later stages interestingly are perinuclear.

In these early stages there is no evidence of dorsi-ventrality. When a late globular stage is reached the differences between the monocots and dicots become apparent. There is no well developed heart shape, a single cotyledon being formed by a series of cell divisions. In grasses, the most conspicuous part of the embryo is the scutellum and the coleoptile. The scutellum-coleoptile have various morphological interpretations, but are in some manner related to the cotyledon.

V. CELL DIVISION AND CELL DETERMINATION IN PLANT EMBRYOS

The classical plant embryologists were struck by the apparent precision of the sequence of divisions during early stages of embryo development. This gave rise to the concept of lines of cell determination that held that the early divisions determined the fate of the cells subsequently formed. Consequently, it was reasonable to claim that at the octant stage the apical four cells were destined to give rise to the stem tip and cotyledons and the basal four to the hypocotyl. Presumably the first cells of the embryo inherit different cytoplasmic potentialities from different regions of the zygote, and their course of development is determined from the beginning (Wardlaw, 1955a).

Later analyses and the accumulation of exceptions to the classical patterns of plant embryogenesis led to a weakening of the concept of cell line determination. Electron microscope studies have failed to detect any basic cytoplasmic differences among the various cells of the octant embryo. Such observations support the conclusion that more subtle differences, such as the position of a cell in the embryo and its relationship to centers of hormone production, will determine its pattern of differentiation.

What seems particularly interesting in the morphological and ultrastructural work is the decrease in cell volume as the embryo undergoes early development. No matter what the condition of the egg, whether highly vacuolate or densely cytoplasmic, after a few divisions of the embryo the cells are small and dense with a simple ultrastructure and relatively few organelles. Moreover, the nuclear:cytoplasmic ratio changes from the zygote to the early embryo. The zygotic nucleus occupies a proportionately smaller share of the cell volume than does the nucleus of the cell of the young embryo. This is not because the embryonic cell nucleus is larger but because the cell volume is decreased. The implication of these observations is that information in the cytoplasm must be reduced before a new developmental sequence from the nucleus can affect the course of cell differentiation. This suggestion is supported by data from the cell culture studies where differentiated cells can repeat embryonic development, but only after differentiation has occurred. Clearly what are needed in this area are detailed analyses of nucleic acid and protein synthesis.

Plant embryogenesis is an area of research poorly explored by modern techniques. Unfortunately, plant embryos are not easily treated experimentally or cultured en masse in the early stages. Nonetheless, we can expect to see continued development in this field as more investigators become interested in this important phenomenon.

12. Disorganized Systems

A. W. DAVIDSON

Department of Biological Sciences, Thames Polytechnic, London, England

P. A. AITCHISON and M. M. YEOMAN

Department of Botany, University of Edinburgh, Scotland

I. INTRODUCTION

The intact plant is an ordered structure in which the component organs and tissues exhibit a high degree of organization. The control exerted by individual meristematic regions over the rest of the intact plant, especially other meristems, and including the feedback from the rest of the organism to each meristem, is the main factor in determining the morphological complexity of an individual. The complexity of this interrelationship increases with the overall complexity of the system. Although environmental factors modify many aspects of growth, including cell division, the dominating influence on the pattern of development is intrinsic to the plant. The integrated physical structure and physiological action shown by individually specialized cells results in what we may describe as an organized system. The influence of one part of a plant on the growth and development of other parts is transmitted by growth hormones and these act as regulators of the genetically determined patterns of development. The immediate effect of such hormones is on the growth of individual cells, but ultimately the whole growth form may be affected. For instance the striking difference between the characteristic shapes of a spruce and an elm tree results from the different interrelations of the growing points in each. In the first, the primary apex retains a strong dominance over meristems of lateral branches, the resultant size of which is dependent on distance from the primary meristem. In the second, no single meristem retains such an influence, and a highly ramified pattern develops. A high degree of organization of growth at all levels is the rule in intact plants and unorganized growth occurs only when induced by external agents or within the confines of plant tissue culture.

II. LEVELS OF ORGANIZATION

Cell division in the angiosperm is concentrated in the apical and lateral meristems, although divisions occur, albeit less frequently, in other regions: in the developing files of cells derived from apical meristems, the derivatives of the cambium, the epidermis of the root, stem and leaf and the reproductive apparatus. The majority of divisions, therefore, take place in the small, essentially non-vacuolate, densely cytoplasmic cells at the tips of roots and shoots. Divisions in other parts of the plant, e.g. in cambium initials, often occur in much larger vacuolate cells. In both cell types there is a higher degree of organization of events during the cell cycle which is especially apparent at division. The events are similar, but not identical, in transformed tumour cells and isolated cells in culture, even those from which the wall has been removed. This does not mean that unorganized growth at the subcellular level never occurs. Such phenomena as polyploidy, non-disjunction of chromosomes, loss of chloroplasts and syncitium formation may in some cases be regarded as evidence of disorganization within the growth cycle of cells (although in others they are highly regulated processes). Such irregularities of cell division are discussed by Brown and Dyer (1972) and will not be covered here.

The beginnings of organization in the intact plant at the supracellular level occur at the first division of the zygote and lead to the formation of the embryo (see Chapter 11). The zygote is situated at one end of the embryo sac, partly in contact with the base of that structure. Thus, even before it divides it is in a state of polarity, in which one pole of the cell is probably subjected to a higher concentration of nutrient substances and growth factors than the other pole. Therefore, although the fertilized egg cell has no cytoplasmic connections with the embryo sac, the plane of the first division is likely to be determined by its immediate environment.

The first division reinforces the polarity of the pro-embryo, and subsequent divisions, together with cell expansion, establish the shape of the organism. It is the pattern in which cells divide and the subsequent network of cell walls that provide the basic organization of the plant. It follows that the factors which determine and influence the plane of cell division are of supreme importance in the establishment of an ordered structure and, conversely, the breakdown of organization at the tissue, organ and plant level may be attributed to a disturbance in the mechanism which determines the plane of division in each cell in relation to its neighbours. The examination of a section of a plant stem, leaf or root reveals the high degree of structural organization within any organ. The maintenance of this integrated structure and its subsequent growth and development depend on an interaction between the permanently embryonic regions, the meristems, and the rest of the plant, sometimes over long distances. Plant growth hormones undoubtedly play a central role in this overall control. On the other hand the integrity or unity of an apical meristem depends on an interaction taking place, probably continuously, between adjacent cells. This more local interaction may involve proteins associated with the wall, or the movement of macromolecular components across the plasmalemma or through the plasmodesmata. This interaction or molecular conversation is an important aspect of organization as yet relatively unexplored.

III. THE DISTURBANCE OF ORGANIZATION

The removal of a piece of a plant and its subsequent culture in isolation induces a change within the fragment. In organ cultures organization may remain at almost the level present in the intact plant. Root tips, for instance, grown in isolation follow a pattern of development characteristic of intact roots (Street, 1966a, b). Excised shoot apices regenerate roots and become intact plants (Ball, 1946). Only when tissue fragments are employed as the initial inoculum does the subsequent growth become disorganized. Perhaps significantly, the tendency is that the smaller the fragment the greater the degree of disorganization. Under appropriate nutrient conditions a callus may be formed (Yeoman, 1973) and sub-culture of this mass can give rise to a mature callus, made up of a population of many rapidly dividing cells, which exhibits little differentiation and eventually a point may be reached when the callus is unable to differentiate (Gautheret, 1946). A particularly well known example of a callus which has been in culture for many years and has lost the ability to differentiate under conditions in which differentiation would have occurred is that of the *Acer* callus isolated from the cambium by Northcote in 1958 and kept in many laboratories throughout the world. This strain is extremely friable and grows rapidly both on agar and in agitated liquid media. Much of the pioneer work by Street and his co-workers on the development of continuous culture techniques for the growth of plant cells has been conducted with this particular strain of sycamore callus (Street, 1973). These cells will not differentiate visibly though they still retain the ability to synthesize lignin, which is normally a product of differentiated tissue and which in this case accumulates in the culture medium (Henshaw and Pearce, 1969). The only degree of organization at the tissue level retained by this system is that the daughter cells produced at successive cytokineses exhibit a tendency to stay together and form clusters. Other isolates of *Acer* callus which have been in culture for shorter periods have not lost the ability to differentiate.

We may speculate on the reasons for breakdown in organization in these cases in terms of the two levels of information exchange, or cell to cell conversation mentioned above. The high degree of structural organization in a meristem presumably demands a high level of interaction between component cells. This interaction is sufficiently precise for the tight organization to be maintained when the meristem is isolated from the rest of the plant even though a high proportion of the cells are actively engaged in cell division. The structural and physiological characteristics are determined by local conversation which cannot be overridden by interaction with hormones in the culture medium. In the intact plant the long-distance conversation received by apical meristems, once formed, seems to determine only the growth rate of the meristem as a whole.

If, on the other hand, we suppose that the maintenance of the organization of other parts of the intact plant is regulated both by long-distance and local conversation, i.e. the degree of self-specified organization is less, we can see why fragments from these regions produced unorganized growth when cut off from their long-distance conversation.

Further, it is reasonable to suppose that as smaller fragments are taken, the more likely is the level of conversation to fall below a minimum necessary to maintain the organization present.

Another unorganized system encountered in culture situations is that of proliferating protoplasts. The preparation and culture of higher plant protoplasts has been developed intensively during the past few years and it is now common practice to maintain protoplasts in culture as reproducing units (Cocking and Evans, 1973). Protoplasts grown in isolation develop a new cell wall early in the culture period, and the initial period of cell division forms a callus mass, without any of the organization characteristic of the original tissue from which the protoplasts were derived.

Other than in tissue cultures, unorganized systems are encountered only when the normal morphogenetic patterns characteristic of development are disturbed, such as in the formation of crown galls. These tumours are initiated by a tumour-inducing principle (TIP) which is derived from the bacterium *Agrobacterium tumefaciens*. Similar conditions are initiated by other agents, e.g. viruses, bacteria and insects, which cause irritation and wounding. The cells of the host plant acquire, as a result of their transformation by the TIP, a capacity to undergo rapid random divisions. Although they are presumably still supplied with nutrients by the host plant, the growth of tumours does not seem to be regulated in any way by the rest of the plant. They either do not hear, or at least do not respond to the molecular conversation of neighbouring cells. Furthermore, once this disorganized system is established it represents a permanently altered growth form of the host plant and can be cultured indefinitely on a simple medium consisting of mineral salts and a carbon source. Such cultures rarely differentiate or exhibit signs of becoming organized.

However, most unorganized systems eventually assume some degree of organization, and in many cases via highly organized intermediate stages, whole plants may even be regenerated. Whether a disorganized group of cells may assume the capacity to become ordered with respect to each other and thereby produce recognizable structures, as distinct from such structures arising from the descendants of a single cell in an ordered sequence, has not been unequivocally demonstrated. The possible mechanisms involved in this reversal of an unorganized to an organized system will be considered later.

IV. CHARACTERISTICS OF DIVISION IN UNORGANIZED SYSTEMS

From a study of the characteristics of cell division in systems becoming disordered and those reorganizing we may gain some insight into the basis of organization. Four major situations will be examined in some detail: (1) the development of a callus from an explant; (2) the growing callus culture; (3) the dividing plant cell in isolation without a recognizable wall—the protoplast; (4) the plant tumour and tissue cultures that may be derived from it.

A. The Developing Callus

A callus (Fig. 1) may be produced from a piece of tissue removed from within a plant, a fragment with at least one damaged surface, or occasionally from an intact plant, usually a young seedling. Wounding alone can promote proliferation at the damaged surface and result in the formation of a callus. If left on the parent plant, the callus is usually short-lived and rapidly becomes covered and infiltrated with phenolic

Fig. 1. An actively proliferating callus derived from the storage of the carrot, *Daucus carota*. (×1·5.)
Photograph by courtesy of Professor H. E. Street.

substances which seal off the wound from the contaminating environment. In order to promote and maintain cell proliferation it is necessary to isolate the young callus and to culture it in the presence of growth-promoting substances. Removal of the fragment and handling of the material must be carried out under conditions of strict asepsis if a successful culture is to be established.

Mechanical wounding is not, it would seem, always a necessary condition for callus initiation, for callus can be produced from intact seedlings by application of certain chemical stimulants. From this it would appear that a marked change in the chemical environment of the constituent cells promoted by wounding and/or the addition of growth stimulants is the overriding factor in the initiation and establishment of a callus. This seems to act as a disorientation which overcomes the control exerted by the rest of the plant.

The development of a callus from cylindrical explants isolated from Jerusalem artichoke tubers has been studied in this laboratory (Yeoman *et al.*, 1968; Yeoman, 1970). The removal of the tissue and exposure to a nutrient medium containing 2,4-D and coconut milk results in rapid proliferation of the constituent cells and the establishment of a callus. Callus induction does occur in the absence of coconut milk, but subsequent proliferation is limited. Cell division is restricted to the periphery of the cylinder, excluding the outermost layer or two of cells which autolyse and eventually collapse. The cells at the centre of the explant do not divide and remain relatively inert metabolically. The general distribution of divisions at the periphery of the explant is due to the release from the cells at the surface of autolytic products which diffuse inwards. The plane of the first division appears to be random but is followed by successive divisions in which the plane is predominantly periclinal and which produce a characteristic wound cambium. From these observations, which appear to be representative of the early development of a callus in other tissues, it would seem that the formation of an actively dividing mass from a quiescent tissue begins with a brief disordered phase which is quickly replaced by the reorganization of the dividing cells into a new meristem. The diffusion inwards of autolysis products cannot be invoked as the sole determinant for the plane of division, and other factors, including mechanical stress (Yeoman and Brown, 1971), contribute to what is a complex situation.

A variety of theories have been proposed to account for the positioning of the new cell wall in plants (Sinnott, 1960). Sachs (1878) claimed that the position of the new wall was such that it tended to divide the cell into two equal parts. Errera (1888) proposed the generalization that the new wall tends to be of minimum surface and therefore is formed across the narrowest width of the cell. Another generalization is that the new wall intersects the old at right angles. These rules are not in any sense explanatory and it is therefore not surprising that there are numerous exceptions to each of them. A mechanism has been proposed by Yeoman and Brown (1971) which explains the pattern of division during early callus development, and may also be used in the interpretation of the distribution of divisions in the mature stem of a dicotyledon. This proposal invokes the internal mechanical stresses which arise during growth of multicellular structures. These forces bear directly upon the cell wall but are transmitted to the plasmalemma and it is changes that occur within this limiting membrane which determine the point at which the phragmosome is formed and hence the position of the new cell wall. The nature of the changes which occur within or at the surface of the plasmalemma are a matter of speculation and these are fully discussed elsewhere (Yeoman and Brown, 1971). Clearly the magnitude and direction of mechanical stress forces encountered by cells within an excised piece of tissue are very different from those operating on the same cells before excision. It is possible that the initially disorganized growth is a result of freeing the cells from some of these stress relationships (components of local conversation). Of course cells at the centre of an explant are probably subjected to essentially the same stress conditions as in the intact plant. It is perhaps significant that these cells do not contribute to the subsequent disorganized growth pattern.

The wound cambium which is quickly established at the interface of tissue and environment is a meristem only for a limited period (Yeoman *et al.*, 1965). After a period of exponential growth, cell division within the wound cambium slows down and eventually stops. The initial proliferation is replaced by another in which the products of division cut off to the outside of the fragment expand rapidly and become recognizable callus cells, while the new cells to the inside become involved in a process of differentiation which results in the formation of vascular bundles and nodular meristems. It cannot always be excluded in these young calluses that some of the cells at the centre of the explant, which do not initially divide, do not participate in this subsequent differentiation, or at least influence differentiation in adjacent, newly formed cells (Yeoman *et al.*, 1968). Again it might be significant that it is the cells to the inside of the wound cambium that are first involved in differentiation. These cells are in closer contact with the original core of apparently inactive tissue which retains its initial level of organization and may impart some degree of entrainment

FIG. 2. The emergence of organized structures from (a) tobacco callus (*Nicotiana tabacum*) ×1·5 and (b) chicory callus (*Cichorium intybus*) ×2.

on neighbouring areas. There are, of course, other differences, e.g. gradients of gaseous components and nutrients, but these are probably of secondary importance (Yeoman *et al.*, 1968). Thus, apparently two major developmental phases are proceeding together, one in which an organized state is emerging and a second in which the products of division of a recognizable meristem develop into a disordered mass of callus cells (Fig. 2). It is from this situation that a callus and intact plants may emerge together in the same chemical and physical environment.

The callus cells formed to the outside of the tissue fragment are large, highly vacuolate and generally not in contact with their neighbours over a large part of their surface area. These cells tend to be spherical with a diameter of up to 100 μm (Street, 1968b; Sutton-Jones and Street, 1968; Roberts and Northcote, 1970). Cytoplasmic streaming can be observed in the transvacuolar strands by following the movement of mitochondria and spherosomes along these strands. Ribosomes are abundant, both free and in helical and spiral arrangements (Yeoman and Street, 1973). The nucleus tends to be rounded with an entire margin and often contains more than one nucleolus. The fibrillar and granular regions are indistinct and the nucleolus as a whole is less compact than in the quiescent state. The structure and arrangement of the plastids, endoplasmic reticulum, mitochondria and golgi all suggest that the cells are metabolically active. Division of these vacuolated cells produces a mass of callus without any apparent organization. The plane of division in this callus is more or less random and the products can be separated from the initial explant and sub-cultured either on a solidified medium or in an agitated liquid medium. The lack of intimate contact between adjacent cells in a callus may be responsible for the disordered development of these cultures. T. W. Goodwin (personal communication) has demonstrated that electrical continuity even between adjacent cells of a callus is poor and is almost non-existent between cells separated by several others. In contrast it may be noted that electrical conductivity between cells in an ordered tissue such as an *Elodea* leaf is excellent and decreases much less markedly with distance. Goodwin has also shown that the diffusion of a dye which can only move from cell to cell via plasmodesmata is rapid in an *Elodea* leaf but extremely slow in a callus culture. He concludes from these data that molecular continuity between cells in a disordered system is poor, i.e. the molecular conversation is attenuated. This will in turn reinforce the lack of organization as the callus proliferates. It would be interesting to compare the electrical conductivity between cells of crown gall tissue on the intact plant with that between adjacent untransformed cells. If the cell to cell conductivity changes markedly at the boundary of the gall this might confirm that the tumour is in poor molecular continuity with the host tissue, and can thereby continue an independent development.

However, not all callus cultures are friable and easily disintegrated, many are compact and quite hard. A friable callus is the most suitable starting point for the establishment of a cell suspension culture for which maximum dispersion of the tissue in response to mechanical agitation is the aim. Blakely and Steward (1961) have demonstrated that friable and compact calluses derived from *Haplopappus gracilis* are inter-convertible. The change from one form to another may be achieved by changing the levels of coconut milk and naphthalene-acetic acid in the culture medium. Torrey and Shigemura (1957) have shown the concentrations of yeast extract which are high relative to 2,4-D concentration induce friability in pea callus and that compact cultures may give rise to friable cultures but not the reverse. Friable and compact forms of callus from the root of *Vicia faba* may also be inter-converted (Grant and Fuller, 1968). This occurs spontaneously during growth and may be associated with the exhaustion of nutrients in the culture medium. Both Blakely and Steward (1961) and Grant and Fuller (1968) have investigated the basic histology of the different forms and have demonstrated pronounced differences. In *Haplopappus* cultures the major observable difference is in the packing of the cells. Friable callus is composed of loosely arranged cells, whereas the non-friable callus is made up of tightly packed cells with evidence of non-random divisions in localized areas, indicative of the onset of organization, and again suggesting that intimacy of contact is important in organized tissues. In bean callus the friable variety exhibits a large number of organized meristematic nodules separated by large vacuolated cells, whereas the compact type is free from organized centres and consists almost completely of large vacuolated cells with a high degree of "wall to wall" contact. Grant and Fuller (1968) have also compared the chemical composition of the cell walls of friable with non-friable *Vicia* callus. Non-friable callus contains a greater total amount of cell wall polysaccharides. Although the total amount of each particular cell wall fraction per unit dry weight is greater, the relative amount of cellulose compared with pectic substances and hemicelluloses is lower. The overall increase in amount of cellulose will presumably increase the rigidity of the cells, and the increased amount of pectic substances will hold the cells together more firmly and resist fragmentation.

B. Cultured Plant Cells

Callus cultures and cell suspensions are made up of vacuolated cells. The process of division in these cells is rather different from that observed in the small iso-diametric cells which constitute the apical meristems and is more closely allied to the process of division in the cambium and cells within the zones of root and shoot elongation (Sinnott and Bloch, 1940, 1941; Esau and Gill, 1965). A large vacuole is the most prominent feature of these cells and restricts the movement of the nucleus. This organelle may lie close to the wall in a thin layer of peripheral cytoplasm or be suspended in the centre of the vacuole by cytoplasmic strands. Between mitoses there is considerable metabolic activity resulting in the synthesis and accumulation of the components necessary for the production of a new cell. These metabolic changes are reflected in the ultra-structure of the cell (Yeoman and Street, 1973). There is a net synthesis of cytoplasm and a consequent decrease in vacuolar volume as total cell volume does not change. The numbers of several organelles increase, including plastids, mitochondria and golgi. The overall number of ribosomes increases and there is a marked rise in the proportion of ribosomes present in helical and spiral arrangements. Immediately after a division, the nucleus of one daughter lies embedded in a layer of cytoplasm against the new wall opposite the nucleus in the other daughter cell. If there is a temporal gap between divisions then the nucleus may migrate to another part of the cell still lying against the wall or become suspended within the vacuole. Frequently a second division follows quickly after the first, which does not necessitate the movement of the nucleus, providing the new division is at right angles to the previous division. The formation of a file of cells in which a succession of new walls are formed parallel to one another requires that the nucleus migrates at each division.

At the onset of division the nucleus lies within the peripheral cytoplasm midway along the cell. Cytoplasm accumulates around the nucleus, producing a bulge which extends into the vacuole. At about the same time the nucleus assumes a more rounded shape. This change in the nucleus is accompanied by an alteration to the structure of the nucleolus, the fibrillar and granular regions becoming indistinct and the nucleolus less compact. A large, more electron-transparent region may appear in the centre of the nucleolus. This region contains, and is surrounded by, granular particles similar to ribosomes. Fibrillar material similar to the chromatin outside the nucleolus is also found within this body, and in addition smaller electron-transparent regions are present in the fibrillar zone surrounding the central area. Similar nucleolar changes in cells of ageing artichoke tuber discs have been reported (Jordan and Chapman, 1971; Rose et al., 1972).

The cytoplasm continues to extend through the vacuole until a complete baffle is formed. This structure, the phragmosome, separates the vacuole into two compartments. Other trans-vacuolar strands are formed, but it is into the phragmosome that the nucleus migrates (Jones et al., 1960; Mota et al., 1964; Das et al., 1966; Roberts and Northcote, 1970; Yeoman et al., 1970). The cytoplasmic baffle containing the nucleus is rich in a variety of organelles. Often the nucleus takes up a position nearer the wall than the centre of the cell. Yeoman et al. (1970) have suggested that the movement of the nucleus into the phragmosome in cultured artichoke cells may be assisted by nuclear extensions which appear during the period between the formation of the phragmosome and the breakdown of the nuclear envelope at the end of prophase. These structures may be up to 10 μm long and 100 nm in diameter. They differ from the endoplasmic reticulum, which may be continuous with the nuclear envelope, in that they are surrounded by and continuous with the karyolymph, although chromatin has not been detected within them. Unlike the rest of the nuclear envelope, these extensions do not have any pores on their surfaces. While the nucleus is moving from its original position to the centre of the phragmosome, microtubules are often found in association with the nuclear extensions (Yeoman et al., 1970), suggesting that microtubules may be involved in the movement of the nucleus as well as in determining the point at which the phragmosome is formed (Pickett-Heaps and Northcote, 1966b; Burgess and Northcote, 1967; Cronshaw and Esau, 1968; Bagshaw, 1969).

Once the nucleus has taken up a position within the phragmosome it can begin to divide. This is usually accompanied by a virtual cessation of cytoplasmic streaming (Das et al., 1966; Roberts and Northcote, 1970). Mitosis is normal apart from anaphase disjunction which is restricted by the width of the phragmosome, since the direction of the spindle is always at right angles to the phragmosome. Cytokinesis in highly vacuolated cells differs only in detail from this event in essentially non-vacuolated cells. The phragmoplast, which heralds cell plate formation, appears at the equator of the mitotic spindle and moves centrifugally outwards along the phragmosome and eventually reaches the wall of the parent cell. There is some evidence that the phragmoplast fuses first with the side of the cell from which the nucleus originated (Roberts and Northcote, 1970). Cell plate formation begins at the centre of the phragmoplast and moves outwards in the conventional manner (Esau and Gill, 1965; Pickett-Heaps and Northcote, 1966; Hepler and Jackson,

1969) with microtubules playing an integral part in the movement and positioning of the golgi vesicles (Mollenhauer and Morré, 1966; Pickett-Heaps, 1967; Cronshaw and Esau, 1968). Fusion of these vesicles gives rise to the cell plate forming a middle lamella and new plasmalemmae. It has also been suggested that the endoplasmic reticulum may contribute vesicles to the cell plate (Pickett-Heaps, 1967a, b, c). This sequence of events is completed by the deposition of a microfibrillar mesh of cellulose which gives rise to the new primary cell wall.

C. Isolated Protoplasts

It is appropriate to consider cell division in isolated protoplasts separately since some of the events leading up to mitosis and cytokinesis are peculiar to cultured protoplasts and have no counterparts in other cells preparing for division. Indeed the methods which have been developed for preparation, isolation and culture all reflect the differences that exist between plant cell protoplasts and tissues in culture (Cocking, 1972).

The first viable protoplasts were isolated from actively growing tomato roots (Cocking, 1960) and the placental tissue of the tomato fruit (Gregory and Cocking, 1963). It is now possible to prepare viable protoplasts from a wide range of plants (Cocking and Evans, 1973), and several plant tissue cultures have also been used as a source. Cell cultures have proved particularly popular, including suspensions of *Haplopappus gracilis* (Eriksson and Jonasson, 1969), soybean (Keller *et al.*, 1970), 'Paul's Scarlet' rose (Cocking and Evans, 1973) and carrot (Grambow *et al.*, 1972; Wallin and Eriksson, 1973), though callus cultures on agar have also been used (Chupeau and Morel, 1970; Reinert and Hellmann, 1971; Moytoyoshi, 1971; Horine and Ruesink, 1972).

Protoplasts maintained in culture tend to regenerate new cell walls and this appears to be a pre-requisite for cell division (Cocking, 1961; Binding, 1966). A simple hypertonic culture medium appears to be quite satisfactory for the regeneration of a new wall, and the addition of plant growth substances is apparently unnecessary (Horine and Ruesink, 1972). Indeed, auxins may have a deleterious effect on wall regeneration since it has been established that indolyl acetic acid (IAA) promotes swelling and bursting of protoplasts isolated from the tomato root (Cocking, 1961), tobacco leaf (Power and Cocking, 1970) and *Avena* coleoptile (Hall and Cocking, 1971) by changing the permeability characteristics of the limiting membrane. Wall regeneration by protoplasts from the tomato fruit can be detected after a few hours of culture in White's medium plus sucrose (Pojnar *et al.*, 1967). Two developmental stages may be recognized. Initially a multilamellar network is formed; beneath this are deposited other materials, of which cellulose microfibrils make up a major component, and these are laid down in close association with the outer surface of the plasmalemma. After about 2 weeks the new wall is sufficiently strong to withstand the osmotic forces produced within the protoplast and the sucrose concentration in the culture medium may be gradually reduced until it is equivalent to that in routine callus culture media (Willison and Cocking, 1972). Similar observations have been made on cell wall synthesis by tobacco mesophyll protoplasts (Nagata and Yamaki, 1973).

Once cell regeneration has taken place, division may follow (Fig. 3). The composition of the medium is critical in determining whether or not division will take place. The nutrient requirements are more complex than for wall regeneration and resemble those for cell and callus culture. Conditioned medium has been used to facilitate division (Kao *et al.*, 1970), but more recently simple, fully defined media have been employed (Kao *et al.*, 1971; Nagata and Takebe, 1971; Bui-Dang-Ha and Mackenzie, 1973). The density of the inoculum is also important, with high initial protoplast densities essential to promote widespread division. Nagata and Takebe (1970) used protoplasts isolated from tobacco leaf palisade cells at a concentration of 10^5 per ml of medium.

If the protoplasts are derived from highly vacuolated cells, for example the mesophyll of toacco leaves (Nagata and Takebe, 1970; Takebe and Nagata, 1973; Nagata and Yamaki, 1973), several of the features of cell division described earlier in this chapter which are characteristic of cultured cells can be observed. A cytoplasmic baffle similar to the phragmosome of dividing callus cells appears and the nucleus migrates into this structure immediately prior to mitosis. The nucleus takes up a more or less central position within the vacuole supported by a number of cytoplasmic strands. The positioning of the nucleus, supported by cytoplasmic strands, is nearly always central, unlike the situation in callus cells, presumably because of the nearly radial symmetry of a protoplast. Disjunction of the daughter chromosomes at anaphase is usually followed by cell plate formation and the newly regenerated wall also participates in the formation of the new dividing wall by centripetal growth. The chloroplasts of tobacco mesophyll protoplasts undergo distinct changes during the period of culture. The division of chloroplasts may be commonly observed in protoplasts cultured for about 5 days. Filamentous structures appear in the matrix of the chloroplasts before division. These structures have been observed

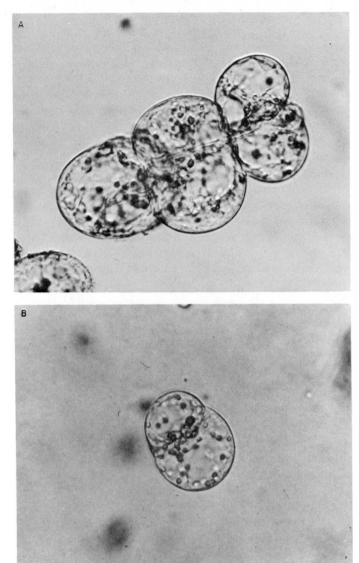

Fig. 3. Protoplasts isolated from the mesophyll of the pea leaf, *Pisum sativum*, dividing in culture. (a) A group of cells derived from one protoplast. (b) An initial division. (× 1100.)
Photograph by courtesy of Dr F. Constabel.

in dividing chloroplasts in Swiss chard cells and are believed to be DNA fibres (Kislev *et al.*, 1965). Similar fibres have also been seen in some of the mitochondria of tobacco mesophyll protoplasts. The rate of division of chloroplasts does not keep pace with the division of the cultured protoplast, resulting in a marked reduction in the number of chloroplasts per cell. With successive divisions the cytological characteristics of the protoplast change from those of a differentiated mesophyll cell to those of an apical meristem cell. By now the nucleus occupies a large proportion of the total cell volume and the large central vacuole has disappeared. The cytoplasm is densely populated with ribosomes and the endoplasmic reticulum is well developed; the chloroplast population shows marked signs of degeneration.

An unusual feature of protoplasts dividing in culture is that nuclear division is not always followed by cytokinesis. The appearance of binucleate protoplasts is usually ascribed to the failure of cell wall formation rather than spontaneous fusion of adjacent protoplasts (Ericksson and Jonasson, 1969). Of course, multinucleate protoplasts may also arise by protoplast fusion (Motoyoshi, 1971; Miller *et al.*, 1971). Reinert and Hellmann (1973) have demonstrated that multinucleate carrot protoplasts may arise both by karyokinesis without cytokinesis and by protoplast fusion. However, they claim that nuclear divisions rather than protoplast fusion are the main source of multinucleate protoplasts from established carrot callus cultures. These nuclear divisions are at least partially synchronous. Protoplasts derived from recently established carrot callus, embryos, and leaves, aggregated regularly and fused more frequently during the isolation procedure giving rise to multinucleate protoplasts. These protoplasts regenerated cell walls much more rapidly than those from long-term cultures. Temporal discrepancies between mitosis and cytokinesis are only commonly observed when nuclear division precedes wall regeneration (Pearce and Cocking, 1973; Frearson *et al.*, 1973).

Normal division in isolated protoplasts was first reported by Nagata and Takebe (1970) and Kao *et al.* (1970), using isolated tobacco mesophyll and soybean suspension cells, respectively. Since then many reports have been published of division in a variety of protoplasts. Nagata and Takebe (1970) showed that most of the protoplasts which regenerated cell walls underwent division and the cycle of mitosis and cytokinesis was repeated two or three times during 2 weeks of culture. The first report of sustained division in protoplasts with regenerated walls was published by Kao *et al.* (1970). This has since been repeated with *Haplopappus* and *Glycine* protoplasts cultured in a completely defined medium (Kao *et al.*, 1971). Repeated cell divisions have also been reported in sugar cane protoplasts (Maretzki and Nickell, 1973), although difficulty has been encountered in persuading protoplasts from other monocotyledons to divide (Evans *et al.*, 1972, 1973). Sustained cell divisions have also been described recently in fused protoplasts obtained from soybean and barley hybrids (Kao and Michayluk, 1974).

Sustained division of protoplasts with regenerated cell walls frequently gives rise to calluses. Nagata and Takebe (1971) incorporated isolated protoplasts into a nutrient medium solidified with agar, and were able to detect the formation of colonies of cells from individual protoplasts within one month of inoculation. Furthermore, the plating efficiency was of the order of 60%, which is high for isolated plant cells. Once the colonies have attained a diameter of 0·5–1·0 mm, they are transferred to a fresh solid culture medium on which they grow into recognizable callus cultures. Other callus cultures have been initiated using this technique, e.g. carrot by Wallin and Eriksson (1973) and *Petunia* by Potrykus and Durand (1972). Protoplasts may also be cultured successfully in a liquid medium without agitation (Cocking and Evans, 1973; Donn *et al.*, 1973) with the formation of a callus, though the use of a solidified medium appears to lead to greater success. Establishment of a callus from dividing protoplasts may be followed by differentiation and organogenesis leading to the formation of new plants. As with cell and callus cultures there are two major paths of development: (1) production of embryoids which then give rise directly to plantlets; and (2) production of roots and/or shoots directly from the callus mass. The regeneration of plantlets from protoplasts in liquid culture via embryoids has been reported for the carrot (Grambow *et al.*, 1972). The sustained division of protoplasts with regenerated walls appears to give rise to organized embryoids within a month. It is also possible to produce embryoids from cell clusters of carrot on a medium solidified with agar (Kameya and Uchimiya, 1972). Between 5 and 30 embryoids were formed from each cell cluster, 4–8 weeks after plating on to a solid medium.

The first report of the regeneration of whole plants from protoplasts via an intermediate callus mass was made by Takebe *et al.* (1971). Tobacco protoplasts were either plated directly on to agar or grown in a liquid medium and subsequently transferred to a solidified medium. After 2 months, the individual colonies of tobacco cells were transferred to a regeneration medium where roots and shoots were developed directly from the callus. *Petunia* plants have been regenerated from leaf protoplasts using similar techniques (Donn *et al.*, 1973; Frearson *et al.*, 1973). Protoplasts isolated from *Asparagus* cladodes can regenerate normal plantlets under similar conditions (Bui-Dang-Ha and Mackenzie, 1973).

D. Formation of Tumours

The relevance of plant neoplasms to a discussion of cell division lies in a consideration of the conversion of a normal differentiated cell to a tumour cell and the consequent initiation and emergence of a rapidly dividing cell type. The changes which accompany this conversion are presumably similar at least in some respects to those described for the initiation of division in developing calluses (Yeoman, 1970; Yeoman and Street, 1973) and protoplast cultures (Cocking and Evans, 1973), since all involve conversion of relatively inactive vacuolated cells to units of intense metabolic activity. The most intensively studied neoplasm which may be initiated in the intact plant and subsequently cultured in the absence of the causative organism is the crown gall. Other conditions such as the wound tumour disease induced by the virus *Aureogenus magnivena* Black and the genetic tumour disease which occurs in certain interspecific hybrids, e.g. *Nicotiana glauca* × *N. langsdorfii*, have been reviewed elsewhere (Black, 1965, 1972; Smith, 1972; Butcher, 1973) and require no further elaboration in this chapter.

Early work on tumour formation in the crown gall disease distinguishes two stages in the process of tumour formation (Braun and Laskaris, 1942). The inception stage is believed to involve conditioning and induction (Braun, 1952) and results in the transformation of differentiated cells into tumour cells by *Agrobacterium tumefaciens*. The development stage is concerned with the autonomous proliferation of the transformed cells into a tumour.

The conditioning process involves the wounding of the tissue by mechanical means, which seems necessary to render the host cells susceptible to transformation. The significance of the conditioning requirement remains unresolved. Wounding does not merely allow the bacteria access to the host cells, as was originally suggested (Smith *et al.*, 1911), since it has now been shown that the bacteria remain outside the cells of the host tissue (Beardsley, 1972). It has also been suggested (Braun, 1947; Klein, 1955) that products of the damaged cells may activate the wound-healing process (Lipetz, 1970) and that the undamaged underlying cells become susceptible to transformation once they have reached a particular stage in their subsequent preparation for division (Braun, 1962). More recent investigations (Therman and Kupila-Ahvenniemi, 1971) also suggest that transformation occurs at a particular point in the wound-healing process. However, it is believed that the situation may be more complex than this since washing of the wound promotes tumour growth in some conditions (Therman and Kupila-Ahvenniemi, 1971). Furthermore those observations imply that the accepted view of transformation which results from an interaction between the bacterium and wound substances to produce the TIP which actually transforms the cell may be an over-simplification.

The identification of TIP has been the subject of numerous investigations and review articles (Braun and Stonier, 1958; Braun, 1962; Klein, 1965). Much of the evidence indicates that a nucleic acid fraction is involved. The rise in DNA content of host cells within 48 h of wounding was initially believed to be due to the uptake of bacterial DNA which was equated with TIP (Klein, 1954). This increase in DNA has now been shown to be a consequence of the wound response and is completely independent of *Agrobacterium tumefaciens* (Kupila and Stern, 1961). Nevertheless, the fact that transformation results in changes in properties and behaviour which are perpetuated in all subsequent cell generations suggests that DNA is involved. Technical problems have prevented rapid progress towards an unequivocal characterization of TIP. Studies have been carried out on the effects of inhibitors of DNA synthesis and functions on the induction of tumours (Beiderbeck, 1970, 1971). However, this approach to the problem has so far proved unsuccessful because of the difficulty in distinguishing between the effects on tumour induction and the independent effects on the bacteria and host cells. Attempts have also been made to induce tumours with cell-free preparations of *A. tumefaciens* (see Butcher, 1973, for review). However, it is difficult to obtain preparations of bacterial DNA or RNA which are free from viable bacteria and this complicates the interpretation of experiments.

Some progress has been made towards the identification of TIP despite these difficulties. After obtaining evidence for partial homology between DNA of *A. tumefaciens* and tumour tissue of *Scorzonera hispanica* and *Nicotiana tabacum*, Quetier *et al.* (1969) have suggested that part of the bacterial DNA is TIP and is incorporated into the genome of the host cell during transformation. Srivastava and Chadha (1970) have produced evidence to suggest that bacterial DNA is incorporated into the DNA of the host cell and maintain that wounding may trigger off the preparation for cell division which involves the unfolding of DNA. This would furnish the necessary conditions for the incorporation of foreign DNA. Such an interpretation is consistent with the results from the investigations of Kupila-Ahvenniemi and Therman (1971) in which they demonstrated that the time of initiation and extent of DNA

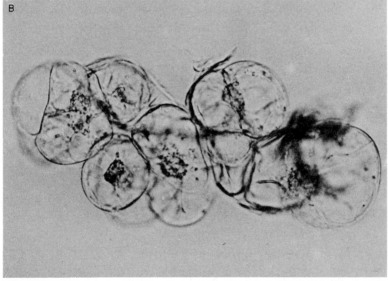

FIG. 4. Crown gall tissue of Virginia creeper, *Parthenocissus tricuspidata* × 1·25. (a) Proliferating tissue on an agar medium. (b) Cells from a culture showing a random plane of division and absence of differentiation. (× 140.)

synthesis in sterile wounds and wounds infected with *A. tumefaciens* were similar.

Other evidence suggests that viruses or bacteriophages of *A. tumefaciens* may be instrumental in the transformation process (Tourneur and Morel, 1971). Little attention is now paid to the possibility that TIP may be the bacterium which enters the wounded cells and is subsequently undetectable.

Whatever the nature of TIP, it would appear that the inception stage in tumorigenesis is complete before the first cell division takes place. It has been suggested (Kupila-Ahvenniemi and Therman, 1968) that the first round of DNA synthesis may be the critical phase and that once this point is reached the fate of the host cell is irreversibly determined. Other investigations with transformed cells have shown increases in nuclear and nucleolar volume, DNA, RNA and non-histone protein before division takes place (Braun, 1962; Rasch *et al.*, 1959; Therman, 1956). It is quite likely that after the inception stage is complete the subsequent pattern of events is similar to that in other disorganized systems preparing for division.

Crown galls freed from contaminating organisms can be cultured indefinitely on a simple mineral salts mixture with sucrose but without auxin or cytokinin. Such tissue cultures grow extremely rapidly to produce friable masses of large vacuolated cells (Fig. 4a). The cells frequently turn green when illuminated but are always dependent on an external source of carbon. The plane of division within the culture is random (Fig. 4b) and organized structures are not observed.

The crown gall culture and the habituated callus represent the most advanced forms of the unorganized state. They are strikingly similar when fully developed, despite the different agencies involved in their initiation, and have similar nutritional requirements. Gautheret (1946) noted that *Scorzonera hispanica* callus tissue which had been sub-cultured for considerable periods was able to grow indefinitely on a simple nutrient medium lacking a source of auxin. More recently, other callus cultures have been described which have simple nutrient requirements. These cultures are described as habituated. Although the phenomenon of habituation is not fully understood there is evidence to suggest that such cultures, like crown gall, show an increased capacity for endogenous auxin synthesis and a decreased capacity for auxin breakdown (Kulescha and Gautheret, 1948; Kulescha, 1952). Furthermore both habituated callus and fully transformed crown gall tissues exhibit an exceedingly low capacity for differentiation.

V. THE GENERATION OF FORM

Some tissue cultures or tumours may proliferate indefinitely as disordered structures. Such systems are the exceptions to the general rule that tissues disturbed by an external agency tend to return to an ordered state after a short period of disorganized growth. As we have seen, the tendency is to return to a situation in which a high degree of organization exists and it is more difficult to maintain a proliferating mass in a disordered state (Fig. 2). Cultures induced and maintained by the continuous provision of growth-promoting substances, in time may lose the ability to regenerate morphogenetic structures. Eventually the culture may become habituated and then exhibits an exceedingly low capacity for differentiation. It would seem that in such cultures, as in crown gall tissue, there is a loss of polarity and a lack of control over the plane of cell division. This leads to a random loose arrangement of cells in which there is low degree of intercellular communication.

The appearance of recognizable structures in younger cultures may occur spontaneously or may be induced by the transfer of a fragment of callus to a medium designed to promote organogenesis. The manipulation of the balance and content of growth substances can produce dramatic changes in the organization of the tissue (Skoog and Miller, 1957). Regeneration of roots is particularly common in both callus and cell cultures and many examples of shoot regeneration from callus cultures have been reported (Reinert, 1973). The derivation of these organized structures from completely unorganized systems has been difficult to establish unequivocally. The formation of vascular elements is always associated with incipient morphogenesis. In some instances it may be that vascular cells from the original explant have been retained during sub-culture and these are responsible for the initiation and development of organized structures. It has been established that cells from the xylem can induce adjacent cells to differentiate into vascular tissue (Camus, 1949; Wetmore and Sorokin, 1955). However, in older, frequently sub-cultured calluses where there has been a considerable period of unorganized growth the situation is rather different. The complete elimination of mature and differentiating vascular elements is extremely difficult to achieve by sub-culture and possibly the instances of habituated cultures are an indication of success which is comparatively rare. Clearly, the presence of one differentiating vascular element can result in a dramatic change to the system. Gradients of growth substances will become rapidly established around the vascular cell and may induce division in neighbouring cells. A dividing group of cells within a tissue will be subjected to considerable mechanical stress and respond accordingly (Yeoman and Brown, 1971). The result is the formation of a nodular or cyclical structure composed of small, essentially non-vacuolate cells which resembles a vascular bundle or a primordium (Gautheret, 1966). Whether such structures subsequently develop into roots or shoots may be related to the particular environmental conditions in which the tissue is cultured. Frequently, the further development of these nodular structures can only be encouraged by exposing the culture to a sequential change of nutritional, hormonal and environmental regimes. Clearly many similarities exist between the formation of lateral root primordia in the intact plant and the development of nodular meristems in tissue cultures.

The production of adventitious embryos in callus and cell cultures is certainly easier to observe and perhaps theoretically simpler to explain. In this situation, a single cell becomes "embryogenic", begins to divide and its descendants give rise to an independent plant. An understanding of the conversion of a cell from an unorganized callus or cell culture to a state in which it can give rise to a plantlet must centre around the question, "What makes a somatic cell behave like a zygote?" Steward has proposed that all somatic cells may be capable of becoming embryogenic when they are isolated from surrounding cells and placed in medium closely resembling that present in the ovule (Steward et al., 1966; Steward, 1970). This hypothesis was formulated as a result of extensive studies of cell cultures from the carrot and water parsnip growing in a liquid medium containing coconut milk. (Embryogenesis has since been demonstrated in fully defined media.) It has also been shown that embryogenic cells in callus tissue from *Ranunculus sceleratus* have many protoplasmic connections with neighbouring cells during the early stages of embryogenesis (Thomas et al., 1972). Epidermal cells of the stem of *R. sceleratus* which are connected with surrounding cells are also capable of becoming embryogenic (Konar et al., 1972). It can be argued that the presence of an apparent cytoplasmic connection does not ensure intercellular communication, for the deposition of a substance (e.g. callose) across the plasmodesmata could prevent the movement of molecules along the cytoplasmic connections from adjacent cells. At present there is no evidence known to the authors to support this point.

Cells at the periphery of tissue masses in suspension cultures and individuals floating in the medium can develop into embryoids (Fig. 5). The pattern of cell division from the unicell (free or attached to the tissue mass) to the mature embryo is similar to that observed in the developing embryo within the ovule of the intact plant. Clearly, the cell in culture inherits its polarity from a previous cellular association.

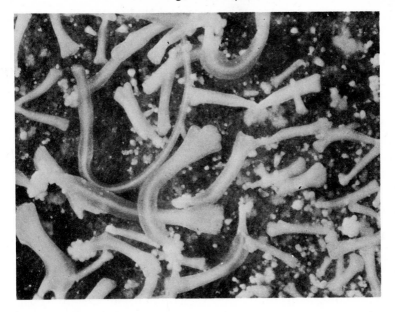

Fig. 5. Embryoids from a suspension culture of *Daucus carota*. ($\times$ 10.)
Photograph by courtesy of Professor H. E. Street.

The first division is usually across the short axis of the cell and establishes the polarity of the developing embryoid; subsequent divisions
result in the formation of a structure similar to the embryo formed
within the ovule of the intact plant. It may be suggested that the critical
step in the formation of an embryo is that the products of the first
division remain in close contact and intimate intercellular conversations can take place. As a result the patterns of subsequent divisions,
in what is a favourable chemical environment, give rise to an
embryoid.

VI. CONCLUDING REMARKS

Various agents can promote disorder in a tissue. Usually the effects are
temporary, but occasionally a permanently disorganized system is produced. All of the agents, whether living or inanimate, induce cell
division and it is the products of division which give rise to a recognizable disordered state. It is possible that under some circumstances a
breakdown in the regulatory mechanism of an organized system occurs,
producing a physiologically disorganized tissue, but this will not

become recognizably disorganized unless cell division occurs. An agent causing a disturbance must modify or overcome two major regulatory controls within the tissue explant or intact plant: (1) the overall chemical environment of the tissue, provided by the long-distance transport of substances from other parts of the plant; (2) the local molecular conversations which take place between adjacent cells. The release of wound substances and/or the effects of added chemical stimulants tend to obliterate the chemical environment of the cells before perturbation and attenuate or over-rule local conversations between cells leading to a breakdown in organization. Separation of cells by enzymic means as in protoplast isolation will of course bring about a similar change in the cellular environment, again the trigger to disorganized development being an effective isolation from the regulatory system of the intact plant.

Usually, the plant quickly reasserts control over the affected region and order is restored. Tissue explants and separated cells are, however, at the mercy of the chemical environment provided by the culture medium and the restoration of order is more difficult. Indeed the maintenance of disorganization depends on the continuous supply of chemicals which neutralize the chemical regime within the tissue explant and interfere with the molecular conversations between cells. Tissue systems maintained and sub-cultured in a disorganized state for extended periods by the provision of growth-promoting substances may eventually become independent of these added substances and lose the ability to differentiate. Such cultures behave in a similar fashion to tissue isolated from plant tumours. A permanent change in the cells, as in these two cases, depends on a seemingly irreversible change in metabolic status coupled with a marked decrease in the extent of communication between adjacent cells, i.e. on the maintenance of a degree of isolation. Conversely the emergence of organized structures from apparently disordered callus or cell cultures, which is a commonly observed phenomenon, results from divisions in which the products remain in close contact, enabling an intimate local conversation or exchange of substances to become established and maintained. Subsequent regulated divisions lead to the formation of organized structures.

D. Summary and Perspectives

13. Summary and Perspectives

M. M. YEOMAN

Department of Botany, University of Edinburgh, Scotland

The preceding chapters in this volume cover the main aspects of cell division in higher plants and it is clear from these contributions that much has been achieved, although much remains to be done. The major achievements fall into two general areas. Firstly, the detailed examination of the visible events of division in somatic and reproductive cells which have resulted in our extensive knowledge of the processes of mitosis (Chapter 2) and meiosis (Chapter 5); secondly, a comprehensive analysis of the cyto-histological changes which take place within the meristematic regions of the green plant. Here most has been achieved with the root apex (Chapter 7), an ideal system for such studies, while the shoot apex (Chapter 8) and the cambia (Chapter 10) have received less attention. From these investigations the concept of a heterogeneous meristem has emerged and considerable amounts of information have been made available on the rates of division and cytological characteristics of cells within the meristem.

Throughout the text a number of themes recur which have received varying amounts of attention over the years but are largely the unanswered problems of the subject. It is by a closer examination of these that we can anticipate future research into cell division and attempt to define the areas which might be productive. Clearly a list of these areas cannot be comprehensive and must to some extent reflect the personal views of the editor; however, the following themes are readily discernible: organelle replication, control of the plane of division, cell communication, regulation of the rate of division, and synchrony, both as a natural phenomenon and as a means to the investigation of the process of division.

While considerable attention has been paid to the division of the nucleus (Chapter 2), less attention has been directed towards the replication of other cellular organelles (Chapters 2 and 4). Present interest in the division of plastids has been stimulated by the general enthusiasm for chloroplasts and photosynthesis and a fair amount is known about the mode of replication of plastids in leaves during the phase of cell expansion. Also, some interest has been accorded to the replication of the mitochondrion but little is known of the means by which endoplasmic reticulum, golgi and lysosomes are produced. It is perhaps salutary to note that, despite the upsurge of interest in plastid replication, the relatively simple question, when do extra-nuclear organelles replicate during the cell cycle? is still unanswered.

Clearly the shape of the green plant is related to the plane of division (Chapter 12), but the factors which determine the plane of division remain largely unresolved. Various attempts have been made to provide an answer to this question, a theme fully explored in the introductory chapter to this book. Mechanical pressure, gradients of growth substances and various other agencies have been postulated as determinants of the plane of division. The tight control exerted by the plant over the plane and rate of division begins in the pro-embryo (Chapter 11) and continues through the growth and development of the adult plant. Only in special circumstances such as the application of growth-promoting substances, tumour formation or the culture of excised tissues does the rate and plane of division become disturbed (Chapter 12) and then in most cases control is quickly reinstated.

The spatial relationships of cells within a multicellular structure, such as a meristem, raises another theme, that of cellular communication. The general lack of information on cell–cell communication is at least partly due to the difficulty of investigating such a phenomenon. Here our basic knowledge is restricted to the facts that cytoplasmic connections or plasmodesmata exist between adjacent cells and that these channels are large enough to allow the transfer of macromolecules. Apart from plasmodesmata, it must also be remembered that the plasmalemmas of opposing cells are separated only by relatively porous walls across which molecular exchange may take place. Presumably, even relatively large proteins can pass unimpeded through the microfibrillar structure of the cell wall. There is of course a general acceptance that transfer of materials does occur via plasmodesmata and that this exchange may be affected by agencies which control the plasmodesmata, perhaps by the deposition of materials which impede movement. Removal of these deposits plugging the plasmodesmata will restore the channel of communication. Such exchange of molecules could provide the basis for the tight control of division and organization observed in meristems.

Interest in the control of division inevitably leads the investigator to the molecular aspects of the process. Here a study of the dividing populations within meristems (Chapters 7 and 8) has provided a firm foundation from which to pursue further studies. However, such systems have severe limitations for investigations at the molecular level and in an attempt to overcome these difficulties another approach has been tried. This approach is relatively new and exploits the use of synchronous or synchronized cell populations (Chapter 3). The amplification provided in these systems allows the investigator to pursue biochemical parameters impossible to study in asynchronous cell populations using the techniques presently available. Synchrony is not observed naturally in meristems, and indeed cannot be induced satisfactorily (Chapters 3 and 7). There are, however, a number of natural situations in which the synchronous division of cells or nuclei is the rule, e.g. developing microspores in anthers of angiosperms (Chapter 5) and the nuclei of developing endosperm. Both of these situations have been exploited, the endosperm for time lapse studies on mitosis and the anther and its contents for the physiological and biochemical investigation of meiosis (Chapter 5). However, in order to discover the molecular basis to the mitotic cell cycle we must look to other systems. Here a promising start has been made using plant tissue and cell cultures (Chapter 12), one of which is naturally synchronous while others are synchronized quite simply without apparently upsetting cellular metabolism. While such systems can and have been used effectively to study the cytological and molecular events which precede mitosis in plant cells, they provide information about cells in an unnatural situation. However, this approach has already provided information about the interphase which can be used to gain a better understanding of the molecular events which take place within the constituent cells of the meristems of the intact plant and help to unravel the problem of the control of division.

From a brief consideration of the recurrent themes, two major problems may be identified. The first relates to the process of cell division. Here, our general lack of knowledge about the structure of the eukaryotic chromosome hinders the development of an understanding as to how transcription may be controlled. Unfortunately the electron microscope has not been able to provide us with the answers. That a large measure of control of the cell cycle takes place at the level of transcription seems fairly certain, but until a detailed picture of chromosome structure emerges our hypotheses must remain unconfirmed (Chapter 1). The second problem, which is basic to an understanding of the part played by division in the generation of form, concerns the factors which determine the plane of division. These have already received attention in this summary and elsewhere in this volume. However, I believe that the emergence of an explanation as to how form is generated must be preceded by the solution to the problem of how the plane of division is controlled.

Despite these major, rather forbidding problems there are a number of defined "growing points" within the general subject. The study of the dynamic properties of apical meristems and the structures derived from them (Chapters 7 and 9) continues most effectively, although the experimental approaches that work so well on these structures have not been fully exploited with the vascular cambium. Most of our knowledge of the cambium tends to be of a descriptive nature and there is a need for a new approach to investigate the dynamic properties of these secondary meristems. The molecular basis of division is already receiving attention (Chapter 3), but here we are only at the very beginning of a line of study, as with organelle replication. The study of sporogenesis and development of the gametophyte in angiosperms (Chapter 5) and ferns (Chapter 6) proceeds apace, as does that of embryo development (Chapter 11). Lastly, the tremendous interest accorded to the whole study of plant tissue and cell culture has contributed not only in the provision of systems to study cell division but also to the whole subject of regeneration (Chapter 12).

References

Numbers in square brackets indicate pages where references occur in the text

ABBÉ, E. C. and PHINNEY, B. O. (1951). The growth of the shoot apex in maize: external features. *Am. J. Bot.* **38,** 737–744. [285, 311, 321]

ABBÉ, E. C., PHINNEY, B. O. and BAER, D. F. (1951). The growth of the shoot apex in maize: internal features. *Am. J. Bot.* **38,** 744–751. [287, 311]

ABBÉ, E. C., RANDOLPH, L. F. and EINSET, J. (1941). The developmental relationship between shoot apex and growth pattern of leaf blade in diploid maize. *Am. J. Bot.* **28,** 778–784. [311]

ABBÉ, E. C. and STEIN, O. L. (1954). The growth of the shoot apex in maize: embryogeny. *Am. J. Bot.* **41,** 287–293. [286]

ADAMSON, D. (1962). Expansion and division in auxin-treated plant cells. *Can. J. Bot.* **40,** 719–744. [116, 201, 202, 204]

AITCHISON, P. A. (1974). Unpublished observations. [131]

AITCHISON, P. A., and YEOMAN, M. M. (1973). The use of 6-methylpurine to investigate the control of glucose-6-phosphate dehydrogenase levels in cultured artichoke tissue. *J. exp. Bot.* **24,** 1069–1083. [117, 127, 128, 130]

AITCHISON, P. A. and YEOMAN, M. M. (1974). Control of periodic enzyme synthesis in dividing plant cells. *In* "Cell Cycle Controls". (Eds G. M. Padilla, I. L. Cameron and A. Zimmerman), pp. 251–263. Academic Press, New York. [83, 117, 127.]

ALDRICH, H. C., (1968). The development of flagella in swarm cells of the myxomycete *Physarum flaviconum. J. gen. Microbiol.* **50,** 217–222. [107]

ALEXANDER, M. P. (1968). Spindle abnormalities in *Saccharum. Indian J. Genet. Pl. Breed.* **28,** 359–364. [248]

ALFERT, M. and DAS, N. K. (1969). Evidence for control of the rate of nuclear DNA synthesis by the nuclear membrane in eukaryotic cells. *Proc. natn. Acad. Sci. U.S.A.* **63,** 123–128. [122]

ALFIERI, I. R. and EVERT, R. F. (1968). Analysis of meristematic activity in the root tip of *Melilotus alba* Desr. *New Phytol.* **67,** 641–648. [62, 266]

ALVIM, P. DE T. (1964). Tree growth and periodicity in tropical climates. *In* "The Formation of Wood in Forest Trees" (Ed. M. H. Zimmermann), pp. 479–495. Academic Press, New York and London. [371, 373, 378]

AMOORE, J. F. (1961). Arrest of mitosis in roots by oxygen lack or cyanide. *Proc. R. Soc.* **154,** 95–108. [28]

ANTON-LAMPRECHT, I., (1966). Beitrage zum Problem der Plastidenabänderung. III. Über der Verkommen von "Rüchmutationen" in einer spontan entstandenen Plastidenschecke von *Epilobium hirsutum. Z. PflPhysiol.* **54,** 417–445. [94]

ARISZ, W. H. (1969). Intercellular polar transport and the role of the plasmodesmata in coleoptiles and *Vallisneria* leaves. *Acta bot. neerl.* **18,** 14–38. [99]

ASHBY, E. (1948). Studies in the morphogenesis of leaves. 1. An essay in leaf shape. *New Phytol.* **47,** 153–176. [321]

ASHLEY, T. (1972). Zygote shrinkage and subsequent development in some *Hibiscus* hybrids. *Planta* **108,** 303–317. [398]

ATKINSON, A. W., JR., JOHN, P. C. L. and GUNNING, B. E. S. (1974). The growth and division of the single mitochondrion and other organelles during the cell cycle of *Chlorella* studied by quantitative stereology and three dimensional reconstruction. *Protoplasma* 81, 77–109. [80, 88]

AUDUS, L. J. (1959). Correlations. *J. Linn. Soc. (Bot.)* **56,** 177–187. [375]

AUDUS, L. J. (1969). Geotropism. *In* "The Physiology of Plant Growth and Development" (Ed. M. B. Wilkins), pp. 203–242. McGraw-Hill, London. [385]

AVANZANI, M. G. (1950). Endomitosi e mitosi a diplochromosomi nelle sviluppe delle cellule del tappeto di *Solanum tuberosum* L. *Caryologia* **2,** 205–222. [219]

AVANZI, S., BRUNORI, A. and D'AMATO, F. (1969). Sequential development of meristems in the embryo of *Triticum durum*. A DNA autoradiographic and cytophotometric analysis. *Devl Biol.* **20,** 368–377. [275]

AVANZI, S., BRUNORI, A., D'AMATO, F. (1970). Nuclear conditions in the meristems of resting seeds of *Triticum durum*. *Wheat Inf. Serv. Kyoto Univ.* **30,** 5–6. [168, 265]

AVANZI, S., BRUNORI, A., D'AMATO, F., RONCHI, V. N. and MUGNOZZA, G. T. S. (1963). Occurrence of 2C (G_1) and 4C (G_2) nuclei in the radicle meristems of dry seeds in *Triticum durum* Desf. Its implications in studies on chromosome breakage and on developmental processes. *Caryologia* **16,** 553–558. [275]

AVANZI, S., CROSSINI, P. G. and D'AMATO, F. (1970). Cytochemical and autoradiographic analyses on the embryo suspensor cells of *Phaseolus coccineus*. *Caryologia* **23,** 605–638. [168, 221]

AVANZI, S. and D'AMATO, F. (1967). New evidence on the organisation of the root apex in Leptosporangiate ferns. *Caryologia* **20,** 257–264. [238, 256, 297]

AVANZI, S. and D'AMATO, F. (1970). Cytochemical and autoradiographic analyses on root primordia and root apices of *Marsilea strigosa*. *Caryologia* **23,** 335–345. [265]

AVANZI, S., MAGGINI, F. and INNOCENTI, A. M. (1973). Amplification of ribosomal cistrons during the maturation of metaxylem in the root of *Allium cepa*. *Protoplasma* **76,** 197–210. [203]

AVERS, C. J. (1963). Fine structure studies of *Phleum* root meristem cells. II. Mitotic asymmetry and cellular differentiation. *Am. J. Bot.* **50,** 140–148. [53, 93, 94]

AVERY, G. S. (1933). Structure and development of the tobacco leaf. *Am. J. Bot.* **20,** 565–592. [316, 328]

AVERY, G. S., BURKHOLDER, P. R. and CREIGHTON, H. B. (1937). Production and distribution of growth hormone in shoots of *Aesculus* and *Malus* and its probable role in stimulating cambial activity. *Am. J. Bot.* **24,** 51–58. [383]

AVERY, O. T., McLEOD, C. M. and McCARTY, M. (1944). Studies on the chemical nature of the substance inducing transformation of pneumococcal types. Induction of transformation by a deoxyribonucleic acid fraction isolated from *Pneumococcus* Type III. *J. exp. Med.* **79,** 137–158. [9]

AYONOADU, U. and REES, H. (1968). The regulation of mitosis by B-chromosomes in rye. *Expl Cell Res.* **52,** 284–290. [122, 183, 190]

AYONOADU, U. W. and REES, H. (1971). The effects of B chromosomes on the nuclear phenotype in root meristem of maize. *Heredity* **27,** 365–383. [214]

BACKS-HUSEMANN, D. and REINERT, J. (1970). Embryobildung durch isolierte Einzellen aus Gewebekulturen von *Daucus carota*. *Protoplasma* **70,** 49–60. [101]

BAENZIGER, H. (1962). Supernumerary chromosomes in diploid and tetraploid forms of crested wheat grass. *Can. J. Bot.* **40,** 549–561. [214]

BAGSHAW, V. (1969). "Changes in Ultrastructure during the Development of Callus Cells". Ph.D. Thesis, University of Edinburgh. [419]

BAILEY, I. W. (1923). The cambium and its derivative tissues. IV. The increase in girth of the cambium. *Am. J. Bot.* **10,** 499–509. [358]

BAILEY, I. W. (1954). Contributions to plant anatomy. *Chronica bot. Waltham, Mass., U.S.A.* **15,** 1–262. [354, 355, 356, 357, 358]

BAIN, J. M. and MERCER, F. V. (1966). Subcellular organization of the developing cotyledons of *Pisum sativum* L. *Aust. J. biol. Sci.* **19,** 49–67. [138]

BAJER, A. (1958). Cine-micrographic studies on chromosome movements in β-irradiated cells. *Chromosoma* **9,** 319–331. [115]

BAJER, A. (1968a). Behaviour and fine-structure of spindle fibres during mitosis in endosperm. *Chromosoma* **25,** 249–281. [63, 70]

BAJER, A. (1968b). Fine structure studies on phragmoplast and cell plate formation. *Chromosoma* **24,** 383–417. [63, 98]

BAJER, A. and MOLE-BAJER, J. (1954). Endosperm: Material for study of the physiology of cell division. *Acta Soc. bot. pol.* **23,** 69–110. [115]

BAJER, A. and MOLE-BAJER, J. (1971). Architecture and function of the mitotic spindle. *In* "Advances in Cell and Molecular Biology" (Ed. E. J. Du Praw), Academic Press, New York and London. Vol. 1, pp. 213–266. [51, 70, 72]

BAJER, A. S. and MOLE-BAJER, J. (1972). Spindle dynamics and chromosome movements. *Int. Rev. Cytol. Suppl.* **3.** [51, 66, 70, 72]

BAL, A. K. and DEEPESH, N. DE (1961). Developmental changes in sub-microscopic morphology of the cytoplasmic components during microsporogenesis in *Tradescantia*. *Devl Biol.* **3,** 241–254. [86]

BALATINECZ, J. J. and KENNEDY, R. W. (1968). Mechanism of earlywood–latewood differentiation in *Larix decidua*. *J. Tech. Ass. Pulp Pap. Ind.* **51,** 414–422. [385]

BALL, E. (1946). Development in sterile culture of stem tips and subjacent regions of *Tropaeolum majus* L. and *Lupinus albus* L. *Am. J. Bot.* **33,** 301–318. [409]

BALL, E. (1960). Cell divisions in living shoot apices. *Phytomorphology* **10,** 377–396. [288, 296, 366]

BALL, E. (1972). The surface "histogen" of living shoot apices. *In* "The Dynamics of Meristem Cell Populations" (Eds M. W. Miller and C. C. Kuehnert), pp. 75–97. Plenum Press, New York. [288]

BALL, E. and SOMA, K. (1965). Effects of sugar concentration on growth of the shoot apex of *Vicia faba. In* "Proc. Int. Conf. Plant Tissue Culture" (Eds P. R. White and A. R. Grove), pp. 269–285. McCutchan, Berkeley. [296, 302]

BALODIS, V. A. and IVANOV, V. B. (1970). Proliferation of root cells in the basal part of meristem and apical part of elongation zone. *Cytology* **12,** 983–992. [270]

BANNAN, M. W. (1950). Frequency of anticlinal divisions in fusiform cambial cells of *Chamaecyparis. Am. J. Bot.* **37,** 511–517. [358]

BANNAN, M. W. (1955). The vascular cambium and radial growth in *Thuja occidentalis* L. *Can. J. Bot.* **33,** 113–138. [361, 362, 374, 375]

BANNAN, M. W. (1960). Ontogenetic trends in conifer cambium with respect to frequency of anticlinal division and cell length. *Can. J. Bot.* **38,** 795–802. [356]

BANNAN, M. W. (1962). The vascular cambium and tree ring development. *In* "Tree Growth" (Ed. T. T. Kozlowski), pp. 3–21. Ronald Press, New York. [373]

BANNAN, M. W. (1968). Polarity in the survival and elongation of fusiform initials in conifer cambium. *Can. J. Bot.* **46,** 1005–1008. [362]

BARBER, H. N. (1941). Chromosome behaviour in *Uvularia. J. Genet.* **42,** 223–257. [246]

BARBER, H. N. (1942). The experimental control of chromosome pairing in *Fritillaria. J. Genet.* **43,** 359–374. [176, 194]

BARGHOORN, E. S., JR. (1940a). Origin and development of the uniseriate ray in the Coniferae. *Bull. Torrey bot. Club* **67,** 303–328. [358]

BARGHOORN, E. S., JR. (1940b). The ontogenetic development and phylogenetic specialization of rays in the xylem of dicotyledons. I. The primitive ray structure. *Am. J. Bot.* **27,** 918–928. [358]

BARGHOORN, E. S., JR. (1941). The ontogenetic development and phylogenetic specialization of rays in the xylem of dicotyledons. II. Modification of the multiseriate and uniseriate rays. *Am. J. Bot.* **28,** 273–282. [358]

BARLOW, P. W. (1969). "Organisation in Root Meristems". D.Phil. Thesis, University of Oxford. [282]

BARLOW, P. W. (1970). RNA synthesis in the root apex of *Zea mays. J. exp. Bot.* **21,** 292–299. [117, 272]

BARLOW, P. W. (1971). Properties of cells in the root apex. *Revista Fac. Agron.* **47,** 275–301. [283]

BARLOW, P. W. (1972a). The ordered replication of chromosomal DNA: a review and a proposal for its control. *Cytobios* **6,** 55–79. [120]

BARLOW, P. W., (1972b). The effect of additional chromosomes on cell division. *In* "Chromosomes Today" (Eds C. D. Darlington and K. R. Lewis), Vol. 3, p. 299. Oliver and Boyd, Edinburgh. [208, 214]

BARLOW, P. W. (1973). Mitotic cycles in root meristems. *In* "The Cell Cycle in Development and Differentiation" (Eds M. Balls and F. S. Billett), *Symp. Br. Soc. Dev. Biol.* **1,** 133–165. Cambridge University Press. [53, 64, 65, 76, 205, 264, 269, 274, 275]

BARLOW, P. W. and MACDONALD, P. D. M. (1973). An analysis of the mitotic cell cycle in the root meristem of *Zea mays*. *Proc. R. Soc.* B **183,** 385–398. [269, 270, 274]

BARLOW, P. W. and VOSA, C. G. (1970). The effect of supernumerary chromosomes on meiosis in *Puschkinia libanotica* (Liliaceae). *Chromosoma* **30,** 344–355. [208, 214]

BARTELS, F. (1964). Plastidenzählungen bei *Epilobium hirsutum*. 1. Mitteilung. Zählungen in Zellen aus unterschiedlich differenzierten geweben des Laubblattes. *Planta* **60,** 434–452. [81]

BAYLISS, M. W. (1974). Unpublished results. [183, 190]

BEARDSLEY, R. E. (1972). The inception phase in the crown gall disease. *Prog. exp. Tumor Res.* **15,** 1–75. [425]

BEATTY, J. W. and BEATTY, A. V. (1953). Duration of the stages of microspore development and in the first microspore division of *Tradescantia paludosa*. *Am. J. Bot.* **40,** 593–596. [162, 176, 189, 190]

BEIDERBECK, R. (1970). Untersuchungen an crown-gall. IV. Rifampicin und ein resisteuter Klon von *Agrobacterium tumefaciens* bei der Tumor induktion. *Z. Naturforsch. Teil.* B **25,** 1458–1460. [426]

BEIDERBECK, R. (1971). Untersuchungen an crown-gall. V. Der Einfluss von Polyornithin auf die Tumorinduktion durch *Agrobacterium tumefaciens*. *Z. PflPhysiol.* **64,** 199–205. [426]

BELL, P. R. (1970a). The archegoniate revolution. *Sci. Prog.* **58,** 27–45. [103]

BELL, P. R. (1970b). Are plastids autonomous? *In* "The Control of Organelle Development" (Ed. P. Miller), *Symp. Soc. exp. Biol.* **24,** 109–127. Cambridge University Press. [80]

BELL, P. R., FREY-WYSSLING, A. and MÜHLETHALER, K. (1966). Evidence for the discontinuity of plastids in the sexual reproduction of a plant. *J. Ultrastruct. Res.* **15,** 108–121. [80, 86]

BELL, P. R. and RICHARDS, B. M. (1958). Induced apospory in polypodiaceous ferns. *Nature, Lond.* **182,** 1748–1749. [103]

BENNETT, M. D. (1970). Natural variation in nuclear characters of meristems in *Vicia faba*. *Chromosoma* **29,** 317–335. [211]

BENNETT, M. D. (1971). The duration of meiosis. *Proc. R. Soc.* B **178,** 277–299. [162, 175, 177, 184, 194, 198]

BENNETT, M. D. (1972). Nuclear DNA content and minimum generation time in herbaceous plants. *Proc. R. Soc.* B **181,** 109–135. [162, 175, 192, 194, 195, 196, 198, 210, 212]

BENNETT, M. D. (1973). The duration of meiosis. *In* "The Cell Cycle in Development and Differentiation" (Eds M. Balls and F. S. Billett), *Symp. Br. Soc. Devel. Biol.* **1,** 111–131., Cambridge University Press. [55, 162, 176, 177, 180, 183, 184]

BENNETT, M. D. (1974). Nuclear characters in plants. *In* "Basic Mechanisms in Plant Morphogenesis", *Brookhaven Symposia in Biology* **25,** 344–366. [182, 192, 194, 195, 198]

BENNETT, M. D., CHAPMAN, V. C., and RILEY, R. (1971). The duration of meiosis in pollen mother cells of wheat, rye and *Triticale. Proc. R. Soc.* B **178,** 259–275. [162, 176, 183]

BENNETT, M. D. and FINCH, R. A. (1971). The duration of meiosis in barley. *Genet. Res.* **17,** 209–214. [176]

BENNETT, M. D., FINCH, R. A., SMITH, J. B., RAO, M. K. (1973a). The time and duration of female meiosis in wheat, rye and barley. *Proc. R. Soc.* B **183,** 301–319. [162, 185, 187, 188]

BENNETT, M. D. and HUGHES, W. G. (1972). Additional mitosis in wheat pollen induced by Ethrel. *Nature, Lond.* **240,** 566–568. [169, 171, 173]

BENNETT, M. D. and KALTSIKES, P. J. (1974). Unpublished observations. [176]

BENNETT, M. D., RAO, M. K., SMITH, J. B. and BAYLISS, M. W. (1973b). Cell development in the anther, the ovule, and the young seed of *Triticum aestivum* L. var. Chinese Spring. *Phil. Trans. R. Soc.* B **266,** 39–81. [162, 164, 166, 168, 172, 183, 189, 191, 193]

BENNETT, M. D. and REES, H. (1969). Induced and developmental variation in chromosomes of meristematic cells. *Chromosoma* **27,** 226–244. [211]

BENNETT, M. D. and SMITH, J. B. (1972). The effects of polyploidy on meiotic duration and pollen development in cereal anthers. *Proc. R. Soc.* B **181,** 81–107. [162, 176, 177, 179, 183, 184, 190, 212, 216]

BENNETT, M. D. and SMITH, J. B. (1973). Genotypic, nucleotypic and environmental effects on meiotic time in wheat. *In* "4th International Wheat Genetics Symposium—University of Missouri" (Eds. E. R. Sears and L. M. S. Sears, pp. 673–644. [162]

BENNETT, M. D., SMITH, J. B. and KEMBLE, R. (1972). The effect of temperature on meiosis and pollen development in wheat and rye. *Can. J. Genet. Cytol.* **14,** 615–624. [162, 176, 180, 184, 185, 189, 190]

BENNICI, A., BUIATTI, M. and D'AMATO, F. (1968). Nuclear conditions in haploid *Pelargonium in vivo* and *in vitro. Chromosoma* **24,** 194–201. [217, 233]

BERGER, C. A., CASIMIR HEROLD, M. and WITKUS, E. R. (1966). Diazouracil and the mitotic spindle. *Caryologia* **19,** 255–259.

BERGER, C. A., McMAHON, R. M. and WITKUS, E. R. (1955). The cytology of *Xanthisma texanum* D.C. III. Differential somatic reduction. *Bull. Torrey bot. Club* **82,** 377–382. [208, 214]

BERGER, C. A. and WITKUS, E. R. (1954). The cytology of *Xanthisma texanum* D.C. I. Differences in the chromosome number of root and shoot. *Bull. Torrey bot. Club* **81,** 489–491. [214]

BERI, S. M. and ANAND, S. C. (1971). Factors affecting pollen shedding capacity in wheat. *Euphytica* **20,** 327–332. [164]

BERLYN, G. P. (1961). Factors affecting the incidence of reaction tissue in *Populus deltoides* Bartr. *Iowa St. J. Sci.* **35,** 367–424. [380]

Bernier, G. (1963). *Sinapis alba* L., a new long-day plant requiring a single photoinductive cycle. *Naturwissenschaften* **50**, 101. [173]

Bernier, G. (1971). Structural and metabolic changes in the shoot apex in transition to flowering. *Can. J. Bot.* **49**, 803–819. [313]

Bernier, G., Bronchart, R., Jacquard, A. and Sylvestre, G. (1967a). Acide gibbérellique et morphogénèse caulinaire. *Bull. Soc. r. Bot. Belg.* **100**, 51–71. [301]

Bernier, G., Bronchart, R. and Kinet, J-M. (1970). Nucleic acid synthesis and mitotic activity in the apical meristem of *Sinapis alba* during floral induction. *In* "Cellular and Molecular Aspects of Floral Induction" (Ed. G. Bernier, pp. 51–79). Longman, London. [313]

Bernier, G., Kinet, J-M. and Bronchart, R. (1967b). Cellular events at the meristem during floral induction in *Sinapis alba* L. *Physiol. Veg.* **5**, 311–324. [313]

Bertagnolli, B. L. and Nadakavurkaren, M. J. (1970). An ultrastructural study of pyrenoids from *Chlorella pyrenoidosa*. *J. Cell Sci.* **7**, 623–630. [83, 88]

Binding, H. (1966). Regeneration und Verschmelzung nackter Laubmoosprotoplasten. *Z. PflPhysiol.* **55**, 305–321. [420]

Bird, A. P. and Birnstiel, M. L. (1971). A timing study of DNA amplification in *Xenopus laevis* oocytes. *Chromosoma* **35**, 300–309. [169]

Birnstiel, M. (1967). The nucleolus in cell metabolism. *A Rev. Pl. Physiol.* **18**, 25–58. [125]

Bisalpultra, T., Ashton, F. M. and Weier, T. E. (1966). Role of dictyosomes in wall formation during cell division of *Chlorella vulgaris*. *Am. J. Bot.* **53**, 213–216. [79, 83]

Bisalpultra, T. and Bisalpultra, A. A. (1969). The ultrastructure of chloroplast of a brown alga *Sphacelaria* sp. 1. Plastid DNA configuration—the chloroplast genophore. *J. Ultrastruct. Res.* **29**, 151–170. [159]

Bisalpultra, T. and Bisalpultra, A. A. (1970). The ultrastructure of chloroplast of a brown alga *Sphacelaria* sp. III. The replication and segregation of chloroplast genophore. *J. Ultrastruct. Res.* **32**, 417–429. [93]

Black, L. M. (1965). Physiology of virus induced tumors in plants *In* "Handbuch der Pflanzenphysiologie (Ed. W. Ruhland), Vol. XV (2), pp. 236–266. Springer-Verlag, Berlin. [425]

Black, L. M. (1972). Plant tumors of viral origin. *Progr. exp. Tumor Res.* **15**, 110–137. [425]

Black, M. and Vlitos, A. J. (1972). Possible interrelationships of phytochrome and plant hormones. *In* "Phytochrome" (Eds K. Mitrakos and W. Shropshire, Jr.), pp. 515–550. Academic Press, London and New York. [389]

Blakely, L. M. and Steward, F. C. (1961). Growth induction in cultures of *Haplopappus gracilis*. I. The behaviour of cultured cells. *Am. J. Bot.* **48**, 351–358. [45, 417]

Blakely, L. M. and Steward, F. C. (1964). Growth and organised development of cultured cells. VII. Cellular variation. *Am. J. Bot.* **51**, 809–820. [249]

BLOCH, R. (1965). Histological foundations of differentiation and development in plants. *In* "Encyclopedia of Plant Physiology" (Ed. W. Ruhland), Vol. XV/I, pp. 146–188. Springer-Verlag, Berlin. [363]

BOASSON, R. and LAETSCH, W. M. (1969). Chloroplast replication and growth in tobacco. *Science, N.Y.* **166**, 749–751. [84, 155, 156]

BOASSON, R., BONNER, J. J. and LAETSCH, W. M. (1972a). Induction and regulation of chloroplast replication in mature tobacco leaf tissue. *Pl. Physiol., Lancaster* **49**, 97–101. [81, 139, 145, 146, 147, 148, 152, 153]

BOASSON, R., LAETSCH, W. M. and PRICE, I. (1972b). The etioplast-chloroplast transformation in tobacco: correlation of ultrastructure, replication and chlorophyll synthesis. *Am. J. Bot.* **59**, 217–223. [81, 145, 146, 147, 148, 149, 150]

BOCHER, T. W. and LEWIS, M. C. (1962). Experimental and cytological studies on plant species. VII. *Geranium sanguineum*. *Skrifter K. Dansk Videnskab Selskab* **11**, 1–25. [321]

BODSON, M. (1975). Variation in the rate of cell division in the apical meristem of *Sinapis alba* during transition to flowering. *Ann. Bot.* **39**, 547–554. [313]

BOLLUM, F. J. and POTTER, V. R. (1959). Nucleic acid metabolism in regenerating rat liver. VI. Soluble enzymes which convert thymidine to thymidine phosphates and DNA. *Cancer Res.* **19**, 561–565. [128]

BORNER, R. VON and RAMSHORN, K. (1968). Untersuchungen über die Veränderung der Basenzusammensetzung der RNS an unterschiedlich differnziertem Wurzelgewebe von *Vicia faba*. *Biol. Zbl.* **87**, 411–418. [275]

BOSEMARK, N. O. (1956). On accessory chromosomes in *Festuca uratensis*. III. Frequency and geographical distribution of plants with accessory chromosomes. *Hereditas* **42**, 189–210. [208, 213, 214]

BOSEMARK, N. O. (1957). Further studies on accessory chromosomes in grasses. *Hereditas* **43**, 236–297. [208, 214]

BOSTOCK, C. (1971). Repetitious DNA. *In* "Advances in Cell Biology" (Eds D. M. Prescott, L. Goldstein and E. McConkey **2**, pp. 153–223. Academic Press, New York and London. [210]

BOUCK, G. B. (1965). Fine structural and organelle associations in brown algae. *J. Cell Biol.* **26**, 523–537. [83]

BOULTER, D., ELLIS, R. J. and YARWOOD, A. (1972). Biochemistry of protein synthesis in plants. *Biol. Rev.* **47**, 113–175. [154]

BOURQUE, D. P. and WILDMAN, S. G. (1973). Evidence that nuclear genes code for several chloroplast ribosomal proteins. *Biochem. biophys. Res. Commun.* **50**, 532–537. [155]

BRACKER, C. E. and GROVE, S. N. (1971). Continuity between cytoplasmic endomembrane and outer mitochondrial membranes in fungi. *Protoplasma* **73**, 15–34. [79]

BRADBEER, J. W., IRELAND, H. M. M., SMITH, J. W., REST, J. and EDGE, H. J. (1974). Plastid development in primary leaves of *Phaseolus vulgaris*, VII. Development during growth in continuous darkness. *New Phytol.* **73**, 263–270. [84]

BRADLEY, M. W. and CRANE, J. C. (1957). Gibberellin-stimulated cambial activity in stems of apricot spur shoots. *Science, N.Y.* **126,** 972–973. [383]

BRÅTEN, T. (1973). Autoradiographic evidence for the rapid disintegration of one chloroplast in the zygote of the green alga *Ulva mutabilis. J. Cell Sci.* **12,** 385–389. [81]

BRAUN, A. C. (1947). Thermal studies on the factors responsible for tumour initiation in crown gall. *Am. J. Bot.* **34,** 234–240. [425]

BRAUN, A. C. (1952). Conditioning of the host cell as a factor in the transformation process in crown gall. *Growth* **16,** 65–74. [425]

BRAUN, A. C. (1962). Tumour inception and development in the crown gall disease. *A. Rev. Pl. Physiol.* **13,** 533–558. [425, 426, 428]

BRAUN, A. C. and LASKARIS, T. (1942). Tumour formation by attenuated crown gall bacteria in the presence of growth promoting substances. *Proc. natn. Acad. Sci. U.S.A.* **28,** 468–477. [425]

BRAUN, A. C. and STONIER, T. (1958). Morphology and physiology of plant tumours. *Protoplasmatologia* **10** (5a), 1–93. [426]

BRENT, T. P., BUTLER, J. A. V. and CRATHORN, A. R. (1965). Variations in phosphokinase activities during the cell cycle in synchronous populations of Hela cells. *Nature, Lond.* **207,** 176–177. [128]

BREWER, E. N. and RUSCH, H. P. (1968). Effect of elevated temperature shocks on mitosis and on the initiation of DNA replication in *Physarum polycephalum. Expl Cell Res.* **49,** 79–86. [225]

BRIGGS, W. R. and SIEGELMAN, H. W. (1965). Distribution of phytochrome in etiolated seedlings. *Pl. Physiol., Lancaster* **40,** 934–941. [333]

BRINKLEY, B. R. (1965). The fine structure of the nucleolus in mitotic divisions of Chinese Hamster Hamster cells *in vitro. J. Cell Biol.* **27,** 411–422. [69, 125]

BRISTOW, J. M. (1962). The controlled *in vitro* differentiation of callus derived from a fern, *Pteris cretica* L., into gametophytic or sporophytic tissue. *Devl Biol.* **4,** 361–375. [103]

BRITTON, D. M. and HULL, J. W. (1957). Mitotic instability in *Rubus. J. Hered.* **48,** 11–20. [248]

BROWN, C. L. (1964). The influence of external pressure on the differentiation of cells and tissues cultured *in vitro. In* "The Formation of Wood in Forest Trees" (Ed. M. Zimmermann), pp. 389–404. Academic Press, New York and London. [367]

BROWN, C. L. and SAX, K. (1962). The influence of pressure on the differentiation of secondary tissues. *Am. J. Bot.* **49,** 683–691. [367]

BROWN, F. A. M. and GUNNING, B. E. S. (1967). Distribution of ribosome-like particles in *Avena* plastids. *In* "Biochemistry of Chloroplasts" (Ed. T. W. Goodwin), Vol. I, pp. 365–373. Academic Press, London and New York. [154]

BROWN, J. M. and BERRY, R. J. (1968). Effects of Radiation on Cellular Proliferation and Differentiation. *Int. atom. Energy Ag. Bull.* 475–491. [271]

BROWN, J. M. and BERRY, R. J. (1969). Effects of X-irradiation on the cell population kinetics in a model tumour and normal tissue system: implications for the treatment of human malignancies. *Br. J. Radiol.* **42,** 372–377. [271]

BROWN, M. S. (1947). A case of spontaneous reduction of chromosome number in somatic tissue of cotton. *Am. J. Bot.* **34,** 384–388. [228]

BROWN, M. S. (1949). Polyploids and aneuploids derived from species hybrids in *Gossypium.* Proc. 8th Internat. Congr. Genetics. *Hereditas* Suppl. Volume, 543–545. [218, 237, 248]

BROWN, R. (1951). The effects of temperature on the durations of the different stages of cell division in the root tip. *J. exp. Bot.* **2,** 96–110. [180, 289]

BROWN, R. (1963). Protein synthesis during cell growth and differentiation. *Brookhaven Symp. Biol.* **16,** 157–169. [8]

BROWN, R. (1975). Unpublished observations.

BROWN, R. and BROADBENT, D. (1950). The development of cells in the growing zones of the root. *J. exp. Bot.* **1,** 249–263.

BROWN, R. and DYER, A. F. (1972). Cell division in higher plants. *In* "Plant Physiology" (Ed. F. C. Steward) Vol. VI C, pp. 49–90. Academic Press, New York and London. [20, 51, 54, 79, 89, 92, 98, 123, 201, 210, 216, 237, 408]

BROWN, R. and RICKLESS, P. A. (1949). A new method for the study of cell division and cell extension with preliminary observations on the effect of temperature and nutrients. *Proc. R. Soc.* B **136,** 110–125. [261]

BROWN, R. and SUNDERLAND, N. (1975). Unpublished data. [32]

BROWN, R. and WIGHTMAN, F. (1952). The influence of mature tissue on division in the meristem of the root. *J. exp. Bot.* **3,** 253–263. [29]

BROWN, S. J. and KLEIN, R. M. (1973). Effects of near ultraviolet and visible radiations on cell cycle kinetics in excised root meristems of *Pisum sativum. Am. J. Bot.* **60,** 554–560. [65]

BROWN, S. W. (1949). Endomitosis in the tapetum of tomato. *Am. J. Bot.* **36,** 703–716. [218, 237, 248]

BROWN, W. V. and EMERY, W. H. P. (1957). Persistent nucleoli and grass systematics. *Am. J. Bot.* **44,** 585–590. [69]

BRUMFIELD, R. T. (1941). Asymmetrical spindles in the first microspore division of certain Angiosperms. *Am. J. Bot.* **35,** 713–722. [56]

BRUMFIELD, R. T. (1943). Cell lineage studies in root meristems by means of chromosome rearrangements induced by X-rays. *Am. J. Bot.* **30,** 101–110. [239, 259, 267, 280]

BRUNORI, A. (1967). Relationship between DNA synthesis and water content during ripening of *Vicia faba* seed. *Caryologia* **20,** 333–338. [275]

BRUNORI, A. (1971). Synthesis of DNA and mitosis in relation to cell differentiation in the roots of *Vicia faba* and *Lactuca sativa. Caryologia* **24,** 209–215. [233, 276]

BRUNORI, A. and D'AMATO, F. (1967). The DNA content of nuclei in the embryos of dry seeds of *Pinus pinea* and *Lactuca sativa. Caryologia* **20,** 153–161. [275]

BRYAN, J. H. D. (1951). DNA protein relations during microsporogenesis of *Tradescantia. Chromosoma* **4,** 369–392. [162, 189]

BUI-DANG-HA, D. and MACKENZIE, I. A. (1973). The division of protoplasts from *Asparagus officinalis* L. and their growth and differentiation. *Protoplasma* **78,** 215–221. [421, 425]

BÜNNING, E. (1952). Weitere Untersuchungen über die Differenzierungsuorgänge in Wurzeln. *Z. Bot.* **40,** 385–406. [303]

BÜNNING, E. (1956). General processes of differentiation. *In* "The Growth of Leaves" (Ed. F. L. Milthorpe), pp. 18–30. Butterworths, London. [102]

BÜNNING, E. (1957). Polarität und Inequele Teilung des Pflanzlichen Protoplasten. *Protoplasmatologia* **VIII,** 1–86. [40]

BURGESS, J. (1970a). Microtubules and cell division in the microspore of *Dactylorchis fuchsii*. *Protoplasma* **69,** 253–264. [56, 70, 94, 107]

BURGESS, J. (1970b). Interactions between microtubules and the nuclear envelope during mitosis in a fern. *Protoplasma* **71,** 77–89. [101, 107]

BURGESS, J. (1970c). Cell shape and mitotic spindle formation in the generative cell of *Endymion non-scriptus*. *Planta* **95,** 72–85. [60]

BURGESS, J. (1971). Observations on structure and differentiation in plasmodesmata. *Protoplasma* **73,** 83–95. [98]

BURGESS, J. (1972). The occurrence of plasmodesmata-like structure in a non-dividing wall. *Protoplasma* **74,** 449–458. [98, 99]

BURGESS, J. and NORTHCOTE, D. H. (1967). A function of the pre-prophase band of microtubules in *Phleum pratense*. *Planta,* **75,** 319–326. [419]

BURHOLT, D. R. and VAN'T HOF, J. (1971). Quantitative thermal-induced changes in growth and cell population kinetics of *Helianthus* roots. *Am. J. Bot.* **58,** 386–393. [180, 289, 304]

BURNS, J. A. (1972). Pre-leptotene chromosome contraction in *Nicotiana* species. *J. Hered.* **63,** 175–178. [184]

BURR, F. A. (1969). Reduction in chloroplast number during gametophyte regeneration in *Megaceros flagellaris*. *Bryologist* **72,** 200–209. [83]

BURR, F. A. (1970). Phylogenetic transitions in the chloroplasts of the Anthocerotales. I. The number and ultrastructure of the mature plastids. *Am. J. Bot.* **57,** 97–110. [83, 86]

BURTON, K. (1956). A study of the conditions and mechanisms of the diphenylamine reactions for the colorimetric estimation of Deoxyribonucleic Acid. *Biochem. J.,* **62,** 315–323. [118]

BUTCHER, D. N. (1973). The origins, characteristics and culture of plant tumour cells. *In* "Plant Tissue and Cell Culture" (Ed. H. E. Street), Botanical Monographs **11,** 356–391. Blackwell Scientific Publications, Oxford [425, 426]

BUTLER, R. D. (1963). The effect of light intensity on stem and leaf growth in broad bean seedlings. *J. exp. Bot.* **14,** 142–152. [332]

BUTTERFASS, T. (1969). Die plastidenverteilung bei der mitose der Schlierzellenmutterzellen. von haploiden Schwedenklee (*Trifolium hybridum* L.). *Planta* **84,** 230–234. [81, 85]

BUTTERFASS, T. (1973). Control of plastid division by means of nuclear DNA amount. *Protoplasma* **76,** 167–195. [81, 85, 93]

Buvat, R. (1952). Structure, evolution et fonctionnement du méristème apical de quelques dicotylédonées. *Annls Sci. nat. (Bot.) Ser. II* **13**, 199–300. [288]

Byrne, J. M. (1973). The root apex of *Malva sylvestris*. III. Lateral root development and the quiescent center. *Am. J. Bot.* **60**, 657–662. [267]

Byrne, J. M. and Heimsch, C. (1970). The root apex of *Malva sylvestris*. I. Structural development. *Am. J. Bot.* **57**, 1170–1178. [266]

Cairns, J. (1962). A minimum estimate for the length of the DNA of *Escherichia coli* obtained by autoradiography. *J. molec. Biol.* **4**, 407–409. [118]

Callan, H. G. (1972). Replication of DNA in the chromosomes of eukaryotes. *Proc. R. Soc.* B **181**, 19–41. [167, 169]

Callan, H. G. (1973). Replication of DNA in eukaryotic chromosomes. *Br. med. Bull.* **29**, 192–195. [118]

Camus, G. (1949). Recherches sur le côte des bourgeons dans les phénomènes de morphogénèse. *Revue Cytol. Biol. vég.* **9**, 1–99. [429]

Carlson, E. M. and Stuart, B. C. (1936). Development of spores and gametophytes in certain New World species of *Salvia*. *New Phytol.* **35**, 68–91. [184]

Carothers, Z. B. (1972). Membrane continuity between plasmalemma and nuclear envelope in spermatogenic cells of *Blasia*. *Science, N.Y.* **175**, 652–654. [79]

Carr, D. J. and Burrows, W. J. (1966). Evidence for the presence in xylem sap of substances with kinetin-like activity. *Life Sci.* **5**, 2061–2077. [385]

Caspersson, T., Farber, S., Foley, G. E., Kudynowski, J., Modest, E. J., Simmonsson, E., Wagh, U. and Zech, L. (1968). Chemical differentiation along metaphase chromosomes. *Expl Cell Res.* **49**, 219–222. [209]

Catesson, A. M. (1964). Origine, fonctionnement et variations cytologiques saisonnieres du cambium de l'*Acer pseudoplatanus* L. (Aceracées). *Annls Sci. nat. (Bot.) 12e ser.* **5**, 229–498. [352, 362, 375]

Catesson, A. M. (1974). Cambial cells. *In* "Dynamic Aspects of Plant Ultrastructure" (Ed. A. W. Robards), pp. 358–390. McGraw-Hill, London. [54, 98]

Cheadle, V. I. and Esau, K. (1964). Secondary phloem of *Liriodendron tulipifera*. *Univ. Calif. Pubs. Bot.* **36**, 143–252. [359]

Chen, C. H. and Ross, J. G. (1963). Colchicine-induced somatic chromosome reduction in *Sorghum*. 1. Induction of diploid plants from tetraploid seedlings. *J. Hered.* **54**, 96–100. [230]

Chi, E. Y., (1971). Brown algae pyrenoids. *Protoplasma* **72**, 101–104. [83, 88]

Chlyah, H. (1974). Formation and propagation of cell division centres in the epidermal layer of internodal segments of *Torenia fournieri* grown *in vitro*. Simultaneous surface observations of all the epidermal cells. *Can. J. Bot.* **52**, 867–872. [102]

Chouinard, A. L. (1955). Nuclear differences in *Allium cepa* root tissues as revealed through induction of mitosis with indoleacetic acid. *Can. J. Bot.* **33**, 628–646. [217]

Chowdhury, K. A. (1961). Growth rings in tropical trees and taxonomy. *Pacif. Sci. Congr. 10th, Abstracts*, 280. [371]

CHRISTENSEN, J. E. and HORNER, JR., H. T. (1970). Microsporogenesis in *Sorghum vulgare*. I. Tapetum. *Am. J. Bot.* **57,** 740. [99]

CHUPEAU, Y. and MOREL, G. (1970). Obtention de protoplastes de plantes supérieur a partir de tissus cultivés *in vitro*. *C.r. hebd. Séanc. Acad. Sci. Paris Ser. D* **270,** 2659–2662. [420]

CHURCH, K. and WIMBER, D. E. (1969). Meiosis in *Ornithogalum virens* (Liliaceae) meiotic timing and segregation of ³H-thymidine labelled chromosomes. *Can. J. Genet. Cytol.* **11,** 573–581. [176]

CLOWES, F. A. (1950). Root apical meristems of *Fagus sylvatica*. *New Phytol.* **49,** 248–268. [260]

CLOWES, F. A. (1953). The cytogenerative centre in roots with broad columellas. *New Phytol.* **52,** 48–57. [268]

CLOWES, F. A. (1954). The promeristem and the minimal constructional centre in grass root apices. *New Phytol.* **53,** 108–116. [265]

CLOWES, F. A. L. (1956). Nucleic acids in root apical meristems of *Zea*. *New Phytol.* **55,** 29–34. [265]

CLOWES, F. A. L. (1958a). Development of quiescent centres in root meristems. *New Phytol.* **57,** 85–88. [266]

CLOWES, F. A. L. (1958b). Protein synthesis in root meristems. *J. exp. Bot.* **9,** 229–238. [272]

CLOWES, F. A. L. (1959a). Reorganization of root apices after irradiation. *Ann. Bot.* **23,** 205–210. [260, 268, 276]

CLOWES, F. A. L. (1959b). Apical meristems of roots. *Biol. Rev.* **34,** 501–529. [260]

CLOWES, F. A. L. (1961a). Effects of β-radiation on meristems. *Expl Cell Res.* **25,** 529–534. [268, 276]

CLOWES, F. A. L. (1961b). Duration of the mitotic cycle in a meristem. *J. exp. Bot.* **12,** 283–293. [268, 269, 289]

CLOWES, F. A. L. (1962). Rates of mitosis in a partially synchronous meristem. *New Phytol.* **61,** 111–118. [62]

CLOWES, F. A. L. (1963a). X-irradiation of root meristems. *Ann. Bot.* **27,** 343–352. [278]

CLOWES, F. A. L. (1963b). Micronuclei in irradiated meristems. *Radiat. Bot.* **3,** 223–229. [278]

CLOWES, F. A. L. (1965a). Synchronization in a meristem by 5-amino-uracil. *J. exp. Bot.* **16,** 581–586. [12, 115]

CLOWES, F. A. L. (1965b). The duration of the G_1 phase of the mitotic cycle and its relation to radio-sensitivity. *New Phytol.* **64,** 355–359. [120, 122, 184, 269]

CLOWES, F. A. L. (1967). Synthesis of DNA during mitosis. *J. exp. Bot.* **18,** 740–745. [20, 64, 269]

CLOWES, F. A. L. (1968). The DNA content of the cells of the quiescent centre and root cap of *Zea mays*. *New Phytol.* **67,** 631–639. [270, 275]

CLOWES, F. A. L. (1970). The proportion of cells that divide in root meristems. of *Zea mays* L. *Ann. Bot.* **35,** 249–261. [184]

CLOWES, F. A. L. (1970a). The immediate response of the quiescent centre to X-rays. *New Phytol.* **69,** 1–18. [278]

CLOWES, F. A. L. (1970b). Nutrition and the quiescent centre of root meristems. *Planta* **90**, 340–348. [282]

CLOWES, F. A. L. (1971). The proportion of cells that divide in root meristems of *Zea mays* L. *Ann. Bot.* **35**, 249–261. [256, 269, 271]

CLOWES, F. A. L. (1972a). Non-dividing cells in meristems. *In* "Chromosomes Today" (Eds C. D. Darlington and K. R. Lewis), Vol. 3, pp. 110–117. Oliver and Boyd, Edinburgh. [239]

CLOWES, F. A. L. (1972b). Cell cycles in a complex meristem after X-irradiation. *New Phytol.* **71**, 891–897. [278]

CLOWES, F. A. L. and JUNIPER, B. E. (1964). The fine structure of the quiescent centre and neighbouring tissues in root meristems. *J. exp. Bot.* **15**, 622–630. [272]

CLOWES, F. A. L. and JUNIPER, B. E. (1968). "Plant Cells", Botanical Monographs **8**, Blackwell Scientific Publications, Oxford. [81, 112, 117, 120, 138, 261]

CLOWES, F. A. L. and STEWART, H. E. (1967). Recovery from dormancy in roots. *New Phytol.* **66**, 115–123. [224, 276, 279]

CLUTTER, M. E. (1960). Hormonal induction of vascular tissue in tobacco pith *in vitro*. *Science, N.Y.* **132**, 548–549. [368]

CLUTTER, M. E. and SUSSEX, I. M. (1968). Ultrastructural development of bean embryo cells containing polytene chromosomes. *J. Cell Biol.* **39**, 26a. [400]

COCKING, E. C. (1960). A method for the isolation of plant protoplasts and vacuoles. *Nature, Lond.* **187**, 962–963. [420]

COCKING, E. C. (1961). Properties of isolated plant protoplasts. *Nature, Lond.* **191**, 780–2. [420]

COCKING, E. C. (1972). Plant cell protoplasts—isolation and development. *A. Rev. Pl. Physiol.* **23**, 29–50. [420]

COCKING, E. C. (1974). Ultrastructure of cultured plant cells. *In* "Dynamic Aspects of Plant Ultrastructure" (Ed. A. W. Robards), pp. 310–330. McGraw-Hill, London. [101]

COCKING, E. C. and EVANS, P. K. (1973). The isolation of protoplasts. *In* "Plant Tissue and Cell Culture" (Ed. H. E. Street), Botanical Monographs **11**, 100–120. Blackwell Scientific Publications, Oxford. [411, 420, 424, 425]

COCUCCI, A. and JENSEN, W. A. (1969). Orchid embryology: the mature megagametophyte of *Epidendrum scutella*. *Kurtziana* **5**, 23–28. [393]

COGGINS, L. N. and GALL, J. G. (1972). The timing of meiosis and DNA synthesis during early oogenesis in the toad, *Xenopus laevis*. *J. Cell Biol.* **52**, 569–576. [169]

COHEN, S. S., FLAKS, J. G., BARNER, H. D., LOEB, M. R. and LICHTENSTEIN, J. (1958). The mode of action of 5-fluorouracil and its derivatives. *Proc. natn. Acad. Sci. U.S.A.* **44**, 1004–1012. [155]

COMINGS, D. E. (1968). The rationale for an ordered arrangement of chromatin in the interphase nucleus. *Am. J. hum. Genet.* **20**, 440–460. [107]

COMINGS, D. E. (1970). The distribution of sister chromatids at mitosis in Chinese Hamster cells. *Chromosoma* **29**, 428–433. [67]

COMINGS, D. E. (1971). Isolabelling and chromosome strandedness. *Nature New Biol.* **229,** 24–25. [67]

COOK, J. R. (1973). Unbalanced growth and replication of chloroplast populations in *Euglena gracilis. J. gen. Microbiol.* **75,** 51–60. [83, 86, 88]

COOPER, D. C. (1952). The transfer of desoxyribose nucleic acid from the tapetum to the microsporocytes at the onset of meiosis. *Am. Nat.* **86,** 219–229. [184]

COOPER, H. L. (1971). Biochemical alterations accompanying initiation of growth in resting cells. *In* "The Cell Cycle and Cancer" (Ed. R. Baserga) pp. 197–226. Dekker, New York. [203]

COOPER, L. S., COOPER, D. C., HILDEBRANDT, A. C. and RIKER, A. J. (1964). Chromosome numbers in single cell clones of Tobacco tissue. *Am. J. Bot.* **51,** 284–290. [216, 249]

CORNMAN, I. and CORNMAN, M. E. (1948). The action of podophyllin and its fractions on marine eggs. *Ann. N.Y. Acad. Sci.* **51,** 1443–1481. [222]

CORSI, G. and AVANZI, S. (1970). Cytochemical analyses on cellular differentiation in the root tip of *Allium cepa. Caryologia* **23,** 381–394. [281]

CORSON, G. E. (1969). Cell division studies of the shoot apex of *Datura stramonium* during transition to flowering. *Am. J. Bot.* **56,** 1127–1134. [290, 291, 292, 305, 313]

CORTI, E. F. and CECCHI, A. F. (1970). The behaviour of the cytoplasm during the megasporogenesis in *Paphiopedilum spicerianum* (Rehb. F.) Pfitzer. *Caryologia* **23,** 715–727. [53]

CÔTÉ, W. A. and DAY, A. C. (1965). Anatomy and ultrastructure of reaction wood. *In* "Cellular Ultrastructure of Woody Plants" (Ed. W. A. Côté), pp. 391–418. Syracuse University Press, New York. [380]

CRAN, D. G. (1970). "The Fine Structure of Fern Protonemata". Ph.D. Thesis, Edinburgh University. [79, 84, 85, 89, 92, 93]

CRAN, D. G. and DYER, A. F. (1973). Membrane continuity and association in the fern *Dryopteris borreri. Protoplasma* **76,** 103–108. [79]

CRAN, D. G. and POSSINGHAM, J. V. (1972a). Two forms of division profile in spinach chloroplasts. *Nature, Lond.* **235,** 142. [84, 89, 92, 138, 145, 146]

CRAN, D. G. and POSSINGHAM, J. V. (1972b). Variation of plastid types in spinach. *Protoplasma* **74,** 345–356. [142, 145]

CRONE, M., LEVY, E. and PETERS, H. (1965). The duration of the premeiotic DNA synthesis in mouse oocytes. *Expl Cell Res.* **39,** 678–688. [169]

CRONSHAW, J. and ESAU, K. (1968). Cell division in leaves of *Nicotiana. Protoplasma* **65,** 1–24. [70, 93, 98, 419, 420]

CROSS, G. L. and JOHNSON, T. J. (1941). Structural features of the shoot apices of diploid and colchicine-induced tetraploid strains of *Vinca rosea* L. *Bull. Torrey bot. Club* **68,** 618–635. [309]

CROTTY, W. J. and LEDBETTER, M. C. (1973). Membrane continuities involving chloroplasts and other organelles in plant cells. *Science N. Y.* **182,** 839–841. [79]

CUMBIE, B. G. (1967). Development and structure of the xylem in *Canavalia* (Leguminosae). *Bull. Torrey bot. Club* **94,** 162–175. [353]

CUMMINS, J. E. and DAY, A. W. (1973). Cell cycle regulation of mating type alleles in the smut fungus *Ustilago violacea*. *Nature, Lond.* **245,** 259–260. [169]

CUTTER, E. G. (1958). Studies of morphogenesis in the Nymphaeaceae. III. Surgical experiments on leaf and bud formation. *Phytomorphology* **8,** 74–95. [318]

CUTTER, E. G. (1971). "Plant Anatomy: Experiment and Interpretation". Part 2. "Organs". Edward Arnold, London. [316, 349]

CUTTER, E. G. and CHING-YUAN HUNG (1972). Symmetric and asymmetric mitosis and cytokinesis in the root-tip of *Hydrocharis morsus-ranae* L. *J. Cell Sci.* **11,** 723–737. [94]

CUTTER, E. G. and FELDMAN, L. J. (1970). Trichoblasts in *Hydrocharis*. I. Origin, differentiation, dimensions and growth. *Am. J. Bot.* **57,** 190–201. [53]

CZERNIK, C. A. and AVERS, C. J. (1964). Phosphatase activity and cellular differentiation in *Phleum* root meristem. *Am. J. Bot.* **51,** 424–431. [53]

DALE, J. E. (1964). Leaf growth in *Phaseolus vulgaris*. I. Growth of the first pair of leaves under constant conditions. *Ann. Bot.* **28,** 579–589. [327, 328, 329]

DALE, J. E. (1965). Leaf growth in *Phaseolus vulgaris*. II. Temperature effects and the light factor. *Ann. Bot.* **29,** 293–308. [335, 338]

DALE, J. E. (1966). The effect of nutritional factors and certain growth substances on the growth of disks cut from young leaves of *Phaseolus*. *Physiologia Pl.* **19,** 385–396. [334, 338, 343]

DALE, J. E. (1967). Growth changes in disks cut from young leaves of *Phaseolus*. *J. exp. Bot.* **18,** 660–671. [334]

DALE, J. E. (1970). Models of cell number increase in developing leaves. *Ann. Bot.* **34,** 267–273. [326, 327]

DALE, J. E. and FELIPPE, G. M. (1968). The gibberellin content and early seedling growth of plants of *Phaseolus vulgaris* treated with the growth retardant CCC. *Planta* **80,** 288–298. [338]

DALE, J. E. and HEYES, J. K. (1970). A *virescens* mutant of *Phaseolus vulgaris*; growth, pigment and plastid characters. *New Phytol.* **69,** 733–742. [336]

DALE, J. E. and MURRAY, D. (1968). Photomorphogenesis, photosynthesis and early growth of primary leaves of *Phaseolus vulgaris*. *Ann. Bot.* **32,** 767–780. [332, 334, 336]

DALE, J. E. and MURRAY, D. (1969). Light and cell division in primary leaves of *Phaseolus*. *Proc. R. Soc.* B **173,** 541–555. [29, 332]

DALESSANDRO, G. (1973). Hormonal control of xylogenesis in pith parenchyma explants of *Lactuca*. *Ann. Bot.* **37,** 375–382. [368, 370]

D'AMATO, F. (1952). Polyploidy in the differentiation and function of tissues and cells in plants. *Caryologia* **4,** 311–358. [203, 206, 216, 217, 219, 233, 236]

D'AMATO, F. (1964). Endopolyploidy as a factor in plant tissue development. *Caryologia* **17,** 41–52. [206, 218, 219, 220, 233, 249]

D'AMATO, F. (1965). Endopolyploidy as a factor in plant tissue development. *In* "Int. Conf. Plant Tissue Culture" (Eds P. R. White and A. R. Grove), pp. 449–462. McCutchan, Berkeley. [217]

D'Amato, F. (1972). Morphogenetic aspects of the development of meristems in seed embryos. *In* "Dynamics of Meristem Cell Populations" (Eds M. W. Miller and C. C. Kuehnert), pp. 149–163. Plenum Press, New York. [275]

D'Amato, F. and Avanzi, S. (1965). DNA content, DNA synthesis and mitosis in the root apical cell of *Marsilea strigosa. Caryologia* **18**, 383–394. [265, 270]

D'Amato, F. and Avanzi, S. (1968). The shoot apical cell of *Equisetum arvense*, a quiescent cell. *Caryologia* **21**, 83–89. [297]

D'Amato, F. and Ronchi, V. N. (1968). The response to colchicine of meristems of roots of *Vicia faba. Caryologia* **21**, 53–64. [267]

Darlington, C. D. (1965). "Cytology". Churchill, London. [163]

Darlington, C. D. (1971). Axiom and process in genetics. *Nature, Lond.* **234**, 521–525. [162]

Darlington, C. D. and La Cour, L. (1941). The genetics of embryo-sac development. *Ann. Bot.* **5**, 547–562. [185]

Darlington, C. D. and Thomas, P. T. (1941). Morbid mitosis and activity of inert chromosomes in *Sorghum. Proc. R. Soc.* B **130**, 127–150. [208, 214]

Darlington, C. D. and Vosa, C. G. (1963). Bias in the internal coiling direction of chromosomes. *Chromosoma* **13**, 609–622. [68]

Das, T. M., Hildebrandt, A. C. and Riker, A. J. (1966). Cine-photomicrography of low temperature effects on cytoplasmic streaming, nucleolar activity and mitosis in single tobacco cells in microculture. *Am. J. Bot.* **53**, 253–259. [419]

Davey, M. R., Frearson, E. M., Withers, L. A. and Power, J. B. (1974). Observations on the morphology, ultrastructure and regeneration of tobacco leaf epidermal protoplasts. *Pl. Sci. Lett.* **2**, 23–27. [101]

Davidson, A. W. and Yeoman, M. M. (1974). A phytochrome mediated sequence of reactions regulating cell division in developing callus cultures. *Ann. Bot.* **38**, 545–554. [116]

Davidson, D. (1959). Changes in the chromosome complements of cells of *Vicia faba* roots following irradiation. *J. exp. Bot.* **10**, 391–398. [239, 240, 241, 242]

Davidson, D. (1960). Meristem initial cells in irradiated roots of *Vicia faba. Ann. Bot.* **24**, 287–295. [239, 242]

Davidson, D. (1965a). Cytological chimaeras in roots of *Vicia faba. Bot. Gaz.* **126**, 149–154. [245]

Davidson, D. (1965b). A differential response to colchicine of meristems of roots of *Vicia faba. Ann. Bot.* **29**, 254–264. [239, 245, 267]

Davidson, D. (1969). The differential response of meristems of *Vicia faba* to colchicine. *Caryologia* **22**, 213–221. [267]

Davidson, D. (1972). Morphogenesis of primordia of lateral roots. *In* "The Dynamics of Meristem Cell Populations" (Eds M. W. Miller and C. C. Kuehnert), pp. 165–185. Plenum Press, New York. [215]

Davidson, D., Macleod, R. D. and Webster, P. L. (1968). Labelling of pericycles with ³H-thymidine. *Naturwissenschaften* **7**, 351–352. [267]

Davies, P. J. and Mitchell, E. K. (1972). Transport of indoleacetic acid in intact roots of *Phaseolus coccineus. Planta* **105**, 139–154. [367]

DEELEY, E. M., DAVIES, H. G. and CHAYEN, J. (1957). The DNA content of cells in the root of *Vicia faba*. *Expl Cell Res.* **12**, 582–591. [201]

DE LA TORRE, C. and CLOWES, F. A. L. (1972). Timing of nucleolar activity in meristems. *J. Cell Sci.* **11**, 713–721. [68, 76, 274]

DE LA TORRE, C. and CLOWES, F. A. L. (1974) Thymidine and the measurement of rates of mitosis in meristems. *New Phytol.* **73**, 919–925. [262]

DE MAGGIO, A. E. (1966). Phloem differentiation: induction stimulation by gibberellic acid. *Science, N.Y.* **152**, 370–372. [383]

DE MAGGIO, A. E. and WETMORE, R. H. (1961a). Growth of fern embryos in sterile culture. *Nature, Lond.* **191**, 94–95. [105]

DE MAGGIO, A. E. and WETMORE, R. H. (1961b). Morphogenetic studies on the fern *Todea barbara*. III. Experimental embryology. *Am. J. Bot.* **48**, 551–565. [105]

DENNE, M. P. (1966a). Morphological changes in the shoot apex of *Trifolium repens* L. 1. Changes in the vegetative apex during the plastochron. *N.Z. Jl. Bot.* **4**, 300–314. [290, 291, 292, 305, 328, 331]

DENNE, M. P. (1966b). Morphological changes in the shoot apex of *Trifolium repens* L. 2. Diurnal changes in the vegetative apex. *N.Z. Jl. Bot.* **4**, 434–443. [303, 328, 331]

DENNE, M. P. (1966c). Diurnal and plastochronal changes in the shoot apex of *Tradescantia fluminensis* Vell. *N.Z. Jl Bot.* **4**, 444–454. [303, 328, 331]

DENNE, M. P. (1966d). Leaf development in *Trifolium repens*. *Bot. Gaz.* **127**, 202–210. [328, 331]

DERMEN, H. (1941). Intra-nuclear polyploidy in bean induced by naphthalene acetic acid. *J. Hered.* **32**, 133–138. [219]

DERMEN, H. and STEWART, R. N. (1973). Ontogenetic study of floral organs of peach (*Prunus persica*) utilizing cytochimerical plants. *Am. J. Bot.* **60**, 283–291. [239]

DE ROPP, R. S. (1945). Studies in the physiology of leaf growth. I. The effect of various accessory growth factors on the growth of the first leaf of isolated stem tips of rye. *Ann. Bot.* **9**, 369–381. [344]

DE ROPP, R. S. (1946a). Studies in the physiology of leaf growth. II. Growth and structure on the first leaf of rye when cultivated in isolation or attached to the intact plant. *Ann. Bot.* **10**, 31–40. [344]

DE ROPP, R. S. (1946b). Studies in the physiology of leaf growth. III. The influence of roots on the growth of leaves and stems in rye. *Ann. Bot.* **10**, 353–359. [344]

DERR, W. F. and EVERT, R. F. (1967). The cambium and seasonal development of the phloem in *Robinia pseudacacia*. *Am. J. Bot.* **54**, 147–153. [374, 375]

DETCHON, P. and POSSINGHAM, J. V. (1972). Ribosomal-RNA distribution during leaf development in spinach. *Phytochemistry* **11**, 943–947. [156]

DETCHON, P. and POSSINGHAM, J. V. (1973). Chloroplast ribosomal ribonucleic acid synthesis in cultured spinach leaf tissue. *Biochem. J.* **136**, 829–836. [156]

DE TOROK, D. and RODERICK, J. (1962). Associations between growth rate, mitotic frequency and chromosome number in a plant tissue culture. *Cancer Res.* **22**, 174–181. [249]

DE TOROK, D. and WHITE, P. R. (1959). Cytological instability in tumours of *Picea glauca. Science, N.Y.* **131,** 730–732. [249]

DE VRIES, A. Ph. (1971). Flowering biology of wheat, particularly in view of hybrid seed production—a review. *Euphytica* **20,** 152–170. [164]

DEWSE, C. D. (1974). Observations on the technique for differential staining of the cell cycle using safranin and indigo-picrocarmine. *Stain Technol.* **49,** 57–64. [63]

DEYSSON, G. (1968). Antimitotic substances. *Int. Rev. Cytol.* **24,** 99–148.

DEYSSON, G. and BENBADIS, M. C. (1966). Modalitiés de l'action inhibitrice de la caféine sur la cytodierese dans les cellules méristèmatiques radiculaires d'*Allium sativum.* L'étude en microscopie électronique. *J. Microscopie* **5,** 511–518. [223, 224]

DHILLON, T. S. and GARBER, E. D. (1962). The genus *Collinsia.* XVI. Supernumerary chromosomes. *Am. J. Bot.* **49,** 168–170. [208, 213]

DIBOLL, A. G. and LARSEN, D. A. (1966). An electron microscopic study of the mature megagametophyte in *Zea mays. Am. J. Bot.* **53,** 391–402. [393]

DICKINSON, H. G. and HESLOP-HARRISON, J. (1970). The ribosome cycle, nucleoli and cytoplasmic nucleoloids in the meiocytes of *Lilium. Protoplasma* **69,** 187–200. [68]

DICKMAN, D. I. (1971). Chlorophyll, ribulose-1,5-diphosphate carboxylase, and Hill reaction activity in developing leaves of *Populus deltoides. Pl. Physiol., Lancaster* **48,** 143–145. [336]

DIERS, L. (1965). Elektronen mikroskopische Beobachtungen zur Archegonium entwichlung des Lebermooses *Sphaerocarpos donnellii* Aust. Die Entwicklung des jungen Archegons bis zum Stadium der fertig ausgebildeten sekundaren Zentralzelle. *Planta* **66,** 165–190. [93, 99]

DIERS, L. (1966). On the plastids, mitochondria and other cell constituents during oogenesis of a plant. *J. Cell Biol.* **28,** 527–544. [86, 93]

DIERS, L. (1970). Origin of plastids: cytological results and interpretations including genetical aspects. *In* SEB Symposium **24,** "The Control of Organelle Development" (Ed. P. Miller), pp. 129–146. Cambridge University Press. [80, 86]

DIETZ, L. (1969). Bau und Funktion des Spindelapparats. *Naturwissenschaften* **56,** 237–248. [72]

DIGBY, J. and WAREING, P. F. (1966a). The effect of applied growth hormones on cambial division and the differentiation of the cambial derivatives. *Ann. Bot.* **30,** 539–548. [383, 384]

DIGBY, J. and WAREING, P. F. (1966b). The relationships between endogenous hormone levels in the plant and seasonal aspects of cambial activity. *Ann. Bot.* **30,** 607–622. [383, 385, 389, 390]

DITTMER, H. J. (1937). A quantitative study of the roots and root hairs of a winter rye plant (*Secale cereale*). *Am. J. Bot.* **24,** 417–420. [239]

DITTMER, H. J. (1948). A comparative study of the number and length of roots produced in nineteen Angiosperm species. *Bot. Gaz.* **109,** 354–360. [239]

DIXON, H. H. (1946). Evidence for a mitotic hormone. Observations on the mitoses of the embryo sac of *Fritillaria imperialis*. *Scient. Proc. R. Dubl. Soc.* **24,** 119–124. [61]

DODD, J. D. (1948). On the shape of cells in the cambial zone of *Pinus sylvestris* L. *Am. J. Bot.* **35,** 666–682. [354]

DODGE, A. D. and WHITTINGHAM, C. P. (1966). Photochemical activity of chloroplasts isolated from etiolated plants. *Ann. Bot.* **30,** 711–719. [336]

DODGE, J. D. (1973). The Fine Structure of Algal Cells. Academic Press, London and New York. [69, 87]

DOLEY, D. and LEYTON, L. (1968). Effects of growth regulating substances and water potential on the development of secondary xylem in *Fraxinus*. *New Phytol.* **67,** 579–594. [378]

DONN, G., HESS, D. and POTRYKUS, I. (1973). Wachstum und Differenzierung in aus isolierten Protoplasten von *Petunia hybrida* entstandenem Kallus. *Z. PflPhysiol.* **69,** 423–437. [424]

DORFFLING, K. (1963). Über das Wuchsstoff-Hemmstoffsystem Von *Acer pseudoplatanus* L. I. Der Jahresgang der Wuchs- und Hemmstoffe in Knospen, Blattern und im Kambium. *Planta* **60,** 390–412. [383]

DORFFLING, K. (1964). Uber Das Wuchstoff-Hemmstoffsystem Von *Acer pseudoplatanus* L. II. Die Bedentung von "Inhibitor β" fur die korrelative Knospenhemmung und fur die regulation der Kambiumtätigkeit. *Planta* **60,** 413–433. [384]

DOVER, G. A. and RILEY, R. (1973). The effect of spindle inhibitors applied before meiosis on meiotic chromosome pairing. *J. Cell Sci.* **12,** 143–161.

DOWNS, R. J. (1955). Photoreversibility of leaf and hypocotyl elongation of dark grown red kidney bean seedlings. *Pl. Physiol., Lancaster* **30,** 468–473. [333]

DUNCAN, R. E. (1945). Production of variable aneuploid numbers of chromosomes within the root tips of *Paphiopedilum wardii*. *Am. J. Bot.* **32,** 506–509. [248]

DUNWELL, J. M. and PERRY, E. (1973). The influence of *in vivo* growth conditions of *N. tabacum* plants on the *in vitro* embryogenic potential of their anthers. *John Innes Annual Report* 69–76. [219]

DU PRAW, E. J. (1970). DNA and chromosomes. Molecular and Cellular Biology Series. Holt, Rinehart and Winston, New York. [69, 71, 73, 231]

DURAND, J., POTRYKUS, I. and DONN, G. (1973). Plantes issues de protoplastes de *Petunia*. *Z. PflPhysiol.* **69,** 26–34. [328]

DYER, A. F. (1963). Allocyclic segments of chromosomes and the structural heterozygosity that they reveal. *Chromosoma* **13,** 545–576. [209]

DYER, A. F. and CRAN, D. G. (1974). The ultrastructure of rhizoids of protonemata of *Dryopteris borreri*. *Ann. Bot.*, in press. [95, 201]

DYER, A. F., JONG, K. and RATTER, J. A. (1970). Aneuploidy: a redefinition. *Notes R. bot. Gdn., Edin.* **30,** 177–182. [241]

EAMES, A. J. (1951). Leaf ontogeny and treatments with 2,4-D. *Am. J. Bot.* **38,** 777–780. [343]

EGGLER, W. A. (1955). Radial growth in nine species of trees in southern Louisiana. *Ecology* **36**, 130–136. [388]

EHRENDORFER, F. (1959). Unterschiedliche Storungssyndrome der Meiose bei diploiden und polyploiden Sippen der *Achillea-millefolium*-Komplexes und ihre Bedeutung für die Mikro-evolution. *Chromosoma* **10**, 482–496. [248]

EHRENDORFER, F. (1961). Akzersorische Chromosomen bei *Achillea* Struktur, cytologisches Verhalten, zahlenmyssige Instabilitat und Enstehung. *Chromosoma* **11**, 523–552. [208, 213, 214]

EIGSTI, O. J. and DUSTIN, P. (1955). "Colchicine—in Agriculture, Medicine, Biology and Chemistry". Iowa State College Press, Ames, Iowa. [222]

EKBERG, I. and ERICKSON, G. (1967). Development and fertility of pollen in three species of *Larix*. *Hereditas* **57**, 303–311. [175, 177, 189]

ELLIS, R. J. (1974). The biogenesis of chloroplast protein synthesis by isolated chloroplasts. *Biochem. Soc. Trans. Meet., Lond.* **2**, 179–182. [154]

ENGELKE, A. L., HAMZI, H. Q. and SKOOG, F. (1973). Cytokinin–gibberellin regulation of shoot development and leaf form in tobacco plantlets. *Am. J. Bot.* **60**, 491–495. [325]

ENZENBERG, U. (1961). Bëiträge zur karyologie des endosperms. *Öst. bot. Z.* **108**, 245–285. [221]

ERICKSON, R. O. (1948). Cytological and growth correlations in the flower bud and anther of *Lilium longiflorum*. *Am. J. Bot.* **35**, 729–739. [164, 166, 183, 189, 190, 192]

ERICKSON, R. O. (1964). Synchronous cell and nuclear division in tissues of the higher plants. *In* "Synchrony in Cell Division and Growth" (Ed. E. Zeuthen), pp. 11–37. John Wiley, New York. [12, 15, 61, 62, 115, 116]

ERICKSON, R. O. and SAX, F. (1956). Rates of cell division and cell elongation in the growth of the primary root of *Zea mays*. *Proc. Am. phil. Soc.* **100**, 499–514. [257, 274, 275, 276]

ERIKSSON, T. (1966). Partial synchronization of cell division in suspension cultures of *Haplopappus gracilis*. *Physiologia Pl.* **19**, 900–910. [115]

ERIKSSON, T. and JONASSON, K. (1969). Nuclear division in isolated protoplasts from cells of higher plants grown *in vitro*. *Planta* **89**, 85–89. [420, 423]

ERNST, H. (1938). Meiosis und crossing over. Zytologische und genetische Untersuchungen an *Antirrhinum majus* L. *Z. Bot.* **33**, 241–294. [176]

ERRERA, L. (1888). Über Zell formen und Seifenblasen. *Bot. Zbl.* **34**, 395–398. [34, 38, 413]

ESAU, K. (1938). Ontogeny and structure of the phloem of tobacco. *Hilgardia* **11**, 343–406. [331, 351]

ESAU, K. (1953). "Plant Anatomy". John Wiley, New York. [351]

ESAU, K. (1954). Primary vascular differentiation in plants. *Biol. Rev.* **29**, 46–86. [350]

ESAU, K. (1960). "Anatomy of Seed Plants". John Wiley, New York. [351]

ESAU, K., (1965a). "Plant Anatomy". John Wiley, New York. [316, 350, 351, 352]

Esau, K. (1965b). "Vascular Differentiation in Plants". Biology Studies. Holt, Rinehart and Winston, New York. [316, 350, 351, 352]

Esau, K. (1972). Apparent temporary chloroplast fusion in leaf cells of *Mimosa pudica*. *Z. PflPhysiol.* **67,** 244–254. [89, 92, 93, 143]

Esau, K. and Gill, R. H. (1965). Observations on cytokinesis. *Planta* **67,** 168–181. [52, 418, 419]

Evans, G. M. and Rees, H. (1971). Mitotic cycles in dicotyledons and mono-cotyledons. *Nature, Lond.* **233,** 350–351. [177, 264]

Evans, H. J., Neary, G. J. and Tonkinson, S. M. (1957). The use of colchicine as an indicator of mitotic rate in broad bean root meristems. *J. Genet.* **55,** 487–502. [263, 290]

Evans, L. S. and Van't Hof (1973). Cell arrest in G_2 in root meristems: a control factor from the cotyledons. *Expl Cell Res.* **82,** 471–473. [122]

Evans, L. T. (1969). "The Induction of Flowering". Macmillan, Australia. [342]

Evans, L. V., (1966). Distribution of pyrenoids among some brown algae. *J. Cell Sci.* **1,** 449–454. [87, 88]

Evans, M. J. (1964). Uptake of ^{3}H-thymidine and patterns of DNA replication in nuclei and chromosomes of *Vicia faba*. *Expl Cell Res.* **35,** 381–393. [117]

Evans, P. K. (1967). "Studies on Cell Division during Early Callus Develop-ment in Tissue isolated from Jerusalem Artichoke Tubers". Ph.D. Thesis University of Edinburgh.

Evans, P. K., Keates, A. G. and Cocking, E. C. (1972). Isolation of proto-plasts from cereal leaves. *Planta* **104,** 178–181. [424]

Evans, P. K., Woodcock, J. and Keates, A. G. (1973). Steps towards cell fusion: studies on cereal protoplasts and protoplasts of haploid tobacco. *In* "Protoplastes et Fusion de Cellules Somatiques Végétales, pp. 469–477. *Coll Int. CNRS* **212.** [424]

Evert, R. F. (1963). The cambium and seasonal development of the phloem in *Pyrus malus*. *Am. J. Bot.* **50,** 149–159. [374, 375]

Evert, R. F. and Deshpande, B. P. (1970). An ultrastructural study of cell division in the cambium. *Am. J. Bot.* **57,** 942–961. [54, 70, 79, 98, 107]

Falk, H. and Sitte, P. (1962). Zellfeinbau bei Plasmolyse. I. Dei Feinbau der *Elodea* Blatt. *Protoplasma* **57,** 290–303. [101]

Fasse-Fransisket, U. (1955). Die Teilung der Proplastiden und Chloroplasten bei *Agapanthus umbellatus* L'Herit. *Protoplasma* **45,** 194–227. [93, 139, 140]

Favard, P. (1969). The golgi apparatus. *In* "Handbook of Molecular Cyto-logy" (Ed. A. Lima-de-Faria, pp. 1130–1155. North-Holland, Amsterdam. [79]

Fayle, D. C. F. (1968). "Radial Growth in Tree Roots". Fac. For. Tech. Rept. No. 9, Univ. of Toronto, 1–183. [361, 374, 381]

Feldmann, J. (1969). *Pseudobryopsis myura* and its reproduction. *Am. J. Bot.* **56,** 691–695. [83, 87]

Felippe, G. M. (1967). "Effects of a Quaternary Ammonium Compound and Gibberellic Acid on the Growth of *Phaseolus*". Ph.D. Thesis, University of Edinburgh. [343, 344]

FELIPPE, G. M. and DALE, J. E. (1968a). Effects of a growth retardant, CCC, on leaf growth in *Phaseolus vulgaris*. *Planta* **80,** 328–343. [343, 344]

FELIPPE, G. M. and DALE, J. E. (1968b). Effects of CCC and gibberellic acid on the progeny of treated plants. *Planta* **80,** 344–348. [343]

FERNÁNDEZ-GÓMEZ, M. A., DE LA TORRE, C. and GIMÉNEZ-MARTÍN, G. (1972). Accelerated nucleolar reorganisation with shortened anaphase and telophase during cycloheximide inhibition of protein synthesis in onion root cells. *Cytobiologie* **5,** 117–124. [274]

FEULGEN, R. and ROSENBECH, H. (1924). Mikroskopisch-chemischer Nachweis einer Nucleinsaure von Typus der Thymonucleinsaure und die darauf Beruhende Elektive Farbung von Zellkernen in Mikroskopisch Praparaten. *Z. Physiol. Chem.* **135,** 203–252. [10]

FILNER, B. and KLEIN, A. O. (1968). Changes in enzymatic activities in etiolated bean seedling leaves after a brief illumination. *Pl. Physiol., Lancaster* **43,** 1587–1596. [334]

FINCH, R. A. and BENNETT, M. D. (1972). The duration of meiosis in diploid and autotetraploid barley. *Can. J. Genet. Cytol.* **14,** 507–515. [176, 185]

FLAMM, W. G. (1972). Highly repetitive sequences of DNA in chromosomes. *Int. Rev. Cytol.* **32,** 1–51. [208, 210]

FLAMM, W. G. and BIRNSTIEL, M. L. (1964). Inhibition of DNA replication and its effect on histone synthesis. *Expl Cell Res.* **33,** 616–619. [155]

FOGWILL, M. (1958). Differences in crossing-over and chromosome size in the sex cells of *Lilium* and *Fritillaria*. *Chromosoma* **9,** 493–504. [188]

FORER, A. (1969). Chromosome movement during cell division. *In* "Handbook of Molecular Cytology" (Ed. A. Lima-de-Faria), pp. 553–601. North-Holland, Amsterdam. [51, 70]

FOSKET, D. E. (1968). Cell division and the differentiation of wound vessel members in cultured stem segments of *Coleus. Proc. natn. Acad. Sci. U.S.A.* **59,** 1089–1096. [106, 217]

FOSKET, D. E. (1972). Meristem activity in relation to wound xylem differentiation. *In* "The Dynamics of Meristem Cell Populations" (Eds M. W. Miller and C. C. Kuehnert), pp. 33–50. Plenum Press, New York. [106]

FOSTER, A. S. (1936). Leaf differentiation in angiosperms. *Bot. Rev.* **2** 349–372. [319]

FOSTER, T. S. and STERN, H. (1959). The accumulation of soluble deoxyribosidic compounds in relation to nuclear division in anthers of *Lilium longiflorum. J. biophys. biochem. Cytol.* **5,** 187–192. [168]

FOURCADE, M. F., BERGER, C. A. and WITKUS, E. R. (1963). Cytological effects of amino-pyrine. *Caryologia* **16,** 347–351. [224]

FOWKE, L. C., BECH-HANSEN, C. W., CONSTABEL, F. and GAMBORG, O. L. (1974). A comparative study of the ultrastructure of cultured cells and protoplasts of soybean during cell division. *Protoplasma* **81,** 189–203. [70, 73, 93]

FOWKE, L. C. and PICKETT-HEAPS, J. D. (1969). Cell division in *Spirogyra*. I. Mitosis. *J. Phycol.* **5,** 240–259. [93]

Fox, J. E. (1963). Growth factor requirements and chromosome number in tobacco tissue cultures. *Physiologia Pl.* **16,** 793–803. [249]

Franke, W. W. and Kartenbeck, J. (1971). Outer mitochondrial membrane continuous with endoplasmic reticulum. *Protoplasma* **73,** 35–41. [79]

Frankel, O. H. (1937). The nucleolar cycle in some species in *Fritillaria. Cytologia* **8,** 37–47 [69]

Fraser, D. A. (1956). Ecological studies of forest trees at Chalk River, Ontario, Canada. II. Ecological conditions and radial increment. *Ecology* **37,** 777–789. [379]

Fraser, R. S. S. (1968). "The Synthesis and Properties of Ribonucleic Acid in Dividing Plant Cells". Ph.D. Thesis, University of Edinburgh. [124]

Fraser, R. S. S., Loening, U. E. and Yeoman, M. M. (1967). Effect of light on cell division in plant tissue cultures. *Nature, Lond.* **215,** 873. [116]

Frazer, T. W. and Gunning, B. E. S. (1969). The ultrastructure of plasmodesmata in the filamentous green alga, *Bulbochaete hiloensis* (Nordst.) Tiffany. *Planta* **88,** 244–254. [98]

Frearson, E. M., Power, J. B. and Cocking, E. C. (1973). The isolation, culture and regeneration of *Petunia* leaf protoplasts. *Devl Biol.* **33,** 130–137. [423, 424]

Freeberg, J. A. and Wetmore, R. H. (1967). The Lycopsida—a study in development. *Phytomorphology* **17,** 78–91. [350, 351, 367]

Frey-Wyssling, A. and Mühlethaler, K. (1965). "Ultrastructural Plant Cytology". Elsevier, Amsterdam. [139]

Friedberg, S. H. and Davidson, D. (1970). Duration of S-phase and cell cycles in diploid and tetraploid cells of mixoploid meristems. *Expl Cell Res.* **61,** 216–218. [122]

Fritts, H. C. (1958). An analysis of radial growth of beech in a central Ohio forest during 1954–1955. *Ecology* **39,** 705–720. [379]

Fröst, S. (1948). B and ring chromosomes in *Centaurea scabiosa. Hereditas* **34,** 255–256. [208]

Fröst, S. (1958). Studies on the genetical effects of accessory chromosomes in *Centaurea scabiosa. Hereditas* **44,** 112–122. [214]

Fröst, S. (1959). The cytological behaviour and mode of transmission of accessory chromosomes in *Plantago serraria. Hereditas* **45,** 191–210. [208, 213]

Fröst, S. (1960). A new mechanism for numerical increase of accessory chromosomes in *Crepis pannonica. Hereditas* **46,** 497–503. [214]

Fröst, S. (1963). Number of accessory chromosome and protein content in rye seeds. *Hereditas* **50,** 150–160. [214]

Fröst, S. and Ostegren, G. (1959). *Crepis pannonica* and *C. conyzaefolia*—two more species having accessory chromosomes. *Hereditas* **45,** 211–214. [208]

Fryxell, P. A. (1957). The mode of reproduction in higher plants. *Bot. Rev.* **23,** 135–233. [61]

Fuchs, M. C. (1966). Observation sur l'extension en largeur du limbe foliaire du *Lupinus albus* L. *C.r. hebd. Séanc. Acad. Sci. Paris Ser. D* **263,** 1212–1215. [320]

FUCHS, M. C. (1968). Localisation des divisions dans le meristeme marginal des feuilles des *Lupinus albus* L., *Tropaeolum peregrinum* L., *Limonium sinuatum* (L) Miller et *Nemophila maculata* Benth. *C.r. hebd. Séanc. Acad. Sci. Paris Ser. D* **267,** 722–725. [321]

FUCHS, M. C. (1972a). Growth of the leaf and establishment of the shape in *Tropaeolum peregrinum* L. I. The mitotic activity. *C.r. hebd. Séanc. Acad. Sci. Paris Ser. D* **274,** 3206–3209. [321]

FUCHS, M. C. (1972b). Growth of the leaf and establishment of the shape in *Tropaeolum peregrinum* L. The mitotic polarity. *C.r. hebd. Séanc. Acad. Sci. Paris Ser. D* **274,** 3375–3378. [321]

FUCHS, M. C. (1972c). Growth of the leaf and establishment of the shape in *Tropaeolum peregrinum* L. III. The cell expansion. *C.r. hebd. Séanc. Acad. Sci. Paris Ser. D* **275,** 345–348. [321]

GALLINSKY, I., (1949). The effect of certain phosphates on mitosis in *Allium* roots. *J. Hered.* **40,** 289–295. [230]

GATTI, M., RIZZONI, M., PALITTI, F. and OLIVIERI, G. (1973). Studies on induced aberrations in diplochromosomes of Chinese hamster cells. *Mutation Res.* **20,** 87–99. [219]

GANTT, E. and ARNOTT, H. J. (1963). Chloroplast division in the gametophyte of the fern *Matteuccia struthiopteris* (L) Todaro. *J. Cell Biol.* 19, 446–448. [93]

GAROT, G., LEBRUN-PEREMANS, L., MOREAU, L. and GILES, A. (1968). Effects des microirradiations sur les microsporocytes et microspores de *Tradescantia paludosa*. *Can. J. Genet. Cytol.* **10,** 44–49. [162]

GATES, R.R. and REES, E. M. (1921). A cytological study of pollen development in *Lactuca*. *Ann. Bot.* **35,** 366–396. [184]

GAUTHERET, R. J. (1946). Comparison entre l'actions de l'acide indoleacetique et celle du *Phytomonas tumefaciens* sur la croissance des tissus végétaux. *C.r. Séanc. Soc. Biol.* **140,** 169–171. [410, 428]

GAUTHERET, R. J. (1966). Factors affecting differentiation of plant tissues grown *in vitro*. *In* "Cell Differentiation and Morphogenesis" (Ed. W. Beerman), pp. 55–71. North-Holland, Amsterdam. [430]

GEITLER, L. (1939). Die Entstehung der polyploiden somakerne der Heteropteran durch Chromosomenteilung ohne Kernteilung. *Chromosoma* **1,** 1–22. [218]

GEITLER, L. (1953). Endomitose und endomitotische polyploidisierung. *Protoplasmatologia* **VIC,** 1–89. [216, 217, 218, 221, 235]

GEITLER, L. (1955). Riesenkerne im Endosperm von *Allium ursinum*. *Ost. bot. Z.* **102,** 460–475. [221]

GENTSCHEFF, G. and GUSTAFSSON, A. (1939). The double chromosome reproduction in *Spinacea* and its causes. II. An X-ray experiment. *Hereditas* **25,** 371–386. [219]

GEORGE, L. and NARAYANASWAMY, S. (1973). Haploid *Capsicum* through experimental androgenesis. *Protoplasma* **78,** 467–470. [103]

GHOSAL, S. K. and MUKHERJEE, B. B. (1971). The chronology of DNA synthesis, meiosis and spermiogenesis in the male mouse and golden hamster. *Can. J. Genet. Cytol.* **13,** 672–682. [169]

GIBBONS, G. S. B. and WILKINS, M. B. (1970). Growth inhibitor production by root caps in relation to geotropic responses. *Nature, Lond.* **226,** 558–559. [283]

GIFFORD, E. M. (1950). The structure and development of the shoot apex in certain woody Ranales. *Am. J. Bot.* **37,** 595–611. [309]

GIFFORD, E. M. (1951). Early ontogeny of the foliage leaf in *Drimys winteri* var. *chilensis. Am. J. Bot.* **38,** 93–105. [308, 316]

GIFFORD, E. M. (1953). Effect of 2,4-D upon the development of the cotton leaf. *Hilgardia* **21,** 605–644. [343]

GIFFORD, E. M. (1954). The shoot apex in angiosperms. *Bot. Rev.* **20,** 477–529. [309]

GIFFORD, E. M. and CORSON, G. E. (1971). The shoot apex in seed plants. *Bot. Rev.* **37,** 143–229. [290, 291]

GIFFORD, E. M. and STEWART, K. D. (1967). Ultrastructure of the shoot apex of *Chenopodium album* and certain other seed plants. *J. Cell Biol.* **33,** 131–142. [302]

GILES, K. L. (1971). The control of chloroplast division in *Funaria hygrometrica*. II. The effects of kinetin and indole acetic acid on nucleic acids. *Pl. Cell Physiol. Tokyo* **12,** 447–450. [99, 102]

GILES, K. L. and SARAFIS, V. (1972). Chloroplast survival and division *in vitro*. *Nature New Biology* **236,** 56–58. [83, 88]

GILES, K. L. and TAYLOR, A. O. (1971). The control of chloroplast division in *Funaria hygrometrica*. I. Patterns of nucleic acid, protein and lipid synthesis. *Pl. Cell Physiol. Tokyo* **12,** 437–445. [93, 99]

GIMÉNEZ-MARTÍN, G., GONZÁLES-FERNÁNDEZ, A. and LÓPEZ-SÁEZ, J. F. (1964). Bimitosis. *Phyton. Rev. Int. Bot. Exp. B. Aires* **21,** 77–84. [224]

GIRBARDT, M. (1971). Ultrastructure of the fungal nucleus. II. The kineto-chore equivalent. *J. Cell Sci.* **9,** 453–473. [107]

GIROLAMI, G. (1954). Leaf histogenesis in *Linum usitatissimum. Am. J. Bot.* **41,** 264–273. [308, 316]

GLÄSS, E. (1961). Weitere untersuchungen zur genomsonderung. II. Die Anordnung der Chromosomen in den Wurzelspitzenmitosen von *Bellevalia romana. Chromosoma* **12,** 422–432. [229]

GLOCK, W. S. (1955). Tree growth. II. Growth rings and climate. *Bot. Rev.* **21,** 73–188. [379]

GNATT, E. and ARNOTT, H. J. (1963). Chloroplast division in the gametophyte of the fern *Matteuccia struthiopteris* L. Todaro. *J. Cell Biol.* **19,** 446–448. [93]

GODLEWSKI, M. and OLSZEWSKA, M. J. (1973). Comparison of the duration of the cell cycle in successive generations of synchronously dividing antheridial filaments of *Chara vulgaris* L. as measured with [³H]-thymidine. *Acta Soc. bot. polon.* **42,** 121–131, [123]

GOIN, O. B., GOIN, C. J. and BACKMANN, K. (1968). DNA and amphibian life history. *Copeia* 532–540. [195]

GOLDSMITH, M. H. M. (1968). The transport of auxin. *A. Rev. Pl. Physiol.* **19,** 347–360. [99]

GOLDSMITH, M. H. M. (1969). Transport of plant growth regulators. *In* "The Physiology of Plant Growth and Development" (Ed. M. B. Wilkins), pp. 125–162. McGraw-Hill, London. [366]

GONZÁLES, A. (1967). Formacion y desarrollo de celulas binucleadas: bimitosis. *Genet. iber.* **19,** 1–98. [224]

GONZÁLES-FERNÁNDEZ, A., GIMÉNEZ-MARTÍN, G., DIEZ, J. L., DE LA TORRE, C. and LÓPEZ-SÁEZ, J. F. (1971). Interphase development and beginning of mitosis in the different nuclei of polynucleate homokaryotic cells. *Chromosoma* **36,** 100–111. [64]

GONZÁLES-FERNÁNDEZ, A., LÓPEZ-SÁEZ, J. F., MORENO, P. and GIMÉNEZ-MARTÍN, G. (1968). A model for dynamics of cell division cycle in onion roots. *Protoplasma* **65,** 263–276. [276, 281]

GOODENOUGH, V. W. (1970). Chloroplast division and pyrenoid formation in *Chlamdomonas reinhardi. J. Phycol.* **6,** 1–6. [83, 88]

GORE, J. R. (1973). "RNA Metabolism in Plants during the Initiation of Cell Division". Ph.D. Thesis, University of Edinburgh. [124]

GORI, P., SARFATTI, G. and CRESTI, M. (1971). Development of spherical organelles from the endoplasmic reticulum in the nucellus of some *Euphorbia* species. *Planta* **99,** 133–143. [80]

GOSSELIN, A. (1940). Action sur la mitose des végétaux, de deux alcaloides puriques. *C.r. hebd. Séanc. Acad. Sci. Paris* **210,** 544–546. [223, 224]

GOUWENTAK, C. A. (1941). Cambial activity as dependent on the presence of growth hormone and the non-resting conditions of stems. *Proc. K. ned. Akad. Wet.* **44,** 654–663. [390]

GRAFL, I. (1940). Ueber das Wachstum der Antipodenkerne von *Caltha palustris. Chromosoma* **2,** 1–11. [219]

GRAMBOW, H. J., KAO, K. N., MILLER, R. H. and GAMBORG, O. L. (1972). Cell division and plant development from protoplasts of carrot cell suspension cultures. *Planta* **103,** 348–355. [420, 424]

GRANICK, S. (1938). Chloroplast nitrogen of some higher plants. *Am. J. Bot.* **25,** 561–567. [139, 140]

GRANICK, S. (1961). The chloroplasts: inheritance, structure and function. *In* "The Cell" (Eds J. Brachet and A. E. Mirsky), Vol. II, pp. 489–602. Academic Press, New York and London. [83, 86]

GRANICK, S. and GIBOR, A. (1967). DNA of chloroplasts, mitochondria and centrioles. *In* "Progress in Nucleic Acid Research and Molecular Biology" **6.** (Eds J. N. Davidson and W. E. Cohn), Vol. 6, pp. 143–186. Academic Press, New York and London. [72]

GRANT, C. J. (1965). Chromosome aberrations and the mitotic cycle in *Trillium* root tips after X-irradiation. *Mutation Res.* **2,** 247–262. [66, 223, 224]

GRANT, M. E. and FULLER, K. W. (1968). Tissue culture of root cells of *Vicia faba. J. exp. Bot.* **19,** 667–680. [417]

GRANT, W. F. and HARNEY, P. M. (1960). Cytogenetic effects of maleic hydrazide treatment on tomato seed. *Can. J. Genet. Cytol.* **2,** 162–174. [222]

GRAY, L. H. and SCHOLES, M. E. (1951). The effect of ionizing radiations on the broad bean root. VIII. Growth rate studies and histological analyses. *Br. J. Radiol.* **24,** 82–92, 176–180, 228–236, 285–291, 348–352. [261, 263]

GREEN, P. B. (1964). Cinematic observations on the growth and division of chloroplasts in *Nitella. Am. J. Bot.* **51,** 334–342. [83, 93, 143]

GREEN, P. B., ERICKSON, R. O. and RICHMOND, P. A. (1970). On the physical basis of cell morphogenesis. *Ann. N.Y. Acad. Sci.* **175,** 712–731. [298]

GREGORY, D. W. and COCKING, E. C. (1963). The use of polygalacturonase for the isolation of plant protoplasts and vacuoles. *Biochem. J.* **88,** 40P. [420]

GREGORY, F. G. (1928). Studies in the energy relations of plants. II. The effect of temperature on increase in area of leaf surface and in dry weight of *Cucumis sativus*. Part 1. The effect of temperature on the increase in area of leaf surface. *Ann. Bot.* **42,** 469–507. [336, 342]

GREGORY, F. G. (1956). General aspects of leaf growth. *In* "The Growth of Leaves" (Ed. F. L. Milthorpe), pp. 3–17. Butterworths, London. [336]

GREGORY, R. A. and ROMBERGER, J. A. (1972). The shoot apical ontogeny of the *Picea abies* seedling. II. Growth rates. *Am. J. Bot.* **59,** 598–606. [312]

GRELL, S. M. (1946). Cytological studies in *Culex.* 1. Somatic reduction divisions. *Genetics* **31,** 60–94. [230, 231]

GRIFFITHS, D. J. (1970). The pyrenoid. *Bot. Rev.* **36,** 29–58. [88]

GRUN, P. (1959). Variability of accessory chromosomes in native populations of *Allium cernuum. Am. J. Bot.* **46,** 218–224. [208]

GULLVAG, B. (1968). Fine structure of the plastids and possible ways of distribution of the chloroplast products in some spores of Archegoniatae. *Phytomorphology* **18,** 520–535. [79, 93]

GUNNING, B. E. S. (1965). The greening process in plastids. I. The structure of the prolamellar body. *Protoplasma* **60,** 111–130. [151]

GUNNING, B. E. S. and JAGOE, M. P. (1967). The prolamellar body. *In* "Biochemistry of Chloroplasts" (Ed. T. W. Goodwin), Vol. 2, pp. 656–676. Academic Press, London and New York. [151]

GUPTA, S. B. (1969). Duration of mitotic cycle and regulation of DNA replication in *Nicotiana plumbaginifolia* and a hybrid derivative of *N. tabacum* showing chromosome instability. *Can. J. Genet. Cytol.* **11,** 133–142. [249]

GUSTAFSSON, A. (1946). Apomixis in higher plants. *Lunds. Univ. Arsskrift* **42,** 1–67. [227, 232]

GUTTENBERG, H. VON (1947). Studien über die Entwicklung des Wurzelvegetationspunktes der dikotyledonen. *Planta* **35,** 360–396. [256, 267]

GUTTENBERG, H. VON (1955). Die Entwicklung des Wurzelvegetationspunkpes. *Naturw. Rdsch., Stuttg.* **10,** 383–388. [260]

GUTTENBERG, H. VON (1960). "Grundzüge der Histogenese Höher Pflanzen. I. Die Angiospermen". Gebrüder Borntraeger, Berlin. [265]

GUTTENBERG, H. VON, BURMEISTER, J. and BROSELL, H-J. (1955). Studien über die Entwicklung des Wurzelvegetationspunktes der dikotyledonen. II. *Planta* **46,** 179–222. [260]

GYLDENHOLM, A. O. (1968). Macromolecular physiology of plants. V. On the nucleic acid metabolism during chloroplast development. *Hereditas* **59,** 142–168. [139, 147]

GYLDENHOLM, A. O. and WHATLEY, F. R. (1968). The onset of photophosphorylation in chloroplasts isolated from developing bean leaves. *New Phytol.* **67,** 461–468. [336]

HABER, A. H. (1962). Non-essentiality of concurrent cell divisions for degree of polarization of leaf growth. I. Studies with radiation-induced mitotic inhibition. *Am. J. Bot.* **49,** 583–589. [26, 322, 323]

HABER, A. H. and FOARD, D. E. (1963). Non-essentiality of concurrent cell divisions for degree of polarization of leaf growth. II. Evidence from untreated plants and from chemically induced changes of the degree of polarization. *Am. J. Bot.* **50,** 937–944. [322, 324, 328]

HABER, A. H. and FOARD, D. E. (1964a). Interpretation concerning cell division and growth. *In* "Regulateurs Naturels de la Croissance Végétale" (Ed. J. P. Nitsch), pp. 491–503. Edition du Centre National de la Recherche Scientifique, Paris. [51, 53, 322]

HABER, A. H. and FOARD, D. E. (1964b). Further studies of gamma-irradiated wheat and their relevance to use of mitotic inhibition for developmental studies. *Am. J. Bot.* **51,** 151–159. [51, 33, 322, 328]

HABERLANDT, G. (1921). Wundhormone als Erreger von Zellteilungen. *Beitr. allgem. Bot.* **2,** 1–53. [363]

HACCIUS, B. and MASSFELLER, D. (1961). Untersuchungen zur biologischen aktivatat der phenylborsaure. *Planta* **56,** 174–188. [343]

HAGEMANN, R. (1965). Advances in the field of plastid inheritance in higher plants. *Genetics Today* **3,** 613–625. [94]

HÅKANSSON, A. (1945). Uberzahlige chromosomen in einer Rasse von *Godetia nutans* Hiorth. *Bot. Notiser* **98,** 1–19. [208]

HÅKANSSON, A. (1948). Behaviour of accessory rye chromosomes in the embryo sac. *Hereditas* **34,** 35–59. [213]

HÅKANSSON, A. (1950). Spontaneous chromosome variation in the roots of a species hybrid. *Hereditas* **36,** 39–59. [248]

HÅKANSSON, A. (1957). Notes on the giant chromosomes of *Allium nutans. Bot. Notiser* **110,** 196–204. [221]

HÅKANSSON, A. and LEVAN, A. (1957). Endo-duplicational meiosis in *Allium odorum. Hereditas* **43,** 179–200. [219]

HALDAR, D., FREEMAN, K. and WORK, T. S. (1966). Biogenesis of mitochondria. *Nature, Lond.* **211,** 9–12. [80]

HALL, M. D. and COCKING, E. C. (1971). The bursting response of isolated *Avena* coleoptile protoplasts to indole-3-yl-acetic acid. *Biochem. J.* **124,** 33P. [420]

HALL, O. (1972). Oxygen requirement of root meristems in diploid and autotetraploid rye. *Hereditas* **70,** 69–74. [245]

HALLET, J-N. (1969). Cytologie végétale-Durée du cycle mitotique dans le point végétative du *Polytrichum formosum* Hedw. *C.r. hebd. Séanc. Acad. Sci. Paris* **269,** 2088–2090. [291]

HAMERTON, J. L., RICHARDSON, B. J., GEE, P. A., ALLEN, W. R. and SHORT, R. V. (1971). Non-random X chromosome expression in female mules and hinnies. *Nature, London.* **232,** 312–315. [210]

HAMMOND, D. (1941). The expression of genes for leaf shape in *Gossypium hirsutum* I. and *Gossypium arboreum* L. I. The expression of genes for leaf shape in *Gossypium hirsutum* L. *Am. J. Bot.* **28,** 124–150. [321]

HANNAM, R. V. (1968). Leaf growth and development in the young tobacco plant. *Aust. J. biol. Sci.* **21,** 855–870. [325, 329, 330]

HANSTEIN, E. (1962). Die Kontinuität der Plastiden und die Beobachtungen von Mühlethaler und Bell. *Z. Vererblehre* **93,** 531–533. [79, 86]

HANSTEIN, J. (1870). Die Entwicklung des keimes der Monokotylen und Dikotylen. *Bot. Adhandl., Bonn* **1,** 1–112. [400]

HANZELY, L. and GILULA, N. B. (1970). Fine structural observations on mitochondrial division in meristematic cells of *Allium sativum. Am. J. Bot.* **57,** 737. [79]

HAQUE, A. (1953). Non-synchronised mitosis in a common cytoplasm. *Heredity* **7,** 429–431. [61, 232]

HARADA, H. (1973). A new method for obtaining protoplasts from mesophyll cells. *Z. PflPhysiol.* **69,** 77–80. [328]

HARKES, P. A. A. (1973). Structure and dynamics of the root cap of *Avena sativa* L. *Acta bot. neerl.* **22,** 321–328. [257]

HARLAND, J., JACKSON, J. F. and YEOMAN, M. M. (1973). Changes in some enzymes involved in DNA biosynthesis following induction of division in cultured plant cells. *J. Cell Sci.* **13,** 121–138. [16, 117, 118, 127, 128, 129]

HARTIG, T. (1853). Ueber die Entwicklung des Jahrringes der Holzpflanzen. *Bot. Ztg* **11,** 553–566, 569–579. [361]

HARTLEY, M. R. and ELLIS, R. J. (1973). Ribonucleic acid synthesis in chloroplasts. *Biochem. J.* **134,** 249–262. [155]

HARTMANN, J. F. and ZIMMERMAN, A. M. (1974). The mitotic apparatus. *In* "The Cell Nucleus" (Ed. H. Busch), Vol. II, pp. 459–486. Academic Press New York and London. [51, 70]

HASITSCHKA, G. (1956). Bildung von chromosomenbündeln nach art der Speicheldrüsenchromosomen, spiralisierte ruhekernchromosomen und andere Struktureigentümlichkeiten in den endopolyploiden riesenkernen der antipoden von *Papaver rhoeas. Chromosoma* **8,** 87–113. [236]

HASKELL, D. A. and POSTLETHWAIT, S. N. (1971). Structure and histogenesis of the embryo of *Acer saccharinum.* 1. Embryosac and pro-embryo. *Am. J. Bot.* **58,** 595–603. [51]

HAYFLICK, L. and MOORHEAD, P. S. (1964). The limited *in vitro* lifetime of human diploid cell strains. *Symp. int. Soc. Cell Biol.* **3,** 155–173. [276]

HEAD, G. V. (1973). Effect of moisture stress on abscisic acid levels in *Ricinus communis* L. with particular reference to phloem exudate. *Planta* **113,** 367–372. [378, 385]

HEATH, I. B. (1974). Genome separation mechanisms in prokaryotes, Algae and Fungi. *In* "The Cell Nucleus" (Ed. H. Busch), Vol. II, pp. 487–515. Academic Press, New York and London. [70, 107]

HECHT, N. B. (1972). DNA polymerase activity during mitosis. *Expl Cell Res.* **70**, 248–250. [128]

HEDDLE, J. A. and TROSKO, J. E. (1966). Is the transition from chromosome to chromatid aberrations the result of single-stranded DNA? *Expl Cell Res.* **42**, 171–177. [66]

HEGWOOD, M. P. and HOUGH, L. F. (1958). A mosaic pattern of chromosome numbers in the White Winter Pearmain apple and six of its seedlings. *Am. J. Bot.* **45**, 349–353. [248]

HEINZ, D. J. and MEE, G. W. P. (1971). Morphologic, cytogenetic and enzymatic variation in *Saccharum* species hybrid clones derived from callus tissue. *Am. J. Bot.* **58**, 257–262. [248, 249]

HEINZ, D. J., MEE, G. W. P. and NICKELL, L. G. (1969). Chromosome numbers of some *Saccharum* species hybrids and their cell suspension cultures. *Am. J. Bot.* **56**, 450–456. [248]

HELDER, R. J. and BOERMA, J. (1969). An electron microscopical study of the plasmodesmata in the roots of young barley seedlings. *Acta bot. neerl.* **18**, 99–107. [99]

HENDERSON, S. A. (1969). Chromosome pairing, chiasmata and crossing over. *In* "Handbook of Molecular Cytology" (Ed. A. Lima-de-Faria), pp. 326–357. North-Holland, Amsterdam. [51]

HENSHAW, G. G. and PEARCE, R. S. (1969). The relationship between growth and the synthesis of phenolics in suspension cultures of *Acer pseudoplatanus* L. *Bot. Cong. (Seattle Abstr.)* p. 89. [410]

HEPLER, P. K. and JACKSON, W. T. (1968). Microtubules and early stages of cell-plate formation in the endosperm of *Haemanthus katherinae* (Baker). *J. Cell Biol.* **38**, 437–446. [419]

HEPLER, P. K. and JACKSON, W. T. (1969). Isopropyl-*N*-phenylcarbamate affects spindle microtubule orientation in dividing endosperm cells of *Haemanthus katherinae* Baker. *J. Cell Sci.* **5**, 727–743. [63]

HERREROS, B. and GIANELLI, F. (1967). Spatial distribution of old and new chromatid sub-units and frequency of chromatid exchanges in induced human lymphocyte endo-reduplications. *Nature, Lond.* **216**, 286–287. [225]

HERRMANN, R. G. (1969). Are chloroplasts polyploid? *Expl Cell Res.* **55**, 414–416 [84]

HERRMANN, R. G. (1970). Multiple amounts of DNA related to the size of chloroplasts. 1. An autoradiographic study. *Planta* **90**, 80–96. [84]

HERRMANN, R. G. and KOWALLIK, K. V. (1970). Multiple amounts of DNA related to the size of the chloroplast. II. Comparison of electronmicroscopic and autoradiographic data. *Protoplasma* **69**, 365–372. [79, 84, 156, 158]

HERSKOWITZ, I. H. (1969). Recombination of nuclear genes—crossing over. *In* "Basic Principles of Molecular Genetics" (Ed. I. H. Herskowitz), pp. 165–166. Nelson, London. [173, 174]

HESLOP-HARRISON, J. (1957). The experimental modification of sex expression in flowering plants. *Biol. Rev.* **32**, 38–90. [173, 175]

HESLOP-HARRISON, J. (1962). Effect of 2-thiouracil on cell differentiation and leaf morphogenesis in *Cannabis sativa*. *Ann. Bot.* **26**, 375–387. [328, 343]

HESLOP-HARRISON, J. (1968). Synchronous pollen mitosis and the formation of the generative cell in masculate orchids. *J. Cell Sci.* **3**, 457–466. [53, 55, 62, 94]

HESLOP-HARRISON, J. (1971a). Wall pattern formation in angiosperm microsporogenesis. *In* "Control Mechanisms of Growth and Differentiation" (Ed. D. D. Davies and M. Balls), *Symp. Soc. exp. Biol.* **25**, 277–300. Cambridge University Press.

HESLOP-HARRISON, J. (1971b). The cytoplasm and its organelles during meiosis. *In* "Pollen Development and Physiology" (Ed. J. Heslop-Harrison), pp. 16–31. Butterworths, London. [69, 86]

HESLOP-HARRISON, J. (1972). Sexuality of angiosperms. *In* "Plant Physiology. A Treatise" (Ed. F. C. Steward), pp. 133–289. Academic Press, New York and London. [55, 62, 64, 80, 86, 99, 101, 103, 182]

HESS, T. and SACHS, T. (1972). The influence of a mature leaf on xylem differentiation. *New Phytol.* **71**, 903–914. [383, 384]

HEYÈS, J. K. (1963). The effects of 8-azaguanine on growth and metabolism in the root. *Proc. R. Soc.* B **158**, 208–221. [27]

HEYES, J. K. and VAUGHAN, D. (1968). The effects of 2-thiouracil on growth and metabolism in the root. *Proc. R. Soc.* B **169**, 77–88. [14]

HIMES, M. (1967). An analysis of heterochromatin in maize root tips. *J. Cell Biol.* **35**, 175–181. [208]

HOAGE, T. R. and KESSEL, R. G. (1968). An electron microscope study of the process of differentiation during spermatogenesis in the drone Honey Bee (*Apis mellifera* L.) with special reference to centriole replication and elimination. *J. Ultrastruct. Res.* **24**, 6–32. [71]

HOLDSWORTH, R. H. (1971). The isolation and partial characterisation of the pyrenoid protein of *Eremosphaera viridis. J. Cell Biol.* **51**, 499–513. [88]

HOLLIDAY, R. (1970). The organization of DNA in eukaryotic chromosomes. *In* "Organization and Control in Prokaryotic and Eukaryotic Cells" (Eds H. P. Charles and B. C. J. E. Knight), *Symp. Soc. Gen. Microbiol.* **20**, 359–380. Cambridge University Press. [51]

HOLLINGSHEAD, L. (1932). The occurrence of unpaired chromosomes in hybrids between varieties of *Triticum vulgare. Cytologia* **3**, 119–141. [248]

HONDA, S. I., HONGLADAROM-HONDA, T., KWANYUEN, P. and WILDMAN, S. G. (1971). Interpretations on chloroplast reproduction derived from correlations between cells and chloroplasts. *Planta* **97**, 1–15. [84]

HOPKINSON, J. M. (1964). Studies on the expansion of the leaf surface. IV. The carbon and phosphorus economy of a leaf. *J. exp. Bot.* **15**, 125–137. [337, 340]

HORINE, R. K. and RUESINK, A. W. (1972). Cell wall regeneration around protoplasts isolated from *Convolvulus* tissue culture. *Pl. Physiol., Lancaster* **50**, 438–445. [420]

HORNER, H. T., JR. and LERSTEN, N. R. (1971). Microsporogenesis in *Citrus limon* (Rutaceae). *Am. J. Bot.* **58**, 72–79. [99]

HOTTA, Y. and STERN, H. S. (1961). Transient phosphorylation of deoxyribosides and regulation of deoxyribonucleic acid synthesis. *J. biophys. biochem. Cytol.* **11**, 311–319. [127]

HOTTA, Y. and STERN, H. (1963a). Inhibition of protein synthesis during meiosis and its bearing on intracellular regulation. *J. Cell Biol.* **16**, 259–279. [162]

HOTTA, Y. and STERN, H. (1963b). Synthesis of messenger-like ribonucleic acid and protein during meiosis in isolated cells of *Trillium erectum*. *J. Cell Biol.* **19**, 45–58. [162, 174, 175, 180, 183, 196]

HOTTA, Y. and STERN, H. S. (1963c). Molecular facets of mitotic regulation. I. Synthesis of thymidine kinase. *Proc. natn. Acad. Sci. U.S.A.* **49**, 648–654. [127]

HOTTA, Y. and STERN, H. S. (1963d). Molecular facets of mitotic regulation. II. Factors underlying the removal of thymidine kinase. *Proc. natn. Acad. Sci. U.S.A.* **49**, 861–865. [127]

HOTTA, Y. and STERN, H. S. (1965). Inducibility of thymidine kinase by thymidine as a function of interphase time. *J. Cell Biol.* **25**, 99–108. [127]

HOWARD, A. and PELC, S. R. (1951a). Nuclear incorporation of ^{32}P as demonstrated by autoradiographs. *Expl Cell Res.* **2**, 178–187. [117, 120]

HOWARD, A. and PELC, S. R. (1951b). Synthesis of nucleoprotein in bean root cells. *Nature, Lond.* **167**, 599–600. [117]

HOWARD, A. and PELC, S. R. (1951c). Synthesis of DNA and nuclear incorporation of ^{35}S as shown by autoradiographs. *In* "Isotopes in Biochemistry", *Ciba* **7**, 138–148. [117]

HOWARD, A. and PELC, S. R. (1953). Synthesis of DNA in normal and irradiated cells and its relation to chromosome breakage. *Heredity, Lond.* (Suppl.) **6**, 261–273. [111, 117]

HSU, T. C. (1961). Chromosomal evolution in cell populations. *Int. Rev. Cytol.* **12**, 69–161. [233, 249]

HSU, T. C. and MOORHEAD, P. S. (1956). Chromosome anomalies in human neoplasms with special reference to the mechanics of polyploidization and aneuploidization in the Hela strain. *Ann. N.Y. Acad. Sci.* **63**, 1083–1094. [219, 237, 249]

HUBERMAN, J. A. and RIGGS, A. D. (1968). On the mechanism of DNA replication in mammalian cells. *J. Molec. Biol.* **32**, 327–341. [118]

HUMPHRIES, E. C. and FRENCH, S. A. W. (1963). The effects of nitrogen, phosphorus, potassium and gibberellic acid on leaf area and cell division in Majestic potato. *Ann. appl. Biol.* **52**, 149–162. [341]

HUMPHRIES, E. C. and WHEELER, A. W. (1963). The physiology of leaf growth. *A. Rev. Pl. Physiol.* **14**, 385–410. [342]

HUMPHRIES, E. C. and WHEELER, A. W. (1964). Cell division and growth substances in leaves. *In* "Regulateurs Naturels de la Croissance Végétale" (Ed. J. P. Nitsch), pp. 505–515. Edition du Centre National de la Recherche Scientifique, Paris. [342]

HUSKINS, C. L. (1948a). Segregation and reduction in somatic tissue. I. Initial observations in *Allium cepa*. *J. Hered.* **39**, 311–325. [229, 230, 233]

HUSKINS, C. L. (1948b). Chromosome multiplication and reduction in somatic tissues. *Nature, Lond.* **161,** 80–83. [228, 229, 230, 233]

HUSKINS, C. L. and CHENG, K. C. (1950). Segregation and reduction in somatic tissues. IV. Reductional groupings induced in *Allium cepa* by low temperature. *J. Hered.* **41,** 13–18. [230]

HUSKINS, C. L. and STEINITZ, L. M. (1948). The nucleus in differentiation and development. II. Induced mitosis in differentiated tissues of *Rhoeo* roots. *J. Hered.* **39,** 66–77. [217]

HUSSEY, G. (1971). Cell division and expansion and resultant tissue tensions in the shoot apex during the formation of a leaf primordium in the tomato. *J. exp. Bot.* **22,** 702–714. [306, 308]

HUSSEY, G. (1972). The mode of origin of a leaf primordium in the shoot apex of the pea (*Pisum sativum*). *J. exp. Bot.* **23,** 675–682, [305, 306, 308]

HUSSEY, G. (1973). Mechanical stress in the shoot apices of *Euphorbia, Lycopersicon* and *Pisum* under controlled turgor. *Ann. Bot.* **37,** 57–64. [307]

IKEDA, M. (1965). Behaviour of sulfhydryl groups of sea urchin eggs under the blockage of cell divisions by UV and heat shock. *Expl Cell Res.* **40,** 282–291. [224]

INGLE, J., POSSINGHAM, J. V., WELLS, R., LEAVER, C. J. and LOENING, U. E. (1970). The properties of chloroplast ribosomal-RNA. *In* "Control of Organelle Development" (Ed. P. L. Miller), *Symp. Soc. exp. Biol.* **24,** 303–325. Cambridge University Press. [156]

ITAI, C., RICHMOND, A. and VAADIA, Y. (1968). The role of root cytokinins during water and salinity stress. *Israel J. Bot.* **17,** 187–195. [378]

ITAI, C. and VAADIA, Y. (1965). Kinetin-like activity in root exudate of water stressed sunflower plants. *Physiologia Pl.* **18,** 941–944. [378]

ITAI, C. and VAADIA, Y. (1971). Cytokinin activity in water stressed shoots. *Pl. Physiol., Lancaster* **47,** 87–90. [378]

ITO, M. (1962). Studies on the differentiation of fern gametophytes. I. Regeneration of single cells isolated from cordate gametophyte of *Pteris vittata. Bot. Mag., Tokyo* **75,** 19–27. [99]

ITO, M. and STERN, H. (1967). Studies of meiosis *in vitro.* I. *In vitro* culture of meiotic cells. *Devl Biol.* **16,** 36–53. [162, 171, 176, 180, 182, 184]

IVANOV, V. B. (1971). Critical size of the cell and its transition to division 1. Sequence of transition to mitosis for sister cells in the corn seedling root tip. *Akademia Nauk CCCP Ontogenez* **2,** 524–535. [261]

IZHAR, S. and FRANKEL, R. (1973). Duration of meiosis in *Petunia* anthers *in vivo* and in floral bud culture. *Acta bot. neerl.* **22,** 14–22. [175, 176]

JACCARD, P. (1939). Tropisme et bois de réaction provoqué par la force centrifuge. *Ber. schweiz. bot. Ges.* **49,** 135–147. [385]

JACKSON, R. C. and NEWMARK, P. (1960). Effects of supernumerary chromosomes on production of pigment in *Haplopappus gracilis. Science, N.Y.* **132,** 1316–1317. [214]

JACKSON, W. T. (1969). Regulation of mitosis. II. Interaction of isopropyl N-phenylcarbamate and melatonin. *J. Cell Sci.* **5,** 745–755. [63]

JACOBS, W. P. (1952). The role of auxin in differentiation of xylem around a wound. *Am. J. Bot.* **39,** 301–309. [368]

JACOBS, W. P. and MORROW, I. B. (1957). A quantitative study of xylem development in the vegetative shoot apex of *Coleus. Am. J. Bot.* **44,** 823–842. [351]

JACOBS, W. P. and MORROW, I. B. (1961). A quantitative study of mitotic figures in relation to development in the apical meristem of vegetative shoots of *Coleus. Devl Biol.* **3,** 569–587. [303]

JACOBS, W. P. and MORROW, I. B. (1967). A quantitative study of sieve-tube differentiation in vegetative shoot apices of *Coleus. Am. J. Bot.* **54,** 524–531. [351]

JACOBSEN, P. (1957). The sex chromosomes in *Humulus* L. *Hereditas* **43,** 357–370. [210]

JACQMARD, A. (1970). Duration of mitotic cycle in the apical bud of *Redbeckia bicolor. New Phytol.* **69,** 269–271. [289, 291, 300]

JACQMARD, A. and MIKSCHE, J. P. (1971). Cell population and quantitative changes of DNA in the shoot apex of *Sinapis alba* during floral induction. *Bot. Gaz.* **132,** 364–367. [313]

JACQMARD, A., MIKSCHE, J. P. and BERNIER, G. (1972). Quantitative study of nucleic acids and proteins in the shoot apex of *Sinapis alba* during transition from the vegetative to reproductive condition. *Am. J. Bot.* **59,** 714–721. [172]

JAGELS, R. (1970). Photosynthetic apparatus in *Selaginella.* I. Morphology and photosynthesis under different light and temperature regimes. *Can. J. Bot.* **48,** 1843–1860. [83, 86, 87]

JAKOB, K. M. (1972). RNA synthesis during the DNA synthesis period of the first cell cycle in the root meristem of germinating *Vicia faba. Expl Cell Res.* **72,** 370–376. [62, 125]

JAKOB, K. M. and TROSKO, J. E. (1965). The relation between 5-amino-uracil-induced mitotic synchronization and DNA synthesis. *Expl Cell Res.* **40,** 56–67. [115]

JANKIEWICZ, L. S., SZPUNAR, H., BARANASKA, R., RUMPLOWA, R. and FIUTOWSKA, K. (1961). The use of auxin to widen crotch angles in young apple trees. *Acta agrobot.* **10,** 151. [387]

JEFFREY, W. R., STUART, K. D. and FRANKEL, J. (1970). The relationship between deoxyribonucleic acid replication and cell division in heat-synchronised *Tetrahymena. J. Cell Biol.* **46,** 533–543. [225]

JENSEN, C. and BAJER, A. (1973). Spindle dynamics and arrangement of microtubules. *Chromosoma* **44,** 73–90. [51, 70]

JENSEN, W. A. (1963). Cell development during plant embryogenesis. *Brookhaven Symp. Biol.* **16,** 179–202. [395, 398]

JENSEN, W. A. (1964). Observations on the fusion of nuclei in plants. *J. Cell Biol.* **23,** 669–672. [92, 394]

JENSEN, W. A. (1965a). The ultrastructure and composition of the egg and central cell of cotton. *Am. J. Bot.* **52,** 781–797. [61, 393]

JENSEN, W. A. (1965b). The composition and ultrastructure of the nucellus in cotton. *J. Ultrastruct. Res.* **13,** 112–128. [79]

JENSEN, W. A. (1968a). Cotton embryogenesis: the zygote. *Planta* **79,** 346–366. [394]

JENSEN, W. A. (1968b). Cotton embryogenesis: the tube containing endoplasmic reticulum. *J. Ultrastruct. Res.* **22,** 296–302. [395]

JENSEN, W. A. (1968c). Cotton embryogenesis; polysome formation in the zygote. *J. Cell Biol.* **36,** 403–406. [395]

JENSEN, W. A. (1973). Fertilisation in flowering plants. *Biol. Sci. Tokyo* **23,** 21–27. [392]

JENSEN, W. A. (1974). Reproduction in flowering plants. *In* "Dynamic Aspects of Plant Ultrastructure" (Ed. A. W. Robards), pp. 481–503. McGraw-Hill, London. [392]

JENSEN, W. A. and KAVALJIAN, L. (1958). An analysis of cell morphology and the periodicity of division in the root tip of *Allium cepa. Am. J. Bot.* **45,** 365–372. [52, 53, 62]

JOHN, B. and LEWIS, K. R. (1965). The meiotic system. "Protoplasmatologia Handbuch der Protoplasmaforschung VIF₁". Springer-Verlag, Wien, New York. [55, 232, 243]

JOHN, B. and LEWIS, K. R. (1968). The chromosome complement. *Protoplasmatologia* **VI A.** [68, 208, 211, 214, 221, 233, 238, 243]

JOHN, B. and LEWIS, K. R. (1969). The chromosome cycle. "Protoplasmatologia Handbuch der Protoplasmaforschung. VI. Kernund Zellteilung. B". [51]

JOHNSON, C., ATTRIDGE, T. and SMITH, H. (1973). Advantages of the fixed angle rotor for the separation of density-labelled from unlabelled proteins by isopycnic equilibrium centrifugation. *Biochim. biophys. Acta* **317,** 219–230. [127]

JOHNSON, R. T. and RAO, P. N. (1971). Nucleo-cytoplasmic interactions in the achievement of nuclear synchrony in DNA synthesis and mitosis in multinucleate cells. *Biol. Rev.* **46,** 97–155. [61, 66]

JOHNSSON, H. (1944). Meiotic aberrations and sterility in *Alopecurus myosuroides* Huds. *Hereditas* **30,** 469–566. [225]

JONES, H. and EAGLES, J. E. (1962). Translocation of ¹⁴carbon within and between leaves. *Ann. Bot.* **26,** 505–510. [337]

JONES, L. E., HILDEBRANDT, A. C., RIKER, A. J. and WU, J. H. (1960). Growth of somatic tobacco cells in microculture. *Am. J. Bot.* **47,** 468–475. [419]

JONES, L. E. and HOOK, P. W. (1970). Growth and development in microculture of gametophytes from stored spores of *Equisetum. Am. J. Bot.* **57,** 430–435. [83, 93]

JONES, R. E. and BANFORD, R. (1941). Chromosome number in the progeny of triploid *Gladiolus* with special reference to the contribution of the triploid. *Am. J. Bot.* **29,** 807–813. [248]

JONES, R. F. (1970). Physiological and biochemical aspects of growth and gametogenesis in *Chlamydomonas reinhardtii. Ann. N.Y. Acad. Sci.* **175,** 648–659. [303]

JONES, R. L. and PHILLIPS, I. D. J. (1966). Organs of gibberellin synthesis in light-grown sunflower plants. *Pl. Physiol.*, *Lancaster* **41**, 1381–1386. [385]

JONES, R. N. and REES, H. (1968). The influence of B-chromosomes upon the nuclear phenotype in rye. *Chromosoma* **24**, 158–176. [214]

JORDAN, E. G. and CHAPMAN, J. M. (1971). Ultrastructural changes in the nucleoli of Jerusalem artichoke (*Helianthus tuberosus*) tuber discs. *J. exp. Bot.* **22**, 627–634. [125, 419]

JOSHI, P. C. and BALL, E. (1968). Growth of isolated palisade cells of *Arachis hypogaea in vitro*. *Devl Biol.* **17**, 308–325. [328]

JOST, L. (1901). Ueber einige Eigenthümlichkeiten des Cambiums der Baüme. *Bot. Ztg* **59**, 1–24. [360]

JOUANNEAU, J. P. (1971). Contrôle par les cytokinines de la synchronisation des mitoses dans les cellules de tabac. *Expl Cell Res.* **67**, 329–337. [115]

JUNIPER, B. E. and BARLOW, P. W. (1969). The distribution of plasmodesmata in the root tip of maize. *Planta* **89**, 352–360. [99]

JUNIPER, B. E. and CLOWES, F. A. L. (1965). Cytoplasmic organelles and cell growth in root caps. *Nature, Lond.* **208**, 864–865. [81]

JUNIPER, B. E., GROVES, S., LANDAU-SCHACHAR, B. and AUDUS, L. J. (1966). Root cap and the perception of gravity. *Nature, Lond.* **209**, 93–94. [283]

KADEJ, A. R. (1966). Organisation and development of apical root meristem in *Elodea canadensis* (Rich) Casp. and *Elodea densa* (Planck) Casp. *Acta Soc. bot. pol.* **35**, 143–158. [268]

KADEJ, F. (1963). Interpretation of the pattern of the cell arrangement in the root apical meristem of *Cyperus gracilis* L. var. *alternifolius*. *Acta. Soc. bot. pol.* **32**, 295–301. [268]

KÄFER, E. (1961). The processes of spontaneous recombination in vegetative nuclei of *Aspergillus nidulans*. *Genetics* **46**, 1581. [229]

KALTSIKES, P. J. (1972). Duration of the mitotic cycle in Triticale. *Caryologia* **25**, 537–542. [190]

KAMEYA, T. (1972). Cell elongation and division of chloroplasts. *J. exp. Bot.* **23**, 62–64. [81, 145, 150]

KAMEYA, T. and TAKAHASHI, N. (1971). Division of chloroplast *in vitro*. *Jap. J. Genet.* **46**, 153–157. [143, 145]

KAMEYA, T. and UCHIMIYA, H. (1972). Embryoids derived from isolated protoplasts of carrot. *Planta*, **103**, 356–360. [424]

KAO, K. N., GAMBORG, O. L., MILLER, R. A. and KELLER, W. A. (1971). Cell divisions in cells regenerated from protoplasts of soybean and *Haplopappus gracilis*. *Nature New Biol.* **232**, 124. [421, 423]

KAO, K. N., KELLER, W. A. and MILLER, R. A. (1970). Cell division in newly formed cells from protoplasts of soybean. *Expl Cell Res.* **62**, 338–340. [421, 423]

KAO, K. N. and MICHAYLUK, M. R. (1974). A method for high frequency intergeneric fusion of plant protoplasts. *Planta* **175**, 355–367. [424]

KAPLAN, D. R. (1970). Comparative foliar histogenesis in *Acorus calamus* and its bearing on the phyllode theory of monocotyledonous leaves. *Am. J. Bot.* **57**, 331–361. [309, 317]

KAPOOR, B. M. and TANDON, S. L. (1963). Contributions to the cytology of endosperm in some angiosperms. III. *Amaryllis belladonna* L. *Cytologia* **28**, 399–408. [211]

KARSTEN, G. (1915). Über embryonales Wachstum und seine Tagesperiode. *Z. Bot.* **7**, 1–34. [303]

KASS, L. B. and PAOLILLO, D. J. (1972). Chloroplast replication in *Polytrichum* spores. *Am. J. Bot.* **59**, 652. [93]

KATES, J. R., CHIANG, K. S. and JONES, R. F. (1968). Studies on DNA replication during synchronised vegetative growth and gametic differentiation in *Chlamydomonas reinhardtii. Expl Cell Res.* **49**, 121–135. [225, 231]

KATO, Y. (1955). Polyploid mitoses, extranuclear bodies and mitotic aberrations in seedlings of *Lilium maximowiczii* Regel. *Cytologia* **20**, 1–10. [219]

KATO, Y. (1965). Physiological and morphogenetic studies of fern gametophytes and sporophytes in aseptic culture. IV. Controlled differentiation in leaf callus tissues. *Cytologia* **30**, 67–74. [103]

KATO, Y. (1970). Physiological and morphogenetic studies of fern gametophytes and sporophytes in aseptic culture. XII. Sporophyte formation in the dark cultured gametophytes of *Pteris vittata* L. *Bot. Gaz.* **131**, 205–210. [103]

KATTERMAN, G. (1933). Ein Beitag zur Frage der Dualitat der Bestandteile des Bastardkernes. *Planta* **18**, 751–785. [248]

KAUFMAN, P. B. (1959). Development of the shoot of *Oryza sativa* L. II. Leaf histogenesis. *Phytomorphology* **9**, 277–311. [317]

KAUFMANN, P. B., CASSELL, S. J. and ADAMS, P. A. (1965). On nature of intercalary growth and cellular differentiation in internodes of *Avena sativa. Bot. Gaz.* **126**, 1–13. [52, 53]

KAUFMAN, P. B., PETERING, L. B. and SONI, S. L. (1970). Ultrastructural studies on cellular differentiation in the internodal epidermis of *Avena sativa. Phytomorphology* **20**, 281–309. [53, 99]

KAUFMAN, P. B., PETERING, L. B., YOCUM, C. S. and BAIC, D. (1970). Ultrastructural studies on stomata development in internodes of *Avena sativa. Am. J. Bot.* **57**, 33–49. [53, 99]

KAYANO, H. (1956). Cytogenetic studies in *Lilium callosum.* II. Preferential segregation of supernumerary chromosome. *Mem. Fac. Sci. Kyushu Univ. Ser. E* **2**, 53–60. [208, 213]

KAYANO, H. (1957). Cytogenetic studies in *Lilium callosum.* III. Preferential segregation of a supernumerary chromosome in EMS's. *Proc. Japan Acad.* **33**, 553–557. [208, 213]

KEINHOLZ, R. (1934). Leader, needle, cambial and root growth of certain conifers and their interrelations. *Bot. Gaz.* **96**, 73–92. [375]

KELLER, W. A., HARVEY, B., GAMBORG, O. L., MILLER, R. A. and EVELEIGH, D. E. (1970). Plant protoplasts for use in somatic cell hybridization. *Nature, Lond.* **226**, 280–282. [420]

KEMP, C. L. (1964). The effects of inhibitors of RNA and protein synthesis on cytological development during meiosis. *Chromosoma* **15**, 652–665. [162, 180]

KENDE, H. (1965). Kinetin-like factors in the root exudate of sunflowers. *Proc. natn. Acad. Sci. U.S.A.* **53**, 1302–1307. [385]

KENDE, H. (1971). The cytokinins. *Int. Rev. Cytol.* **31**, 301–338. [225]

KIERMAYER, O. and JAROSCH, R. (1962). Die Formbildung von *Micrasterias rotata* Ralfs und ihre experimentelle Beeinflussung. *Protoplasma* **54**, 382–420. [101]

KIHLMANN, B. A. (1949). The effects of purine derivations on chromosomes. *Hereditas* **35**, 393–396. [223, 224]

KIHLMANN, B. A. (1967). "Actions of Chemicals on Dividing Cells". Prentice-Hall, New York. [223, 224]

KIHLMANN, B. A. and LEVAN, A. (1949). The cytological effect of caffeine. *Hereditas* **35**, 109–111. [223, 224]

KINET, J.-M., BERNIER, G. and BRONCHART, R. (1967). Sudden release of the meristematic cells from G_2 as a primary effect of flower induction in *Sinapis*. *Naturwissenschaften* **13**, 351. [313]

KING, B. and CHAPMAN, J. M. (1972). The effect of inhibitors of protein and nucleic acid synthesis on nucleolar size and enzyme induction in Jerusalem artichoke tuber slices. *Planta* **104**, 306–315. [125]

KING, P. J. and STREET, H. E. (1973). Growth patterns in cell cultures. *In* "Plant Tissue and Cell Culture" (Ed. H. E. Street). Botanical Monographs **11**, 269–337. Blackwell Scientific Publications, Oxford. [115, 128]

KIRK, J. T. O. (1972). Genetic control of plastid formation; recent advances and strategies for the future. *Sub-cell. Biochem.* **1**, 333–361. [158]

KIRK, J. T. O. and TILNEY-BASSETT, R. A. E. (1967). "The Plastids". Freeman, San Francisco. [94, 137, 138, 154, 246, 247]

KIRKHAM, M. B., GARDNER, W. R. and GERLOFF, G. C. (1972). Regulation of cell division and cell enlargement by turgor pressure. *Pl. Physiol., Lancaster* **49**, 961–962. [378]

KISLEV, N., SWIFT, H. and BOGORAD, L. (1965). Nucleic acids of chloroplasts and mitochondria in Swiss chard. *J. Cell Biol.* **25**, 327–344. [158, 423]

KLEIN, H. D. (1972). Timing anomalies during meiosis. *Pisum Newsletter* **4**, 14–15. [180]

KLEIN, R. M. (1954). Mechanisms of crown-gall induction. *Brookhaven Symp. Biol.* **6**, 97–114. [426]

KLEIN, R. M. (1955). Resistance and susceptibility of carrot roots to crown-gall tumor formation. *Proc. natn. Acad. Sci. U.S.A.* **41**, 271–274. [425]

KLEIN, R. M. (1965). The physiology of bacterial tumors in plants and of habituation. *In* "Handbuch der Pflanzenphysiologie" (Ed. W. Ruhland), Vol. XV/2, pp. 209–235. Springer-Verlag, Berlin. [426]

KLEINMANN, A. (1923). Ueberkern und Zellteilungen in Cambium. *Bot. Arch.* **4**, 113–147. [362]

KOCHER, P. B. and LEONARD, O. A. (1971). Translocation and metabolic conversion of ^{14}C-labelled assimilates in detached and attached leaves of *Phaseolus vulgaris* L. in different phases of leaf expansion. *Pl. Physiol., Lancaster* **47**, 212–216. [338]

KODANI, M. (1948). Sodium ribose nucleate and mitosis. *J. Hered.* **39**, 327–335. [230]

KOEHLER, P. G. (1973). The roles of cell division and cell expansion in the growth of alfalfa leaf mesophyll. *Ann. Bot.* **37**, 65–68. [328]

KOFMAN-ALFARO, S. and CHANDLEY, A. C. (1970). Meiosis in the male mouse. An autoradiographic investigation. *Chromosoma* **31**, 404–420. [169]

KOHLENBACH, H. W. (1965). Über organisierte bildungen aus *Macleaya cordata* kallus. *Planta* **64**, 37–40. [328]

KONAR, R. N., THOMAS, E. and STREET, H. E. (1972). Origin and structure of embryoids arising from epidermal cells of the stem of *Ranunculus sceleratus* L. *J. Cell Sci.* **11**, 77–93. [102, 430]

KORIBA, K. (1958). On the periodicity of tree-growth in the tropics, with reference to the mode of branching, the leaf fall, and the formation of the resting bud. *Gdns Bull. Straits Settl.* **17**, 11–81. [371]

KORN, R. W. (1969). Chloroplast inheritance in *Cosmarium turpinii* Breb. *J. Phycol.* **5**, 332–336. [83]

KOVACS, C. J. and VAN'T HOF, J. (1970). Synchronization of a proliferative population in a cultured plant tissue. *J. Cell Biol.* **47**, 536–539. [115]

KOWALLIK, K. V. and HERRMANN, R. G. (1972). Variable amounts of DNA related to the size of chloroplasts. IV. Three-dimensional arrangement of DNA in fully differentiated chloroplasts of *Beta vulgaris* L. *J. Cell Sci.* **11**, 357–377. [79, 84, 158]

KOZLOWSKI, T. T. (1971). "Growth and Development of Trees. Vol. II. Cambial Growth, Root Growth, and Reproductive Growth". Academic Press, New York and London. [363, 370, 371, 376, 377, 379, 380]

KOZLOWSKI, T. T., WINGET, C. H. and TORRIE, J. H. (1962). Daily radial growth of oak in relation to maximum and minimum temperature. *Bot. Gaz.* **124**, 9–17. [379]

KUEHNERT, C. C. (1969). Developmental potentialities of leaf primordia of *Osmunda cinnamomea*. II. Further studies on the influence of determined leaf primordia on undetermined leaf primordia. *Can. J. Bot.* **47**, 59–63. [319]

KUEHNERT, C. C. (1972). On determination of leaf primordia in *Osmunda cinnamomea* L. *In* "The Dynamics of Meristem Cell Populations" (Eds M. W. Miller and C. C. Kuehnert), pp. 101–118. Plenum Press, New York. [319]

KULESCHA, Z. (1952). Recherches sur l'élaboration de substances de croissance par les tissus végétaux. *Revue gén. Bot.* **59**, 241–264. [428]

KULESCHA, Z. and GAUTHERET, R. J. (1948). Sur l'élaboration de substances de croissance par trois types de cultures de tissus de Scorsonère cultures normales, cultures de crown-gall et cultures accontumées a l'hetero-auxine. *C.r. hebd. Séanc. Acad. Sci. Paris* **227**, 292–294. [428]

Kupila, S. and Stern, H. (1961). DNA content of broad bean (*Vicia faba*) internodes in connection with tumor induction by *Agrobacterium tumefaciens*. *Pl. Physiol., Lancaster* **36**, 216–219. [426]

Kupila-Ahvenniemi, S. and Therman, E. (1968). Morphogenesis of crown gall. *In* "Advances in Morphogenesis" (Eds M. Abercrombie and J. Brachet), Vol. 7, pp. 45–78. Academic Press, London and New York. [428]

Kupila-Ahvenniemi, S. and Therman, E. (1971). First DNA synthesis around sterile and crown-gall inoculated wounds in *Vicia faba*. *Physiologia Pl.* **24**, 23–26. [426]

Kurabayashi, M., Lewis, H. and Raven, P. H. (1962). A comparative study of mitosis in the Onagraceae. *Am. J. Bot.* **49**, 1003–1026. [208]

Kuroiwa, T. and Tanaka, N. (1970). DNA replication pattern in somatic chromosomes of *Crepis capillaris*. *Cytologia* **35**, 271–279. [119]

Kusunoki, S. and Kawasaki, Y. (1936). Beobachtung über die Chloroplastenteilung bei einigen Blutenpflanzen. *Cytologia* **7**, 530–534. [93]

La Cour, L. F. (1949). Nuclear differentiation in the pollen grain. *Heredity* **3**, 319–338. [246]

Ladefoged, K. (1952). The periodicity of wood formation. *K. denske Vidensk. Selsk. Skr.* **7**, 1–98. [373, 379]

Laetsch, W. M. (1974). The C_4 syndrome: a structural analysis. *A. Rev. Pl. Physiol.* **25**, 27–52. [142, 153]

Laetsch, W. M. and Boasson, R. (1971). Effect of growth regulators on organelle development. *In* "Hormonal Regulation in Plant Growth and Development" (Eds H. Kaldeway and Y. Vardar), pp. 453–465. *Proc. Adv. Study Inst. Izmir.* Verlag Chemie, Weinheim. [147]

Laetsch, W. M. and Stetler, D. A. (1967). Regulation of chloroplast development in cultured plant tissues. *In* "Le Chloroplaste" (Ed. C. Sironval), pp. 291–297. Marvu, Paris. [151]

Lafontaine, J. G. (1974). The nucleus. *In* "Dynamic Aspects of Plant Ultrastructure (Ed. A. W. Robards), pp. 1–51. McGraw-Hill, London. [51, 63, 67, 68, 107, 208, 210]

Lafontaine, J. G. and Chouinard, L. A. (1963). A correlated light and electron microscope study of the nucleolar material during mitosis in *Vicia faba*. *J. Cell Biol.* **17**, 167–201. [125]

La Fountain, K. L. and Mascarenhas, J. P. (1972). Isolation of vegetative and generative nuclei from pollen tubes. *Expl Cell Res.* **73**, 233–236. [60]

Lala, P. K. (1968). Cytokinetic control mechanisms in Ehrlich Ascites tumour growth. *In* "Effects of Radiation on Cellular Proliferation and Differentiation", pp. 463–474. *Int. atomic Energy Ag. Bull.*, Vienna. [271]

Lamerton, L. F. and Steel, G. G. (1968). Cell population kinetics in normal and malignant tissues. *In* "Progress in Biophysics and Molecular Biology" (Eds J. A. V. Butler and D. Noble), pp. 245–283. Pergamon Press, Oxford. [271]

Lance, A. (1952). Sur la structure et le fonctionnement du point végétatif de *Vicia faba* L. *Annls Sci. nat (Bot.) Ser. XI* **13**, 301–339. [303]

LANCE, A. (1958). Récherches cytologiques sur l'évolution de quelques méristèmes apicaux et sur ses variations provoquées par des traitements photopériodiques. *Annls Sci. nat. (Bot.) Ser. XI* **19,** 165–202. [139]

LANG, A. (1965). Progressiveness and contagiousness in plant differentiation and development. *In* "Encyclopedia of Plant Physiology" (Ed. W. Ruhland), Vol. XV/1, pp. 409–423. Springer-Verlag, Berlin. [368]

LANGDON, O. G. (1963). Growth patterns of *Pinus elliottii* var. *densa. Ecology* **44,** 825–827. [371]

LARK, K. G. (1967). Non-random segregation of sister chromatids in *Vicia faba* and *Tricicum boeoticum. Proc. natn. Acad. Sci. U.S.A.* **58,** 352–359. [67]

LARSEN, K. A. I. (1960). Cytological and experimental studies in *Koeleria*. I. *Koeleria pubescens. Hereditas* **46,** 312–318. [208, 215]

LARSON, P. R. (1962a). The indirect effect of photoperiod on tracheid diameter in *Pinus resinosa. Am. J. Bot.* **49,** 132–137. [377, 379]

LARSON, P. R. (1962b). Auxin gradients and the regulation of cambial activity. *In* "Tree Growth" (Ed. T. T. Kozlowski), pp. 97–117. Ronald Press, New York. [379, 383]

LARSON, P. R., ISEBRANDS, J. G. and DICKSON, R. E. (1972). Fixation pattern of ^{14}C within developing leaves of Eastern Cottonwood. *Planta* **107,** 301–314. [337]

LAWTON, E. (1932). Regeneration and induced polyploidy in ferns. *Am. J. Bot.* **19,** 303–333. [103]

LEACH, R. W. A. and WAREING, P. F. (1967). Distribution of auxin in horizontal woody stems in relation to gravimorphism. *Nature, Lond.* **214,** 1025–1027. [385, 386, 387]

LEAVER, C. J. (1974). The biogenesis of plant mitochondria. *In* "The Chemistry and Biochemistry of Plant Proteins" (Eds J. B. Harborne and C. F. Van Sumere), pp. 137–165. Academic Press, London and New York. [136]

LEECH, R. M., RUMSBY, M. G. and THOMSON, W. W. (1973). Plastid differentiation, acyl lipid and fatty acid changes in developing green maize leaves. *Pl. Physiol., Lancaster* **52,** 240–245. [140]

LEECH, R. M., RUMSBY, M. G., THOMSON, W. W., CROSBY, W. and WOOD, P. (1972). Lipid changes during plastid differentiation in developing maize leaves. *In* "Proc. 2nd Int. Cong. Photosynthesis" (Eds G. Forti, M. Avron and A. Melandri), Vol. II, pp. 2479–2497. Dr Junk, The Hague. [140]

LEEDALE, G. F. (1959). Periodicity of mitosis and cell division in the Euglenineae. *Biol. Bull.* **116,** 162–174. [303]

LEEDALE, G. F. (1970). Phylogenetic aspects of nuclear cytology in the algae. *Ann. N.Y. Acad. Sci.* **175,** 429–453. [231]

LEHMAN, H. and SCHULZ, D. (1969). Elektron Mikroskopische Untersuchungen von differenzierungs-vorgangen bei Moosen. II. Die Zellplatten und Zellwandbildung. *Planta* **85,** 313–325. [95]

LEROUX, R. (1954). Recherches sur les modifications anatomiques de trois

espèces d'osiers (*Salix viminilas* L., *Salix purpurea* L., *Salix fragilis* L.) provoquées par l'acide naphthalèneacétique. *C.r. Séanc. Soc. Biol.* **148**, 284–286.
[363]

LERSTEN, N. (1965). Histogenesis of leaf venation in *Trifolium wormskioldii*
(Leguminosae). *Am. J. Bot.* **52**, 767–774. [317, 331]

LESHEM, B. and CLOWES, F. A. L. (1972). Rates of mitosis in shoot apices of
potatoes at the beginning and end of dormancy. *Ann. Bot.* **36**, 687–691. [290,
291]

LEVAN, A. (1948). Polyploidy in flax, sugar beets and timothy. *Proc. 8th Int.
Congr. of Genetics*, 46–47. [248]

LEWIS, K. R. and JOHN, B. (1961). Hybridisation in a wild population of *Eleo-
charis palustris*. *Chromosoma* **12**, 433–448. [55]

LEWIS, K. R. and JOHN, B. (1963). "Chromosome Marker". Churchill, London. [243]

LEWIS, K. R. and JOHN, B. (1964). "The Matter of Mendelian Heredity".
Churchill, London. [51]

LEWIS, W. H. (1962). Aneusomaty in aneuploid populations of *Claytonia virginica*. *Am. J. Bot.* **49**, 918–928. [248]

LICHTENSTEIN, L. M. and MARGOLIS, S. (1968). Histamine release *in vitro*. Inhibition by catecholamines and methylxanthines. *Science, N.Y.* **161**, 902–
903. [224]

LIEBERMAN, I., ABRAMS, R., HUNT, N. and OVE, P. (1963). Levels of enzyme
activity and deoxyribonucleic acid synthesis in mammalian cells cultured
from the animal. *J. biol. Chem.* **238**, 3955–3962. [128]

LIMA-DE-FARIA, A. (1947). Disturbances in microscope cytology of *Anthoxanthum*. *Hereditas* **33**, 539–551. [214]

LIMA-DE-FARIA, A. (1964). Seriation of meiotic stages and spindle orientation
in *Gazania*. *Port. Acta biol.* **8**, 147–152. [164]

LIMA-DE-FARIA, A. (1969). DNA replication and gene amplification in Heterochromatin. *In* "Handbook of Molecular Cytology" (Ed. A. Lima-de-Faria),
pp. 277–325. North-Holland, Amsterdam. [120, 208]

LINDER, A. (1959). Cytochemical effects of 5-fluorouracil on sensitive
and resistant Ehrlich ascites tumor cells. *Cancer Res.* **19**, 189–194.
[225]

LIPETZ, J. (1970). Wound-healing in higher plants. *Int. Rev. Cytol.* **27**, 1–28.
[425]

LIST, A., JNR. (1963). Some observations on DNA content and cell and nuclear
volume growth in the developing xylem cells of certain higher plants. *Am.
J. Bot.* **50**, 320–329. [53, 216, 236]

LOEB, L. A. and AGARWAL, S. S. (1971). DNA polymerase correlation with
DNA replication during transformation of human lymphocytes. *Expl Cell
Res.* **66**, 299–304. [128]

LOEWENBERG, J. R. (1955). The development of bean seeds. (*Phaseolus vulgaris*
L.) *Pl. Physiol., Lancaster* **30**, 244–250. [327]

LOISEAU, J-E. (1962). Activité mitotique des cellules superficielles du sommet
végétatif caulinaire. *Mém. Soc. bot. Fr.* 14–23. [295, 298]

LOMBARDO, G. and GEROLA, F. M. (1968a). Ultrastructure of the pollen grain and taxonomy. *Giorn. bot. ital.* **102,** 353–380. [56]

LOMBARDO, G. and GEROLA, F. M. (1968b). Cytoplasmic inheritance and ultrastructure of the male generative cell of higher plants. *Planta* **82,** 105–110. [56]

LOOMIS, R. S. and TORREY, J. G. (1964). Chemical control of vascular cambium initiation in isolated radish roots. *Proc. natn. Acad. Sci.. U.S.A.* **52,** 3–11. [383, 384]

LÓPEZ-SÁEZ, J. F., RISUEÑO, M. C. and GIMÉNEZ-MARTÍN, G. (1966). Inhibition of cytokinesis in plant cells. *J. Ultrastruct. Res.* **14,** 85–94. [95, 223, 224]

LOVE, R. M. (1938). Somatic variation of chromosome numbers in hybrid wheats. *Genetics, Princeton* **23,** 517–522. [248]

LØVLIE, A. and BRÅTEN, T. (1970). On mitosis in the multicellular alga *Ulva mutabilis* Føyn. *J. Cell Sci.* **6,** 109–129. [79, 83, 88]

LUXOVÁ, M. and MURÍN, A. (1973). The extent and differences in mitotic activity of the root tip of *Vicia faba* L. *Biologia Pl.* **15,** 37–43. [276]

LUYKX, P. (1970). Cellular mechanisms of chromosome distribution. *Int. Rev. Cytol. Suppl.* **2.** [51, 66, 70, 71]

LUYKX, P. (1974). The organization of meiotic chromosomes. *In* "The Cell Nucleus" (Ed. H. Busch), Vol. 2, pp. 163–207. Academic Press, New York and London. [51, 66, 67, 107]

LYNDON, R. F. (1967). The growth of the nucleus in dividing and non-dividing cells of the pea root. *Ann. Bot.* **31,** 133–146. [112]

LYNDON, R. F. (1968). Changes in volume and cell number in the different regions of the shoot apex of *Pisum* during a single plastochron. *Ann. Bot.* **32,** 371–390. [325, 330]

LYNDON, R. F. (1970a). Rates of cell division in the shoot apical meristem of *Pisum. Ann. Bot.* **34,** 1–17. [290, 291, 292, 293, 305, 308, 310]

LYNDON, R. F. (1970b). Planes of cell division and growth in the shoot apex of *Pisum. Ann. Bot.* **34,** 19–28. [298, 299, 306, 308, 309, 310]

LYNDON, R. F. (1970c). DNA, RNA and protein in the pea shoot apex in relation to leaf initiation. *J. exp. Bot.* **21,** 286–291. [302]

LYNDON, R. F. (1972a). Leaf formation and growth at the shoot apical meristem. *Physiol. Veg.* **10,** 209–222. [287, 293, 305, 308, 309, 310]

LYNDON, R. F. (1972b). Nucleic acid synthesis in the pea shoot apex. *Symp. Biol. Hung.* **13,** 345–353. [302]

LYNDON, R. F. (1973). The cell cycle in the shoot apex. *In* "The Cell Cycle in Development and Differentiation" (Eds M. Balls and F. S. Billett), pp. 167–183. Cambridge University Press. [55, 62, 287, 289, 292, 297, 300, 301, 310]

LYNDON, R. F. (1975). Unpublished observations. [84, 310, 314]

LYNDON, R. F. and ROBERTSON, E. S. (1975). The quantitative ultrastructure of the pea shoot apex in relation to leaf initiation. *Protoplasma*, in press. [302]

McArthur, I. C. S. (1967). "Experimental Vascular Differentiation in *Geum chiloense Balbis*". Ph.D. Thesis, University of Saskatchewan. [366]

MacDaniels, L. H. and Cowart, F. F. (1944). The development and structure of the apple leaf. *N.Y. (Cornell) Agr. Expt. Sta. Mem.* 258. [316, 328]

Macdonald, P. D. M. (1970). Statistical inference from the fraction labelled mitoses curve. *Biometrika* **57**, 489–503. [262]

Macdougal, D. T. (1903). "The Influence of Light and Darkness upon Growth and Development". The New Era Printing Company, Lancaster. [332]

McDermott, I. (1971). Human male meiosis: chromosome behaviour at pre-meiotic and meiotic stages of spermatogenesis. *Can. J. Genet. Cytol.* **13**, 536–549. [172, 174]

Mache, R., Rozier, C., Loiseaux, S. and Vial, A. M. (1973). Synchronous division of plastids during the greening of cut leaves of maize. *Nature New Biology* **242**, 158–160. [81, 153, 155]

McKenzie, A., Heslop-Harrison(J. and Dickinson, H. G. (1967). Elimination of ribosomes during meiotic prophase. *Nature, Lond.* **215**, 997–999. [183]

McLaughlin, D. J. (1971). Centrosomes and microtubules during meiosis in the mushroom *Boletus pubinellus*. *J. Cell Biol.* **50**, 737–745. [73]

McLeish, J. (1969). Changes in the amount of nuclear RNA during interphase in *Vicia faba*. *Chromosoma* **26**, 312–325. [117, 123]

McLeish, J. and Sunderland, N. (1961). Measurements of deoxyribosenucleic acid (DNA) in higher plants by Feulgen photometry and chemical methods. *Expl Cell Res.* **24**, 527–540. [278]

MacLeod, A. (1974). Unpublished observations.

MacLeod, R. D. (1968). Changes in the mitotic cycle in lateral root meristems of *Vicia faba* following kinetin treatment. *Chromosoma* **24**, 177–187. [270]

MacLeod, R. D. (1971). Thymidine kinase and thymidylate synthetase in meristems of roots of *Vicia faba*. *Protoplasma* **73**, 337–348. [127, 267]

MacLeod, R. D. (1972). Colchicine induced changes in the rate of entry of cells into prophase in roots of *Vicia faba* and *Hyacinthus orientalis*. *Öst. bot. Z.* **120**, 15–28. [263]

MacLeod, R. D. (1973a). The emergence and early growth of the lateral root in *Vicia faba* L. *Ann. Bot.* **37**, 69–75. [267]

MacLeod, R. D. (1973b). The response of root meristems to colchicine and indol-3-yl-acetic acid in *Vicia faba* L. *Ann. Bot.* **37**, 687–697. [267]

MacLeod, R. D. and Davidson, D. (1968). Delayed incorporation of ³H-thymidine by primordial cells. *Chromosoma* **24**, 1–9. [267]

MacLeod, R. D. and Davidson, D. (1970). Incorporation of ³H-deoxynucleosides: changes in labelling indices during root development. *Can. J. Bot.* **48**, 1659–1663. [267]

Macnutt, M. M. and von Maltzahn, K. E. (1960). Cellular dedifferentiation in *Splachnum ampullaceum* (L) Hedw. *Can. J. Bot.* **38**, 895–908. [93, 99]

McWilliam, A. A., Smith, S. M. and Street, H. E. (1974). The origin and development of embryoids in suspension cultures of carrot (*Daucus carota* L.). *Ann. Bot.* **38**, 243–250. [101]

Maheshwari, P. (1950). "An Introduction to the Embryology of Angiosperms". McGraw-Hill, London. [61, 161, 192, 236, 392]

Mahlberg, P. G. and Sabharwal, P. S. (1967). Mitosis in the non-articulated laticifer of *Euphorbia marginata. Am. J. Bot.* **54**, 465–472. [53, 61, 216]

Mahmood, A. (1968). Cell grouping and primary wall generations in the cambial zone, xylem, and phloem in *Pinus. Aust. J. Bot.* **16**, 177–195. [362]

Mak, S. (1965). Mammalian cell cycle analysis using microspectrophotometry combined with autoradiography. *Expl Cell Res.* **39**, 286–289. [300]

Maksymowych, R. (1959). Quantitative analysis of leaf development in *Xanthium pennsylvanicum. Am. J. Bot.* **46**, 635–644. [325]

Maksymowych, R. (1963). Cell division and cell elongation in leaf development of *Xanthium pennsylvanicum. Am. J. Bot.* **50**, 891–901. [239, 328]

Maksymowych, R. (1973). Analysis of leaf development. *In* "Developmental and Cell Biology" (Eds M. Abercrombie, D. R. Newth and J. G. Torrey), Vol. 1, pp. 1–109. Cambridge University Press. [319, 330]

Maksymowych, R. and Blum, M. K. (1966). Incorporation of ^{3}H-thymidine in leaf nuclei of *Xanthium pennsylvanicum. Am. J. Bot.* **53**, 134–142. [328, 330]

Maksymowych, R., Blum, M. K. and Devlin, R. G. (1966). Autoradiographic studies of the synthesis of nuclear DNA in various tissues during leaf development of *Xanthium pennsylvanicum. Devl Biol.* **13**, 250–265. [332]

Maksymowych, R. and Erickson, R. O. (1960). Development of the lamina in *Xanthium italicum* represented by the plastochron index. *Am. J. Bot.* **47**, 451–459. [319, 320]

Malamud, D. (1971). Differentiation and the cell cycle. *In* "The Cell Cycle and Cancer" (Ed. R. Baserga), pp. 132–142. Dekker, New York. [64]

Manton, I. (1950). "Problems of Cytology and Evolution in the Pteridophytes". Cambridge University Press. [219]

Manton, I. (1966a). Som possibly significant structural relations between chloroplasts and other cell components. *In* "Biochemistry of Chloroplasts" (Ed. T. W. Goodwin), pp. 23–47. Academic Press, London and New York. [88]

Manton, I. (1966b). Further observations on the fine structure of *Chrysochromulina chiton*, with special reference to the pyrenoid. *J. Cell Sci.* **1**, 187–192. [88]

Marchant, H. J. and Pickett-Heaps, J. D. (1972). Ultrastructure and differentiation of *Hydrodictyon reticulatum*. III. Formation of the vegetative daughter net. *Aust. J. biol. Sci.* **25**, 265–278. [72, 107]

Maretzki, A. and Nickell, L. G. (1973). Formation of protoplasts from sugarcane cell suspensions and the regeneration of cell cultures from protoplasts. *In* "Protoplastes et Fusion de Cellules Somatiques Végétales". *Coll. Int. CNRS* **212,** 51–61. [424]

MARGULIES, M. M. (1962). Effects of chloramphenicol on light dependent development of seedlings of *Phaseolus vulgaris* var. Black Valentine, with particular reference to development of photosynthetic activity. *Pl. Physiol., Lancaster* **37,** 473–480. [334]

MARGULIES, M. M. (1965). Relationship between red light mediated glyceraldehyde-3-phosphate dehydrogenase formation and light dependent development of photosynthesis. *Pl. Physiol., Lancaster* **40,** 57–61. [334]

MARITHAMU, K. M. and THRELKELD, S. F. H. (1966). The distribution of tritiated thymidine in tetrad nuclei of *Haplopappus gracilis*. *Can. J. Genet. Cytol.* **8,** 603–612. [176, 183]

MARQUARDT, H. (1951). Die Wirkung der Röntgenstrahlen auf die Chiasma frequenz in der Meiosis von *Vicia faba*. *Chromosoma* **4,** 232–238. [176]

MARUYAMA, K. (1968). Electron microscopic observation of plastids and mitochondria during pollen development in *Tradescantia paludosa*. *Cytologia* **33,** 482–498. [81, 86]

MATTHYSE, A. G. and TORREY, J. G. (1967). DNA synthesis in relation to polyploid mitoses in excised pea root segments cultured *in vitro*. *Expl Cell Res.* **48,** 484–498. [203]

MATTINGLY, E. (1966). Synchrony of cell division in root meristems following treatment with 5-aminouracil. *In* "Cell Synchrony" (Eds I. L. Cameron and G. M. Padilla), pp. 256–268. Academic Press, New York and London. [62, 115]

MAZIA, (1961). Mitosis and the physiology of cell division. *In* "The Cell" (Eds J. Brachet and A. E. Mirsky), Vol. III, pp. 77–412. Academic Press, New York and London. [50, 68, 72, 224]

MAZIA, D. (1974). The cell cycle. *Scient. Am.* **230,** 54–68. [106]

MECHELKE, F. (1953). Die Entstehung der polyploiden Zelkerne des Antherentapetums bei *Antirrhinum majus*. *Chromosoma* **5,** 246–295. [216, 218, 221, 237, 238]

MEHRA, P. N. and SULKLYAN, D. S. (1969). *In vitro* studies on apogamy, apospory and controlled differentiation of a rhizome segment of the fern, *Ampelopteris prolifera* (Retz.) Copel. *J. Linn. Soc. (Bot.)* **62,** 431–443. [103]

MELANDER, Y. (1948). Cytological studies on Scandinavian flat-worms belonging to Tricladida, Paludicola. *Proc. 8th Int. Congr. of Genetics,* 625–626. [219]

MENDELSOHN, M. L. (1962). Autoradiographic analysis of cell proliferation in spontaneous breast cancer of C_3H mouse. III. The growth fraction. *J. natn. Cancer Inst.* **28,** 1015–1029. [271]

MENDELSOHN, M. L. and TAKAHASHI, M. (1971). A critical evaluation of the fraction of labelled mitoses method as applied to analysis of tumor and other cell cycles. *In* "The Cell Cycle and Cancer" (Ed. R. Baserga), pp. 58–95. Dekker, New York. [64]

MENZEL, M. Y. and BROWN, M. S. (1952). Polygenomic hybrids in *Gossypium*. II. Mosaic formation and somatic reduction . *Am. J. Bot.* **39,** 59–69. [248]

MEPHAM, R. H. and LANE, G. R. (1970). Observations on the fine structure of developing microspores of *Tradescantia bracteata*. *Protoplasma* **70,** 1–20. [55, 79]

MERRITT, C. (1968). Effect of environment and heredity on the root-growth pattern of red pine. *Ecology* **49,** 34–40. [376]

MICHAELIS, P. (1962). Uber zahlengesetsmässigkeiten Plasmatischer Erbträger insbesondere der Plastiden. *Protoplasma* **55,** 177–231. [139, 140]

MICHAUX, N. (1969). Cytologie végétale. Durée des phases du cycle mitotique dans le méristeme apical de *l'Ixtes setocea* Lam. *C.r. hebd. Séanc. Acad. Sci. Paris.* **269,** 1396–1399. [291]

MIHARA, S. and HASE, E. (1971). Studies on the vegetative life cycle of *Chlamydomonas reinhardi* Dangeard. in synchronous culture. 1. Some characteristics of the cell cycle. *Pl. Cell Physiol. Tokyo* **12,** 225–236. [83]

MILINKOVIC, V. (1957). Accessory chromosomes in the roots of *Poa alpina*. *Hereditas* **43,** 583–588. [214]

MILLER, M. B. (1975). "The Transition from Vegetative to Floral Development in the Shoot Apex". Ph. D. Thesis, University of Edinburgh. [314]

MILLER, M. B. and LYNDON, R. F. (1975). The cell cycle in vegetative and floral shoot meristems measured by a double labelling technique. *Planta,* **126,** 37–43. [296]

MILLER, R. A., GAMBORG, O. L., KELLER, W. A. and KAO, K. N. (1971). Fusion and division of nuclei in multinucleated soybean protoplasts. *Can. J. Genet. Cytol.* **13,** 347–353. [423]

MILTHORPE, F. L. and NEWTON, P. (1963). Studies on the expansion of the leaf surface. III. The influence of radiation on cell division and leaf expansion. *J. exp. Bot.* **14,** 483–495. [325, 326, 329, 336, 337, 340, 341]

MITCHELL, J. P. (1967). DNA synthesis during the early division cycles of Jerusalem artichoke callus cultures. *Ann. Bot.* **31,** 427–435. [64, 116, 118, 202, 203]

MITCHELL, J. P. (1968). The pattern of protein accumulation in relation to DNA replication in Jerusalem artichoke callus cultures. *Ann. Bot.* **32,** 315–326. [116, 118, 119, 126]

MITCHELL, J. P. (1969). RNA accumulation in relation to DNA and protein accumulation in Jerusalem artichoke callus cultures. *Ann. Bot.* **33,** 25–34. [116, 123, 124]

MITCHISON, J. M. (1969). Enzyme synthesis in synchronous cultures. *Science, N.Y.* **165,** 657–663. [128]

MITCHISON, J. M. (1971). "The Biology of the Cell Cycle". Cambridge University Press. [51, 64, 73, 79, 114, 123, 125, 129, 130, 203]

MITCHISON, J. M. and VINCENT, W. S. (1965). Preparation of synchronous cell cultures by sedimentation. *Nature, Lond.* **205,** 987–9. [114, 225]

MITCHISON, J. M. and WALKER, P. M. B. (1965). RNA synthesis during the cell life cycle of a fission yeast, *Schizosaccharomyces pombe*. *Expl Cell Res.* **16,** 49–58. [124]

MITRA, J. and STEWARD, F. C. (1961). Growth induction in cultures of *Haplo-*

pappus gracilis II. The behaviour of the nucleus. *Am. J. Bot.* **48,** 358–368. [174, 228, 249]

MITTERMAYER, C., BRAUN, R. and RUSCH, H. P. (1964). RNA synthesis in the mitotic cycle of *Physarum polycephalum*. *Biochim. biophys. Acta* **91,** 399–405. [123]

MOENS, P. B. (1969). The fine structure of meiotic chromosome polarization and pairing in *Locusta migratoria* spermatocytes. *Chromosoma* **28,** 1–25. [107]

MOHR, H. (1966). Differential gene activation as a mode of action of phytochrome 730. *Photochem. Photobiol.* **5,** 469–483. [333]

MOHR, (1972). "Lectures on Photomorphogenesis". Springer-Verlag, Berlin and Heidelberg. [333, 334]

MOLDER, M. and OWENS, J. N. (1972). Ontogeny and histochemistry of the vegetative apex of *Cosmos bipinnatus* "sensation". *Can. J. Bot.* **50,** 1171–1184. [309]

MOLLENHAUER, H. H. and MORRE, D. J. (1966). Golgi apparatus and plant secretions. *A. Rev. Pl. Physiol.* **17,** 27–46. [420]

MONESI, V. (1962). Autoradiographic study of DNA synthesis and the cell cycle in spermatogonia and spermatocytes of mouse testis using tritiated thymidine. *J. Cell Biol.* **14,** 1–18. [166, 169]

MONESI, V., CRIPPA, M. and ZITO-BIGNAMI, R. (1967). The stage of chromosome duplication in the cell cycle as revealed by X-ray breakage and ³H-thymidine labelling. *Chromosoma* **21,** 369–386. [66]

MOORING, J. S. (1960). A cytogenetic study of *Clarkia unguiculata*. II. Supernumerary chromosomes. *Am. J. Bot.* **47,** 847–854. [208]

MOREL, G. (1963). Leaf regeneration in *Adiantum pedatum*. *J. Linn. Soc. (Bot.)* **58,** 381–383. [103]

MORGAN, T. H. (1919). "The Physical Basis of Heredity." Lippincott, Philadelphia. [5]

MORRÉ, D. J., MERRITT, W. D. and LAMBI, C. A. (1971). Connections between mitochondria and endoplasmic reticulum in rat liver and onion stem. *Protoplasma* **73,** 43–49. [79]

MORRÉ, D. J. and MOLLENHAUER, H. H. (1974). The endomembrane concept: a functional integration of endoplasmic reticulum and Golgi apparatus. *In* "Dynamic Aspects of Plant Ultrastructure" (Ed. A. W. Robards), pp. 84–137. McGraw-Hill, London. [79]

MOSER, J. W. and KREITNER, G. L. (1970). Centrosome structure in *Anthoceros laevis* and *Marchantia polymorpha*. *J. Cell Biol.* **44,** 454–458. [72]

MOSES, M. J. and TAYLOR, J. H. (1955). Deoxypentose nucleic acid synthesis during microsporogenesis in *Tradescantia*. *Expl Cell Res.* **9,** 474–488. [64]

MOSHKOV, B. S. (1933). Photoperiodicity of certain woody species. *Biol. Abstr.* **7,** 20678. [388]

MOSHKOV, B. S. (1934). Photoperiodicity of trees and its practical importance. *Biol. Abstr.* **8,** 1680. [388]

MOSHKOV, B. S. (1935). Photoperiodismus und Frostharte ausdauernder Gewachse. *Planta* **23,** 774–803. [388]

MOSS, G. I. and HESLOP-HARRISON, H. (1967). A cytochemical study of DNA,

RNA and protein in the developing maize anther. II. Observations. *Ann. Bot.* **31**, 555–572. [164, 167]

Moss, J. P. (1966). The adaptive significance of B-chromosomes in rye. *In* "Chromosomes Today" (Eds C. D. Darlington and K. R. Lewis). Vol. 1, pp. 15–23. Oliver and Boyd, Edinburgh. [214]

Most, B. H. (1971). Abscisic acid in immature apical tissue of sugar cane and in leaves of plants exposed to drought. *Planta* **101**, 67–75. [378]

Mota, M., Hildebrandt, A. C. and Riker, A. J. (1964). Movements of cytoplasm between and during nuclear divisions in living tobacco cells of tissue cultures. A motion picture film. *Agronomia Lusit.* **26**, 205–212. [419]

Motoyoshi, F. (1971). Protoplasts isolated from callus cells of maize endosperm. *Expl Cell Res.* **68**, 452–456. [420, 423]

Mueller, G. C. (1971). Biochemical perspectives of the G_1 and S intervals in the replication cycle of animal cells: a study in the control of cell growth. *In* "The Cell Cycle and Cancer" (Ed. R. Baserga), pp. 269–307). Dekker, New York. [106]

Mueller, D. M. J. (1974). Spore wall formation and chloroplast development during sporogensis in the moss *Fissidens limbatus*. *Am. J. Bot.* **61**, 525–534. [83, 85, 86, 87, 99]

Mughaz, S. and Godward, M. B. E. (1973). Kinetochore and microtubules in two members of *Chlorophyceae*, *Cladophora tracta* and *Spirogyra majuscula*. *Chromosoma* **44**, 213–229. [69]

Mühlethaler, K. (1960). Die struktur de Grana- und Stromalamellen in Chloroplasten. *Z. wiss. Mikrosk.* **64**, 444–452. [145]

Munroe, M. H. and Bell, P. R. (1970). The fine structure of fern root cells showing apospory. *Devl Biol.* **23**, 550–562. [103]

Muntzing, A. (1966). Some recent data on accessory chromosomes in *Secale* and *Poa*. *In* "Chromosomes Today" (Eds C. D. Darlington and K. R. Lewis), vol. 1, pp. 7–14. Oliver and Boyd, Edinburgh. [208, 213, 214]

Murashige, T. (1974). Plant propagation through tissue culture. *A. Rev. Pl. Physiol.* **25**, 135–166. [102]

Murashige, T. and Skoog, F. (1962). A revised medium for rapid growth and bioassays with tobacco tissue cultures. *Physiologia Pl.*, **15** 473–497. [145]

Murray, D. (1968). "Light and Leaf Growth in *Phaseolus*". Ph.D. Thesis, University of Edinburgh. [332, 333, 334, 344]

Nachtwey, D. S. and Cameron, I. L. (1968). Cell cycle analysis. *In* "Methods in Cell Physiology" (Ed. D. M. Prescott), vol. 3, pp. 213–259. Academic Press, New York and London. [64, 301]

Naf, U. (1962). Developmental physiology of lower archegoniates. *A. Rev. Pl. Physiol.* **13**, 507–532. [103]

Nagata, T. and Takebe, I. (1970). Cell wall regeneration and cell division in isolated tobacco mesophyll protoplasts. *Planta* **92**, 301–308. [421, 423]

Nagata, T. and Takebe, I. (1971). Plating of isolated tobacco mesophyll protoplasts on agar medium. *Planta* **99**, 12–20. [101, 421, 424]

Nagata, T. and Yamaki, T. (1973). Electron microscopy of isolated tobacco mesophyll protoplasts cultured *in vitro*. *Z. PflPhysiol.* **70**, 452–459. [421]

NAGL, W. (1969). Banded polytene chromosomes in the legume *Phaseolus vulgaris*. *Nature, Lond.* **221**, 70–71. [400]

NAGL, W. (1970). The mitotic and endomitotic nuclear cycle in *Allium carinatum*. II. Relations between DNA replication and chromatin structure. *Caryologia* **23**, 71–78. [63]

NAGL, W. (1970a). Inhibition of polytene chromosome formation in *Phaseolus* by polyploid mitoses. *Cytologia* **35**, 252–258. [238]

NAGL, W. (1970b). On the mercaptoethanol induced polyploidisation in *Allium cepa*. *Protoplasma* **70**, 349–359. [222, 224, 225]

NAGL, W. (1972). Molecular and structural aspects of the endomitotic chromosome cycle in angiosperms. *In* "Chromosomes Today" (Eds C. D. Darlington and K. R. Lewis), vol. 3, pp. 17–23. Oliver and Boyd, Edinburgh. [63, 221]

NAGL, W. (1973). The mitotic and endomitotic nuclear cycle in *Allium carinatum*. IV. ³H-uridine incorporation. *Chromosoma* **44**, 203–212. [221, 235]

NAKAZAWA, S. (1963). Role of the protoplasmic connections in the morphogenesis of fern gametophytes. *Sci. Rep. Tokoku Univ. Ser. 4* **29**, 247–255. [101]

NASJILETI, C. E. and SPENCER, H. H. (1968). Effects of chloramphenicol on mitosis of phytohemagglutinin-stimulated human leukocytes. *Expl Cell Res.* **53**, 11–17. [222]

NATROVKA, Z. (1968). Studium velikosti pylovych zrn jarniho jecmene v klasech odnozi ruzneho radu [Investigation of the size of the pollen grains of summer barley in the ears of shoots of different order] *Genetika a Slechteni* **4**, 85–94. [191]

NAYLOR, J. M. (1958). Control of nuclear processes by auxin in auxillary buds of *Tradescantia paludosa*. *Can. J. Bot.* **36**, 221–232. [203]

NAYLOR, J., SANDER, G. and SKOOG, F. (1954). Mitosis and cell enlargement without cell division in excised tobacco pith tissue. *Physiologia Pl.* **7**, 25–29. [225]

NEILSEN, E. L. and NATH, J. (1961). Somatic instability in derivatives from *Agroelymus turneri* resembling *Agropyron repens*. *Am. J. Bot.* **48**, 345–349. [248]

NEUBAUER, B. F. (1972). The development of the achene of *Polygonum pennsylvanicum:* embryo, endosperm and pericarp. *Am. J. Bot.* **58**, 655–664. [51]

NEWCOMB, E. H. (1969). Plant microtubules. *A. Rev. Pl. Physiol.* **20**, 253–288. [51]

NEWCOMB, W. and FOWKE, L. C. (1973). The fine structure of the change from free nuclear to cellular condition in the endosperm of chickweed *Stellaria media*. *Bot. Gaz.* **134**, 236–241. [61]

NEWMAN, I. V. (1956). Pattern in meristems of vascular plants. I. Cell partition in living apices and in the cambial zone in relation to the concepts of initial cells and apical cells. *Phytomorphology* **6**, 1–19. [296, 362]

NEWMAN, I. V. (1961). Pattern in the meristems of vascular plants. II. A review of shoot apical meristems of gymnosperms, with comments on apical biology and taxonomy, and a statement of some fundamental concepts. *Proc. Linn. Soc. N.S.W.* **86**, 9–59. [297]

NICKLAS, R. B. (1971). Mitosis. *In* "Advances in Cell Biology" (Eds D. M. Prescott, L. Goldstein and E. McConkey) Vol. 2, pp. 225–297. Academic Press, New York and London. [51]

NIELSON-JONES, W. (1969). "Plant Chimeras". Methuen, London. [94, 246, 247]

NIITSU, T. (1958). Effects of chemicals on mitosis studied in *Tradescantia* cells *in vivo* IV. 2-methyl amino 1:3 diaza-azulene with special reference to its effect on cytokinesis. *Cytologia* **23**, 372–382. [224]

NINNEMAN, H. and EPEL, B. (1973). Inhibition of cell division by blue light. *Expl Cell Res.* **79**, 318–326. [173]

NISHI, A. and SUGANO, N. (1970). Growth and division of carrot in suspension culture. *Pl. Cell Physiol., Tokyo* **11**, 757–765. [110]

NJOKU, E. (1956). Studies in the morphogenesis of leaves. XI. The effect of light intensity on leaf shape in *Ipomea caerulea. New Phytol.* **55**, 91–110. [321]

NJOKU, E. (1956). The effect of defoliation on leaf shape in *Ipomea caerulea. New Phytol.* **55**, 213–228. [321]

NORSTOG, K. (1972). Early development of the barly embryo: fine structure. *Am. J. Bot.* **59**, 123–132. [51, 99, 404]

NORSTOG, K., WALL, W. E. and HOWLAND, G. P. (1969). Cytological characteristics of ten-year-old rye-grass endosperm tissue culture. *Bot. Gaz.* **130**, 83–86. [63, 249]

NORTHCOTE, D. H. (1971). Organisation of structure, synthesis and transport within the plant during cell division and growth. *In* "Control Mechanisms and Differentiation of Growth" (Eds D. D. Davies and M. Balls), *Symp. Soc. exp. Biol.* **25**, 51–70. Cambridge University Press. [79, 99, 106, 108]

NORTHCOTE, D. H. (1972). Chemistry of the plant cell wall. *A. Rev. Pl. Physiol.* **23**, 113–132. [370]

NOUGARÈDE, A. (1967). Experimental cytology of the shoot apical cells during vegetative growth and flowering. *Int. Rev. Cytol.* **21**, 203–351. [288, 292, 294, 296, 313]

NUR, U. (1968). Endomitosis in the mealy bug; *Planococcus citri* (Homoptera: Coccoidea). *Chromosoma* **24**, 202–209. [218]

O'BRIEN, T. P. and THIMANN, K. V. (1967). Observations on the fine structure of the oat coleoptile. II. The parenchyma cells of the apex. *Protoplasma* **63**, 417–442. [98]

OGIHARA, R. (1962). B-chromosomes of *Lilium auratum* Lindl. 1. Meiotic behaviour of the B chromosomes in PMC. *La Kromosomo* **53/54**, 1778–1784. [208, 215]

OHNUKI, Y. (1968). Structure of chromosomes. 1. Morphological studies of the spiral structure of human somatic chromosomes. *Chromosoma* **25**, 402–428. [67]

OKA, H. I. and MORISHIMA, H. (1967). Variations in the breeding systems of a wild rice, *Oryza perennis. Evolution, Lancaster, Pa.* **21**, 249–258. [164]

OLDEN, E. J. (1953). Sexual and apomictic seed formation in *Malus sieboldii* Rehd. *Bot. Notiser* **106**, 105–128. [248]

OMODEO, P. (1952). Cariologia di Lumbricidae. *Caryologia* **4**, 173–275. [219]

OOTAKI, T. (1967). Branchings and regeneration patterns of isolated single cells of a fern protonema. *Bot. Mag., Tokyo* **80**, 1–10. [99]

OOTAKI, T. (1968). Polarity in the branching of *Pteris vittata* protonemata. *Embryologia* **10**, 152–163. [99]

OOTAKI, T. and FURUYA, M. (1969). Experimentally induced apical dominance in protonemata of *Pteris vittata. Embryologia* **10**, 284–296. [99]

ÖSTEGREN, G. and FRÖST, S. (1962). Elimination of accessory chromosomes from the roots in *Haplopappus gracilis. Hereditas* **48**, 363–366. [208, 214]

ÖSTEGREN, G. KOOPMANS, A. and REITALU, J. (1953). The occurrence of the amphiastral type of mitosis in higher plants and the influence of aminopyrin on mitosis. *Bot. Notiser* **106**, 417–419.

PACKER, L. (1966). Evidence of contractility in chloroplasts. *In* "Biochemistry of Chloroplasts" (Ed. T. W. Goodwin) Vol. 1, pp. 233–242. Academic Press, London and New York. [88]

PALIWAL, R. L. and HYDE, B. L. (1959). The association of a single B-chromosome with male sterility in *Plantago coronopus. Am. J. Bot.* **46**, 460–466. [208, 214]

PAOLILLO, D. J., JR. (1962). The plastids of *Isoetes howellii. Am. J. Bot.* **49**, 590–598. [83, 86]

PAOLILLO, D. J., JR. (1974). Motile male gametes of plants. *In* "Dynamic Aspects of Plant Ultrastructure" (Ed. A. W. Robards), pp. 504–531. McGraw-Hill, London. [69, 71, 72, 83, 86]

PARKHURST, D. F. (1972). Conductive capacities of veins in expanding leaves of *Quercus. Aust. J. biol. Sci.* **25**, 425–428. [341]

PARTANEN, C. R. (1963). Plant tissue culture in relation to developmental cytology. *Int. Rev. Cytol.* **15**, 215–243. [217, 219, 233, 249]

PATAU, K. and DAS, N. K. (1961). The relation of DNA synthesis and mitosis in tobacco pith tissue cultured *in vitro. Chromosoma* **11**, 553–572. [202, 203, 225]

PATAU, K., DAS, N. K. and SKOOG, F. (1957). Induction of DNA synthesis by kinetin and indolacetic acid in excised tobacco pith tissue. *Physiologia Pl.* **10**, 949–966. [202, 203, 225]

PATEL, R. N. (1965). A comparison of the anatomy of the secondary xylem in roots and stems. *Holzforschung* **19**, 72–79. [361]

PAULEY, S. S. (1958). Photoperiodism in relation to tree improvement. *In* "The Physiology of Forest Trees" (Ed. K. V. Thimann), pp. 557–571. Ronald Press, New York. [388]

PEARCE, R. S. and COCKING, E. C. (1973). Behaviour in culture of isolated protoplasts from "Paul's Scarlet" rose suspension culture cells. *Protoplasma* **77**, 165–180. [423]

PEREIRA, A. S. R. and LINSKENS, H. F. (1963). The influence of glutathione and glutathione antagonists on meiosis in excised anthers of *Lilium henryi. Acta bot. neerl.* **12**, 302–314. [176, 184]

PETERSEN, D. F., ANDERSON, E. C. and TOBEY, R. A. (1968). Mitotic cells as a source of synchronized cultures. *In* "Methods in Cell Physiology" (Ed. D. M. Prescott) vol. 3, pp. 347–370. Academic Press, New York and London. [114]

PETERSON, R. L. (1973). Control of cambial activity in roots of turnip (*Brassica rapa*). *Can. J. Bot.* **51,** 475–480. [383, 384]

PHILIPSON, W. R. and WARD, J. M. (1965). The ontogeny of the vascular cambium in the stem of seed plants. *Biol. Rev.* **40,** 534–579. [361]

PHILIPSON, W. R., WARD, J. M. and BUTTERFIELD, B. G. (1971). "The Vascular Cambium". Chapman and Hall, London. [349, 358, 359, 360, 361, 374, 375, 376, 377, 380, 383]

PHILLIPS, H. L., JR. and TORREY, J. G. (1971a). Deoxyribonucleic acid synthesis in root cap cells of cultured roots of *Convolvulus*. *Pl. Physiol., Lancaster* **48,** 213–218. [266, 275]

PHILLIPS, H. L., JR. and TORREY, J. G. (1971b). The quiescent center in cultured roots of *Convolvulus arvensis*. L. *Am. J. Bot.* **58,** 665–671. [266]

PHILLIPS, H. L., JR. and TORREY, J. G. (1972). Duration of cell cycles in cultured roots of *Convolvulus*. *Am. J. Bot.* **59,** 183–188. [53, 122]

PHILLIPS, I. D. J. (1969). Apical dominance. *In* "The Physiology of Plant Growth and Development" (Ed. M. B. Wilkins), pp. 163–202. McGraw-Hill, London. [375, 387, 388]

PHILLIPS, I. D. J. (1971). "Introduction to the Biochemistry and Physiology of Plant Growth Hormones". McGraw-Hill, New York. [381]

PHILLIPS, I. D. J. (1975). Apical dominance. *A. Rev. Pl. Physiol.* **26,** 341–367.

PICKETT-HEAPS, J. D. (1967a). The effects of colchicine on the ultrastructure of dividing plant cells, xylem wall differentiation and distribution of cytoplasmic microtubules. *Devl Biol.* **15,** 206–236. [72, 420, 222]

PICKETT-HEAPS, J. D. (1967b). Further observations of the Golgi apparatus and its functions in cells of the wheat seedling. *J. Ultrastruct. Res.* **18,** 287–303. [420]

PICKETT-HEAPS, J. D. (1967c). Ultrastructure and differentiation in *Chara*. sp. II. Mitosis. *Aust. J. biol. Sci.* **20,** 883–894. [72, 93, 222, 420]

PICKETT-HEAPS, J. D. (1968). Ultrastructure and differentiation in *Chara fibrosa*. IV. Spermatogenesis. *Aust. J. biol. Sci.* **21,** 655–690. [72]

PICKETT-HEAPS, J. D. (1969a). The evolution of the mitotic apparatus: an attempt at comparative ultrastructural cytology in dividing plant cells. *Cytobios* **1,** 257–280. [71, 72, 73, 107, 108, 223, 224]

PICKETT-HEAPS, J. D. (1969b). Pre-prophase microtubule bands in some abnormal mitotic cells of wheat. *J. Cell Sci.* **4,** 397–420. [70, 95. 107, 223, 224]

PICKETT-HEAPS, J. D. (1969c). Pre-prophase microtubules and stomatal differentiation in *Commelina cyanea*. *Aust. J. biol. Sci.* **22,** 375–391. [70, 107, 223, 224]

PICKETT-HEAPS, J. D. (1969d). Pre-prophase microtubules and stomatal differentiation: some effects of centrifugation on symmetrical and asymmetrical cell division. *J. Ultrastruct. Res.* **27,** 24–44. [70, 223, 224]

PICKET-HEAPS, J. D. (1970). Mitosis and autospore formation in the green alga *Kirchneriella lunaris*. *Protoplasma* **70,** 325–347. [88]

PICKET-HEAPS, J. D. (1974). Plant microtubules. *In* "Dynamic Aspects of Plant Ultrastructure" (Ed. A. W. Robards), pp. 219–255. McGraw-Hill, London. [51, 69, 70, 71, 73]

PICKETT-HEAPS, J. D. and NORTHCOTE, D. H. (1966a). Cell division in the formation of the stomatal complex of the young leaves of wheat. *J. Cell Sci.* **1**, 121–128. [53]

PICKETT-HEAPS, J. D. and NORTHCOTE, D. H. (1966b). Organisation of microtubules and endoplasmic reticulum during mitosis and cytokinesis in wheat meristems. *J. Cell Sci.* **1**, 109–120. [53, 419]

PIENIAZEK, J. (1964). Kinetin-induced breaking of dormancy in 8-month-old apple seedlings of 'Antonovka' variety. *Acta Agrobot.* **16**, 157–169. [383]

PILET, P-E. (1972). Root cap and root growth. *Planta* **106**, 169–171. [283]

PILET, P-E. (1973). Growth inhibitor from the root cap of *Zea mays*. *Planta* **111**, 275–278. [283]

PILET, P-E. and LANCE-NOUGARÈDE, A. (1965). Quelques characteristiques structurales et physiologiques du méristème radiculaire du *Lens culinaris*. *Bull. Soc. fr. Physiol. vég.* **11**, 187–201. [272]

PITELKA, D. R. (1969). Centriole replication. *In* "Handbook of Molecular Cytology" (Ed. A. Lima-de-Faria), pp. 1199–1218. North-Holland, Amsterdam. [71]

POJNAR, E., WILLISON, J. H. M. and COCKING, E. C. (1967). Cell wall regeneration by isolated tomato-fruit protoplasts. *Protoplasma* **64**, 460–480. [420]

POLLOCK, E. G. and JENSEN, W. A. (1964). Cell development during early embryogenesis in *Capsella* and *Gossypium*. *Am. J. Bot.* **51**, 915–921. [51, 398, 400, 402]

POPHAM, R. A. (1958). Cytogenesis and zonation in the shoot apex of *Chrysanthemum morifolium*. *Am. J. Bot.* **45**, 198–206. [303]

PORTER, K. R. and MACHADO, R. D. (1960). Studies on the endoplasmic reticulum. IV. Its form and distribution during mitosis of cells of onion root tip. *J. biophys. biochem. Cytol.* **7**, 167–180. [93]

POSSINGHAM, J. V. (1973). Chloroplast growth and division during the greening of spinach leaf discs. *Nature New Biol.* **245**, 93–94. [149, 153]

POSSINGHAM, J. V. and SAURER, W. (1969). Changes in chloroplast number per cell during leaf development in spinach. *Planta* **86**, 186–194. [81, 84, 93, 138, 139, 140, 142]

POSSINGHAM, J. V. and SMITH, J. W. (1972). Factors affecting chloroplast replication in spinach. *J. exp. Bot.* **23**, 1050–1059. [81, 142, 149, 150, 153]

POTRYKUS, I. and DURAND, J. (1972). Callus formation from single protoplasts of *Petunia*. *Nature New Biol.* **237**, 286–287. [424]

POWER, J. B. and COCKING, E. C. (1970). Isolation of leaf protoplasts: macro-molecule uptake and growth substance response. *J. exp. Bot.* **21**, 64–70. [420]

PRASAD, A. B. and GODWARD, M. B. E. (1965). Comparison of the developmental response of diploid and tetraploid *Phalaris* following irradiation of the dry seed. I. Determination of mitotic cycle time, mitotic time and phase time. *Radiat. Bot.* **5**, 465–474. [122]

PRESCOTT, D. M. (1961). The growth-duplication cycle of the cell. *Int. Rev. Cytol.* **11**, 255–282. [51, 77]

PRESCOTT, D. M. (1966). The synthesis of total macronuclear protein, histone and DNA during the cell cycle in *Euplotes eurystomus*. *J. Cell Biol.* **31**, 1–9. [118]

PRESTON, A. P. and BARLOW, H. W. B. (1950). The use of growth substances to widen crotch angles. *Rep. E. Malling Res. Stn* No. 26. [387]

PRIESTLEY, J. H. (1930). Studies in the physiology of cambial activity. III. The seasonal activity of the cambium. *New Phytol.* **29**, 316–354. [373, 379, 381, 389]

PRIESTLEY, J. H. and SCOTT, L. I. (1936). A note upon summer wood production in the tree. *Proc. Leeds phil. lit. Soc. (Sci. Sect.)* **3**, 235–248. [374]

PRITCHARD, A. J. and COURT, R. D. (1968). The cytological effects of mimosine. *Cytologia* **33**, 73–77. [222]

PRITCHARD, E. (1968). A cytogenetic study of supernumerary chromosomes in *Haplopappus gracilis*. *Can. J. Genet. Cytol.* **10**, 928–936. [213, 214]

QUASTLER, H. and SHERMAN, F. G. (1959). Cell population kinetics in the intestinal epithelium of the mouse. *Expl Cell Res.* **17**, 420–438. [262]

QUETIER, F., HUGUET, T. and GUILLE, E. (1969). Induction of crown gall: partial homology between tumor cell DNA, bacterial DNA and the G+C-rich DNA of stressed normal cells. *Biochem. biophys. Res. Commun.* **34**, 128–133. [426]

RAATZ, W. (1892). Die Stabbildungen in secondären Holzkörper der Bäume und die Initialentheorie. *Jb. wiss. Bot.* **23**, 567–636. [362]

RADIN, J. W. and LOOMIS, R. S. (1971). Changes in the cytokinins of radish roots during maturation. *Physiologia Pl.* **25**, 240–244. [383, 384, 385, 389]

RAJHATHY, T. (1963). Chromosome mosaics and the recovery of the original strain from octaploid *Hordeum murinum*. *Z. Vererblehre,* **94**, 269–279. [248]

RAJU, M. V. S., STEEVES, T. A. and NAYLOR, J. M. (1964). Developmental studies on *Euphorbia esula* L.: apices of long and short roots. *Can. J. Bot.* **42**, 1615–1628. [266]

RAM, H. (1960). Occurrence of endosperm haustorium in *Cannabis sativa* L. *Ann. Bot.* **24**, 79–82. [61]

RAPER, C. D., JR. and THOMAS, J. F. (1972). Temperatures in early post-transplant growth. Effect on shape of mature *Nicotiana tabacum* L. leaves. *Crop Sci.* **12**, 540–542. [321]

RASCH, E., SWIFT, H. and KLEIN, R. M. (1959). Nucleoprotein changes in plant tumor growth. *J. biophys. biochem. Cytol.* **6**, 11–34. [123, 219, 230, 231, 428]

RASHID, A. and STREET, H. E. (1974). Growth, embryogenic potential and stability of a haploid cell culture of *Atropa belladonna* L. *Pl. Sci. Lett.* **2**, 89–94. [103]

REES, H. (1961). Genotypic control of chromosome form and behaviour. *Bot. Rev.* **27**, 288–318. [232]

REES, H. and JONES, R. N. (1972). The origin of the wide species variation in nuclear DNA content. *Int. Rev. Cytol.* **32**, 53–92. [208, 210, 211, 212]

REES, H. and NAYLOR, B. (1960). Developmental variation in chromosome behaviour. *Heredity* **15**, 17–27. [184]

REEVE, R. M. (1942). Structure and growth of the vegetative shoot apex of *Garrya elliptica* Dougl. *Am. J. Bot.* **29**, 697–711. [309]

REEVE, R. M. (1948). The "Tunica-Corpus" concept and development of shoot apices in certain dicotyledons. *Am. J. Bot.* **35**, 65–75. [309]

REID, D. M. and CARR, D. J. (1967). Effects of a dwarfing compound, CCC, on the production and export of gibberellin-like substances by root systems. *Planta* **73**, 1–11. [343]

REINDERS-GOUWENTAK, C. A. (1949). Cambiumwerkzaamheid en groeistof. *Vakb. Biol.* **29**, 9–17. [390]

REINDERS-GOUWENTAK, C. A. (1965). Physiology of the cambium and other secondary meristems of the shoot. *In* "Encyclopedia of Plant Physiology" (Ed. W. Ruhland), Vol. XV/1, pp. 1077–1105. Springer-Verlag, Berlin. [390]

REINERT, J. (1959). Über die Kontrolle der Morphogenese und die Induktion von adventiv Embryonen an Gewebekulturen aus Karotten. *Planta* **53**, 318–333. [101]

REINERT, J. (1973). Aspects of organisation—organogenesis and embryogenesis. *In* "Plant Tissue and Cell Culture" (Ed. H. E. Street), Botanical Monographs **11**, 338–355. Blackwell Scientific Publications, Oxford. [249, 429]

REINERT, J. and HELLMANN, S. (1971). Mechanism of the formation of polynuclear protoplasts from cells of higher plants. *Naturwissenschaften* **58**, 419. [420]

REINERT, J. and HELLMANN, S. (1973). Aspects of nuclear division and cell wall formation in protoplasts of different origin. *In* "Protoplastes et Fusion de Cellules Somatiques Végétales", *Coll. Int. C.N.R.S.* **212**, 273–279. [423]

RENAUD, F. L. and SWIFT, H. (1964). The development of basal bodies and flagella in *Allomyces arbusculus*. *J. Cell Biol.* **23**, 339–354. [72]

RESCH, A. (1958). Weitere untersuchungen über das Phloem von *Vicia faba*. *Planta* **52**, 121–143. [53, 235]

RETALLAK, B. and BUTLER, R. D. (1970). One development and structure of pyrenoids in *Bulbochaete hiloensis*. *J. Cell Sci.* **6**, 229–242. [88]

RHOADES, M. M. (1961). Meiosis. *In* "The Cell" (Eds J. Brachet and A. E. Mirsky), Vol. 3, pp. 1–75. Academic Press, New York and London. [50]

RICHARDS, F. J. (1948). The geometry of phyllotaxis and its origin. *In* "Growth in Relation to Differentiation and Morphogenesis", *Symp. Soc. exp. Biol.* **2**, 217–245. Cambridge University Press. [312]

RICHARDS, F. J. (1951). Phyllotaxis: its quantitative expression and relation to growth in the apex. *Phil. Trans. R. Soc.* B **235**, 509–564. [286, 311, 312]

RICHARDS, F. J. (1956). Spatial and temporal correlations involved in leaf pattern production at the apex. *In* "Growth of Leaves" (Ed. F. L. Milthorpe), pp. 66–76. Butterworths, London. [312]

RIDLEY, S. M. and LEECH, R. M. (1970a). Division of chloroplasts in an artificial environment. *Nature, Lond.* **227**, 463–465. [143, 144]

RIDLEY, S. M. and LEECH, R. M. (1970b). Light-dependent consumption of oxygen by *Vicia faba* chloroplasts. *Archs Biochem. Biophys.* **139**, 351–360. [143, 144, 155]

RIEGER, R. and MICHAELIS, A. (1958). Cytologische und stoffwechselphysiologische Untersuchungen amaktiven Meristem der Wurzelspitze von *Vicia faba* L. 1. Der Einfluss der Unterwasserquelling der Samen auf die chromosomale Aberationsrate. *Chromosoma* **9**, 238–257. [249]

RIEGER, R., MICHAELIS, A., SCHUBERT, I. and MEISTER, A. (1973). Somatic interphase pairing of *Vicia* chromosomes as inferred from the hom/het ratio of induced chromatid interchanges. *Mutation Res.* **20**, 295–298. [107]

RIJVEN, A. H. G. C. (1968). Determination of uni- and trifoliolate leaf form in fenugreek. *Aust. J. biol. Sci.* **21**, 155–156. [318]

RILEY, R. and BENNETT, M. D. (1971). Meiotic DNA synthesis. *Nature, Lond.* **230**, 182–185. [162, 169, 185]

RIS, H. and PLAUT, W. (1962). Ultrastructure of DNA containing areas in the chloroplast of *Chlamydomonas*. *J. Cell Biol.* **13**, 383–391. [83, 88, 158]

RISUEÑO, M. C., GIMÉNEZ-MARTÍN, G. and LÓPEZ-SÁEZ, J. F. (1968). Experimental analysis of plant cytokinesis. *Expl Cell Res.* **49**, 136–147. [222, 224]

RIZZONI, M. and PALITTI, F. (1973). Regulatory mechanisms of cell division. 1. Colchicine induced endoreduplication. *Expl Cell Res.* **77**, 450–458. [225]

ROBARDS, A. W. (1965). Tension wood and eccentric growth in Crack Willow (*Salix fragilis* L.). *Ann. Bot.* **30**, 513–523. [222, 380]

ROBARDS, A. W. (1966). The application of the modified sine rule to tension wood production and eccentric growth in the stem of Crack Willow (*Salix fragilis* L.). *Ann. Bot.* **30**, 513–523. [385]

ROBARDS, A. W. (1971). The ultrastructure of plasmodesmata. *Protoplasma* **72**, 315–323. [98]

ROBARDS, A. W., DAVIDSON, E. and KIDWAI, P. (1969). Short-term effects of some chemicals on cambial activity. *J. exp. Bot.* **20**, 912–921. [384]

ROBARDS, A. W. and KIDWAI, P. (1969). A comparative study of the ultrastructure of resting and active cambium of *Salix fragilis* L. *Planta* **84**, 239–249. [356, 374]

ROBBELEN, G. (1966). Chloroplastendifferenzierung nach gehindurzierter Plastommutation bei *Arabidopsis thaliana* (L) Heynh. *Z. PflPhysiol.* **55**, 387–403. [94]

ROBBINS, E., JENTZSCH, G. and MICALI, A. (1968). The centriole cycle in synchronised HeLa cells. *J. Cell Biol.* **36**, 329–339. [71]

ROBERTS, K. and NORTHCOTE, D. H. (1970). The structure of sycamore callus cells during division in a partially synchronized suspension culture. *J. Cell Sci.* **6**, 299–321. [52, 93, 98, 115, 416, 419]

ROBERTS, L. W. and FOSKET, D. E. (1966). Interactions of gibberellic acid and indoleacetic acid in the differentiation of wound vessel members. *New Phytol.* **65**, 5–8. [384]

ROBINOW, C. F. and CATEN, C. E. (1969). Mitosis in *Aspergillus nidulans*. *J. Cell Sci.* **5**, 403–431. [73]

RODKIEWICZ, B. (1970). Callose in cell walls during megasporogenesis in angiosperms. *Planta* **93**, 39–47. [99]

ROMBERGER, J. A. (1963). "Meristems, Growth and Development in Woody

Plants". U.S. Dept. of Agriculture, Forest Service, Technical Bulletin No. 1293. [376, 377]

ROMBERGER, J. A. (1966). Developmental biology and the spruce tree. *J. Washington Acad. Sci.* **56,** 69–81. [379]

RONCHI, N. V., AVANZI, S. and D'AMATO, F. (1965). Chromosome endore-duplication (endopolyploidy) in pea root meristems induced by 8-aza-guanine. *Caryologia* **18,** 599–617. [225]

ROSE, R. J., CRAN, D. G. and POSSINGHAM, J. V. (1974). Distribution of DNA in dividing spinach chloroplasts. *Nature, Lond.* **251,** 641–642. [84]

ROSE, R. J., SETTERFIELD, G. and FOWKE, L. C. (1972). Activation of nucleoli in tuber slices and the function of nucleolar vacuoles. *Expl Cell Res.* **71,** 1–16. [125, 419]

ROSENBERGER, R. F. and KESSEL, M. (1968). Non-random sister chromatid segregation and nuclear migration in hyphae of *Aspergillus nidulans. J. Bact.* **96,** 1208–1213. [67]

ROST, T. L. and VAN'T HOF, J. (1973). Radio-sensitivity, RNA and protein metabolism of "leaky" and arrested cells in sunflower root meristems (*Helianthus annuus*). *Am. J. Bot.* **60,** 172–181. [122, 168]

ROTH, I. (1957). Relation between the histogenesis of the leaf and its external shape. *Bot. Gaz.* **118,** 237–245. [317]

ROTTA, H. (1949). Untersuchungen uber tagesperiodische Vorgonge in Spross- und Wurzelvegetationspunkten. *Planta* **37,** 399–412. [303]

ROUFFA, A. S. and GUNCKEL, J. E. (1951). Leaf initiation, origin and pattern of pith development in the Rosaceae. *Am. J. Bot.* **38,** 301–307. [309]

RUTISHAUSER, A. (1956). Chromosome distribution and spontaneous chromo-some breakage in *Trillium grandiflorum. Heredity* **10,** 367–407. [214, 239, 240, 242, 249]*

RUTISHAUSER, A. and ROTHLISBERGER, E. (1966). Boosting mechanism of B-chromosomes in *Crepis capillaris. In* "Chromosomes Today" 1 (Eds C. D. Darlington and K. R. Lewis), Vol. 1, pp. 28–30. Oliver and Boyd, Edin-burgh. [213]

SACHS, J. (1878). Über die Anordnung der Zellen in jüngsten Pflanzerstheilen. *Arb. bot. Inst. Würzburg* **2,** 46–104. [413]

SACHS, J. (1887). "Lectures on the Physiology of Plants". Clarendon Press, Oxford. [6]

SACHS, L. (1952). Chromosome mosaics in experimental amphidiploid in the Triticinae. *Heredity* **6,** 157–170. [248]

SACRISTÁN, M. D. and WENDT-GALLITELLI, M. F. (1973). Tumorous cultures of *Crepis capillaris:* chromosomes and growth. *Chromosoma* **43,** 279–288. [249]

SAINT-CÔME, R. (1969). Cytologie végétale—durée du cycle mitotique dans le point végétatif du *Coleus blumei* Berth. *C.r. hebd. Séanc. Acad. Sci. Paris* **268,** 508–511. [291]

SAINT-CÔME, R. (1971). Cytologie végétale—durée du cycle mitotique chez le *Coleus blumei* Berth. durant les phases préflorale et reproductive. *C.r. hebd. Séanc. Acad. Sci. Paris* **272,** 44–47. [313]

Saint-Côme, R. (1973). Cytologie végétale—détermination de la durée du cycle mitotique dans le point végétatif du *Coleus blumei* Berth. après traitement à la colchicine. *C.r. hebd. Séanc. Acad. Sci. Paris* **277**, 1001–1004. [290, 291]

Salisbury, E. J. (1961). "Weeds and Aliens". Collins, London. [195]

Samejima, J. (1958). Meiotic behaviour of accessory chromosomes and their distribution in natural population of *Lilium medeoloides* A. Gray. *Cytologia* **23**, 159–171. [213]

Sanger, J. M. and Jackson, W. T. (1971). Fine structure study of pollen development in *Haemanthus katherinae* Baker. I. Formation of vegetative and generative cells. *J. Cell Sci.* **8**, 289–301. [56]

Sanio, K. (1873). Anatomie der gemeinen Kiefer (*Pinus sylvestris* L.). *Jb. wiss. Bot.* **9**, 50–126. [361]

Sapehin, A. A. (1915). Untersuchungen uber die Individualität der Plastide. *Arch. Zellt.* **13**, 319–398. [83, 86, 87]

Satina, S. (1959). Chimeras. *In* "Blakeslee: The Genus Datura" (Eds A. G. Avery, S. Satina and J. Rietsema), pp. 132–151. Ronald Press, New York. [233, 235, 239, 245]

Saunders, P. F., Alvim, R. and Harrison, M. A. (1973). Abscisic acid and tree growth. *In* "Plant Growth Substances", pp. 871–881. Hirokawa Publishing Co., Tokyo, Japan. [390]

Saurer, W. and Possingham, J. V. (1970). Studies on the growth of spinach leaves (*Spinacea oleracea*). *J. exp. Bot.* **21**, 151–158. [328, 329]

Sauerland, H. (1956). Quantitative Untersuchungen von Röntgen-effekten nach Bestrahlung verschiedener Meiosisstadien bei *Lilium candidum* L. *Chromosoma* **7**, 627–654. [176]

Savage, J. R. K. and Wigglesworth, D. J. (1972). Lack of uniform radiosensitivity of dormant cells in the root meristem of barley seeds. A preliminary report. *In* "Technical Report Series 141", *Int. atom. Energy Ag. Bull.* **5**, 77–86. [280]

Savelkoul, R. M. H. (1957). Distribution of mitotic activity within the shoot apex of *Elodea densa*. *Am. J. Bot.* **44**, 311–317. [303]

Sax, K. and Edmonds, H. W. (1933). Development of the male gametophyte in *Tradescantia*. *Bot. Gaz.* **95**, 156–163. [176]

Sax, K. and Husted, L. (1936). Polarity and differentiation in microspore development. *Am. J. Bot.* **23**, 606–609. [56]

Schachtschabel, D. O., Killander, D., Zetterberg, A., McCarthy, R. E. and Foley, G. E. (1968). Effects of 4-aminopyrazolo (3,4-D) pyrimidine in combination with guanine on nucleic acid and protein synthesis by Ehrlich ascites cells in culture. *Expl Cell Res.* **50**, 73–80. [222]

Schaffner, M. (1906). The embryology of the Shepherd's purse. *Ohio Naturalist* **7**, 1–34. [400]

Schiff, J. A. and Epstein, H. T. (1965). The continuity of the chloroplast in *Euglena*. *In* "Reproduction: Molecular, Subcellular and Cellular" (Ed. M. M. Locke), 24th Annual Symposium of the Society for Development and Growth, pp. 131–189. Academic Press, New York. [83]

SCHIFF, J. A. and EPSTEIN, H. T. (1966). The replicative aspect of chloroplast continuity in *Euglena. In* "Biochemistry of Chloroplasts" (Ed. T. W. Goodwin), pp. 341–353. Academic Press, London and New York. [83]

SCHNABEL, V. (1941). Der Bau der Sprossvegetationspunkte von *Honckenya peploides, Silene maritima, Dianthus caryophyllus* und *Clematis paniculata. Bot. Arch.* **42,** 461–502. [309]

SCHÖTZ, F. and SENSER, F. (1964). Untersuchungen über die Chloroplastenentwicklung bei *Oenothera.* III. Der *Pictirubata*-typ. *Planta* **63,** 191–212. [93]

SCHRÖDER, K. H. (1962). Mikroskopische Untersuchungen über die Vermehrung der Plastiden in Scheitelmeristem und in den Blättern von *Oenothera albilacta. Z. Bot.* **50,** 348–367. [81, 84]

SCHÜEPP, O. (1917). Untersuchungen über Wachstum und Formwechsel von Vegetationspunkten. *Jb. wiss. Bot.* **57,** 17–79. [257]

SCHÜEPP, O. (1938). Über periodische Formbildung bei Pflanzen. *Biol. Rev.* **13,** 59–92. [287]

SCHULZ, P. and JENSEN, W. A. (1969). *Capsella* embryogenesis: the suspensor and the basal cell. *Protoplasma* **67,** 139–163. [398, 400]

SCHULZ, P. and JENSEN, W. A. (1971). *Capsella* embryogenesis: the chalazal proliferating tissue. *J. Cell Sci.* **8,** 201–227. [99]

SCHULZ, S. R. and JENSEN, W. A. (1968a). *Capsella* embryogenesis: the early embryo. *J. Ultrastruct. Res.* **22,** 376–392. [51, 393, 395, 398, 400]

SCHULZ, S. R. and JENSEN, W. A. (1968b). *Capsella* embryogenesis: the egg, zygote and young embryo. *Am. J. Bot.* **55,** 807–819. [51, 53, 400, 401, 402]

SCHULZ, S. R. and JENSEN, W. A. (1968c). *Capsella* embryogenesis: the synergids before and after fertilization. *Am. J. Bot.* **55,** 541–552. [61]

SCHWABE, W. W. (1971). Chemical modification of phyllotaxis and its implications. *In* "Control Mechanisms of Growth and Differentiation" (Eds D. D. Davies and M. Balls), *Symp. Soc. exp. Biol.* **25,** 301–322. Cambridge University Press. [299, 312]

SCHWANITZ, F. (1967). "The Origin of Cultivated Plants". Harvard University Press, Cambridge, Mass. [211]

SCHWARZACHER, H. G. and SCHNEDLE, W. (1966). Position of labelled chromatids in diplochromosomes of endoreduplicated cells after uptake of tritiated thymidine. *Nature, Lond.* **209,** 107–108. [225]

SCHWEIZER, D. (1973). Differential staining of plant chromosomes with Giemsa. *Chromosoma* **40,** 307–320. [209]

SCURFIELD, G. (1973). Reaction wood: Its structure and function. *Science, N.Y.* **179,** 647–655. [361, 380, 387]

SCURFIELD, G. and WARDROP, A. B. (1962). The nature of reaction wood. VI. The reaction anatomy of seedlings of woody perennials. *Aust. J. Bot.* **10,** 93–105. [380]

SELMAN, G. G. (1966). Experimental evidence for the nuclear control of differentiation in *Micrasterias. J. Embryol. exp. Morph.* **16,** 469–485. [26]

SETTERFIELD, G. (1963). Growth regulation in excised slices of Jerusalem artichoke tuber tissue. *In* "Cell Differentiation", *Symp. Soc. exp. Biol.* **17,** 98–126. Cambridge University Press. [116, 125]

SHANKS, R. (1965). Differentiation in leaf epidermis. *Aust. J. Bot.* **13,** 143–151. [328]

SHARMA, A. K. (1956). A new concept of a means of speciation in plants. *Caryologia* **9,** 93–130. [233, 248, 249]

SHARMA, A. K. and BHATTACHARYYA, B. (1956). Vitamins: their property of inducing chromosome division in adult cells of plants. *Caryologia* **9,** 38–52. [230]

SHARMA, A. K. and MAJUMDAR, A. (1955). Cytological peculiarity of *Pteris longifolia* L and its importance in evolution. *Sci. Cult.* **21,** 338–339. [248]

SHARMA, A. K. and MOOKERJEA, A. (1959). Induction of division in cells. A study of causal factors involved. *Bull. bot. Soc. Bengal* **8,** 24–100. [217, 228]

SHARMA, A. K. and SHARMA, A. (1965). "Chromosome Techniques, Theory and Practice". Butterworths, London. [222]

SHARMAN, B. C. (1942). Developmental anatomy of the shoot of *Zea mays* L. *Ann. Bot.* **6,** 245–282. [316]

SHELDRAKE, A. R. (1973). The production of hormones in higher plants. *Biol. Rev.* **48,** 509–559. [366, 389]

SHELDRAKE, A. R. and NORTHCOTE, D. H. (1968). The production of auxin by tobacco internode tissues. *New Phytol.* **67,** 1–13. [389]

SHEPHARD, D. V. and THURMAN, D. A. (1973). Effect of nitrogen sources upon the activity of L-glutamate dehydrogenase of *Lemna gibba*. *Phytochemistry* **12,** 1937–1946. [127]

SHEPHERD, K. R. (1964). Some observations on the effect of drought on the growth of *Pinus radiata* D. Don. *Aust. Forestry* **28,** 7–22. [378]

SHEPHERD, K. R. and ROWAN, K. S. (1967). Indoleacetic acid in cambial tissue of radiata pine. *Aust. J. biol. Sci.* **20,** 637–646. [383]

SHIMIZU, Y. (1961). Sex chromatin in *Rumex acetosa*. *La Kromosomo* **49,** 1521–1523. [210]

SHUSHAN, S. and JOHNSON, M. A. (1955). The shoot apex and leaf of *Dianthus caryophyllus* L. *Bull. Torrey bot. Club* **82,** 266–283. [309]

SIMCHEN, G., PINON, R. and SALTS, Y. (1972). Sporulation in *Saccharomyces cerevisiae*: premeiotic DNA synthesis, readiness and commitment. *Expl Cell Res.* **75,** 207–218. [166, 169, 172, 177]

SIMPSON, G. M. and SAUNDERS, P. F. (1972). Abscisic acid associated with wilting in dwarf and tall *Pisum sativum*. *Planta* **102,** 272–276. [378]

SINHA, A. K. (1967). Spontaneous occurrence of tetraploidy and near-haploidy in mammalian peripheral blood. *Expl Cell Res.* **47,** 443–448. [228]

SINNOTT, E. W. (1960). "Plant Morphogenesis". McGraw-Hill, New York. [111, 375, 413]

SINNOTT, E. W. and BLOCH, R. (1940). Cytoplasmic behaviour during division of vacuolate plant cells. *Proc. natn. Acad. Sci. U.S.A.* **26,** 223–227. [418]

SINNOTT, W. W. and BLOCH, R. (1941). Division in vacuolated plant cells. *Am. J. Bot.* **28,** 225–232. [52, 418]

SINNOTT, E. W. and BLOCH, R. (1945). The cytoplasmic basis of intercellular pattern in vascular differentiation. *Am. J. Bot.* **32,** 151–156. [369]

SITTE, P. (1963). Zellfeinbau bei plasmolyse. II. Der Feinbau der *Elodea* Blattzellen bei Zucker und Ionenplasmolyse. *Protoplasma* **57**, 304–333. [101]

SITTON, D., ITAI, C. and KENDE, H. (1967). Decreased cytokinin production in the roots as a factor in shoot senescence. *Planta* **73**, 296–300. [344]

SKOOG, F. and MILLER, C. O. (1957). Chemical regulation of growth and organ formation in plant tissues cultured "in vitro". *In* "The Biological Action of Growth Substances". *Symp. Soc. exp. Biol.* **11**, 118–131. Cambridge University Press. [429]

SKULT, H. (1969). Growth and cell population kinetics of tritiated thymidine labelled roots of diploid and autotetraploid barley. *Acta Acad. åbo Ser.* B **29**, 1–15. [264]

SMILLIE, R. M. and STEELE-SCOTT, N. (1969). Organelle biosynthesis: the chloroplast. *Prog. Mol. subcell. Biol.* **1**, 136–202. [154]

SMITH, D. L. (1972). Staining and osmotic properties of young gametophytes of *Polypodium vulgare* L. and their bearing on rhizoid function. *Protoplasma* **74**, 465–479. [101]

SMITH, E. F., BROWN, N. A. and TOWNSEND, C. O. (1911). Crown-gall of plants: its cause and remedy. *U.S. Dept. Agr. Bur. Plant Ind. Bull.* **213**, 1–215. [425]

SMITH, H. H. (1972). Plant genetic tumors. *Progr. exp. Tumor Res.* **15**, 138–164. [425]

SMITH, H., STEWART, G. R. and BERRY, D. R. (1970). The effects of light on plastid ribosomal-RNA and enzymes at different stages of barley etioplast development. *Phytochemistry* **9**, 977–983. [156]

SMITH, R. H. and MURASHIGE, T. (1970). *In vitro* development of the isolated shoot apical meristem of angiosperms. *Am. J. Bot.* **57**, 562–568. [366]

SMITH-WHITE, S. (1955). The life history and genetic system of *Leucopogon juniperinus*. *Heredity* **9**, 79–91. [212]

SNOAD, B. (1955). Somatic instability of chromosomes number in *Hymenocallis calathinum*. *Heredity* **9**, 129–134. [248]

SNOW, M. and SNOW, R. (1931). Experiments on phyllotaxis. I. The effect of isolating a primordium. *Phil. Trans. R. Soc.* B **221**, 1–43. [305]

SNOW, R. (1935). Activation of cambium by pure hormones. *New Phytol.* **34**, 347–360. [368, 381]

SNOW, R. (1942). On the causes of regeneration after longitudinal splits. *New Phytol.* **41**, 101–107. [367, 368]

SÖDING, H. (1937). Wuchstoff und Kambiumtätigkeit der Baume. *Jb. wiss. Bot.* **84**, 639–670. [383]

SOMA, K. (1958). Morphogenesis in the shoot apex of *Euphorbia lathyris* L. *J. Fac. Sci. Tokyo Univ. III* **7**, 199–256. [309]

SOMA, K. and BALL, E. (1963). Studies of the surface growth of the shoot apex of *Lupinus albus*. *Brookhaven Symp. Biol.* **16**, 13–45. [288, 294, 295, 297]

SOROKIN, S. (1962). Centrioles and the formation of rudimentary cilia by fibroblasts and smooth muscle cells. *J. Cell Biol.* **15**, 363–377. [72]

SOROKIN, H. P., MATHUR, S. N. and THIMANN, K. V. (1962). The effects of auxins and kinetin on xylem differentiation in the pea epicotyl. *Am. J. Bot.* **49,** 444–454. [383]

SOUÈGES, E. C. (1919). Les premières divisions de l'œuf et les différenciations du suspenseur chez le *Capsella bursa-pastoris* moench. *Annls Sci. nat. 10 (Bot.)* **1,** 1–28. [400]

SPANG, H. A. and PLATT, R. S. (1972). The effect of plant growth substances on the hyperchromicity of DNA. *Physiologia Pl.* **27,** 321–326. [225]

SPANSWICK, R. M. (1972). Electrical coupling between cells of higher plants: a direct demonstration of intercellular communication. *Planta* **102,** 215–227. [99]

SPARROW, A. H. (1942). Colchicine induced univalents in diploid *Antirrhinum majus* L. *Science, N.Y.* **96,** 363–364.

SPARROW, A. H. and SPARROW, R. C. (1949). Treatment of *Trillium erectum* prior to and during mass production of permanent smear preparations. *Stain Technol.* **24,** 47–55. [162, 171, 189]

SPARVOLI, E., GAY, H. and KAUFMANN, B. P. (1966). Duration of the mitotic cycle in *Haplopappus gracilis. Caryologia* **19,** 65–71. [183]

SPREY, B. (1968). Zum Verhalten DNS—haltiger Areale des Plastidenstromas bei der Plastidenteilung. *Planta* **78,** 115–133. [88, 93, 158]

SRINIVASACHAR, D. and PATAU, K. (1958). Reductional groupings in cold-treated onion roots. *Chromosoma* **9,** 229–237. [229]

SRIVASTAVA, B. I. and CHADHA, K. C. (1970). Liberation of *Agrobacterium tumefaciens* DNA from the crown gall tumor cell DNA by shearing. *Biochem. biophys. Res. Commun.* **40,** 968–972. [426]

SRIVASTAVA, L. M. (1966). On the fine structure of the cambium of *Fraxinus americana* L. *J. Cell Biol.* **31,** 79–93. [355, 363, 374]

SRIVASTAVA, L. M. and O'BRIEN, T. P. (1966). On the ultrastructure of cambium and its vascular derivatives. I. Cambium of *Pinus strobus* L. *Protoplasma* **61,** 257–276. [355, 374]

SRIVASTAVA, L. M. and PAULSON, R. E. (1968). The fine structure of the embryo of *Lactuca sativa*. II. Changes during germination. *Can. J. Bot.* **46,** 1447–1453. [138]

STACK, S. M. and CLARKE, C. R. (1973). Differential Giemsa staining of the telomeres of *Allium cepa* chromosomes: observations related to chromosome pairing. *Can. J. Genet. Cytol.* **15,** 619–624. [68]

STAHL, A. (1972). Nucleoli and chromosome; their relationship during the meiotic prophase of the human fetal oocyte. *Humangenetik* **14,** 269–284.

STALLARD, W. A. (1962). "The Mitotic Pattern in a Root Meristem". Ph.D. Thesis, University of Texas. [268]

STANGE, L. (1964). Regeneration in lower plants. *In* "Advances in Morphogenesis" (Eds M. Abercrombie and J. Brachet), Vol. 4, pp. 111–153. Academic Press, New York and London. [99, 102]

STANGE, L. (1965). Plant cell differentiation. *A. Rev. Pl. Physiol.* **16,** 119–140. [53, 200, 201, 219]

STANGE, L. and KLEINHAUF, H. (1968). Der Zeitpunkt der DNS–Synthese bei der Embryonalisierung von Zellen des Lebermooses *Riella* in bezug auf RNS-Synthese und Kernteilung. *Planta* **80,** 280–287. [99]

STAVITSKY, G. (1970). Embryoid formation in callus tissues of coffee. *Acta. bot. neerl.* **19,** 509–514. [102]

STEBBINS, G. L. (1971). "Chromosome Evolution in Higher Plants". Edward Arnold, London. [195]

STEBBINS, G. L. and JAIN, S. K. (1960). Developmental studies of cell differentiation in the epidermis of monocotyledons. I. *Allium, Rhoeo* and *Commelina. Devl Biol.* **2,** 409–426. [53]

STEBBINS, G. L. and SHAH, S.S. (1960). Developmental studies in cell differentiation in the epidermis of monocotyledons. II. Cytological features of stomatal development in the Gramineae. *Devl Biol.* **2,** 477–500. [53, 245, 328]

STEBBINS, G. L., SHAH, S. S., JAMIN, D. and JURA, P. (1967). Changed orientation of the mitotic spindle of stomatal guard cell divisions in *Hordeum vulgare. Am. J. Bot.* **54,** 71–80. [110, 245]

STEER, B. T. (1971). The dynamics of leaf growth and photosynthetic capacity in *Capsicum frutescens* L. *Ann. Bot.* **35,** 1003–1015. [328, 329]

STEEVES, T. A. (1957). Contribution to discussion—a conference on tissue culture. *J. natn. Cancer Inst.* **19,** 583–585. [319]

STEEVES, T. A. (1963). Morphogenetic studies of *Osmunda cinnamomea* L. The shoot apex. *J. Indian bot. Soc.* **42A,** 225–236. [350, 367]

STEEVES, T. A., HICKS, M. A., NAYLOR, J. M. and RENNIE, P. (1969). Analytical studies on the shoot apex of *Helianthus annuus. Can. J. Bot.* **47,** 1367–1375. [289]

STEEVES, T. A. and SUSSEX, I. M. (1972). "Patterns in Plant Development". Prentice-Hall, New Jersey. [349, 351, 367]

STEFFERSON, D. A. (1966). Synthesis of ribosomal RNA during growth and division of *Lilium. Expl Cell Res.* **44,** 1–12. [64]

STEIL, W. N. (1951). Apogamy, apospory and parthenogenesis in the Pteridophytes II. *Bot. Rev.* **17,** 90–104. [103]

STEINITZ, L. (1944). The effect of lack of oxygen on meiosis in *Tradescantia. Am. J. Bot.* **31,** 428–443. [162, 176]

STEPHEN, J. (1974). Cytological investigations on the endosperm of *Borassus flabellifera* Linn. *Cytologia* **39,** 195–207. [219, 236, 239]

STEPHENS, S. G. (1944). The genetic organization of leaf-shape development in the genus *Gossypium. J. Genet.* **46,** 28–51. [321]

STERLING, C. (1949). The primary body of the shoot of *Dianthera americana. Am. J. Bot.* **36,** 184–193. [309]

STERN, C. (1958). The nucleus and somatic cell variation. *J. Cell comp. Physiol.* **522,** Suppl. *1,* 1–34. [207]

STERN, H. (1966). The regulation of cell division. *A. Rev. Pl. Physiol.* **17,** 345–378. [116]

STERN, H. and HOTTA, Y. (1969). Biochemistry of meiosis. *In* "Handbook of Molecular Cytology" (Ed. A. Lima-de-Faria) pp. 520–539. North-Holland, Amsterdam. [55, 162, 166, 172, 173, 183]

STETLER, D. A. and LAETSCH, W. M. (1969). Chloroplast development in *Nicotiana tabacum* "Maryland Mammoth". *Am. J. Bot.* **56**, 260–270. [93]

STEVENS, B. J. and ANDRÉ, J. (1969). The nuclear envelope. In "Handbook of Molecular Cytology" (Ed. A. Lima-de-Faria), pp. 837–871. North-Holland, Amsterdam. [69, 107]

STEWARD, F. C. (1968). "Growth and Organization in Plants". Addison-Wesley, New York. [375]

STEWARD, F. C. (1970). From cultured cells to whole plants: the induction and control of their growth and morphogenesis. *Proc. R. Soc.* B **175**, 1–30. [430]

STEWARD, F. C., AMMIRATO, P. V. and MAPES, M. O. (1970). Growth and development of totipotent cells. Some problems, procedures and perspectives. *Ann. Bot.* **34**, 761–787.

STEWARD, F. C., KENT, A. E. and MAPES, M. A. (1966). The culture of free plant cells and its signification for embryology and morphogenesis. In "Current Topics in Developmental Biology" (Eds A. A. Moscona and A. Monroy), pp. 113–154. Academic Press, New York and London. [430]

STEWART, R. N. and BURK, L. G. (1970). Independence of tissues derived from apical layers in ontogeny of the tobacco leaf and ovary. *Am. J. Bot.* **57**, 1010–1016. [246, 247]

STEWART, R. N. and DERMEN, H. (1970). Determination of number and mitotic activity of shoot apical initial cells by analysis of mericlinal chimeras. *Am. J. Bot.* **57**, 816–826. [296]

STEWART, R. N., SEMINIUK, P. and DERMEN, H. (1974). Competition and accommodation between apical layers and their derivatives in the ontogeny of chimeral shoots of *Pelargonium* × *hortorum*. *Am. J. Bot.* **61**, 54–67. [246]

STEWART, W. N. (1948). A study of the plastids in the cells of the mature sporophyte of *Isoetes*. *Bot. Gaz.* **110**, 281–300. [83, 86]

STOCKERT, J. C., FERNÁNDEZ-GÓMEZ, M. E., GIMÉNEZ-MARTÍN, G. and LÓPEZ-SÁEZ, J. F. (1970). Organisation of argyrophilic nucleolar material throughout the division cycle of meristematic cells. *Protoplasma* **69**, 265–278. [68]

STONE, W. E. (1932). The origin, development and increase of chloroplasts in the potato. *J. agric. Res.* **45**, 421–435. [93]

STOREY, W. B. (1968a). Somatic reduction in cycads. *Science, N.Y.* **159**, 648–650. [228, 229] 234]

STOREY, W. B. (1968b). Mitotic phenomena in the spider plant. *J. Hered.* **59**, 23–27. [238]

STOW, I. (1930). Experimental studies on the formation of the embryo-sac-like giant pollen grain in the anther of *Hyacinthus orientalis*. *Cytologia* **1**, 417–439. [163]

STOW, I. (1933). On the female tendencies of the embryo-sac-like giant pollen grain of *Hyacinthus orientalis*. *Cytologia* **5**, 88–108. [163]

STRASBURGER, E. (1880). "Zellbildung und Zelltheilung". Gustav Fisher, Jena.

STREET, H. E. (1966a). The physiology of root growth. *A. Rev. Pl. Physiol.* **17**, 315–344. [383, 384, 409]

STREET, H. E. (1966b). Growth, differentiation and organogenesis in plant tissue and organ culture. *In* "The Biology of Cells and Tissues in Culture" (Ed. E. N. Willmer), pp. 631–689. Academic Press, New York and London. [233, 409]

STREET, H. E. (1968a). Factors influencing the initiation and activity of meristems in roots. *In* "Root Growth" (Ed. W. J. Whittington), pp. 20–41. Butterworths, London. [275, 282]

STREET, H. E. (1968b). The induction of cell division in plant cell suspension cultures. *In* "Les Cultures de Tissues de Plantes", pp. 177–193. Colloques Internationaux du C.N.R.S., Strasbourg. [416]

STREET, H. E. (1973). Cell (suspension) cultures—techniques. *In* "Plant Tissue and Cell Culture" (Ed. H. E. Street), Botanical Monographs **11**, 59–99. Blackwell Scientific Publications, Oxford. [29, 410]

STUBBLEFIELD, E. and BRINKELY, B. R. (1967). Architecture and function of the mammalian centriole. *In* "Formation and Fate of Cell Organelles" (Ed. K. B. Warren), pp. 175–218. Academic Press, New York and London. [71, 72]

STUBBLEFIELD, E. and MUELLER, G. C. (1965). Thymidine kinase activity in synchronized HeLa cell cultures. *Biochem. biophys. Res. Commun.* **20**, 535–538. [128]

STUDHALTER, R. A., GLOCK, W. S. and AGERTER, S. R. (1963). Tree growth. Some historical chapters in the study of diameter growth. *Bot. Rev.* **29**, 245–365. [374, 376]

SUBRAHMANYAM, N. C. and KASHA, K. J. (1973). Selective chromosomal elimination during haploid formation in barley following inter-specific hybridization. *Chromosoma* **42**, 111–125. [174]

SUN, C. N. (1961). Sub-microscopic structure and development of the chloroplasts of *Psilotum triquetrum*. *Am. J. Bot.* **48**, 311–315. [83]

SUNDERLAND, N. (1960). Cell division and expansion in the growth of the leaf. *J. exp. Bot.* **11**, 68–80. [239, 325, 329]

SUNDERLAND, N. (1961). Cell division and expansion in the growth of the shoot apex. *J. exp. Bot.* **12**, 446–457. [286, 287, 294, 312, 313]

SUNDERLAND, N. (1973a). Pollen and anther culture. *In* "Plant Tissue and Cell Culture" (Ed. H. E. Street), Botanical Monographs **11**, 205–239. Blackwell Scientific Publications, Oxford. [105]

SUNDERLAND, N. (1973b). Nuclear cytology. *In* "Plant Tissue and Cell Culture" (Ed. H. E. Street), Botanical Monographs **11**, 161–190. Blackwell Scientific Publications, Oxford. [103, 219, 233, 241, 249]

SUNDERLAND, N. and BROWN, R. (1956). Distribution of growth in the apical region of the shoot of *Lupinus albus*. *J. exp. Bot.* **7**, 127–45. [286, 294, 329]

SUNDERLAND, N., HEYES, J. K. and BROWN, R. (1956). Growth and metabolism in the shoot apex of *Lupinus albus*. *In* "The Growth of Leaves" (Ed. F. L. Milthorpe), pp. 77–90. Butterworths, London. [329]

SUNDERLAND, N. and WICKS, F. M. (1971). Embryoid formation in pollen grains of *Nicotiana tabacum*. *J. exp. Bot.* **22**, 213–226. [103]

SUSSEX, I. M. (1955). Morphogenesis in *Solanum tuberosum* L.: experimental investigation of leaf dorsiventrality and orientation in the juvenile shoot. *Phytomorphology* **5**, 286–300. [318]

SUSSEX, I. M. and CLUTTER, M. E. (1968). Differentiation in tissues, free cells and reaggregated plant cells. *In* "Differentiation and Defense in Lower Organisms *in vitro*" (Ed. M. M. Sigel), pp. 3–12. Tissue Culture Association, New York. [101]

SUSSEX, I. and ROSENTHAL, D. (1973). Differential ³H-thymidine labelling of nuclei in the shoot apical meristem of *Nicotiana*. *Bot. Gaz.* **134**, 295–301. [290]

SUTTON-JONES, B. and STREET, H. E. (1968). Studies on the growth in culture of plant cells. III. Changes in fine structure during the growth of *Acer pseudo-platanus* L. cells in suspension culture. *J. exp. Bot.* **19**, 114–118. [416]

SWIFT, H. (1950). The constancy of desoxyribose nucleic acid in plant nuclei. *Proc. natn. Acad. Sci. U.S.A.* **36**, 643–654. [10, 275]

SWIFT, H. and WOLSTENHOLME, D. R. (1969). Mitochondria and chloroplasts: nucleic acids and the problem of biogenesis (genetics and biology). *In* "Handbook of Molecular Cytology" (Ed. A. Lima-de-Faria), pp. 972–1046. North-Holland, Amsterdam. [791]

TÄCKHOLM, G. (1922). Zytologische Studien über Gattung *Rosa*. *Acta. Horti. Bergiani* **7**, 97–381. [212]

TAGEEVA, S. V., GENEROSOVA, I. P., DEREVYANKO, V. G., LADYGIN, V. G. and SEMENOVA, G. A. (1971). Ultrastructural organisation of chloroplasts as effected by the functional state of plant tissues and organs, genetic factors and light conditions. *In* "Photosynthesis and Solar Energy Utilisation" (Ed. O. V. Zalenzky) pp. 144–148. Hayka Academy of Science, U.S.S.R. [136]

TAI, W. (1970). Multipolar meiosis in diploid crested wheat grass *Agropyron cristatum*. *Am. J. Bot.* **57**, 1160–1169. [73, 93]

TAKAHASHI, C. (1962). Cytological study on induced apospory in ferns. *Cytologia* **27**, 79–96. [103]

TAKEBE, I., LABIB, G. and MELCHERS, G. (1971). Regeneration of whole plants from isolated mesophyll protoplasts of tobacco. *Naturwissenschaften* **58**, 318–320. [424]

TAKEBE, I. and NAGATA, T. (1973). Culture of isolated tobacco mesophyll protoplasts. *In* "Protoplastes et Fusion de Cellules Somatiques Végétales", *Coll. Int. C.N.R.S.* **212**, 175–187. [421]

TAKEHISA, S. (1961). Aneusomaty in the leaves of diploid *Petunia*. *Bot. Mag. Tokyo* **74**, 494–497. [248]

TAMIYA, H. (1964). Growth and Cell Division of *Chlorella*. *In* "Synchrony in Cell Division and Growth" (Ed. E. Zeuthen, pp. 247–305. John Wiley, New York [201, 224]

TAYLOR, H. (1949). Increase in bivalent interlocking and its bearing on the chiasma hypothesis of metaphase pairing. *J. Hered.* **40**, 65–69. [162, 176]

TAYLOR, H. (1950). The duration of differentiation in excised anthers. *Am. J. Bot.* **37**, 137–143. [162, 176]

TAYLOR, H. (1953). Autoradiographic detection of incorporation of ^{32}P into chromosomes during meiosis and mitosis. *Expl Cell Res.* **4**, 164–173. [176]

TAYLOR, H. (1958). Incorporation of phosphorus-32 into nucleic acids and protein during microgametogenesis of *Tulbaghia*. *Am. J. Bot.* **45**, 123–131. [64]

TAYLOR, J. H. (1958). The mode of chromosome duplication in *Crepis capillaris*. *Expl Cell Res.* **15**, 350–357. [119]

TAYLOR, J. H. and McMASTER, R. D. (1954). Autoradiographic and microphotometric studies of desoxyribose nucleic acid during microgametogenesis in *Lilium longiflorum*. *Chromosoma* **6**, 489–581. [162, 168, 169, 176]

TERASIMA, T. and TOLMACH, L. J. (1963). Growth and nucleic acid synthesis in synchronously dividing populations of HeLa cells. *Expl Cell Res.* **30**, 344–362. [124]

TERRY, N., WALDRON, L. J. and ULRICH, A. (1971). Effects of moisture stress on the multiplication and expansion of cells in leaves of sugar beet. *Planta* **97**, 281–289. [378]

TEWARI, K. K. and WILDMAN, S. G. (1969). Information content in the chloroplast DNA. *In* "The Control of Organelle Development" (Ed. P. Miller), *Symp. Soc. exp. Biol.* **24**, 147–179. Cambridge University Press. [79, 84]

THERMAN, E. (1951). The effect of indole-3-acetic acid on resting nuclei. 1. *Allium cepa*. *Ann. Acad. Sci. Fenn. Ser.* A **IV**, 4–40. [217]

THERMAN, E. (1956). Dedifferentiation and differentiation of cells in crown gall of *Vicia faba*. *Caryologia* **8**, 325–348. [428]

THERMAN, E. and KUPILA-AHVENNIEMI, S. (1971). Crown gall development stimulated by wound washing. *Physiologia Pl.* **25**, 178–180. [425, 426]

THOMAS, D. R. (1967). Quiescent centre in excised tomato roots. *Nature, Lond.* **214**, 739. [266]

THOMAS, E., KONAR, R. N. and STREET, H. E. (1972). The fine structure of the embryogenic callus of *Rammculus sceleratus* L. *J. Cell Sci.* **11**, 95–109. [430]

THOMAS, P. T. (1936). Genotypic control of chromosome size. *Nature, Lond.* **138**, 402. [210]

THOMPSON, J. and CLOWES, F. A. L. (1968). The quiescent centre and rates of mitosis in the root meristem of *Allium sativum*. *Ann. Bot.* **32**, 1–13. [269, 270, 278]

THOMPSON, M. M. (1962). Cytogenetics of *Rubus*. III. Meiotic instability in some higher polyploids. *Am. J. Bot.* **49**, 575–582. [248]

TILNEY-BASSETT, R. A. E. (1963). The structure of the periclinal chimeras. I. The analysis of periclinal chimeras. *Heredity* **18**, 265–285. [246, 247, 319]

TILNEY-BASSETT, R. A. E. (1965). Genetics and plastid physiology in *Pelargonium*. II. *Heredity* **20**, 451–466. [94]

TILNEY-BASSETT, R. A. E. (1973). The control of plastid inheritance in *Pelargonium*. II. *Heredity* **30**, 1–13. [94]

TOBEY, R. A., PETERSEN, D. F. and ANDERSON, E. C. (1971). Biochemistry of G$_2$ and mitosis. *In* "The Cell Cycle and Cancer" (Ed. R. Baserga) pp. 310–353. Dekker, New York. [202, 203]

TOKUYASU, K. T. (1972). Identification of early S phase nuclei by observation of centriole replication in cultured human lymphocytes. *Expl Cell Res.* **73,** 17–24. [63, 107]

TORREY, J. G. (1957). Auxin control of vascular pattern formation in regenerating pea root meristems grown *in vitro. Am J. Bot.* **44,** 859–870. [367, 368]

TORREY, J. G. (1961). Kinetin as a trigger for mitosis in mature endomitotic plant cells. *Exp Cell Res.* **23,** 281–299. [203, 217, 219]

TORREY, J. G. (1965a). Cytological evidence of cell selection by plant tissue culture media. *In* "Proc. Int. Conf. Plant Tissue Cult." (Eds P. R. White and A. R. Grove), pp. 473–484. McCutchan, Berkeley. [217]

TORREY, J. G. (1965b). Physiological bases of organization and development in the root. *In* "Encyclopedia of Plant Physiology" (Ed. W. Ruhland), Vol. XV/1, pp. 1256–1327. Springer-Verlag, Berlin. [366]

TORREY, J. G. (1967). Morphogenesis in relation to chromosome constitution in long term plant cultures. *Physiologia Pl.* **20,** 265–275. [249]

TORREY, J. G. (1972). On the initiation of organization in the root apex. *In* "The Dynamics of Meristem Cell Populations" (Eds M. W. Miller and C. C. Kuehnert), pp. 1–13. Plenum Press, New York. [282]

TORREY, J. G. and FOSKET, D. E. (1970). Cell division in relation to cytodifferentiation in cultured pea root segments. *Am. J. Bot.* **57,** 1072–1080. [106, 233]

TORREY, J. G. and SHIGEMURA, J. (1957). Growth and controlled morphogenesis in pea root callus tissue grown in liquid media. *Am. J. Bot.* **44,** 334–344. [417]

TOURNEUR, J. and MOREL, G. (1971). Bacteriophages et crown-gall. Resultats et hypothèses. *Physiol. Vég.* **9,** 527–539. [428]

TROY, M. R. and WIMBER, D. E. (1968). Evidence for a constancy of the DNA synthetic period between diploid-polyploid groups in plants. *Expl Cell Res.* **53,** 145–154. [120]

TSCHERMAK-WOESS, E. (1956a). Notizen über die Riesenkerne und "Riesenchromosomen" in den Antipoden von *Aconitum. Chromosoma* **8,** 114–134. [219, 221]

TSCHERMAK-WOESS, E. (1956b). Karyologische pflanzenanatomie. Ein kritischer Überblick. *Protoplasma* **46,** 798–834. [221]

TSCHERMAK-WOESS, E. (1957). Über Kernstrukturen in den endopolyploiden Antipoden von *Clivia miniata. Chromosoma* **8,** 637–649. [221]

TUCKER, C. M. and EVERT, R. F. (1969). Seasonal development of the secondary phloem in *Acer negundo. Am. J. Bot.* **56,** 275–284. [375]

TURGEON, R. and WEBB, J. A. (1973). Leaf development and phloem transport in *Cucurbita pepo*: transition from import to export. *Planta* **113,** 179–191. [337]

TYREE, M. T. (1970). The symplast concept. A general theory of symplastic transport according to the thermodynamics of irreversible processes. *J. theor. Biol.* **26,** 181–214. [99]

UEDA, R., TOMINAGA, S. and TANUMA, T. (1970). Cinematographic observations on the chloroplast division in *Mnium* leaf cells. *Sci. Rep. Tokyo. Kyoiku Daigaku* **14,** 121–128. [93]

UNDERBRINK, A. G. and OLAH, L. V. (1968). Effect of digitonin on cellular division. Part III. Fine structural aspects of early phragmoplast development in the absence of an organised mitotic spindle. *Cytologia* **33,** 155–164. [222]

VAARMA, A. (1949). Spindle abnormalities and variation in chromosome number in *Ribes nigrum*. *Hereditas* **35,** 136–162. [248]

VAN'T HOF, J. (1963). DNA, RNA and protein synthesis in the mitotic cycle of pea root meristem cells. *Cytologia* **28,** 30–35. [117, 123, 126]

VAN'T HOF, J. (1965a). Relationships between mitotic cycle duration, S-period duration and the average rate of DNA synthesis in the root meristem cells of several plants. *Expl Cell Res.* **39,** 48–58. [117, 121, 264]

VAN'T HOF, J. (1965b). Cell population kinetics of excised roots of *Pisum sativum*. *J. Cell. Biol.* **27,** 179–189. [289]

VAN'T HOF, J. (1967). RNA synthesis during G_1, S and G_2 periods of diploid and colchicine-induced tetraploid cells in the same tissue of *Pisum*. *Expl. Cell Res.* **45,** 638–645.

VAN'T HOF, J. (1968a). Some kinetic aspects of recovery of G_1 and G_2 cells in stationary-phase excised pea roots. *Radiat. Res.* **34,** 626–636. [64]

VAN'T HOF, J. (1968b). Control of cell progression through the mitotic cycle by carbohydrate provision. I. Regulation of cell division in excised plant tissue. *J. Cell Biol.* **37,** 773–780. [282]

VAN'T HOF, J. (1968c). Experimental procedures for measuring cell population kinetic parameters in plant root meristems. *In* "Methods in Cell Physiology" (Ed. D. M. Prescott), Vol. 3, pp. 95–117. Academic Press, New York and London. [64]

VAN'T HOF, J. (1974). Control of the cell cycle in higher plants. *In* "Cell Cycle Controls" (Eds G. M. Padilla, K. Cameron and A. Zimmerman), pp. 77–85. Academic Press, New York. [202, 203, 204, 233]

VAN'T HOF, J., HOPPIN, D. P. and YAGI, S. (1973). Cell arrest in G_1 and G_2 of the mitotic cycle of *Vicia faba* root meristems. *Am. J. Bot.* **60,** 889–895. [122]

VAN'T HOF, J. and KOVACS, C. J. (1972). Mitotic cycle regulation in the meristem of cultured roots: the principal control point hypothesis. *In* "The Dynamics of Meristem Cell Populations" (Eds M. W. Miller and C. C. Kuehnert), pp. 15–30. Plenum Press, New York. [203, 281]

VAN'T HOF, J. and MCMILLAN, B. (1969). Cell population kinetics in callus tissues of cultured pea root segments. *Am. J. Bot.* **56,** 42–51. [217]

VAN'T HOF, J. and ROST, T. L. (1972). Cell proliferation in complex tissues: the control of the mitotic cycle of cell populations in the cultured root meristem of sunflower (*Helianthus*). *Am. J. Bot.* **59,** 769–774. [203, 301]

VAN'T HOF, J. and SPARROW, A. A. (1963). A relationship between DNA content, nuclear volume and minimum mitotic cycle time. *Proc. natn. Acad. Sci. U.S.A.* **49,** 897–902. [117, 264]

VASIL, I. K. (1959). Cultivation of excised anthers *in vitro*—effect of nucleic acids. *J. exp. Bot.* **10,** 399–408. [176]

VASIL, I. K. (1967). Physiology and cytology of anther development. *Biol. Rev.* **42,** 327–373. [174, 184]

VASIL, I. K. and ALDRICH, H. C. (1970). A histochemical and ultrastructural study of the ontogeny and differentiation of pollen in *Podocarpus macrophyllus* D. Don. *Protoplasma* **71,** 1–37. [86, 99]

VASIL, I. K. and ALDRICH, H. C. (1971). Histochemistry and ultrastructure of pollen development in *Podocarpus macrophyllus* D. Don. *In* "Pollen Development and Physiology" (Ed. J. Heslop-Harrison), pp. 70–74. Butterworths, London. [99]

VENDRELY, C. (1971). Cytophotometry and histochemistry of the cell cycle. *In* "The Cell Cycle and Cancer" (Ed. R. Baserga), pp. 227–268. Dekker, New York. [64]

VENKETSWARAN, S. (1963). Tissue culture studies on *Vicia faba*. II. Cytology. *Caryologia* **16,** 91–100. [249]

VERNER, L. (1938). The effect of a plant growth substance on crotch angles in young apple trees. *Proc. Am. Soc. hort. Sci.* **36,** 415–422. [387]

VERNER, L. (1955). Hormone relations in the growth and training of apple trees. *Res. Bull. Idaho Agric. Exp. Stn.* No. **28,** 1–31. [387]

VESK, M., MERCER, F. V. and POSSINGHAM, J. V. (1965). Observation on the origin of chloroplasts and mitochondria in the leaf cells of higher plants. *Aust. J. Bot.* **13,** 161–169. [93, 139]

VINIKKA, Y. (1973). The occurrence of B chromosomes and their effect on meiosis in *Najas marina. Hereditas* **75,** 207–212. [214]

VIRGIN, H. J., KAHN, A. and VON WETTSTEIN, D. (1963). The physiology of chlorophyll formation in relation to structural changes in chloroplasts. *Photochem. Photobiol.* **2,** 83–91. [151]

VÖCHTING, H. (1877). Über Theilbarkeit in Pflanzenreich und die Wirkung ausserer und innerer Kräfte auf Organbildung an Pflanzentheilen. *Pflugers Arch. ges. Physiol.* **15,** 153–190. [364]

VÖCHTING, H. (1906). Über Regeneration und Polarität bei höheren an Pflanzen. *Bot. Ztg* **64,** 101–148. [43]

VON WETTSTEIN, D. (1958). The formation of plastid structures. *In* "The Photochemical Apparatus: its Structure and Function". *Brookhaven Symp. Biol.* 138–158. [151]

VON WETTSTEIN, D. (1965). Die Induktion und experimentelle Berinflüssung der Polarität bei Pflanzen. *In* 'Encyclopedia of Plant Physiology" (Ed. W. Ruhland), Vol. XV/1, pp. 275–330. Springer-Verlag, Berlin.

VON WETTSTEIN, D. (1967). Chloroplast Structure and Genetics. *In* Harvesting the Sun" (Eds A. San Pietro, F. Greer and T. J. Army), pp. 153–190. Academic Press, New York and London. [151]

WETTSTEIN, D. VON and ERIKSON, G. (1965). The genetics of chloroplasts. *Genetics Today* **3,** 591–610. [94]

Vosa, C. G. (1966). Seed germination and B-chromosomes in the leek (*Allium porrum*). *In* "Chromosomes Today" (Eds C. D. Darlington and K. R. Lewis), Vol. 1, pp. 24–27. Oliver and Boyd, Edinburgh. [214]

Vosa, C. G. (1969). Heterochromatic B chromosomes in *Puschkinia libanotica*. *In* "Chromosomes Today" (Eds C. D. Darlington and K. R. Lewis), Vol. 2, pp. 189–191. Oliver and Boyd, Edinburgh. [208, 214]

Vosa, C. G. (1972). Two track heredity: differentiations of male and female meiosis in *Tulbaghia*. *Caryologia* **25**, 275–281. [188]

Vosa, C. G. (1973). Heterochromatin recognition and analysis of chromosome variation in *Scilla sibirica*. *Chromosoma* **43**, 269–278. [209]

Vosa, C. G. and Barlow, P. G. (1972). Meiosis and B chromosomes in *Listera ovata* (Orchidaceae). *Caryologia* **25**, 1–8. [188]

Wagenaar, E. B. (1969). End-to-end chromosome attachments in mitotic interphase and their possible significance to meiotic chromosome pairing. *Chromosoma* **26**, 410–426. [107]

Waisel, Y. and Fahn, A. (1965). The effects of environment on wood formation and cambial activity in *Robinia pseudacacia* L. *New Phytol*. **64**, 436–442. [377, 383, 390]

Waisel, Y., Liphschitz, N. and Arzee, T. (1967). Phellogen activity in *Robinia pseudacacia* L. *New Phytol*. **66**, 331–335. [363]

Waisel, Y., Noah, I. and Fahn, A. (1966). Cambial activity in *Eucalyptus camaldulensis* Dehn. II. The production of phloem and xylem elements. *New Phytol*. **65**, 319–324. [383]

Walker, D. A. (1970). Three phases of chloroplast research. *Nature, Lond*. **226**, 1204–1208. [135]

Walker, G. W. R. and Dietrich, J. F. (1961). Abnormal microsporogenesis in *Tradescantia* anthers cultured with sucrose-deficiency and kinetin-supplementation. *Can. J. Genet. Cytol*. **3**, 170–183. [162, 182, 190, 191]

Wallin, A. and Eriksson, T. (1973). Protoplast cultures from cell suspensions of *Daucus carota*. *Physiologia Pl*., **28**, 33–39. [420, 424]

Walters, M. S. (1958). Aberrant chromosome movement and spindle formation in meiosis of *Bromus* hybrids: an interpretation of spindle organization. *Am. J. Bot*. **45**, 271–289. [248]

Walters, M. S. (1972). Preleptotene chromosome contraction in *Lilium longiflorum* "Croft". *Chromosoma* **39**, 311–322. [162, 173]

Wardlaw, C. W. (1944). Experimental and analytical studies of pteridophytes. III. Stelar morphology: The initial differentiation of vascular tissue. *Ann. Bot*. **8**, 173–188. [350, 367]

Wardlaw, C. W. (1947). Experimental investigations of the shoot apex of *Dryopteris aristata*. Druce. *Phil. Trans. R. Soc*. B **232**, 343–384. [365]

Wardlaw, C. W. (1955a). "Embryogenesis in Plants". John Wiley, New York. [405]

Wardlaw, C. W. (1955b). Experimental and analytical studies of pteridophytes. XXVIII. Leaf symmetry and orientation in ferns. *Ann. Bot*. **19**, 389–399. [318]

Wardlaw, C. W. (1956). The inception of leaf primordia. *In* " The Growth of Leaves" (Ed. F. L. Milthorpe), pp. 53–65. Butterworths, London. [321]

WARDLAW, C. W. (1965). The organization of the shoot apex. *In* "Encyclopedia of Plant Physiology" (Ed. W. Ruhland), Vol. XV/I, pp. 966–1076. Springer-Verlag, Berlin. [364]

WARDROP, A. B. (1964). The reaction anatomy of arborescent angiosperms. *In* "The Formation of Wood in Forest Trees" (Ed. M. H. Zimmermann), pp. 405–456. Academic Press, New York and London. [380]

WARDROP, A. B. (1965). The formation and function of reaction wood. *In* "Cellular Ultrastructure of Woody Plants" (Ed. W. A. Cote), pp. 371–390. Syracuse University Press, New York. [380]

WARDROP, A. B. and DADSWELL, H. E. (1952). The cell wall structure of xylem parenchyma. *Aust. J. sci. Res. Ser. B, Biol. Sci.* **5**, 223–236. [380]

WAREING, P. F. (1951). Growth studies in woody species. III. Further photoperiodic effects in *Pinus sylvestris. Physiologia Pl.* **4**, 41–56. [373, 377, 388]

WAREING, P. F. (1954). Growth studies in woody species. VI. The locus of photoperiodic perception in relation to dormancy. *Physiologia Pl.* **7**, 261–277. [388]

WAREING, P. F. (1958a). The physiology of cambial activity. *J. Inst. Wood Sci.* **1**, 34–42. [373, 374, 377, 379, 388]

WAREING, P. F. (1958b). Interaction between indole-acetic and gibberellic acid in cambial activity. *Nature, Lond.* **181**, 1744–1745. [383, 384]

WAREING, P. F. (1969). Germination and dormancy. *In* "Physiology of Plant Growth and Development" (Ed. M. B. Wilkins), pp. 603–644. McGraw-Hill, London. [379, 390]

WAREING, P. F., HANNEY, C. E. A. and DIGBY, J. (1964). The role of endogenous hormones in cambial activity and xylem differentiation. *In* "The Formation of Wood in Forest Trees" (Ed. M. H. Zimmermann), pp. 323–344. Academic Press, New York and London. [382, 383, 386]

WAREING, P. F. and PHILLIPS, I. D. J. (1970). "The Control of Growth and Differentiation in Plants". Pergamon Press, Oxford. [381]

WAREING, P. F. and ROBERTS, D. L. (1956). Photoperiodic control of cambial activity in *Robinia pseudacacia. New Phytol.* **55**, 356–366. [373, 377, 383]

WAREING, P. F. and SAUNDERS, P. F. (1971). Hormones and dormancy. *A. Rev. Pl. Physiol.* **22**, 261–288. [385, 390]

WARR, J. R. (1968). A mutant of *Chlamydomonas reinhardii* with abnormal cell division. *J. gen. Microbiol.* **52**, 243–251. [83]

WARR, J. R. and DURBER, S. (1971). Studies on the expression of a mutant with abnormal cell division in *Chlamydomonas reinhardii. Expl Cell Res.* **64**, 463–469. [232]

WATSON, D. P. (1948). An anatomical study of the modification of bean leaves as a result of treatment with 2,4-D. *Am. J. Bot.* **35**, 543–555. [343]

WEBSTER, B. D. and RADIN, J. W. (1972). Growth and development of cultured radish roots. *Am. J. Bot.* **59**, 744–751. [383]

WEBSTER, P. C. and DAVIDSON, D. (1968). Evidence from thymidine-³H-labelled meristems of *Vicia faba* of two cell populations. *J. Cell Biol.* **39**, 332–338. [184]

WEBSTER, P. L. and LANGENAUER, H. D. (1973). Experimental control of the activity of the quiescent centre in excised root tips of *Zea mays*. *Planta* **112**, 91–100. [279, 283]

WEBSTER, P. L. and VAN'T HOF, J. (1970). DNA synthesis and mitosis in meristems: requirements for RNA and protein synthesis. *Am. J. Bot.* **57**, 130–139. [125]

WEHRMEYER, W. (1965). Zur Kristallgitterstruktur der sogenannten Prolamellarkorper in Proplastiden etiolierter Bohnen. I. Pentagondodekaeder als Mittelpunkt Konzentrischer Prolamellarkörper. *Z. Naturf.* **20B**, 1270–1278. [152]

WEIER, T. E., SJOLAND, R. D. and BROWN, D. L. (1970). Changes induced by low light intensities on the prolamellar body of 8-day, dark-grown seedlings. *Am. J. Bot.* **57**, 276–284. [151]

WENT, F. W. (1938). Transplantation experiments with peas. *Am. J. Bot.* **25**, 44–55. [342, 344]

WENT, H. A. (1966). The behaviour of centrioles and the structure and formation of the achromatic figure. *Protoplasmatologia* **VIG1**, 1–109. [51]

WESTING, A. H. (1965a). Formation and function of compression wood in gymnosperms. *Bot. Rev.* **31**, 381–480. [380, 387]

WESTING, A. H. (1965b). Compression wood in the regulation of branch angle in gymnosperms. *Bull. Torrey bot. Club.* **92**, 62–66. [385, 387]

WETMORE, R. H. (1954). The use of *in vitro* cultures in the investigation of growth and differentiation in vascular plants. *Brookhaven Symp. Biol.* **6**, 22–40. [366]

WETMORE, R. H. and JACOBS, W. P. (1953). Studies on abscission: the inhibiting effect of auxin. *Am. J. Bot.* **40**, 272–276. [366, 385]

WETMORE, R. H. and RIER, J. P. (1963). Experimental induction of vascular tissues in callus of angiosperms. *Am. J. Bot.* **50**, 418–430. [368, 369, 370]

WETMORE, R. H. and SOROKIN, S. (1955). On the differentiation of xylem. *J. Arnold Arbor.* **36**, 305–317. [368, 369, 429]

WETMORE, R. H. and STEEVES, T. A. (1971). Morphological introduction to growth and development. *In* "Plant Physiology, A Treatise" (Ed. F. C. Steward), Vol. VIA, pp. 3–166. Academic Press, New York and London. [349, 350, 352]

WETTSTEIN, R. and SOTELO, J. R. (1971). The molecular architecture of synaptonemal complexes. *In* "Advances in Cell and Molecular Biology" (Ed. E. J. DuPraw), Vol. 1, pp. 109–152. Academic Press, New York and London. [51]

WHATLEY, J. M. (1971). The chloroplasts of *Equisetum telmateia* Erhr: a possible developmental sequence. *New Phytol.* **70**, 1095–1102. [83, 93]

WHATLEY, J. M. (1974). The behaviour of chloroplasts during cell division of *Isoetes lacustris* L. *New Phytol.* **73**, 139–142. [83]

WHEELER, A. W. (1960). Changes in a leaf-growth substance in cotyledons and primary leaves during the growth of dwarf bean seedlings. *J. exp. Bot.* **11,** 217–226. [343]

WHEELER, A. W. (1973). Longitudinal distribution of growth substances in leaves of wheat (*Triticum aestivum* L.). *Planta* **112,** 129–135. [343]

WHITE, D. J. B. (1962). Tension wood in a branch of *Sassafras*. *J. Inst. Wood. Sci.* **10,** 74–80. [380]

WHITE, D. J. B. (1965). The anatomy of reaction tissues of plants. *In* "Viewpoints in Biology" (Eds J. D. Carthy and C. L. Duddington), Vol. IV. Butterworths, London. [380]

WHITE, M. J. D. and WEBB, G. C. (1972). Labelling pattern of the parthenogenetic grasshopper *Moraba vigo*. *In* "Chromosomes Today" (Eds C. D. Darlington and K. R. Lewis), Vol. 3, p. 307. Oliver and Boyd, Edinburgh. [219]

WHITE, R. A. (1968). A correlation between the apical cell and the heteroblastic leaf series in *Marsilea*. *Am. J. Bot.* **55,** 485–493. [321]

WHITE, R. A. (1971). Experimental and developmental studies of the fern sporophyte. *Bot. Rev.* **37,** 509–540. [103]

WHITEHOUSE, H. L. K. (1973). "Towards an Understanding of the Mechanism of Heredity". Edward Arnold, London. [67]

WHITMORE, F. W. and ZAHNER, R. (1967). Evidence for a direct effect of water stress on tracheid cell wall metabolism of cell wall in *Pinus*. *For. Sci., Peking* **13,** 397–400. [378]

WHITTIER, D. P. (1974). Apogamous sporophytes from gametophytes of *Botrychium dissectum* in axenic culture. *Am. J. Bot.* **61,** Supplement 40. [103]

WHITTIER, D. P. and PRATT, L. H. (1971). The effect of light quality on the induction of apogamy in prothalli of *Pteridium aquilinum*. *Planta* **99,** 174–178. [103]

WILCOX, H. (1962). Cambial growth characteristics. *In* "Tree Growth" (Ed. T. T. Kozlowski), pp. 57–88. Ronald Press, New York. [374]

WILCOX, H. E. (1968). Morphological studies of the root of red pine, *Pinus resinosa*. I. Growth characteristics and patterns of branching. *Am. J. Bot.* **55,** 247–254. [376]

WILDMAN, S. G. (1967). The organisation of Grana-containing chloroplasts in relation to location of some enzymatic systems concerned with photosynthesis, protein synthesis and ribonucleic acid synthesis. *In* "The Biochemistry of Chloroplasts" (Ed. T. W. Goodwin), Vol. II, pp. 295–319. Academic Press, London and New York. [139, 147]

WILDMAN, S. G., LU LIAO, C. and WONG STAAL, F. (1973). Maternal inheritance, cytology and macromolecular composition of defective chloroplasts in a variegated mutant of *Nicotiana tabacum*. *Planta* **113,** 293–312. [94]

WILLEMSE, M. Th. M. (1972). Morphological and quantitative changes in the population of cell organelles during microsporogenesis of *Gasteria verrucosa*. *Acta bot. neerl.* **21,** 17–31. [86]

WILLIAMS, R. F. (1960). The physiology of growth in the wheat plant. 1. Seedling growth and the pattern of growth at the shoot apex. *Aust. J. biol. Sci.* **13,** 401–428. [330]

WILLIAMSON, D. H. (1966). Nuclear events in synchronously dividing yeast cultures. *In* "Cell Synchrony" (Eds I. L. Cameron and G. M. Padilla), pp. 81–101. Academic Press, New York and London. [169]

WILLISON, J. H. M. and COCKING, E. C. (1972). The production of microfibrils at the surface of isolated tomato fruit protoplasts. *Protoplasma* **75**, 397–403. [421]

WILSON, G. B. and CHANG, K. C. (1949). Segregation and reduction in somatic tissues. *J. Hered.* **40**, 3–6. [228]

WILSON, G. B., HAWTHORN, M. E. and TSOU, T. M. (1951). Spontaneous and induced variations in mitosis. *J. Hered.* **42**, 183–189. [228]

WILSON, G. L. (1966). Studies on the expansion of the leaf surface. V. Cell division and expansion in a developing leaf as influenced by light and upper leaves. *J. exp. Bot.* **17**, 440–451. [330, 337, 340]

WILSON, G. L. and LUDLOW, M. M. (1968). Bean leaf expansion in relation to temperature. *J. exp. Bot.* **19**, 309–321. [327]

WILSON, H. J. (1970). Endoplasmic reticulum and microtubule formation in dividing cells of higher plants—a postulate. *Planta* **94**, 184–190. [73, 93]

WILSON, J. Y. (1959). Duration of meiosis in relation to temperature. *Heredity* **13**, 263–267. [176, 180]

WILSON, S. B., KING, P. J. and STREET, H. E. (1971). Studies on the growth in culture of plant cells. XII. A versatile system for the large-scale batch or continuous culture of plant cell suspensions. *J. exp. Bot.* **22**, 177–207. [115]

WIMBER, D. E. (1960). Duration of the nuclear cycle in *Tradescantia paludosa* root tips as measured with ^{3}H-thymidine. *Am. J. Bot.* **47**, 828–834. [117, 266]

WIMBER, D. E. (1961). Asynchronous replication of DNA in root tip chromosomes of *Tradescantia paludosa*. *Expl Cell Res.* **23**, 402–407. [117, 119]

WIMBER, D. E. (1966). Duration of the nuclear cycle in *Tradescantia* root tips at 3 temperatures as measured with ^{3}H-thymidine. *Am. J. Bot.* **53**, 21–24. [65, 117, 122, 190, 264]

WIMBER, D. E. and QUASTLER, H. (1963). A ^{14}C- and ^{3}H-thymidine double labelling technique in the study of cell proliferation in *Tradescantia* root tips. *Expl Cell Res.* **30**, 8–22. [262]

WIPF, L. and COOPER, D. C. (1940). Somatic doubling of chromosomes and nodular infection in certain Leguminosae. *Am. J. Bot.* **27**, 821–824. [219]

WITHERS, L. A. and COCKING, E. C. (1972). Fine-structural studies on spontaneous and induced fusion of higher plant protoplasts. *J. Cell Sci.* **11**, 59–75. [98]

WITSCH, V. H. and FLÜGEL, A. (1951). Über photoperiodisch induzierte endomitose bei *Kalanchoe blossfeldiana*. *Naturwissenschaften* **38**, 138–139. [235]

WODZICKI, T. (1960). Investigation on the kind of *Larix polonica* Rac. wood formed under various photoperiodic conditions. I. Plants growing in natural conditions. *Acta Soc. bot. pol.* **29**, 713–730. [388]

WODZICKI, T. (1961a). Investigation on the kind of *Larix polonica* Rac. wood formed under various photoperiodic conditions. II. Effect of different light conditions on wood formed by seedlings grown in greenhouse. *Acta Soc. bot. pol.* **30**, 111–131. [388]

Wodzicki, T. (1961b). Investigation on the kind of *Larix polonica* Rac. wood formed under various photoperiodic conditions. III. Effect of decapitation and ringing on the wood formation and cambial activity. *Acta Soc. bot. pol.* **30**, 293–306. [388]

Wodzicki, T. (1964). Photoperiodic control of natural growth substances and wood formation in larch (*Larix decidua* DC.). *J. exp. Bot.* **15**, 584–599. [383]

Wodzicki, T. (1965). Annual ring of wood formation and seasonal changes of natural growth inhibitors in larch. *Acta Soc. bot. pol.* **34**, 117–151. [383, 384, 385]

Wolff, S. (1969). Strandedness of chromosomes. *Int. Rev. Cytol.* **25**, 279–296. [66, 67, 211]

Woo, K. C., Anderson, J. M., Boardman, N. K., Downton, W. J. S., Osmond, C. B. and Thorne, S. W. (1970). Deficient photosystem II in agranal bundle sheath chloroplasts of C_4 plants. *Proc. natn. Acad. Sci. U.S.A.* **67**, 18–25. [153]

Wood, H. N., Lin, M. C. and Braun, A. C. (1972). The inhibition of plant and animal adenosine 3′:5′-cyclic monophosphate phosphodiesterases by a cell-division-promoting substance from tissues of higher plant species. *Proc. natn. Acad. Sci. U.S.A.* **69**, 403–406. [224, 225]

Woodard, J. W. (1958). Intracellular amounts of nucleic acids and protein during pollen grain growth in *Tradescantia*. *J. biophys. biochem. Cytol.* **4**, 383–390. [64]

Woodard, J., Gelber, B. and Swift, H. (1961). Nucleoprotein changes during the mitotic cycle in *Paramecium aurelia*. *Expl Cell Res.* **23**, 258–264. [14, 112, 113, 117, 118, 119, 123, 124, 126]

Woodard, J., Rasch, E. and Swift, H. (1961). Nucleic acid and protein metabolism during the mitotic cycle of *Vicia faba*. *J. biophys. biochem. Cytol.* **9**, 445–462. [14, 112, 113, 117, 118, 119]

Woodcock, C. L. F. and Bogorad, L. (1971). Nucleic acids and information processing in chloroplasts. *In* "Structure and Function of Chloroplasts" (Ed. M. Gibbs), pp. 90–128. Springer-Verlag, Berlin. [154, 158]

Woods, P. S. (1960). "The Cell Nucleus". Butterworths, London. [125]

Woods, P. S. and Taylor, J. H. (1959). Studies of ribonucleic acid metabolism with tritium-labelled cytidine. *Lab. Invest.* **8**, 309–318. [125]

Wort, D. J. (1962). Physiology of cambial activity. *In* "Tree Growth" (Ed. T. T. Kozlowski), pp. 89–95. Ronald Press, New York. [383]

Wright, S. T. C. (1969). An increase in the "Inhibitor β" content of detached wheat leaves following a period of wilting. *Planta* **86**, 10–20. [378]

Wright, S. T. C. and Hiron, R. W. P. (1969). (+)-Abscisic acid, the growth inhibitor induced in detached wheat leaves by a period of wilting. *Nature, Lond.* **224**, 719–720. [378]

Yang, S. J. (1965). Numerical chromosome instability in *Nicotiana* hybrids. II. Intraplant variation. *Can. J. Genet. Cytol.* **7**, 112–119. [248]

YEOMAN, M. M. (1970). Early development in callus cultures. *Int. Rev. Cyt.* **29,** 383–409. [111, 117, 413, 425]

YEOMAN, M. M. (1973). Tissue (callus) cultures—techniques. *In* "Plant Tissue and Cell Culture" (Ed. H. E. Street), Botanical Monographs **11,** 31–58. Blackwell Scientific Publications, Oxford. [410]

YEOMAN, M. M. (1974). Division synchrony in cultured cells. *In* "Tissue Culture and Plant Science" (Ed. H. E. Street), pp. 1–17. Academic Press, London and New York. [117]

YEOMAN, M. M. and AITCHISON, P. A. (1973a). Changes in enzyme activities during the division cycle of cultured plant cells. *In* "The Cell Cycle in Development and Differentiation" (Eds M. Balls and F. S. Billett), *Symp. Br. Soc. Dev. Biol.* **1,** 185–201. Cambridge University Press. [16, 117, 128]

YEOMAN, M. M. and AITCHISON, P. A. (1973b). Growth patterns in tissue (callus) cultures. *In* "Plant Tissue and Cell Culture" (Ed. H. E. Street), Botanical Monographs **11,** 240–268. Blackwell Scientific Publications, Oxford. [16, 117, 126, 128]

YEOMAN, M. M. and BROWN, R. (1971). Effects of mechanical stress on the plane of cell division in developing callus cultures. *Ann. Bot.* **35,** 1001–12. [35, 102, 110, 318, 413, 414, 430]

YEOMAN, M. M. and DAVIDSON, A. W. (1971). Effect of light on cell division in developing callus cultures. *Ann. Bot.* **35,** 1085–1100. [29, 116]

YEOMAN, M. M., DYER, A. F. and ROBERTSON, A. I. (1965). Growth and differentiation of plant tissue cultures. I. Changes accompanying the growth of explants from *Helianthus tuberosus* tubers: *Ann. Bot.* **29,** 265–276. [13, 116, 414]

YEOMAN, M. M. and EVANS, P. K. (1967). Growth and differentiation of plant tissue cultures. II. Synchronous cell division in developing callus cultures. *Ann. Bot.* **31,** 323–332. [11, 62, 116, 118, 203]

YEOMAN, M. M., EVANS, P. K. and NAIK, G. G. (1966). Changes in mitotic activity during early callus development. *Nature, Lond.* **209,** 1115–1116. [116]

YEOMAN, M. M., NAIK, G. G. and ROBERTSON, A. I. (1968). Growth and differentiation of plant tissue cultures. III. The initiation and pattern of cell division in developing callus cultures. *Ann. Bot.* **32,** 301–313. [116, 124, 413, 414, 416]

YEOMAN, M. M. and STREET, H. E. (1973). General Cytology of Cultured Cells. *In* "Plant Tissue and Cell Culture" (Ed. H. E. Street), Botanical Monographs **11,** 121–160. Blackwell Scientific Publications, Oxford. [52, 125, 416, 418, 425]

YEOMAN, M. M., TULETT, A. J. and BAGSHAW, V. (1970). Nuclear extensions in dividing vacuolated plant cells. *Nature, Lond.* **226,** 557–558. [419]

YOSHIDA, Y. (1962). Nuclear control of chloroplast activity in *Elodea* leaf cells. *Protoplasma* **54,** 476–492. [98]

ZAHNER, R. (1968). Water deficits and growth of trees. *In* "Water Deficits and Plant Growths" (Ed. T. T. Kozlowski), pp. 191–254. Academic Press, New York and London. [378]

ZEEVART, J. A. D. (1971). (+)-Abscisic acid content of spinach in relation to photoperiod and water stress. *Pl. Physiol., Lancaster* **48,** 86–90. [378]

ZEIGER, E. and STEBBINS, G. L. (1972). Developmental genetics in barley: a mutant for stomatal development. *Am. J. Bot.* **59,** 143–148. [53, 94]

ZEPF, E. (1952). Über die Differenzierung des *Sphagnumblattes. Z. Bot.* **40,** 87–118. [53, 101]

ZETTERBERG, A. (1966). Synthesis and accumulation of nuclear and cytoplasmic proteins during interphase in mouse fibroblasts *in vitro. Expl Cell Res.* **42,** 500–511. [126]

ZETTERBERG, A. and KILLANDER, D. (1965). Quantitative cytochemical studies on interphase growth. II. Derivation of synthesis curves from the distribution of DNA, RNA and mass value of individual mouse fibroblasts *in vitro. Expl Cell Res.* **39,** 22–32. [118, 119, 124]

ZIMMERMAN, W. A. (1928). Histologische Studies am Vegetationspunkt von *Hypericum uralum. Jb. wiss. Bot.* **68,** 289–344. [309]

ZINSMEISTER, D. D. and CAROTHERS, Z. B. (1974). The fine structure of oogenesis in *Marchantia polymorpha. Am. J. Bot.* **61,** 499–512. [86]

ZIRKLE, C. (1927). The growth and development of plastids in *Lunularia vulgaris, Elodea canadensis* and *Zea mays. Am. J. Bot.* **14,** 429–445. [93]

Index

For the first time, most of the varied aspects of cell division in higher plants have been brought together in a single volume. This book gives a critical but constructive appraisal of the current theories and research achievements of the subject, and provides a stimulating basis for future investigation. Hence, it is a work aimed at all cell biologists, at senior undergraduates postgraduates, and those established scientists who are interested in cell division or considering research in this area.

In the four part arrangement of the text, special emphasis has been given to the processes of mitotic and meiotic cell division and the relationship of division to growth and the generation of form. In other sections the authors consider the significance of cell division both as a process and as a part of overall growth, and speculate on the possible future direction of research.

The contributors, all distinguished specialists in their particular fields, demonstrate various approaches to the subject, from the descriptive to the physiological and molecular. In editing this book, Dr. Yeoman has specifically aimed at compiling an integrated and coherent survey by allowing some overlap between chapters, by the extensive use of cross references, and by the inclusion of a unified bibliography. In addition, the relationship between the different chapters is reviewed in the concluding summary.